Springer Geophysics

The Springer Geophysics series seeks to publish a broad portfolio of scientific books, aiming at researchers, students, and everyone interested in geophysics. The series includes peer-reviewed monographs, edited volumes, textbooks, and conference proceedings. It covers the entire research area including, but not limited to, geodesy, planetology, geodynamics, geomagnetism, paleomagnetism, seismology, and tectonophysics.

More information about this series at http://www.springer.com/series/10173

Xuetao Hu · Shuyong Hu
Fayang Jin · Su Huang
Editors

Physics of Petroleum Reservoirs

Second Edition

石油工业出版社
PETROLEUM INDUSTRY PRESS

Springer

Editors
Xuetao Hu
School of Oil and Natural Gas Engineering
Southwest Petroleum University
Chengdu, Sichuan
China

Fayang Jin
School of Oil and Natural Gas Engineering
Southwest Petroleum University
Chengdu, Sichuan
China

Shuyong Hu
School of Oil and Natural Gas Engineering
Southwest Petroleum University
Chengdu, Sichuan
China

Su Huang
PetroChina Southwest Oil and Gasfield Company
Chengdu, Sichuan
China

Springer Geophysics
ISBN 978-3-662-55025-0 ISBN 978-3-662-55026-7 (eBook)
DOI 10.1007/978-3-662-55026-7

Jointly published with Petroleum Industry Press, Beijing, China

The print edition is not for sale in China Mainland. Customers from China Mainland please order the print book from Petroleum Industry Press.

Library of Congress Control Number: 2017946950

Printed on acid-free paper

This Springer imprint is published by Springer Nature
The registered company is Springer-Verlag GmbH Germany
The registered company address is: Heidelberger Platz 3, 14197 Berlin, Germany

Foreword

With the acceleration of economic globalization, worldwide competition is intensifying, which brings significant changes in the talent demand of China. Professionals, with strong expertise as well as proficiency in English, are in urgent need. Bilingual teaching, in such case, is an effective way of cultivating application-oriented talents with an international vision. Therefore, it has been widely promoted in colleges and universities across China, taking a higher and higher proportion in the curriculum systems.

Southwest Petroleum University and some other Universities have offered bilingual teaching for petroleum-related majors in succession recent years. Teachers are encouraged to teach both fundamental and compulsory courses using foreign language (English mainly). Some Universities even setup international-talents-training classes. Bilingual teaching has been considered as the major method of cultivating international and inter-disciplinary talents, which helps the students to improve their international awareness as well as communication and cooperation abilities, at the same time, makes them more competitive in the job market.

Physics of Petroleum Reservoirs, as one of the most important fundamental courses in petroleum engineering and related majors, has been taught bilingually in many colleges and universities. However, there are currently too few bilingual textbooks for the teachers and students to choose from; existing teaching materials from abroad, on the other hand, can hardly match with the Chinese materials, be understood or accepted by students, nor keep up with the latest developments of this discipline. Bilingual teaching can only reach its best effect when there are suitable bilingual textbooks, otherwise it cannot even be carried on.

The authors, led by Editor Xuetao Hu, write this textbook based on their years' experience of bilingual teaching in Physics of Petroleum Reservoirs. Not only did they carefully research into massive original teaching materials abroad, but also took the students' levels and capacities into fully consideration. It is a textbook that covers the basic knowledge this course requires, at the same time, illustrates part of the frontier theories and latest developments related. This textbook timely provides bilingual teaching materials in accordance with the current teaching situation. Notably, it also matches well with the Chinese textbook of this course in SWPU, edited by Prof. Gengshen He and Prof. Hai Tang, which has been used for a long time in many colleges and universities. It is believed that the publication of this textbook can help to further improve the effect of bilingual teaching in this course, and play a key role in the education of application-oriented international talents needed by China's petroleum industry; in the meantime, it is expected to benefit to both engineers and technicians in related fields.

Chengdu, China
March 2016

Shilun Li

Preface

Physics of Petroleum Reservoirs, as one of the fundamental courses in petroleum engineering and related majors, is widely used in many fields, such as geological research, geophysical exploration, drilling engineering, oil/gas production and so on. To promote the development of such an important course, bilingual teaching is indispensable so as to help students quickly master the basic technique vocabulary, improve their English application ability, and lay a foundation for their study on the subsequent bilingual courses. For further consideration, bilingual teaching on this course can also broaden the students' horizon on the international trend of this subject and guides them to enter the related research fields as early as possible, therefore in accordance with the urgent need in China's petroleum industry for large numbers of internationalized application-oriented professionals.

However, in the practice of bilingual teaching in this course, there are too few bilingual textbooks for teachers and students to use at present. The lack of an appropriate bilingual textbook of this course has been suffered for years in Southwest Petroleum University. In view of this, the authors, based on their years' experience engaging in the bilingual teaching of this course, come out with this bilingual textbook specially. When composing this textbook, authors not only consult some original teaching materials abroad, but also take the students' knowledge levels and capacities into fully consideration. Besides the basic knowledge, this course requires, the textbook also covers some frontier theories and latest developments in this subject; hence, its strong practicability and pertinence can well satisfy the demands in bilingual teaching. This book is written to be a textbook for undergraduates who major petroleum engineering, resources exploration engineering and other relevant majors, or a reference book for oilfield engineers.

The book focuses the attributes of petroleum reservoirs, the important physical and chemical phenomena, and physical processes occurring in petroleum production. To specify, there are four chapters in this book. Chapter 1 addresses the physical properties of reservoir rocks. Chapter 2 discusses the physical properties of reservoir fluids. Chapter 3 presents the microscopic mechanism of multiphase fluids flowing through rocks. And Chap. 4 is a simple introduction to the principles of enhanced oil recovery. Considering the depth and breadth, the book mainly introduces the basic concepts, primary theories, common test items and methods in oilfield development. Through the book, students can understand the important properties of petroleum reservoirs with their applications on petroleum engineering, at the same time learn the principles and methods of measuring these physical properties as well as the required experimental skills; ultimately, lay a solid foundation for their following courses and practical work in the future.

The textbook is arranged to meet the need of 48–56 class hours. Due to the limitation of space and the adaptability for undergraduates, some topics in this subject cannot be deeply discussed. Interested readers may refer to related literatures and books.

This textbook is organized into five chapters. Xuetao Hu writes most of the book, mainly including the Preface, Chaps. 1, 2, 3 and Sect. 4.1 of Chap. 4. The rest of Chap. 4 is written by Shuyong Hu and Chap. 5 by Fayang Jin. Su Huang (from Southwest Oil and Gas Field Company, CNPC) participates in the writing of the Preface, Chap. 1 and sections of Chap. 2; and help to translate the Foreword and review the whole book in English. Xuetao Hu edits all the drafts.

We are especially indebted to Prof. Shilun Li for taking time to carefully review this book and write the Foreword, also to the leaders and teachers from Southwest Petroleum University and the School of Oil & Natural Gas Engineering, who offered strong support and great assistance in the completion of this book. As the textbook of a vital fundamental course, it inherits the advantages and absorbs the essence of various related textbooks published in different times both at home and abroad; numerous references including published and unpublished are cited, and some of them failed to be listed for some reason. Great appreciation goes to all of them.

International System of Units (SI system) is dominating in this book, whereas the system of imperial units is customarily used in the world. This book retains imperial system in some parts as well, wishing to help readers gain an international view. Therefore, it is advised that readers pay more attention to the units.

The study on Physics of Petroleum Reservoirs is ever developing. While this book is being written, innovations may take place in laboratories and oilfields throughout the world. Limited by time and the authors' level, the latest technologies and methods may not be presented completely, and omissions and shortcomings inevitable. Criticism and suggestions are always welcome.

Chengdu, China Xuetao Hu
June 2016

Contents

Chapter 1
Introduction

Xuetao Hu and Su Huang

1.1 Importance of This Course

As the power source of social developments, energy is of decisive importance to a country's economic performance, competition ability and overall national strength. Among all kinds of energy, petroleum and natural gas, functioning as the key resources, high-quality chemical raw materials and indispensable war materials, are capturing more and more attention nowadays by counties all around the world, that is why they are also called the *black gold*.

During the past decades, China's annual output of oil has been increasing all the way from less than 90,000 tons before liberation, exceeding 100 million tons in 1978, to 160–170 million tons in recent years, ranking into the world's top four major oil producing countries. However, it should not be neglected that, China has become a net importer of crude oil ever since 1997, and now imports more than 100 million tons each year since the beginning of the twenty-first century, making itself the world's second largest oil consumer. According to the national socioeconomic development plan, the future demand for oil and gas resources will continue to rise in China.

Therefore, in order to meet the resources needs for national economy modernization and sustainable development, it is necessary that

---

X. Hu (✉)
School of Oil and Natural Gas Engineering, Southwest Petroleum University, Chengdu, Sichuan, China
e-mail: huxt@swpu.edu.cn

S. Huang
Exploration, PetroChina Southwest Oil and Gasfield Company, Chengdu, Sichuan, China
e-mail: huangsu419@petrochina.com.cn

X. Hu et al. (eds.), *Physics of Petroleum Reservoirs*, Springer Geophysics,
DOI 10.1007/978-3-662-55026-7_1

(a) find out new oil and gas reservoirs to increase resources backup reserves;
(b) apply advanced technologies to maximize the reasonable development of hydrocarbon reservoirs;
(c) enhance the recovery of existing oil and gas fields to increase petroleum production;
(d) develop international cooperation and participate in the development of international oil and gas resources, to satisfy the requirement of China's petroleum industry in exploiting domestic and international resources facing domestic and international markets;
(e) utilize adequately various energy-saving measures, alternative energy, and new energy to reduce the pressure in petroleum demand.

For those who are going to be or already engaged in this career, proficiency in the theories and practices of petroleum exploration and exploitation is highly required, so as to make better contributions to China's oil industry in the future.

Physics of Petroleum Reservoirs studies the physical and physical-chemical phenomena in petroleum geology and petroleum production from the viewpoint of physics and physical chemistry. It is one of the required courses in both petroleum engineering and resources exploration engineering majors. It develops based on the experiments, and is a highly practical course. Meanwhile, it is fundamental to the subsequent major courses in petroleum areas, such as fluid mechanics in porous media, reservoir engineering, reservoir simulation, reservoir production engineering, well test analysis, reservoir protection technology, nature gas engineering, and enhanced oil recovery and so on.

The chief objectives of learning this course are the following:

(a) to understand the basic physical properties of reservoir rocks and reservoir fluids; as well as the mechanism of multiphase fluids flowing through reservoir rocks;
(b) to catch the point of the latest EOR methods and technologies;
(c) to learn the basic skills of experiments related with the attributes of reservoir rocks;
(d) to develop students' ability of analyzing and solving the problems encountered in reservoir production.

1.2 History and Developments of the Discipline

Physics of Petroleum Reservoirs is a relatively young subject. The earliest research related to this area can be dated back to the 1930s. At that time, these engineers, who were engaged in the petroleum production in the United States, the former Soviet Union, and some other countries, noticed the influence of reservoir fluids' properties on the production and started the preliminary studies on the fluid properties and the related determination methods.

Before 1930s, the former Soviet Union experts had the publications on this area. However, this subject came into being from the publication: *Physical Principles of Oil Production* written by Muskat in 1949 [1], U.S. that turned the researches in production data (oil, gas and water) into a subject. Aimed at the properties of reservoir rocks and fluids, this book collected various researches and production data in the first half of the twentieth century, and summarized them to give explanations from a physical view. Another breakthrough took place in the 1950s, when the former Soviet Union professor Kotyakhov from Moscow Petroleum Institute published *Основы физик0438 нефтяного пласта* (*Foundation of Reservoir Physics*) [2], which marks the beginning of Physics of Petroleum Reservoirs as an independent branch of petroleum production engineering.

The following years witnessed extensive and in-depth developments of the subject. A variety of books reflecting the research results related with Physics of Petroleum Reservoirs had been successively published in China and other countries in the world, such as the United States, Soviet Union, Canada, and so on. Despite the different names of these books (e.g., *The Basic Principles of Reservoir Engineering*), their contents are broadly consistent. Some iconic publications at abroad are

(a) In the 1970s, Chilingar et al. [3] published *Oil and Gas Production of Carbonate Reservoirs*, which describes the characteristics of carbonate reservoir from the exploitation point of view, and gives a more systematic and comprehensive depiction of the carbonate reservoir and its physical properties. The book, *The Properties of Petroleum Fluids* written by William et al. [4], explains systematically on the physical and chemical properties of oil and gas with related calculations. In 1977, a new book, *Physics of Oil and Gas Reservoirs*, was created by F.I. Kotyakhov and his assistant, in which the properties of oil and gas reservoirs and the pore structure of the reservoir rocks were reconsidered.

    Физико—Химическая Механика Нефтянотo пласта (*The Physical and Chemical Mechanism of Oil-Bearing Reservoir*) by Markhasin (1982) [5] described the physical-chemical processes of crude oil seepage and migration from the physical and chemical point of view for the first time. His book highlights that the oil recovery can only be enhanced through elaborate studies of the physical and chemical properties of oil-bearing reservoirs based on the traditional reservoir physics.
(b) The studies in the 1980s mainly focused on the characteristics of carbonate reservoirs, clastic rocks, and secondary pores; the non-Newtonian nature of hydrocarbon fluids, the application of phase-equilibrium equation, and so on. At that time, experimental testing techniques and computer technology also made remarkable progresses.
(c) Since the 1990s, the study on Physics of Petroleum Reservoirs has stepped into a new stage. Various research findings and literatures prospered.

In China, Physics of Petroleum Reservoirs was noticed as an individual subject around the 1950s. The signal event was that a former Soviet Union expert being invited to have, for the first time, a lecture on this subject in our first petroleum school (Beijing Petroleum Institute), and to tutor graduate students. Since then, there had been more and more professionals engaged in teaching and researching of this area in step with the increasing numbers of oilfields in China. At the same time, the successive discoveries and developments of the giant oilfields in eastern China also promoted further development of this subject in China. Institutes and laboratories for the research in this area were established one after another in Beijing and the major oilfields in China. Various experiments and pilot tests on petroleum reservoir development and EOR technologies were also carried out. As a result, numerous research achievements were made, and the professional teams have thus been better developed. The study on Physics of Petroleum Reservoirs in China study has progressed significantly. Since the 1960s, many iconic works reflecting the latest research results on this area have been published. To name a few: *Reservoir Physical Basis* by Shiduo Hong (1985) [6], *Reservoir Physics* by Zhetan Luo (1985) [7], *Physics of Oil-gas Reservoir* by Boqun Zhang et al. (1989) [8], *Reservoir Physics* by Gensheng He (1994) [9], *Fundamentals of Petrophysics* by Shenglai Yang (2004, 2011) [10, 11], *Reservoir Physics* by Yuncheng Wang (2006) [12], etc. In addition, a growing number of articles published in a variety of magazines.

Through the constant efforts of generation after generation of scholars both at home and abroad, following achievements have been made in this subject:

(a) A relatively complete theoretical system;
(b) Standardized experimental technologies;
(c) Wide test contents: various physical properties of reservoir rocks and reservoir fluids;
(d) Updated research areas and method: pore structure, multiphase flow, surface properties, capillary pressure, and so on;
(e) Increasing test items: conventional core analysis, special core analysis, EOR tests, flow tests for formation sensitivity, and so on.

In a word, more and more normative and mature research methods have developed, and a number of industry standards have established.

However, we should notice that, due to the heterogeneity of reservoir rocks and fluids, and the complicated physical phenomena and processes occurring in reservoirs, we are not able to fully understand, explain, and describe all phenomena in petroleum reservoirs at present. Some understandings may be relatively immature.

However, the needs of oil-gas production always promote the development of this subject. New problems emerging in petroleum production bring about new research targets. Physics of Petroleum Reservoir will continue to grow, blossom, and evolve with the following features:

(a) Integration: mutual penetration and cooperation between disciplines forms new edge theories;
(b) Innovation: the application of new theories accelerates the innovations in research methods and test technology to solve new problems occurring in petroleum production;

(c) Reproducibility: tests and experiments will be more reliable to simulate actual conditions and development process of reservoirs.

Nowadays, it enters a brand-new research stage of Physics of Petroleum Reservoirs, in not only the breadth and depth of theoretical research, but also the contents of experimental tests and the service areas for oil-gas production. It has become an indispensable theoretical basis for oilfield development in China. With the development of deep hydrocarbon reservoirs, high water-cut oil reservoirs and low-permeability oil/gas reservoirs, increasing difficulties may occur in petroleum production. Physics of Petroleum Reservoirs, offering the indispensable fundamental data, theories and research methods for oil-gas production, will no doubt play a more important role in the future.

1.3 Description and Targets of This Course

The English word *petroleum* is derived from the combination of Greek word *petro* (rocks) and Latin word *Oleum* (fuel oil), meaning the fuel oil that comes from rocks. Physics of Petroleum Reservoirs is here focused on the properties of reservoir rocks and reservoir fluids (oil, gas, and water), as well as the microscopic mechanism of reservoir fluids flowing through rocks. The main contents include the following:

(a) The physical properties of reservoir rocks
Underground reservoir space refers to the void space within the rocks. The physical properties of a rock include the properties of matrix and pores in the rock. The unique properties of reservoirs make them fundamentally different from ordinary liquid containers on the ground. As porous media, the unique properties of rocks determine the distribution and flow patterns of reservoir fluids in rocks. Therefore, the properties of reservoir rocks are the basis of petrophysics for petroleum production, and are the important issues of oilfield development theories.

(b) The physical properties of reservoir fluids
Different from the oil on the ground, the crude oil in reservoirs is under high pressure and temperature, with a large number of dissolved gas in it. The physical properties of reservoir fluids and the phase behavior of oil-gas system are the basis of fluid investigations in petroleum production, and are the important issues of oilfield development theories as well.

(c) The microscopic mechanism of fluids flowing through rocks
This part explains essentially the mechanism of the distribution, flow, and remaining of oil and gas in reservoirs, which lays a solid foundation for the research of EOR and the implementation of EOR measures.
In a word, the knowledge of the physical properties of reservoir rocks and fluids, and the interaction between rock and fluids is always involved in the whole life of reservoir production. At the early stage of a reservoir's life, it can

be used for the determination of oil/gas reservoir types, and serves for the design of development plan. During the reservoir development, it is indispensable for performance analysis and reservoir management. When it comes to the late life of a reservoir, it serves as a vital groundwork of studies on the distribution of remaining oil, EOR methods and so on. Therefore, the main targets of this subject are to provide reliable basic parameters for accurately recognizing, actively reforming and effectively developing hydrocarbon reservoirs.

References

1. Muskat M (1949) Physical principles of oil production. McGraw-Hill, New York
2. Kotyakhov FI (1958) Основы физики нефтяного пласта (Foundation of Reservoir Physics). Petroleum Industry Press, China
3. Chilingar GV, Rieke HH, Mannon R (1972) Oil and gas production from carbonate reservoirs. Elsevier, Amsterdam
4. William D, McCain (1977) The properties of petroleum fluids. PennWell Books
5. Markhasin (1982) The physical and chemical mechanism of oil-bearing reservoir. Scientific Research and Design Institute of Daqing Petroleum Administration, China
6. Hong SD (1985) Reservoir physical basis. Petroleum Industry Press, China
7. Luo ZT (1985) Reservoir physics. Geology Press, China
8. Zhang BQ et al. (1989) Physics of oil (gas) reservoir. China University of Geosciences press, China
9. He GS (1994) Reservoir physics. Petroleum Industry Press, China
10. He GS, Tang H (2004) Reservoir physics, Petroleum Industry Press, China
11. He GS, Tang H (2011) Reservoir physics. Petroleum Industry Press, China
12. Wang YC (2006) Reservoir physics, Sichuan Science and Technology Press, China

Chapter 2
Physical Properties of Reservoir Rocks

Xuetao Hu and Su Huang

A rock capable of producing oil, gas, or water is called *reservoir rock*. A reservoir rock may be any rock with sufficient porosity and permeability to allow oil and gas to accumulate and be produced in commercial quantities [1]. Petroleum generally occurs in sandstones, limestones, dolomites, conglomerates, and shales, but sometimes it is also found in igneous and metamorphic rocks. As a whole, sandstone and carbonate rocks are the most common reservoir rocks. However, reservoir rocks are quite variable in composition and physical properties [2]. The physical properties of reservoir rocks are the vital information for producers to estimate the geological reserve and ultimate recovery of an oilfield and determine the most efficient method of petroleum production.

Reservoir rocks may be investigated from two points of view, namely, microscopic and megascopic. If we study a core or a sample of a reservoir rock containing petroleum and place it under a magnifying glass (binocular or microscope), we can observe that the reservoir rock is made of a framework of minerals (rigid or friable) which fills but part of the void space. The void space in the rock is broadly known as *pore*. Oil, gas, or/and water can enter and fill in the pores if they are interconnected. In general, interconnected pores are interlaced within the mineral framework. The size, quantity, and shape of the interconnected pores in a reservoir rock determine the ability of the rock storing and transmitting fluids. Therefore, the physical properties of reservoir rocks are in essence dependent on the mineral framework of the rock, namely on the mineral composition and the texture of the rock.

X. Hu (✉)
School of Oil and Natural Gas Engineering, Southwest Petroleum University,
Chengdu, Sichuan, China
e-mail: huxt@swpu.edu.cn

S. Huang
Exploration, PetroChina Southwest Oil and Gasfield Company,
Chengdu, Sichuan, China
e-mail: huangsu419@petrochina.com.cn

X. Hu et al. (eds.), *Physics of Petroleum Reservoirs*, Springer Geophysics,
DOI 10.1007/978-3-662-55026-7_2

2.1 Matrix Features of Reservoir Rocks

Although we find oil and gas occasionally in igneous rocks, most of reservoir rocks are sedimentary in origin. Sedimentary rocks that contain oil and gas are generally divided into two categories: clastic sedimentary rocks and carbonate sedimentary rocks. The former chiefly includes conglomerate, sandstone, and mudstone, while the latter usually refers to limestone and dolomite. Sandstone is one of the most common and important petroleum reservoirs. Sandstone is a clastic sedimentary rock composed of more than 50 % sand-sized minerals or rock grains.

Sandstone is an aggregate of various particles. The size and nature of these particles in sandstone determine the characteristics of the rock. Actually, all clastic rocks have two basic properties: composition and texture. Rock texture indicates the way the particles or minerals are put together to constitute the rock. Some other properties, like structure, density, color, electrical properties, porosity, and permeability, are only second- or third-order derivatives of the two basic properties (composition and texture). In order to find the best way to exploit petroleum from underground rocks, it is necessary for the producers to know what a rock is composed of and how it is formed because the knowledge of the mineral composition and rock texture can give people an idea of how easy the petroleum may get into the rock.

In the description of sedimentary rocks, texture is a key content. It can help to interpret the mechanisms and environments of deposition. The texture of clastic sediments includes external characteristics of sediment grains, such as size, shape, and orientation [3]. Sediment texture is dependent on the grain morphology, surface features, and the fabric of the sediment.

Generally, rock's texture can be adequately studied by means of casting thin section and scanning microscope. For sand and silt-sized sediments, the analysis of rock's texture mainly focuses on the size, sorting, shape, and roundness of particles (grains). The grain-size of clastic rocks may vary from clay in shales; silt in siltstones; sand in sandstones; and gravel; cobble; to boulder-sized fragments in conglomerates and breccias (Fig. 2.1). Different rocks consist of different particles. As a result, the size, distribution, and nature of particles in a rock are responsible for many important physical and chemical properties of the rock, such as porosity, permeability, wettability, and so on. The grain-size composition and distribution of a reservoir rock are thus very important information for petroleum engineers to understand the properties of the rock.

2.1.1 Grain-Size Distribution of Rocks

Grain-size is one of the most important characteristics of rock particles. It is customarily defined as the average diameter of grains in sediments or lithified particles in clastic rocks, expressed in millimeter. It is a measure of the particle size of a rock.

Fig. 2.1 A display of grains ranging in size from silt to very coarse sand

Table 2.1 Grain-size classes for sediments and clastic rocks (Zhu [4])

Grain-size (mm)	Class	Sediment	Rock
>1000	Boulder	Gravel	Conglomerate and breccia
100–1000	Cobble		
10–100	Pebble		
2–10	Granule		
1–2	Huge sand	Sand	Sandstone
0.5–1	Coarse sand		
0.25–0.5	Medium sand		
0.1–0.25	Fine sand		
0.05–0.1	Coarse silt	Silt	Siltstone
0.005–0.05	Fine silt		
<0.005		Clay	Claystone (mudstone)

According to grain-size, the grains of clastic rocks can be classified into several broad groups: gravel, sand, silt, and clay (Table 2.1), which are generally called as grain-size classes. The rocks corresponding to each grain-size class above are known as conglomerate/breccia, sandstone, siltstone, and mudstone, respectively. In China, the accepted grain-size scale is classified with decimal system (Table 2.1). However, the common parameter is crystal size rather than grain-size in the description of chemical rocks, such as evaporates, recrystallized limestones, and dolomites.

2.1.1.1 Grain-Size Composition (Granulometric Composition)

A grain-size composition means the constitution of various size particles in a rock. It is customarily defined as the percentage of different size particles in a rock (e.g., sandstone), in weight. Mathematically, it can be expressed as

$$G_i = \frac{w_i}{\sum w_i} \times 100\,\% \tag{2.1}$$

where G_i is the weight percent of ith group of rock particles, %; w_i is the weight of ith group of rock particles, g.

Using the grain-size composition of a rock, the size and distribution of particles in the rock can be quantitatively described. It is helpful to the understanding of the rock texture.

2.1.1.2 Grain-Size Analysis

The measurement of the grain-size distribution of a rock is called *grain-size analysis*. In general, there are various laboratory techniques available for grain-size analysis. However, what technique can be used in the analysis mainly depends on the size of grains, the volume of the rock, and the consolidated degree of the rock. Generally, a systematic method of grain-size analysis includes the following techniques: sieve analysis, sedimentation method, direct measurement, and other methods.

In grain-size analysis, direct measurement is a subsidiary method, just suitable for the samples consisting of larger size particles, e.g., cobble, boulder, etc. Other methods mainly include optical method, electrical method, thin section analysis, image and analysis. These methods are often used for special samples, such as tiny samples and tightly cemented samples. Sieve analysis and sedimentation method are the routine methods in grain-size analysis. The principles of the two methods are simply described as follows.

Sieve analysis

Sieve analysis is a quick method and can provide a reliably and relatively broad grain-size spectrum. It is thus a main method of grain-size analysis and often used for determining the grain-size distribution of conventional sandstones.

In this method, a rock sample should be first broken into dispersed grains or particles. Then the grains are separated into several fractions with a stack of sieves on a shaker (Fig. 2.2). Each fraction of the grains has different average grain-sizes. After that, each fraction of grains is weighed and the grain-size composition of the sample can then be calculated according to the weight of each fraction of grains.

A typical sieve analysis involves a nested column of sieves with wire mesh cloth (screen) which has different mesh sizes (opening size) (Figs. 2.2 and 2.3). The opening size of the nested sieves should be suitable for the grain size of the sample. The nested sieves should be placed in the order of decreasing opening size from top to bottom on a mechanical sieve shaker (Fig. 2.2). A pan should be placed below the stack of sieves to collect the aggregate that passes through the last sieve.

The mesh size of sieves is expressed in two ways. One is the number of openings per inch (linear) of mesh cloth, called mesh, e.g., 200 mesh means 200 openings per inch of mesh cloth. The other one is the size (diameter) of openings of mesh cloth

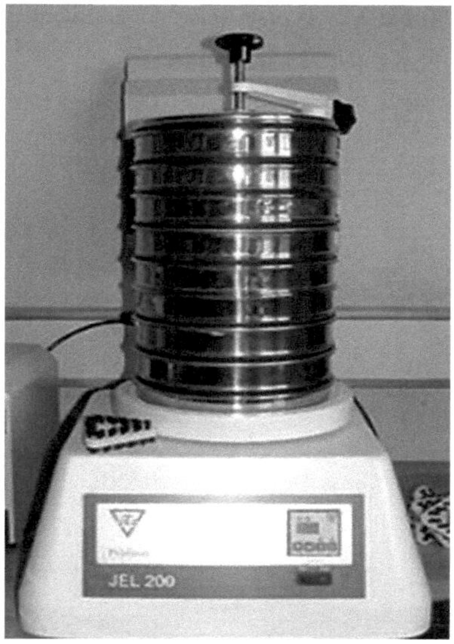

Fig. 2.2 A shaker used in sieve analysis

Fig. 2.3 Sieves used in sieve analysis

expressed in millimeter. In general, the opening size of a sieve is approximately $1/\sqrt{2}$ or $1/\sqrt{42}$ increase than that of the next sieve. The opening size of sieves used in China and USA is listed in Table 2.2. The series of 9, 10, 14, 20, 27, 35, 40, 60, 80, 100 mesh sieves are commonly used for the grain-size analysis of conventional sandstones.

The routine procedure of sieve analysis includes the following steps:

(a) *Sample preparation*: first, select a representative sample from rock cores. Then let the sample be crumbled in a pulverizer to dispersed grains.
(b) *Sieving*: put the dispersed grains of the sample into the top sieve of the stack of sieves on a shaker shown in Fig. 2.2. Turn on the shaker and let it run for enough time for the separation of the grains. During the process of shaking, the grains pass through each sieve from top to bottom. Each of the sieves

Table 2.2 Typical sieve no. and corresponding opening sizes in sieve analysis

Sieve size (China) [5]		Standard sieve size (U.S.) [6]	
Sieve no. (mesh)	Opening (mm)	Sieve no. (mesh)	Opening (mm)
4	4.599	4	4.760
5	3.962	5	4.000
6	3.327	6	3.360
7	2.794	7	2.830
8	2.362	8	2.380
9	1.981	10	2.000
10	1.651	12	1.680
12	1.397	14	1.410
14	1.165	16	1.190
16	0.991	18	1.000
20	0.833	20	0.840
24	0.701	25	0.710
27	0.589	30	0.590
32	0.495	35	0.500
35	0.417	40	0.420
40	0.350	45	0.354
60	0.245	50	0.297
65	0.220	60	0.250
80	0.198	70	0.210
100	0.165	80	0.177
110	0.150	100	0.149
180	0.083	120	0.125
200	0.074	140	0.105
250	0.061	170	0.088
270	0.053	200	0.074
325	0.047	230	0.063
425	0.033	270	0.053
500	0.025	325	0.044
625	0.020	400	0.037

retains part of the grains due to larger grain-size than the opening size of the sieve. The dispersed grains of the sample are thus separated into several fractions.

(c) *Determine the weight of each fraction*: measure weight of the grains retained on each sieve with a balance. The weight of each fraction of grains is expressed as w_1, w_2, …, w_n. w_n is the weight of the grains in the pan. The cumulative weight, W of all fractions is then calculated by the weight of each fraction. Namely, $W = w_1 + w_2 + \cdots + w_n$.

(d) *Determine grain-size distribution*: the percentage of each grain fraction retained on each sieve can be determined by Eq. (2.1), expressed as G_1, G_2, … G_n.

Sieve analysis has been used for decades to monitor the particle size of different particle materials [7]. For coarser materials, the sizes may range down to 0.150 mm; sieve analysis is accurate and the particle size distribution is consistent. However, for materials finer than 0.150 mm, this method is significantly less accurate. For finer particles or grains, it could be harder to make the particles pass through the openings of sieves because of the increasing mechanical energy required as well as effects of interfacial adsorption between particles and between particle and the mesh cloth.

This method assumes that all particles are orbicular or nearly so, and one particle can only pass through the square-shaped openings with larger sizes than its own diameter. For elongated and flat particles, sieve analysis can give a less reliable result because an elongated particle may pass through the screen from end-on, but would be prevented the other way (e.g., if it is side-on). For example, a needle-shaped particle can either pass through a mesh or be retained on the screen, depending on its orientation during sifting. Even so, the density of particles has no effect on the analysis results. Sieve analysis is a favorite method for loose or weakly cemented rocks, but it may not be suitable for soft or tight rocks. Such kind of rocks may generate size-distorted particles during the process of breaking the sample.

In addition, the precision of sieve analysis is also affected by the following factors: the number of sieves used in the analysis and their quality, moisture of grains, and sieving time. Experience of the user is also reflected in the measuring results.

In a word, this method is simple, easily available, and relatively reliable. It is widely used in grain-size analysis of particle materials and reservoir rocks. Sieve analysis can also provide useful information and materials for the study on mineralogy and particle shape in the future.

Sedimentation method

Sedimentation analysis appeared more than a century ago [8]. It was used in laboratories at the end of the nineteenth century and served as the most common method of soil and sediment particle size analysis.

Sedimentation method is a subsidiary method for the grain-size analysis of conventional sandstones. It is awfully appropriate for the grain-size analysis of tiny particles retained in the pan in sieve analysis. It is accurate for particles ranging between 0.053 and 0.074 mm, but it is less accurate for particles in sub-micrometers due to the effects of Brownian motion.

This method is based on the application of Stokes' law. Stokes' law describes the settling velocity of spherical particles in a viscous fluid under gravity (i.e., free falling). Under the conditions of low Reynolds numbers (i.e., laminar flow), the settling velocity of particles depends on the ratio of the particle density to the fluid density, the viscosity of the fluid, and the size of particles. Therefore, different particles have different settling velocities when subsiding in a viscous liquid (Fig. 2.4).

Fig. 2.4 Sketch map of particle settling

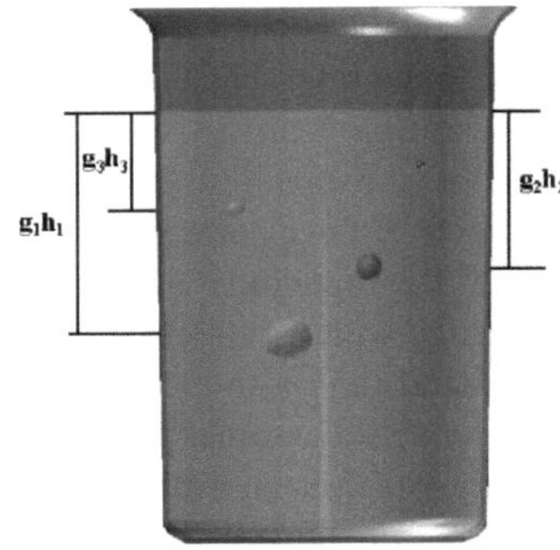

g_i: acceleration of gravity; h_i: distance

In the simplest case, the settling velocity of a particle is proportional to its size. According to the particle's velocity measured, the diameter of particles can be determined according to the Stokes equation of fluid mechanics:

$$d = \sqrt{\frac{18\gamma v}{g(\rho_s/\rho_L - 1)}} \tag{2.2}$$

where γ is the kinematical viscosity of suspension liquid, cm^2/s; v is the terminal settling velocity of a particle in the liquid, cm/s; g is gravity acceleration, cm/s^2; ρ_s is particle density, g/cm^3; ρ_L is liquid density, g/cm^3; d is particle's diameter, cm.

Equation (2.2) is valid for single spherical particle subsiding slowly in a fluid without the interference of other forces or motions. The concentration of particles in the liquid should be low enough to ensure that particles are adequately dispersive in the liquid and no appreciable interaction and interference between particles occur during the experiment. Therefore, the maximum mass concentration, commonly accepted, is usually less than 1 %. In addition, In order to reduce the interference of vessel wall to particle settling, the distance of a particle to the vessel wall should be at least 0.5 cm. In this situation, however, some particles may dissolve partially or fully in the fluid altering the grain-size distribution.

Mean size of particles

In clastic reservoir rocks, the particle size varies greatly in nature. In sieve analysis, each sieve actually retains an aggregate of different size particles. We thus know the size range of particles between the opening sizes of two adjacent sieves but grain-size of each particle. Therefore, it is very necessary to know the mean size of particles in each sieve for the study of rock texture. In an ordinary way, the mean size of particles in each sieve/fraction can be determined by the following formula:

$$\frac{1}{\bar{d}_i} = \frac{1}{2}\left(\frac{1}{\bar{d}'_i} + \frac{1}{\bar{d}''_i}\right) \tag{2.3}$$

where $\bar{d}_i$ is the mean diameter of particles retained on ith sieve, μm; d_i' is the opening size of last sieve (i-1th), μm; d_i'' is the opening size of ith sieve, μm.

2.1.1.3 Grain-Size Distribution

To display the results of the grain-size analysis of a rock, first find the percent of each fraction of grains. To do so, Eq. (2.1) can be used. Next, find the cumulative percent of all grains till ith fraction. The percentage of cumulative grains to ith fraction is determined by Eq. (2.1) using the total weight of all grains to ith fraction.

The result of grain-size analysis can be represented using either a table or a diagram. Table 2.3 shows the result of a grain-size analysis for a typical sandstone sample. Generally, the graphical method is more popular in oilfields, normally including grain-size distribution curve, cumulative grain-size distribution curve, grain-size frequency histogram, etc.

The curve of the weight percent of grains to their diameter is referred to as *grain-size distribution curve* or *grain-size frequency curve* (Fig. 2.5). A curve with cumulative weight percent on the y axis and grain-size (grain diameter or phi value) on the x axis is called *cumulative grain-size distribution curve* or *grading curve* (Fig. 2.6).

The two curves can visually display the grain-size distribution of a rock. Generally, a sharper peak of grain-size distribution or a steeper curve of cumulative grain-size distribution reflects more uniform grain-sizes (good sorting) of a rock; a flat peak or less steep curve reflects less uniform grain-sizes (poor sorting) in opposite manner. Different locations of the curve in the diagram indicate different average grain sizes of rocks (Fig. 2.7). From here we see that both the two curves can characterize illustratively the size and distribution of rock particles.

Table 2.3 The results of a grain-size analysis for a typical sandstone

Opening size of sieve (mm)	0.833	0.701	0.589	0.495	0.417	0.35
Weight percent (%)	2.10	13.11	54.15	18.50	7.44	4.70
Cumulative weight percent (%)	2.10	15.21	69.36	87.86	95.30	100.0

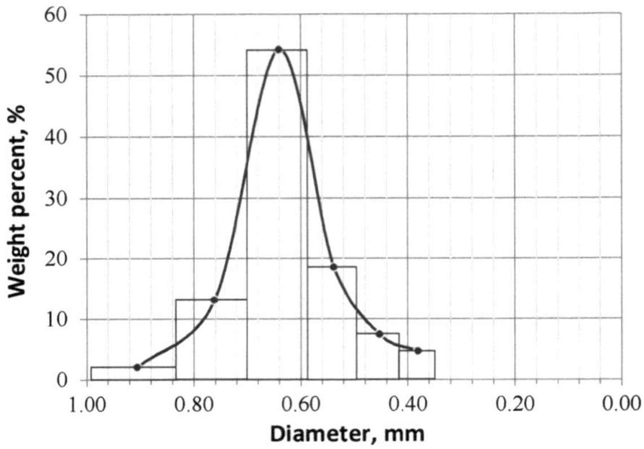

Fig. 2.5 Grain-size distribution curve

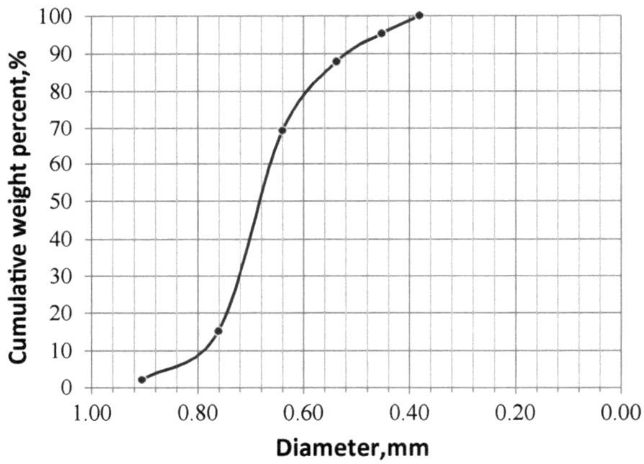

Fig. 2.6 Cumulative grain-size distribution curve

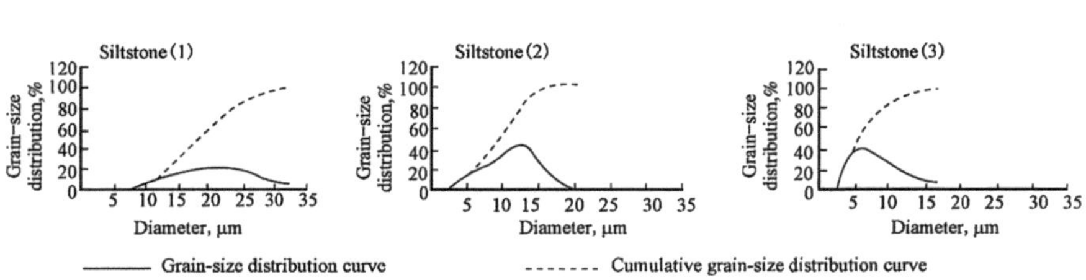

Fig. 2.7 Grain-size distribution curves of different siltstones

2.1.1.4 Grain-Size Statistical Parameters

Sorting of a grain population is a measure of the range of grain-sizes present and the magnitude of the spread or scatter of these sizes around the mean size [9]. It indicates the uniformity of grain-size distribution within a sedimentary rock. In well-sorted sediments, the grain-size of particles is very similar, while the particles consisting of poorly sorted sediments usually are distributed in a wide range of grain-size (Fig. 2.8). In the same way, a rock having a wide range of grain-size is said to be poorly sorting, whereas a well-sorted rock has a relatively narrow grain-size range. Well-sorted rocks are generally porous and have high permeability, but poorly sorted rocks have lower porosity and permeability.

Describing the significant feature of grain-size distributions, grain-size parameters can be used to evaluate the grain-size distribution or sorting of rock particles. Grain-size parameters can be determined in terms of different mathematical methods. Most grain-size parameters are defined based on graphical-statistical method, such as median, mean, standards deviation, skewness, kurtosis, etc. This is to say that these parameters are calculated with the grain diameters read from a graph (cumulative grain-size distribution curve). At present, the common parameters are customarily calculated with the method proposed by Folk and Ward. Here are several common grain-size parameters.

Standard deviation

Standard deviation is widely used to evaluate the sorting of rock particles. It describes the uniformity of grain-sizes in a rock. Folk and Ward (1957) proposed the following expression for calculating graphic standard deviation [10]:

$$\sigma = \frac{\phi_{84} - \phi_{16}}{4} + \frac{\phi_{95} - \phi_5}{6.6} \tag{2.4}$$

where phi (ϕ) is a logarithmic transformation: $\phi_i = -\log_2 d_i$; d_i is grain diameter in millimeters; The subscript of 5, 16, 84, and 95 denotes separately the 5, 16, 84, and 95 % by cumulative weight of a cumulative grain-size distribution curve. So, ϕ_5, ϕ_{16}, ϕ_{84}, and ϕ_{95} represent the phi values at 5, 16, 84, and 95 cumulative percentage (Fig. 2.9).

The standard deviation of grain-size distributions of rocks ordinarily ranges from <0.35 to >4.00. Based on the statistical results of a lot of samples, a verbal

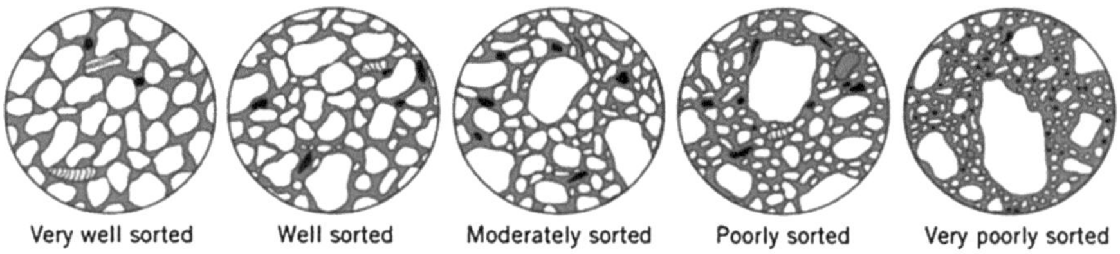

Fig. 2.8 Schematic diagram of sorting levels of sediment particles

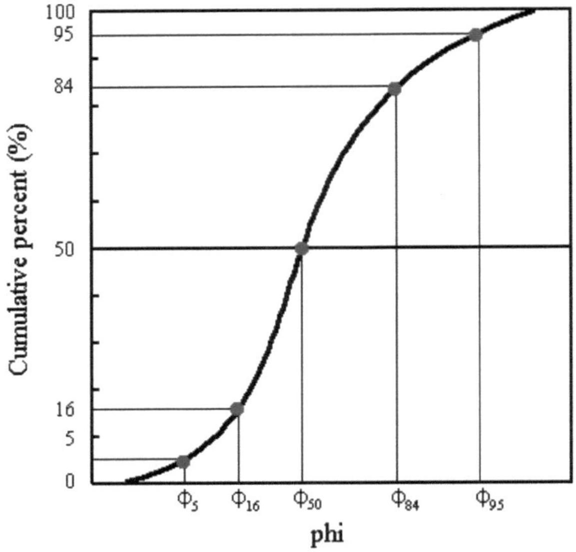

Fig. 2.9 Schematic diagram for determining the values of ϕ_5, ϕ_{16}, ϕ_{84} and ϕ_{95} on the curve of cumulative grain-size distribution

Table 2.4 Classification scale of standard deviation

Standard deviation	<0.35	0.35–0.50	0.50–0.71	0.71–1.00	1.00–2.00	2.00–4.00	>4.00
Scale	Very well sorted	Well sorted	Moderately well sorted	Moderately sorted	Poorly sorted	Very poorly sorted	Extremely poorly sorted

classification scale for sorting corresponding to various values of graphic phi standard deviation is presented by Folk and Ward (1974) (Table 2.4) [10, 11]. Obviously, the smaller the standard deviation is, the more the uniform rock particles are.

Skewness

Skewness is an additional measure of grain-size sorting. Actually, two samples may have the same average grain-size and sorting but may be quite different to their degrees of symmetry. Skewness measures the degree to which a grain-size distribution curve approaches symmetry. The grain-size distributions of most clastic sediments do not yield a perfect normal, or lognormal, distribution curve. Instead, they display an asymmetrical distribution. Folk's inclusive graphic skewness (1957) is calculated by the equation [10]:

$$S_{\mathrm{kp}} = \frac{\phi_{84} + \phi_{16} - 2\phi_{50}}{2(\phi_{84} - \phi_{16})} + \frac{\phi_{95} + \phi_5 - 2\phi_{50}}{2(\phi_{95} - \phi_5)} \tag{2.5}$$

where S_{kp} is skewness of grain-size distribution. The phi values (ϕ_i) represent the same percentages as those for standard deviation, namely ϕ_5, ϕ_{16}, ϕ_{50}, ϕ_{84}, and ϕ_{95} represent the phi values at 5, 16, 50, 84, and 95 cumulative percentage.

In general, the skewness of a grain-size distribution varies between −1 and 1, as $-1 < S_{kp} < 1$. Symmetrical grain-size curves (normal distribution) have a skewness equal to 0; those with a large proportion of fine particles are said to be fine skewed, or positively skewed (fine sediment has positive phi values, $S_{kp} > 0$) (Fig. 2.10a); those with a large proportion of coarse particles are said to be coarse skewed, or negatively skewed ($S_{kp} < 0$) (Fig. 2.10b). A verbal classification scale proposed by Folk (1966) [4] is listed in Table 2.5. Obviously, the more the value of the skewness deviates from zero, the greater the skewness of grain-size distribution.

Kurtosis

Kurtosis is a measure of the *sharpness* of a grain-size distribution curve. The sharpness or peakedness of a grain-size distribution curve is known as Kurtosis. Kurtosis describes the concentrative degree of grain-size distribution in a rock. Generally, a normal distribution curve is mesokurtic; a sharp-peaked curve is said to be leptokurtic; and a flat-peaked curve is platykurtic. Sharp-peaked curves indicate better sorting in the central portion of the grain-size distribution than the tails, and flat-peaked curves indicate the opposite.

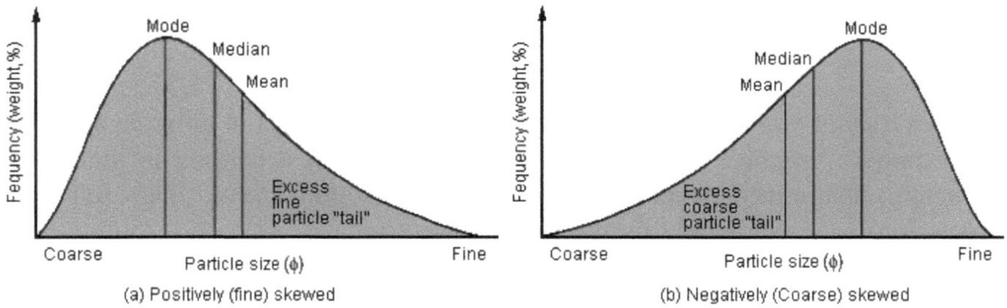

Fig. 2.10 Skewed grain-size frequency curves, illustrating the difference between positive (fine) and negative (coarse) skewness (Sam [9])

Table 2.5 Gradation of skewness (Folk 1966)

Skewness	<−0.30	−0.3 to −1	−0.1 to +0.1	+0.1 to +0.30	>+0.3
Scale	Strongly coarse skewed	Coarse skewed	Nearly symmetrical	Fine skewed	Strongly fine skewed

Folk's formula (1957) for calculating graphic kurtosis is as follows [10]:

$$K_p = \frac{\phi_{95} - \phi_5}{2.44(\phi_{75} - \phi_{25})} \tag{2.6}$$

where K_p is the kurtosis of a grain-size distribution; the phi values represent the same percentages as those of standard deviation.

In general, normal distribution curves have a kurtosis equal to 1; leptokurtic curves have a kurtosis higher than 1, and platykurtic curves have a kurtosis lower than 1. If a curve is very flat or double-peaked, K_p may be less than 0.6; A much sharp-peaked curve may have a K_p ranging from 1.5 to 3.

Mode

Mode (Mo) is the most frequently occurring grain-size in a population of rock particles. The modal diameter is the diameter of grains represented by the peak of a distribution curve or the steepest point of a cumulative curve, as shown in Fig. 2.10.

To a certain extent, it represents the grain-size of the majority and the sorting of rock particles.

Median

Median (Md) is the grain-size corresponding to the 50 % cumulative weight of a cumulative grain-size distribution curve. It represents the midpoint of a grain-size distribution. In a rock sample, half of the grains by weight are coarser than the median, and half are finer.

Mean

Mean of a grain-size distribution refers to the arithmetic mean of all grain-sizes in a rock sample. Actually, a graphic mean can be obtained according to the typical diameters determined from a cumulative distribution curve. The following expression is the commonly used algorithm:

$$M_z = \frac{d_{16} + d_{50} + d_{84}}{3} \tag{2.7}$$

where d_{16}, d_{50}, and d_{84} are particle diameters separately corresponding to 16, 50, and 84 % cumulative weight of the cumulative distribution curve, μm.

2.1.2 Specific Surface Area of Rocks

Specific surface area, or SSA for short, is an essential property of reservoir rocks. It is a measure of the total surface area of grains in a rock. The SSA of reservoir rocks has a particular importance for the study of adsorption and reactions on surfaces in petroleum engineering.

2.1.2.1 Specific Surface Area

Specific surface area is defined either by the total surface area of rock matrix divided by the volume of a rock, or the total internal surface area of void space divided by the volume of a rock, That is:

$$S = \frac{A}{V} \tag{2.8}$$

where S is the specific surface area of a rock, cm^2/cm^3 or 1/cm; A is the total surface area of rock matrix or total internal surface area of the void space in the rock, cm^2; V is the bulk volume or the view volume of the rock, cm^3.

Specific surface area can also be defined either by the total surface area of rock matrix divided by the mass of a rock (with units of cm^2/g). The specific surface area of conventional sandstones is about 500–5000 cm^2/g, and the specific surface area of shale is about 1,000,000 cm^2/g, namely 100 m^2/g.

In practical applications, specific surface area may be expressed as the surface area per unit solid volume of a rock V_s or per unit pore volume in a rock V_p, namely $S_s = A/V_s$, and $S_p = A/V_p$. Here S_s and S_p are the specific surface areas which are defined by V_s and V_p, respectively. Based on the relationship between V and V_p, or between V and V_s: $\phi = V_p/V = 1 - V_s/V$, the following expression can be obtained:

$$S = \phi \times S_p = (1 - \phi)S_s \tag{2.9}$$

where ϕ is the porosity of a rock, f; S is the specific surface area per unit bulk volume of the rock, cm^2/cm^3; S_p is the specific surface area per unit pore volume of the rock, cm^2/cm^3; S_s is the specific surface area per unit solid volume of the rock, cm^2/cm^3.

Suppose an ideal porous medium composed of same size spherical particles. In this medium, the total internal surface area of void space equals the total surface area of all particles without regard to the effect of particle contact. Suppose the radius of the spherical particle is R (Fig. 2.11), the specific surface area of the medium can be easily determined by the formula: $S = 8 \times 4\pi R^2/(4R)^3$. Obviously, the specific surface area of a porous medium is inversely proportional to the particle radius (R) packing the medium. This conclusion also applies to sedimentary rocks. The specific surface areas of clastic rocks generally increase with the decrease in

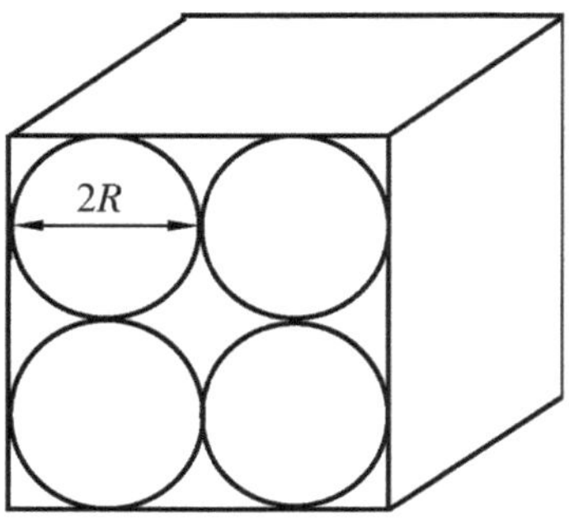

Fig. 2.11 Packing of spherical particles

Table 2.6 The statistical results of the specific surface areas of rocks

Rocks	Grain diameter (mm)	Specific surface area (1/cm)
Coarse, medium sandstone	1.0–0.25	<950
Fine sandstone	0.25–0.1	950–2300
Siltstone	0.1–0.01	>2300

grain-size (Table 2.6). Therefore, specific surface area can characterize the average size and the dispersion degree of rock particles.

Essentially, the SSA of a rock characterizes the total surface area of rock particles exposed to fluids in the rock. When flowing through a rock, the fluid in rock pores is subject to the adsorption of rock particles. Therefore, SSA also represents indirectly the magnitude of the adsorption of rock particles to fluids flowing through the rock. The larger the SSA of a rock, the stronger the adsorption of particles to fluids. SSA can also help in understanding the important reservoir phenomena, such as adsorption, wettability, capillarity, solubility, and free interfacial energy (see Chap. 4). As a result, rock's SSA is highly concerned by petroleum engineering.

2.1.2.2 Measurement of Specific Surface Area

The SSA of a rock can be measured by at least two techniques. One is based on the permeability of rocks, called gas permeability method or penetrant method; and the other is based on the adsorption of gas on particle's surface, called gas adsorption method. Besides, the petrographic image analysis (PIA) and nuclear magnetic resonance (NMR) may also be used to measure the SSA of various particle materials at present [12]. Details of gas permeability method and gas adsorption method are briefly introduced as follows.

Gas permeability method

Based on the permeability of rocks to the gas flowing through the rock, gas permeability method is developed by Kozeny. Kozeny–Carman equation (1927) is widely used for all permeametry modifications [13]. In terms of the theory of Kozeny–Carman [14], the rock pores that fluids flow can be supposed to be a bundle of parallel capillary tubes, the SSA of a rock is then equal to the summation of the areas of all pore-walls when the fluid flows through the rock. Therefore, the SSA of a rock can be directly determined from the measurement of gas permeability (air is often used).

The apparatus of this method is shown in Fig. 2.12. It primarily consists of a Mariotte bottle, a core in core holder, and a piezometer. The porosity of the core is given.

In the measurement, first of all, close valve 3 and open valve 1, 2; then inject water to Mariotte bottle through the funnel until water level reaches the specified height. After that, close valve 1, 2 and let air flow into the bottle through the core until the water level of U-table reaches the specified pressure drop. Then, open valve 3 and let water flow into the water collector. When the pressure drop of

Fig. 2.12 Apparatus of SSA measurement

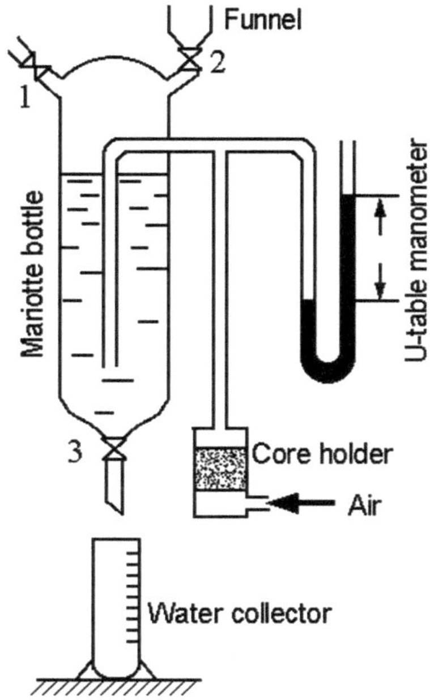

U-table is unchanged, begin recording and measure the volume of water flowing out from valve 3 in a given period of time. Under the condition of stable flow, the flow rate of air flowing through the core is exactly equal to the flow rate of water flowing out from the bottle. Therefore, the flow rate of air through the core can be determined according to the water volume measured in a given time. Finally, the SSA of the rock core can be calculated using Kozeny–Carman equation:

$$S = 14\sqrt{\phi^3}\sqrt{\frac{AH}{Q_0 \mu L}} \tag{2.10}$$

where S is the SSA of a rock core, cm^2/cm^3; ϕ is the porosity of the core, f; A is the cross-sectional area of the core, cm^2; L is the length of the core, cm; Q_0 is the flow rate of air through the core, cm^3/s; H is pressure drop, cm water column; μ is the viscosity of air at room temperature, mPa s.

This method is simple, fast, and easy to use. It is widely used in oilfields. However, it fails to measure the surface area of a deep texture. That is to say, the surface area of tiny pores can hardly be measured by this method. Generally, it is appropriate for the rocks whose particle sizes range between 0.001 and 0.1 mm.

Gas adsorption method

Gas adsorbing on the surface of solids is a common phenomenon. Gas adsorption is often influenced by surface imperfections and irregularities [15]. In porous media, the influence is superposed by their internal structure, such as shape and dimensions of the pores. On the surface of a porous medium, the adhesion of gas on the surface of the medium occurs due to the interaction between molecules of gas and the surface of the medium by the molecular interaction forces (van der Waals forces).

The principle of this method is based on the Brunauer–Emmett–Teller (BET) model. The BET model was built for determining the amount of gas adsorbtion on the surface of a sample at low temperatures [13]. The BET theory was developed by Brunauer, Emmett, and Teller in 1938 [16]. The BET theory aims to explain the physical adsorption of gas molecules on a solid surface and serves as the basis for the measurement of the SSA of porous media.

The gas adsorption technique involves determining the amount of inert gas, typically nitrogen (N_2), argon (Ar), krypton (Kr), or xenon (Xe). This method requires a single layer adsorption of inert gas on the solid surface at low temperatures. The surface area of the solid can be determined according to the area occupied by gas molecules.

In the measurement, the inert gas can enter not only the large-size rock pores through which fluids can freely flow, but also the tiny pores/fractures or dead-end pores in which fluids can hardly flow. As a result, the SSA measured by this method is frequently larger than that measured by gas permeability method. In gas permeability method, the surface of particles, though "smoothed out," is determined without regard to particle's roughness and dead-end pores.

At present, gas adsorption method is popular in the measurement of the SSA of reservoir rocks, especially in low-permeable rocks and shale. The reason is that the SSA and pore-size distribution of a rock can be time determined at a time using this method. Moreover, this method has a particular advantage of measuring the surface of fine structures and deep texture on particles. It should be, however, limited to porous media that do not have large specific surfaces [12].

In addition, the nuclear magnetic resonance (NMR) method appears to be currently the most accurate technique for estimating the specific surface area [12]. In this case, the specific surface areas S_{vg} and S_{pv} are obtained from [12]:

$$S_{\text{Vg}} = A_{\text{NMR}}\rho_m \tag{2.11}$$

$$S_{\text{pv}} = \left(\frac{1-\phi}{\phi}\right)A_{\text{NMR}}\rho_m \tag{2.12}$$

where A_{NMR} is the NMR surface area of dry material, m^2/g; ρ_m is grain-matrix density, g/cm^3; S_{vg} is the SSA per unit grain volume, m^2/cm^3; S_{pv} is the SSA per unit pore volume, m^2/cm^3.

Note that values of SSA obtained from NMR are generally higher than values obtained by the gas adsorption technique.

2.1.2.3 Estimation of Specific Surface Area

Besides the methods introduced above, the SSA of a rock can also be directly estimated according to the physical properties (porosity and permeability) of the rock.

Based on the simplified capillary model of rocks, Kozeny developed the following equation:

$$K = \frac{\phi^3}{kS^2} \times 10^8 = \frac{\phi^3}{2\tau^2 S^2} \times 10^8 \tag{2.13}$$

where k is Kozeny constant, $k = 2\tau^2$; τ is the average tortuosity of rock pores, f; ϕ is rock porosity, f; K is rock permeability, D; S is the SSA of the rock, 1/m.

Generally let $\tau = 1.4$, Eq. (2.13) is expressed as follows:

$$S \approx 5051\phi\sqrt{\phi/K} \tag{2.14}$$

Equation (2.12) is the final expression of estimating the SSA of a rock.

2.2 Porosity of Reservoir Rocks

The first element of an oil/gas reservoir is the reservoir rock. The essential feature of a reservoir rock is *porosity:* the rock must contain pores or voids, which should be of such size and character as to permit the storage of oil and gas in reservoirs that are large enough to justify exploitation. Porosity, however, is not enough; the pores must be interconnected to permit the passage of oil and gas through the rock. In other words, the rock must be permeable (it is said to have *permeability*); otherwise there would be little if any accumulation into reservoirs, nor could any petroleum that accumulated be produced by drilling wells, as the accumulation of petroleum would not move into the wells fast enough. A pumice rock, for example, would not make a good reservoir even though the greater part of it might consist of pore space, for the pores are not interconnected and the porosity is not effective. The conventional shale generally cannot become a reservoir rock, for the pores are so tiny that the capillary attraction of the fluids for the mineral grains effectively holds the fluids in the rock. Trying to get petroleum out of a shale would be like trying to remove ink from a blotter. Therefore, the porosity and the pore structure of reservoir rocks are the key factors to affect the quality of reservoirs and control the off-take potentials of oil or gas wells.

2.2.1 Pore and Pore Structure of Reservoir Rocks

Pore refers to the void space not occupied by minerals within a rock (Fig. 2.13). In a rock, most pores are fully interconnected (Fig. 2.13b). From the interconnected point of view, the idea of "throats" can be important. If the void space in a rock is shaped like the cartoon shown in Fig. 2.13c, it can be loosely defined as the "throat." A "throat" means the narrow void space connected with larger void space

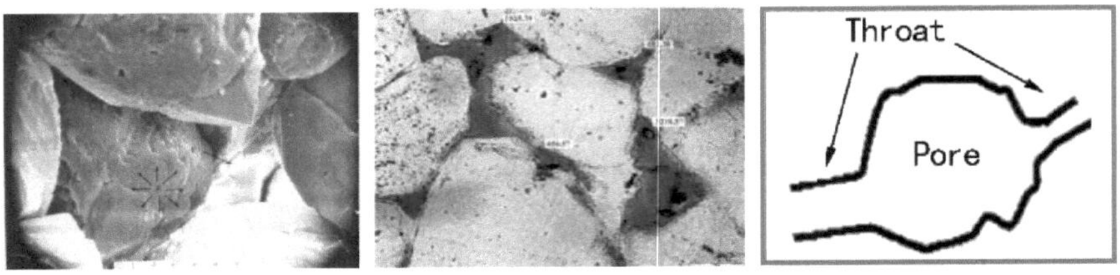

(a) (b) (c)

(a) SEM microphotograph of bore body and pore throat (Jordon, 17); (b) Intergranular pores (outlined in red) in a andstone; (c) sketch map defining a throat and pore.

Fig. 2.13 Images and sketch map of rock pores

as shown in Fig. 2.13c. In a narrow sense, a "pore" means the larger space connected with narrow space in a void system. In the void system of a rock, the size of the "throat" limits the accessibility of the larger "pore."

In sedimentary rocks, there is a wide variation in the size, shape, and arrangement of pores with respect to one another. The pore system of a rock depends essentially on the depositional environment and diagenetic process of the rock. The feature of a pore system is closely related with the microcosmic texture of the rock, namely related with the size, sorting, and arrangement of rock particles.

2.2.1.1 Pore Types of Reservoir Rocks

In petroleum reservoirs, there are various types of rock pores. Rock pores can be classified from different point of view. The common classification of pores in reservoir rocks is simply described as follows.

Classification by genesis

From the genetic point of view, rock pores in sedimentary rocks may be classified into two groups: primary pores and secondary pores, as shown in Table 2.7.

Table 2.7 Pore types classified by genesis (Richard [18])

Type		Origin
Primary/depositional pores	Intergranular or interparticle	Sedimentation
	Intergranular or interparticle	
Secondary/postdepositional pores	Intercrystalline pore	Cementation
	Fenestral pore	
	Moldic pore	Solution
	Vuggy	
	Fracture	Tectonic movement, compaction or dehydration

By definition, primary pores are the void space which formed when sediments first deposit and still remain after the sediments lithify. Primary pores contain two types: interparticle pores and intergranular pores. Interparticle pores are characteristics of sandstones, whereas intergranular pores are usually found in skeletal carbonate rocks.

Secondary pores form after deposition and are a result of diagenetic processes [19]. Recrystallization and notable dolomitization can generate intercrystalline pores. Solution can generate moldic, vuggy, and cavernous pores [18]. Because such pores are often isolated from one another, the permeability of rock matrix may be low. Fractures form in both the unconsolidated and brittle sediments. In the first instance, the fractures remain closed, but in brittle rocks, fracture may be preserved, enlarged by solution, or diminished by cementation [19]. Fracture porosity occurs not just in indurated sediments, but also in igneous and metamorphic rocks.

Intergranular pore [18]
Intergranular or interparticle pores occur in the spaces between the detrital grains that form the framework of a sediment (Fig. 2.14a). Intergranular pores are a main type of pores in sandstone, and are most common.

Intragranular pore [18]
Intragranular pores, those within grains, may exist naturally in skeletal grains or in diagenetically altered grains of any origin (Fig. 2.14b). Usually they occur in porous, skeletal allochems, which make up varying amounts of detrital sediments. Bryozoans, for example, have internal pores, as do corals sponges and many other reef-building organisms. In a bioclastic grainstone reservoir consisting of sand-sized bryozoans fragments, the bryozoans may contribute significantly to total porosity. In that case, intragranular pore corresponds with the distribution of skeletal grain-stones and packstones and, in turn, with facies maps. Common pore sizes are <0.01 to 1 mm [20].

Intercrystalline pores are spaces between more or less equal-sized crystals, often related to early and late diagenetic recrystallization and dolomitization processes (Fig. 2.14c). Common pore sizes are <1–10 μm [20].

Fenestral pore [18]
Fenestral or "bird's-eye" pores result from desiccation or from expulsion of gas during decay of organic matter in muddy sediment. The pores may be millimeter to centimeter in size, they are elongate and planar in a direction parallel to bedding (Fig. 2.14d), and they are especially common in tidal-flat environments where sediments are alternately wet and dry. Fenestral pores may not be in flow communication with each other; consequently, they may have the petrophysical properties of small, separate vugs. If they are touching and are not filled with cement, they may present high permeability values at comparatively low porosity.

Moldic pore [18]
Moldic pores result from the selective removal, commonly by solution of primary depositional grains (Fig. 2.14e), e.g., fossils or ooids. Molds are fabric selective. That is to say, solution is confined to individual particles and does not cross-cut

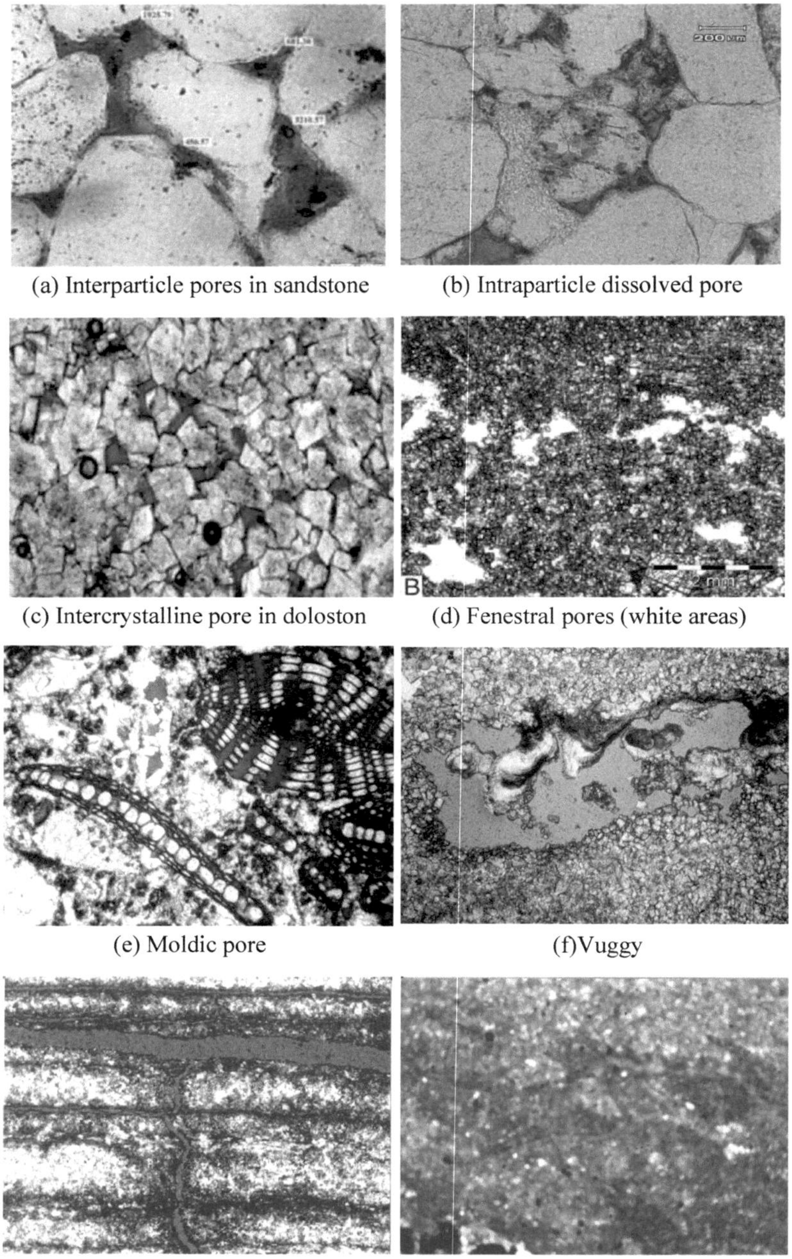

(a) Interparticle pores in sandstone
(b) Intraparticle dissolved pore
(c) Intercrystalline pore in doloston
(d) Fenestral pores (white areas)
(e) Moldic pore
(f)Vuggy
(g) Tectonic fractures
(h) Dissolved fracture

Fig. 2.14 Illustration of pore genetic type in reservoir rocks

cement, matrix, and framework. Typically, in any rock it is all the grains of one particular type that are dissolved. Therefore, it requires a distinctive mineralogical or microstructural difference between the solubility of grains and matrix or cements.

Biomoldic pores refer to pores caused by removal of fossils. Pore sizes <0.5 to several millimeters. Dissolution of (mainly aragonitic) ooids results in the formation of oomoldic pores, particularly in meteoric vadose and meteoric phreatic environments. Common pore sizes <0.20 to >0.5 mm.

In an oomoldic rock, pores will be sub-spherical and of similar size. In biomoldic rocks, by contrast, pores may be very variable in size and shape, ranging from minute apertures to curved planar pores where shells have dissolved, and cylinders where echinoid spines have gone into solution.

Vuggy pore [18]

Vugs are a second type of pore formed by solution and, like molds, they are typically found in carbonates (Fig. 2.14f). Vugs differ from molds because they cross-cut the primary depositional fabric of the rock. Vugs thus tend to be larger than molds. They are often lined by a selvage of crystals. With increasing size vugs grade into what is loosely termed "cavernous pore." People once proposed that the minimum dimension of a cavern is a pore which allows a speleologist to enter or which, when drilled into, allows the drill string to drop by more than half a meter through the rotary table. Large-scale vuggy and cavernous pore is commonly developed beneath unconformities where it is referred to as "paleokarst." This serves as a petroleum reservoir in some fields such as Tahe oilfield.

Fracture pore [18]

Fractures occur in many kinds of rocks other than sediments. It often takes the form of microfaulting caused by slumping, sliding, and compaction. Fractures in plastic sediments are instantaneously scaled, however, and thus seldom preserve porosity. In brittle rocks, fractures may remain open after formation, thus giving rise to fracture porosity. This pore type is generally formed later than the other varieties of pore. It is important to note that fracture can be found in well-cemented sandstones and carbonates, but may also occur in shales and igneous and metamorphic rocks.

Fracture is much more difficult to observe and analyze than most other pore systems. Though fractures range from microscopic to cavernous in size, they are difficult to study in cores.

Fracture can occur in a variety of ways and situations. Tectonic movement can form fracture pores in two ways. Tension over the crests of compressional anticlines and compactional drapes can generate fracture (Fig. 2.14g). Fracture is also intimately associated with faulting and some oil fields show very close structural relations with individual fault systems.

Fracture can also form from atectonic processes. It is often found beneath unconformities. Here fractures, once formed by weathering, may have been enlarged by solution (especially in limestone paleokarst) and preserved without subsequent loss of their pore (Fig. 2.14h). Fracture is extremely important in both aquifers and petroleum reservoirs. This is because a very small amount of fracture can give a very large permeability; the fractures connecting many pores of other types that might otherwise be ineffective.

In a word, sedimentary rocks generally contain more than one type of pores. The combination of open fractures with other types of pores is of particular significance. Fine-grained rocks, such as shales, microcrystalline carbonates and fine sands, have considerable porosity; but they often have very low permeability because of their narrow pore-throats and concomitant higher capillary pressures. However, the presence of fractures in such rocks makes it possible to yield easily the fluids contained in the rocks. The success of many oil and water wells in such formations often depends on whether they happen to penetrate the open fractures. Recognition of the significance of fractures in producing fluids from low-permeability rocks has led to the development of artificial fracturing technology. Similarly, the productivity of fractured carbonate reservoirs can be improved by injecting acid to the rocks to dissolve and enlarge the fractures.

Classification by pore size

Pores may be classified with respect to their size. Generally, the pores found in petroleum reservoirs may be placed, by size, in three groups: supercapillary, capillary, and subcapillary pores (Table 2.8).

Supercapillary pores include tubular openings greater than 0.5 mm in diameter and sheet openings (fractures) more than 0.25 mm in width. Large fractures and dissolved caverns in rocks, and the pores in carbonate rocks or weakly cemented sandstones belong to this category. In such pores, water obeys the ordinary laws of hydrostatics and the fluids can flow freely under gravity. That to say, fluid movement in such pores is determined more by gravity or normal pressure difference than capillary action.

Capillary pores include tubular openings between 0.5 and 0.0002 mm in diameter and sheet pores between 0.25 and 0.0001 mm in width. Pores in conventional sandstones are such openings. In this pore system, the flow of water does not obey the ordinary laws of hydrostatics but is primarily affected by capillary forces. Broadly speaking, fluids in these pores are subject to both gravity and capillary forces, and cannot flow freely due to greater friction along the pore walls.

Table 2.8 Pore types classified by pore size (He [21])

Type	Pore diameter (mm)	Fracture width (mm)	States of fluids in pores	Cases
Supercapillary pore	>0.5	>0.25	Flow freely	Dissolved pores or caverns, big fractures
Capillary pore	0.5–0.0002	0.25–0.0001	Can't flow freely unless $\Delta p > p_c$	Pores in conventional sandstones
Subcapillary pore	<0.0002	<0.0001	Can hardly flow under normal conditions	Pores in conventional mudstones

Under the condition of great pressure drop, however, fluids may be forced through capillary openings. It is possible for fluids to circulate through capillary pores if gravity force or pressure drop is greater than capillary force.

Subcapillary pores are tubular pores less than 0.0002 mm in diameter and sheet pores (fractures) less than 0.0001 mm in width. These pores are so small that great intermolecular force between particles and liquids makes fluids hardly flow in these pores. Fluids may enter such pores, but it tends to remain as if fixed to the walls, prohibiting further entrance of fluids. That is to say, under normal pressure drops of oil reservoirs, there is no circulation of fluids through such openings.

Obviously, the smaller the pore, the harder the flow of fluids in pores. For example, mud stones, clays, and shales, primarily containing minute pores or micropore, are always natural barriers to circulation unless sufficient pressure drops exists, which may force fluids flow through this kind of medium. Therefore, subcapillary pores are usually noneffective for the circulation of fluids in oil reservoirs. In gas reservoirs, however, even subcapillary pores may be effective owing to the small molecule size and strong flowability of natural gas. For example, nanometer-scale pores in shale gas reservoirs are even effective for gas production.

In reservoir evaluation, pores are customarily assorted according to the median radius of pore-throats (R_{50}) of rocks (Table 2.9). The parameter (R_{50}) of rocks can be determined from the capillary curve of the rock.

Classification by pore connectivity

During the process of sediment lithification, sediments underwent various geologic processes. These processes will change rock texture and characterize a rock on microscopic scale, for example the cementation between particles, dissolution of minerals, recrystallization, and so on. After these variations, the void space in reservoir rocks changes greatly in pore-size, pore connectivity, and so on throughout the reservoirs.

In reservoir rocks, generally, most of pores are well interconnected, which provides continuous channels for fluid flow; but there is always a small amount of pores that are not connected or isolated due to various geologic variations in the rocks. In loose sand, all voids are interconnected, whereas part of pores in limestones, dolomites, and sandstones may be completely sealed off from the other pores, or blind (dead-end). Based on the connectivity of pores, consequently, the void space in rocks can be divided into: interconnected pore, isolated pore, and blind pore (Fig. 2.15).

Table 2.9 Pore types classified by the median radius of pore-throats (SYT 6285-2011 China)

Type	Median radius of pore-throats (R_{50}) (μm)
Extra-large pore	$R_{50} \geq 25$
Large pore	$15 \leq R_{50} < 25$
Median pore	$5 \leq R_{50} < 5$
Small pore	$3 \leq R_{50} < 5$
Ultra-small pore	$R_{50} < 3$

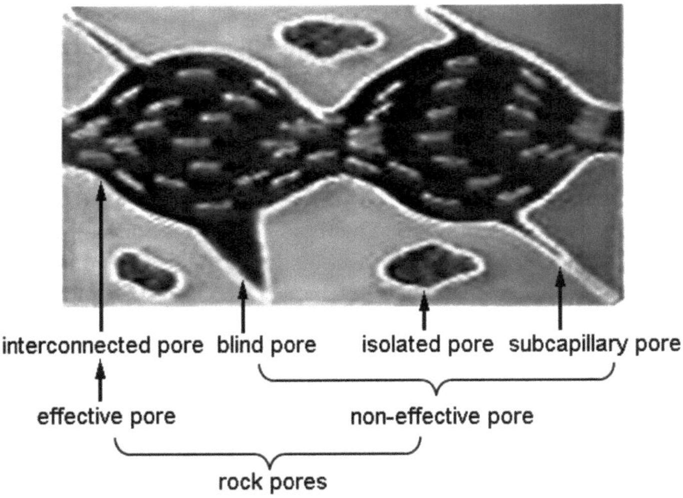

Fig. 2.15 Schematic diagram of pore types

Perceptibly, the pores that are not part of a continuous channel network do not contribute to oil and gas production. These pores are known as noneffective pores (Fig. 2.15), namely, they provide no space for fluid flow in reservoirs. Noneffective pores are not considered in the estimation of petroleum reserves. As a result, the measure of interconnected pore volume as opposed to total pore volume is of more interest to reservoir engineers in estimating producible fluid volume and analyzing reservoir performance.

In a word, interconnected supercapillary and capillary pores are effective for petroleum production, but interconnected subcapillary pores may be effective depending on gas or oil flowing in these pores. Whether it is gas flowing, or oil flowing, isolated pores are noneffective.

Classification by void shape

As viewed from the shapes, the void space of reservoir rocks may be classified to two sorts: pore/cave and fracture (Table 2.10). The former has a lower ratio of length to width, including primary and secondary pores. The latter has a higher ratio of length to width. Pore/cave is more or less tubular while fracture is round about fissuriform.

Table 2.10 Pore types classified by void shape (SYT 6285-2011, China)

Type		Length/width of void	Diameter of void, mm
Pore or cave	Pore	1:1–<10:1	<2
	Cave		≥2
Fracture		≥ 10:1	

In carbonate rocks, various pore types (pore, fracture and dissolved cavern) can be found. Pore means the voids between carbonate particles. It is often primary. Dissolved cavern is a secondary pore formed by the dissolution of groundwater. In general, dissolved caverns are larger than pores in carbonate rocks.

2.2.1.2 Types of Reservoir Space

In petroleum reservoirs, *reservoir space* means the void system which can hold gas or oil. Reservoir space may be composed of one of the three void types: pore, fracture, dissolved cavern, or free combination of the three void types. Reservoir space is thus complicated. Different combinations of void types form different types of reservoir space, in which fluids have different flow patterns. Both the types of reservoir space and flow patterns of fluids are very concerned by reservoir engineers, they are the important foundation of reservoir engineering calculation and have a great effect on the oil recovery.

Based on the combination type of voids, reservoir space of petroleum reservoirs can be classified as: single-porosity medium, dual-porosity medium, and triple-porosity medium.

Single-porosity medium means that the reservoir space of petroleum reservoirs consists of one type of voids. It may be single-pore, single-fracture, or single-dissolved cavern system. In this kind of reservoir, fluids are held and flow in the same void space.

The reservoir of dual-porosity medium has two types of voids. It may be pore-fracture system, dissolved cavern-pore system, or fracture-dissolved cavern system. For example, if a rock contains both fracture network and intergranular pores (e.g., a sandstone), the rock is called dual-porosity medium.

In a dual-porosity reservoir, such as a pore-fracture reservoir, the pores are customarily the space holding oil or gas, whereas the fracture network mainly acts as the flow passage of oil or gas. The flow through the fractures is accompanied by the exchange of fluid from and to the surrounding porous rock matrix. Exchange between the fracture network and the porous block (matrix) is normally represented by a term that describes the rate of mass transfer. The physical parameters of both the fracture and block are needed to be defined at each mathematical point in the flow domain. As a result, dual porosity and dual permeability are necessary to characterize the properties of this kind of reservoir. At present, this kind of reservoir is mainly found in carbonate rock, especially in limestone or dolomite. Such reservoirs exist extensively in carbonate rocks in Sichuan gas field of China.

Triple-porosity medium is the most complicated petroleum reservoir. It consists of three types of voids: intergranular pore, fracture and dissolved cavern/solution cavity. In this reservoir, the flow mechanism of fluids in each type of voids is different. Triple-porosity mediums have been found in Tahe oilfield in Western China.

2.2.1.3 Pore-Size Distribution

The topology of the pore space of a porous medium is very important in determining the properties of the medium. In terms of topology, we may know what size the pores are and how the pores are connected. In general, rock pores can be characterized by the terms: pore size and pore-size distribution.

In oilfield exploitation, pore size and its distribution of a reservoir rock are the significant foundation of recognizing the properties of reservoir rocks, and have a very important effect on reservoir behavior with respect to single-phase or multiphase flow in the reservoir. Therefore, this kind of information is absolutely indispensable to recognize the mechanism of fluids flowing through porous mediums.

The number of discrete pores in a conventional reservoir of a cubic meter is enormous. Since the average diameter of particles in most sandstones is between 0.05 and 0.25 mm, the number of pores in a cubic meter rock may be between 10^9 and 10^{12} because the pores in most sandstones have radii between 20 and 200 μ. Moreover, the pore size in sandstone reservoirs is also extremely heterogeneous. Evidently, it is impossible to analyze the pore size in a rock one by one. The best way to investigate pore size is statistical analysis, by which one can give the distribution and sorting of a pore system.

In the laboratory, several techniques are commonly used for the analysis of the size and distribution of rock pores. They are mercury injection, thin section analysis, image analysis, gas adsorption, and so on.

Pore-size distribution

In engineering researches, pore size is typically described with a parameter, namely diameter. In real rocks, but, pores are much varied either in size or shape. There is scarcely a pore which is typically spherical. One way to define a "pore size" is by defining the pore diameter at a point within the pore space of rock as the diameter of the largest sphere that contains this point while still remaining entirely within the pore space [22]. Generally, the unit of measure typically associated with pore-size values is micrometers (more commonly referred to as microns), whereby 1 μm equals 1×10^{-6} m. For reference, a human hair is approximately 25 μm in diameter.

Pore-size distribution is the fraction of pore volume with respect to pore size. Alternatively, it may be defined by the fraction of pore area with respect to pore size. The former can be obtained from mercury injection analysis; whereas the latter can be determined from thin section analysis.

The measurement techniques (e.g., mercury injection) report pore-size information as a distribution of data points. For this reason, pore-size information is best described graphically, rather than with a single number. The most commonly used is the pore-size distribution curve and pore-size cumulative distribution curve (Fig. 2.16).

Clearly, the pore-size distribution curve, shown in Fig. 2.16, describes visually the size and distribution of rock pores. A sharp peak of distribution curve in

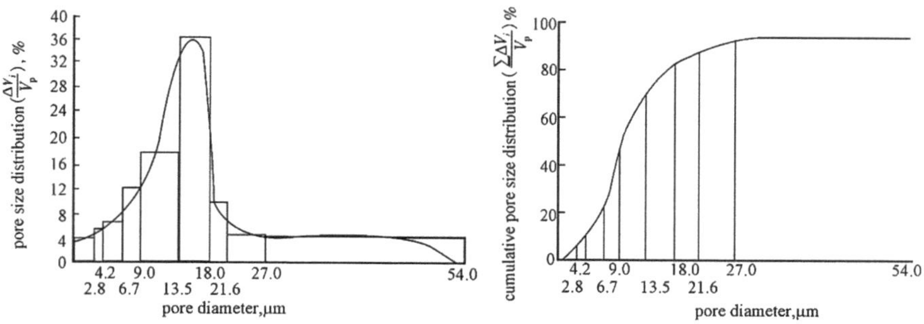

Fig. 2.16 Pore-size distribution curve (**a**) and pore-size cumulative distribution curve (**b**)

Fig. 2.16a reflects a uniform pore size (good sorting), yet a flat curve indicates a nonuniform pore size (poor sorting). Similarly, a steep cumulative curve reflects a uniform pore size. Different locations of curves indicate different average pore sizes. Thereby, the pore-size distribution curves can qualitatively characterize the size and distribution of rock pores.

Pore sorting evaluation

Besides pore-size distribution curves, an alternative method of describing pore-size distribution is to use statistical parameters. These parameters can describe the pore-size distribution in more detail. The common parameters include sorting coefficient, skewness and kurtosis, mean pore diameter, etc., they are easily obtained from the pore-size cumulative distribution curve.

Sorting coefficient

Sorting coefficient (S_p) is a standard deviation measure of pore sizes in a rock. It is often used for describing the uniformity of pore-throat sizes, also named uniformity coefficient, generally expressed as follows (Folk and Ward 1957):

$$S_p = \frac{\phi_{84} - \phi_{16}}{4} + \frac{\phi_{95} - \phi_5}{6.6} \tag{2.15}$$

where $\phi_i = -\log_2 d_i$, subscript i represents 5, 16, 84 and 95; d_i is a mean pore diameter corresponding to the ith accumulative pore volume, μm; The subscript of 5, 16, 84, and 95 separately denotes 5, 16, 84, and 95 % of accumulative pore volume in pore-size cumulative distribution curve.

In general, the smaller the sorting coefficient (S_p), the more uniform the pore-size, and then the better the sorting of the pore-size.

Skewness

Skewness (S_{kp}) is a measure of the asymmetry of pore-size distribution, or the departure from symmetry of a pore-size distribution. It is defined as follows (Folk and Ward 1957):

$$S_{kp} = \frac{\phi_{84} + \phi_{16} - 2\phi_{50}}{2(\phi_{84} - \phi_{16})} + \frac{\phi_{95} + \phi_5 - 2\phi_{50}}{2(\phi_{95} - \phi_5)} \tag{2.16}$$

where S_{kp} is skewness of pore-size distribution; $\phi_i = -\log_2 d_i$, subscript i means 5, 16, 50, 84, and 95; d_i is a mean pore diameter corresponding to the ith accumulative pore volume, μm. The subscript of 5, 16, 50, 84, and 95 denote separately 5, 16, 50, 84, and 95 % of accumulative pore volume.

The asymmetry of a pore-size distribution can be given by its skewness. In general, the skewness of rock pore-size varies between −1 and 1, namely $-1 < S_{kp} < 1$.

If a pore-size distribution curve is symmetrical (normal distribution), the skewness of rock pores is zero. If skewness is positive, it indicates that the curve has a tail in big pore that is "coarse skewed"; or a negative value indicates that the curve is skewed toward small pores, named "fine skewed."

Kurtosis

Kurtosis (K_p) is a measure of the degree of peakedness, that is, the ratio between the spread of the pore diameters in the tails and the spread of the pore diameters in the central portion of the distribution (Folk and Ward 1957):

$$K_p = \frac{\phi_{95} - \phi_5}{2.44(\phi_{75} - \phi_{25})} \tag{2.17}$$

where K_p is the kurtosis of pore-size distribution; $\phi_i = -\log_2 d_i$, subscript i means 5, 25, 75, and 95; d_i is a mean pore diameter corresponding to the ith accumulative pore volume, μm; The subscript of 5, 25, 75, and 95 denote separately 5, 25, 75, and 95 % of accumulative pore volume.

Kurtosis is used for describing the concentration of major pores in a rock. It is also referred to as sharpness.

Generally, normal distribution curve has a K_p of one, called mesokurtic. If the curve is flat or double peaks, K_p may be less than 0.6, called very platykurtic. A curve represented by a high narrow peak (very leptokurtic) may have K_p values ranging from 1.5 to 3.

If rock pores is composed of two or more different void types (such as pore, fracture, even dissolved cavern), double peaks, even multi-peak, will occur on the distribution curve.

Mean pore diameter

Median pore diameter (D_{50}) is the diameter (μm) which is larger than half of the pores in the distribution and smaller than the other half (i.e., the middlemost member of the distribution) [23]. It reflects the overall pore size as influenced by the chemical or physical origin of the rock and any subsequent alteration. It may be a very misleading value, however.

Mean pore diameter (D_m) is the measure of the overall average pore size. It is an arithmetic mean by the pore diameters of a few feature points of pore-size cumulative distribution curve. The following averaging methods are commonly used [23].

$$D_m = \frac{D_5 + D_{15} + D_{25} + \cdots + D_{85} + D_{95}}{10} \tag{2.18}$$

or

$$D_m = \frac{D_{16} + D_{50} + D_{84}}{3} \tag{2.19}$$

or

$$D_m = \frac{D_{25} + D_{50} + D_{75}}{3} \tag{2.20}$$

In Eqs. (2.18)–(2.20), D_i (i = 5, 15, 16, 25, 50, 75, 84, 85, and 90) means the pore diameters corresponding to 5, 15, 16, 25, 50, 75, 84, 85, and 95 % of cumulative pore volume.

2.2.1.4 Pore Structure

The pore structure of reservoir rocks results from the complex interplay of the various factors that influence the porosity of the reservoir rock. Pore structure means the all characters of whole pore system, including the pore size and shape, the connectivity between pores, the surface roughness of pore-throats, the pore-size distribution, and the connecting pattern of pores. The size, shape, distribution, and communication ways of pores control all mechanic and hydraulic properties of a rock, such as conductivity, deformation behavior, anisotropy, etc.

The pore structure of clastic rocks is mainly affected by several petrographic characteristics, which include: (1) grains: size, shape, sorting, chemical composition, mineral composition; (2) matrix: the amount of each mineral, the distribution of minerals and chemical composition; (3) cement: composition, amount, distribution with respect to grains and matrix. The pore structure of chemical rocks is mostly dependent on the following factors: (1) fossil content, (2) fracturing and jointing, (3) solution and redeposition, (4) dolomite content, (5) recrystallization, (6) clay content, (7) bedding planes.

The size of individual pores ranges from subcapillary through capillary openings to solution cavities of all shapes and sizes, including caverns formed in carbonate rocks.

The individual pore may be tubular, like a capillary tube; or it may be nodular and may feather out into the bounding constrictions between nodules; or it may be a thin, intercrystalline, tabular opening that is 50–100 or more times as wide as it is thick [24]. The wall of pore may be clean quartz, chert, or calcite, or it may be coated with clay mineral particles, platy accessory minerals, or rock fragments.

The feature of some individual pores may be observed in well cuttings and cores by the unaided eye. Many pores, however, can be seen only with a binocular or petrographic microscope. More fine detail of pores can be examined through electron micrography because much of the reservoir pores is submicroscopic in size. Pores filled with oil may also be observed under ultraviolet rays. The fluorescence of oil trapped in minute fractures and intercrystalline pores, not visible to the eye, stands out prominently, and pools have been discovered when this was the only way the oil was observed. Casts of the interconnected pores may be made by forcing wax or plastic material under pressure into a core or rock fragment and then dissolving away the surrounding rock material. Such a pore cast of the ordinary reservoir sandstone looks much like a piece of bread, while the cast of a rock with angular grains and crystals has the appearance of rock candy (Fig. 2.17). Photomicrographs of pore casts viewed stereoscopically offer a good means of observing the pore structure (Fig. 2.17b).

Several parameters have been proposed to describe the pore structures of reservoir rocks in the past, such as aspect ratio, coordination number, pore tortuosity, threshold pressure, and so on. The common parameters are described as follows.

Aspect ratio (pore/pore-throat size ratio) is the ratio of pore radius to pore-throat radius [26]. Or it is defined as the ratio of pore diameter to pore-throat size. This is a fundamental control on hydrocarbon displacement [27].

The pore/pore-throat size ratio is a very important factor in estimating recovery efficiency because large pores connected by small pore-throats are difficult to drain [28]. As nonwetting fluids are withdrawn from a system of large pores and small pore-throats, the nonwetting fluid column breaks apart in the narrow throats, leaving a large amount of fluids isolated in the large pores.

Pore-to-pore accessibility is determined by *coordination number*, or the number of pore-throats that connect with a pore [28]. In general, it ranges between 2 and 15.

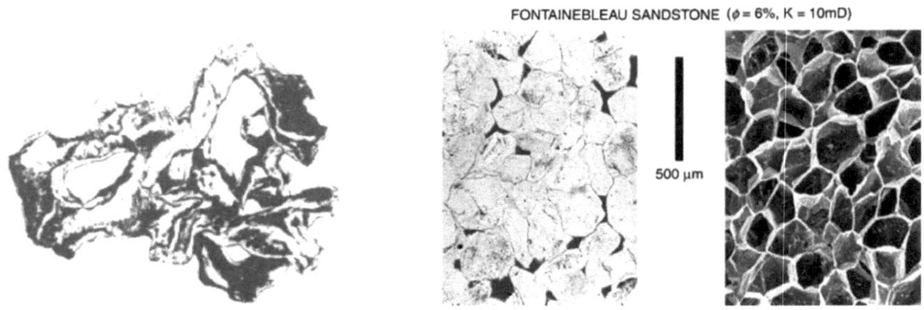

(a) (b)

(a)Cast of pore space insandstone (Collins,25); (b) Thin section(left)and epoxy pore cast (right) of fontainebleau sandstone (after Zinszner and Pellerin, 2007)

Fig. 2.17 Pore castsretort distillation at atmospheric

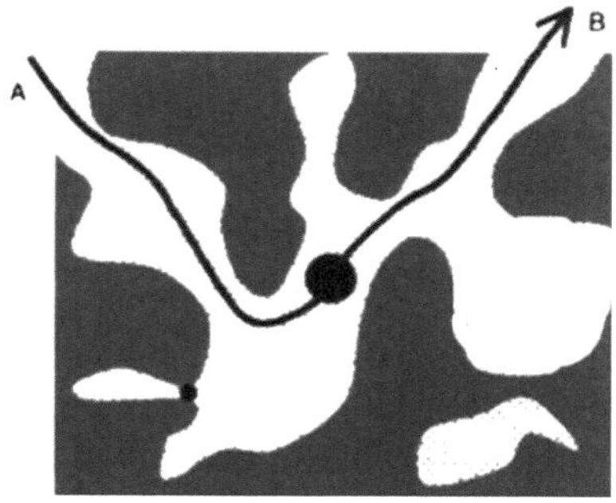

Fig. 2.18 Schematic diagram of tortuosity of a path AB in a porous medium (after Christopher and William [30])

The connectivity of a pore network is generally characterized by a coordination number which represents the number of independent paths through which pores are connected to each other. An average coordination number representing the connectivity between constrictions to chamber is an important parameter. In other words, we would like to know, on an average, how many pore constrictions (namely, throats) are connected to a pore chamber. Likewise, there can be a coordination number which represents an average number of pore chambers connected to a pore constriction. A reasonable estimate of an average coordination number for sandstones is between 4 and 8 [28].

Pore tortuosity is defined as the ratio of the effective average path of connected pores in a rock to the straight-line distance of the two end pores measured (Fig. 2.18). It is actually the ratio of the true length of a pore to its projection [29]. The tortuosity has nothing to do with the size of the pores but entirely depends on the connectivity of the pore system. Tortuosity, a dimensionless factor, expresses the degree of complexity of the sinuous pore path. It is not easy to determine directly the value of the parameter. By experience, it is about in the range of 1.2–2.5.

2.2.2 Porosity of Reservoir Rocks

The development of oil and gas reservoirs shows that the void space holding oil or gas is only small part of whole rock volume. Oil or gas is not in the underground lakes or streams of oil which once was a popular idea. Porosity is a measure of the void space of reservoir rocks, and thus represents the storage capacity of the reservoirs. Porosity is the best-known physical characteristic of petroleum reservoirs.

Porosity is very important rock property because it can be used to measure the potential volume of hydrocarbons in reservoir rocks. All recovery computations must be based on the knowledge of this parameter. It is one of the most important reservoir parameters in the study of petroleum reservoir engineering.

2.2.2.1 Porosity

As shown in Fig, 2.19, the total volume of a rock (V_b), also known as bulk volume, or view volume, is composed of the pore volume (V_p) and the solid volume (matrix volume) of rock particles (V_s) (Fig. 2.19), that is:

$$V_b = V_p + V_s \tag{2.21}$$

The porosity of a rock is defined as the ratio of the void volume of the rock to the total volume of the rock, mathematically expressed as:

$$\phi = \frac{V_p}{V_b} \times 100\ \% \tag{2.22}$$

where ϕ is the porosity of a rock, %; V_p is the volume of void space in the rock, cm^3; V_b is the bulk volume of the rock, cm^3.

The void space includes all pores, cracks, vugs, inter- and intra-crystalline spaces. The porosity is conventionally given the symbol ϕ, and is expressed either as a fraction varying between 0 and 1, or a percentage varying between 0 and 100 %. The fractional form is often used in engineer calculations.

From Eq. (2.21), we know $V_p = V_b - V_s$. Combine it with Eq. (2.22), then:

$$\phi = \left(1 - \frac{V_s}{V_b}\right) \times 100\ \% \tag{2.23}$$

where V_s is the total volume of solid composing the rock matrix, cm^3.

It should be noted that the porosity does not give any information concerning pore size, pore-size distribution, and pore connectivity. Thus, rocks of the same porosity may have widely different physical properties. An example of this might be a carbonate rock and a sandstone. Each could have a porosity of 0.1, but carbonate pores are often much unconnected resulting in its permeability being much lower

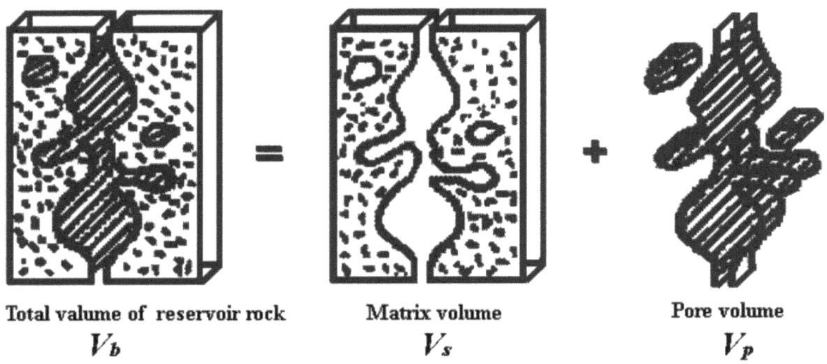

Fig. 2.19 Total volume, matrix volume and pore volume of a reservoir rock

than that of the sandstone. Different definitions of porosity have been recognized and used within the hydrocarbon industry. For example, total porosity, connected porosity, effective porosity, primary porosity, secondary porosity, fracture porosity, etc.

2.2.2.2 Absolute Porosity and Effective Porosity

Absolute porosity is the total porosity. It is defined as the ratio of the total void volume of a rock to the total volume of the rock.

Connected porosity means the ratio of the connected pore volume to the total rock volume.

Effective porosity is the ratio of the effective pore volume to the total rock volume. Here, effective pores means the pores which can transmit fluids in conventional reservoir conditions. Effective pore does not include isolated pores and the tiny pores which are too small to effectively transmit fluids. Even in high quality reservoirs, an appreciable difference between the total porosity and effective porosity often exists. In general, the effective porosity of a rock is less than its absolute porosity in most cases.

It should be emphasized that the effective space of a rock is often not equal to the spaces of fluids flowing actually in the rock due to the effect of adsorption of fluids on the surface of the rock particles. A term "flowing porosity" is thus used to describe the percent of pore volume of fluids really occupying in rock pores when fluids flow through the rock. Flowing porosity may be considered as the fraction of the volume of space occupied by fluids flowing in pores to the rock volume. Flowing pore volume does not contain the space which is occupied by the liquid film adsorbed on the surface of rock particles. Thereby, the flowing porosity is usually less than the effective porosity of a rock; and the flowing porosity is changeable. It often changes with displacement pressure difference, physical properties of fluids, and the surface properties of rock particles.

Effective porosity is of significance in petroleum production. If a rock has a high total porosity, but a low effective porosity (e.g., pumice), the oil in rock could not be recovered even if the rock might conceivably be saturated with oil because the oil could not find effective passages to flow through the rock to a well drilled into the stratum. The effective porosity characterizes the properties of a rock holding and transmitting fluids. In petroleum industry, effective porosity is thus used for calculating the *original oil in place* (OOIP).

Generally, the values of absolute porosity and effective porosity can be determined in the laboratory. It is worth noting that, under normal circumstances, effective porosity is called porosity for short in petroleum engineering.

In addition, the following porosities used in petroleum production should also be known [31]:

Primary porosity: the porosity of the rock resulting from its original deposition.
Secondary porosity: the porosity resulting from diagenesis.

Microporosity: the porosity resident in small pores (<2 mm) commonly associated with detrital and authigenic clays.
Intergranular porosity: the porosity due to pore volume between the rock grains.
Intragranular porosity: the porosity due to voids within the rock grains.
Dissolution porosity: the porosity resulting from dissolution of rock grains.
Intercrystal porosity: microporosity existing along intercrystalline boundaries usually in carbonate rocks.
Moldic porosity: a type of dissolution porosity in carbonate rocks resulting in molds of original grains or fossil remains.
Fenestral porosity: a holey ("bird's-eye") porosity in carbonate rocks usually associated with algal mats.
Vug porosity: porosity associated with vugs, commonly in carbonate rocks.

2.2.2.3 The Range of Porosity Values in Nature

In recently deposited sediments, such as those you might find on the floor of a lake, porosity may be very high (values up to 80 % have been recorded). More common materials, such as loose sands, can have porosities as high as 45 %. These materials are either extremely unstable or stabilized by cements. In reservoir rocks, high porosities can also be found due to dissolution (secondary porosity), particularly in carbonates. Generally, porosities can be very low in carbonate rocks. In some cases, the total porosity of carbonate rocks may be very high, but their permeability can be very low as the pores and vugs that make up the pore structure are unconnected. In massy fractured carbonates, the porosity of matrix can be as low as 1 %. Igneous and metamorphic rocks almost always have porosities less than 1 %. Sandstones, generally, lie in the range 5–25 %. Table 2.11 gives the approximate ranges of porosities for some common lithologies.

Table 2.11 The range of porosity values for rocks (after Paul [31])

Lithology	Porosity range (%)
Unconsolidated sands	35–45
Reservoir sandstones	15–35
Compact sandstones	1–15
Compact carbonate rocks	<1–5
Shales	0–45
Clays	0–45
Massive limestones	5–10
Vuggy limestones	10–40
Dolomite	10–30
Chalk	5–40
Granite	<1
Basalt	<0.5
Gneiss	<2
Conglomerate	1–15

Table 2.12 Gradation of reservoir porosity [23]

Clastic rock		Carbonate rock	
Type of reservoir porosity	Porosity, ϕ (%)	Type of reservoir porosity	Porosity, ϕ (%)
Extremely high porosity	$\phi \geq 30$		
High porosity	$25 \leq \phi < 30$	High porosity	$\phi \geq 20$
Moderate porosity	$15 \leq \phi < 25$	Moderate porosity	$12 \leq \phi < 20$
Low porosity	$10 \leq \phi < 15$	Low porosity	$4 \leq \phi < 12$
Extremely low porosity	$5 \leq \phi < 10$	Extremely low porosity	$\phi < 4$
Ultra-low porosity	$\phi < 5$		

Porosity is a vital parameter of petroleum reserves calculation, reservoir evaluation, and engineering calculation. The classification and evaluation of petroleum reservoirs is also based on the parameter. The gradations of porosity for reservoir rocks are shown in Table 2.12 according to the industry standard of China. Generally, there are different gradation for clastic rocks and carbonate rocks due to different types, size, and scale of pores in the two types of rocks.

2.2.2.4 Dual Porosity*

Dual porosity has been developed in the oil industry to deal with specific problems associated with fractured petroleum reservoirs. In the most general sense, the use of dual porosity indicates that the reservoir has a complex porosity system that can be divided into fracture porosity and a smaller scale porosity that exists in the pore spaces within the rocks [32]. In some cases, this kind of porosity system is referred to as a double porosity system.

In double porosity system, the matrix usually has low permeability but high porosity and thus dominates the storage, while the fracture system has high permeability but low porosity, and dominates fluid flow.

As a general rule, fractured reservoir rocks are made up of two porosity systems: the first intergranular porosity (Fig. 2.20a) formed by void spaces between the

(a) Void space between grains (matrix) (b) Simplified vug and fracture space

Fig. 2.20 Sketch map of pores, vug and fractures

grains of the rock, and the second formed by void spaces of fractures and vugs (Fig. 2.20b). The first type is called primary porosity and is typical of sandstone or limestone. The second type is called secondary porosity, or vugular porosity (referring to only vugs)/fracture porosity (referring to only fractures).

Secondary porosity is generally found in compact, brittle rocks of relatively low intergranular porosity, such as compact limestones, shales, shaly sandstones, siltstones, schists, etc. Secondary porosity is normally caused by rock fracturing, jointing, and dissolution by circulating water.

In a fractured reservoir, the total porosity (ϕ_t) is the result of the simple addition of the primary and secondary porosities:

$$\phi_t = \phi_p + \phi_f \tag{2.24}$$

where ϕ_p = matrix void volume/total bulk volume, called primary porosity; ϕ_f = fracture void volume/total bulk volume, named fracture porosity.

From the experimental results of porosity measurements on various types of rocks, it can be found that the fracture porosity of a rock is generally considerably less than the primary porosity of the rock.

We should note that the two porosities (ϕ_p, ϕ_f) are defined based on the total bulk (matrix + fractures) volume of a rock. However, it is not easy to obtain a typical core of a dual-porosity system in practice. The cores used to measure in the laboratory are usually from the part of the matrix of a fractured rock. The porosity of the rock determined in the laboratory is therefore referred to as matrix porosity (ϕ_m) because it is only related with the matrix volume of the fractured rock. So, the matrix porosity (ϕ_{m}) is defined as follows:

$$\phi_m = \frac{\text{void volume of matrix}}{\text{matrix bulk volume}} \tag{2.25}$$

Suppose the total bulk volume of rock is one unit and the fracture porosity is ϕ_f. The matrix bulk volume is then $(1 - \phi_f)$; and the void volume of matrix is then $(1 - \phi_f)\phi_m$. The fracture and primary porosities are presented schematically in Fig. 2.21, where the unit of bulk is scaled in the upper part. In this case, the primary porosity, as a function of matrix porosity, is expressed by the following equation:

$$\phi_p = \frac{\text{void volume of matrix}}{\text{total bulk volume of rock}} = \frac{(1 - \phi_f)\phi_m}{1} = (1 - \phi_f)\phi_m \tag{2.26}$$

Then the total porosity is:

$$\phi_t = (1 - \phi_f)\phi_m + \phi_f \tag{2.27}$$

By experience, when $\phi_{\mathrm{t}} < 10\ \%$, the maximum of fracture porosity is often lower than $0.1\phi_t (\phi_{\mathrm{fmax}} < 0.1\phi_t)$, whereas the maximum of fracture porosity is probably lower than $0.04\phi_{\mathrm{t}}$ ($\phi_{\mathrm{fmax}} < 0.04\phi_t$) when $\phi_{\mathrm{t}} > 10\ \%$. From the reservoir

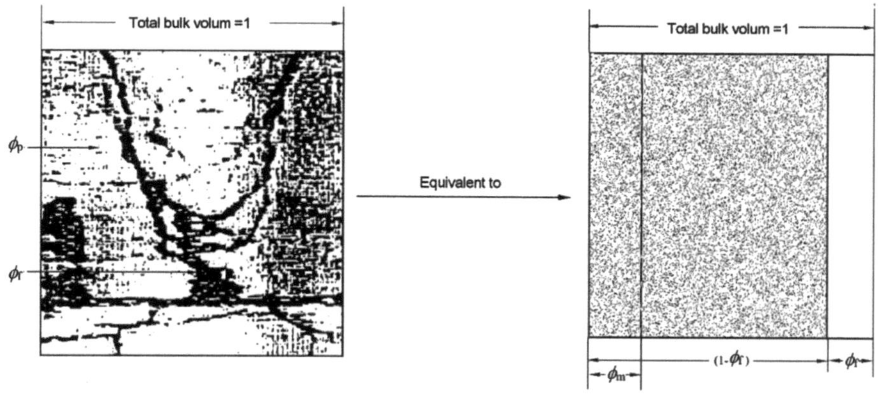

Fig. 2.21 Sketch map of double porosity

space point of view, fracture porosity is much lower than matrix porosity, and sometimes is ignored. However, the existence of fractures is significant for the storage of oil or gas in tight rocks ($\phi_t < 5\ \%$), especially for the flow of oil or gas in compacted rocks.

2.2.3 *Factors Affecting Porosity*

Primary pore is the original void space created when the rock was formed in geological processes. The porosity of a rock is created and altered by geological processes, and is thus affected by a host of factors during deposition, as well as in the long periods of the following diagenesis.

The initial (pre-diagenesis) porosity is affected by the following microstructural parameters. They are grain-size, packing pattern of particles, particle shape, and grain-size distribution. However, the initial porosity in real rocks is usually subsequently affected by secondary factors, such as compaction and geochemical diagenetic processes. This section briefly reviews these factors and controls on porosity.

2.2.3.1 Grain-Size and Packing Pattern

In real rocks, the effect of grain-size and packing pattern to porosity is quite intricate by reason of the heterogeneity of grain-size and shape. To simplify this problem, begin with the analysis of porosity determination of an ideal medium.

The ideal medium is an imaginary rock which is packed with spherical grains in same size or not. In theory, the porosity of an ideal medium can be easily calculated by the grain-size.

Suppose an imaginary rock is a regular cubic packing with same size spherical grains (Fig. 2.22a), and the packing is the least compact arrangement (one grain packing on the top of the other). The porosity of the cubic packing is calculated as follows:

Considering the cube filled with eight grains of radius r.

Volume of the cube is: $(4r)^3$.

Total volume of the eight grains in the cube is: $8 \times \frac{4\pi r^3}{3}$

The pore volume of the cubic packing is: $(4r)^3 - 8 \times \frac{4\pi r^3}{3}$

Then the porosity of the cubic packing is:

$$\phi = \frac{V_b - V_s}{V_b} = 1 - \frac{V_s}{V_b} = 1 - \frac{32\pi r^3}{3 \times 64 r^3} = 1 - \frac{\pi}{6} = 47.46\ \% \tag{2.28}$$

where V_b is the bulk volume of a packing, cm^3; V_s is the total volume of grains in the packing, cm^3.

From here we see that the *maximum* porosity of the cubic packing is 47.6 %, and it is independent of grain-size (Fig. 2.22a). In a similar way, the porosity of the cubic with rhombohedral arrangement (i.e., the most compact arrangement, Fig. 2.22b) can be determined, and is 25.96 %. So, the porosity of an ideal rock may vary from about 26 to 48 % (Fig. 2.22), depending upon the pattern of packing (how the grains are packed together); and the value of the porosity is independent of the grain-size. For real rocks, however, the way of grain packing depends upon the conditions of deposition.

Theoretically, porosity has nothing to do with grain-size for a regular cubic packing with same size spherical grains. This is also true for other ordered packing, but not true for the random arrangement of spherical grains. In real depositional environments, ordered packings cannot be formed because deposition is energetically unstable, and grain-size is heterogeneous. Therefore, it does not prove true in grain assemblages of a real rock due to variable sizes and shapes of rock particles.

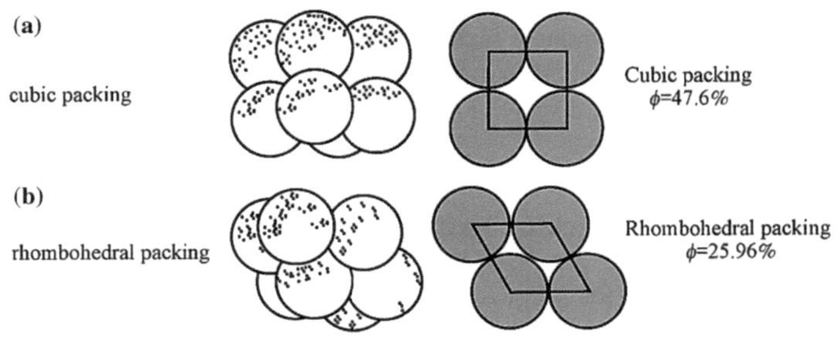

Fig. 2.22 Ideal rock: **a** cubic packing; **b** rhombohcdral packing (after Graton and Fraser)

2.2.3.2 Grain Sorting

Real rocks contain a distribution of various size grains, and the distribution is often multimodal. To understand the effect of grain sorting on porosity, in the same way, consider an ideal rock consisting of two size spherical gains (Fig. 2.23c). It can be figured out that the porosity of the packing does not exceed about 37 %, and the porosity reaches its minimum when packed with the particles of two sizes each in half (Fig. 2.23c).

It can be proved that the porosity of real rocks is always related with the size and sorting of rock particles. In general, well-sorted sediments are more porous and have higher porosity than poorly sorted ones because small grains take up space between large grains (Fig. 2.23). The statistical results of real rocks show that the porosity of rocks is usually decrease with the increase in grain-size. The reason is that fine-grained sediments is well sorted and more spherical than that coarse-grained sediments.

For clastic rocks, a general rule is proved as shown in Fig. 2.24. The increase in porosity only becomes significant when sorting coefficient is lower than 2; and the

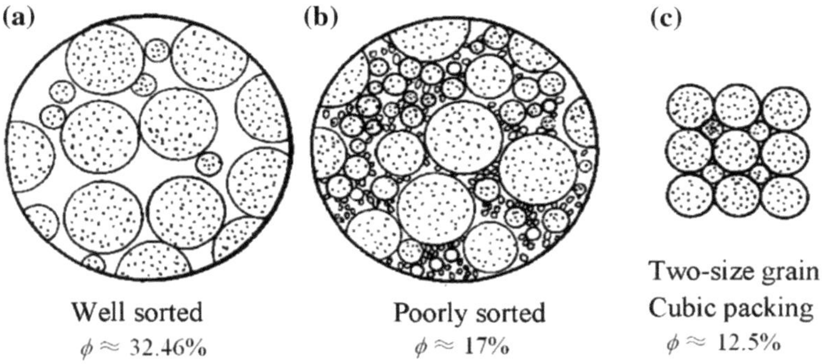

Fig. 2.23 Effect of sorting on porosity (Bear [33])

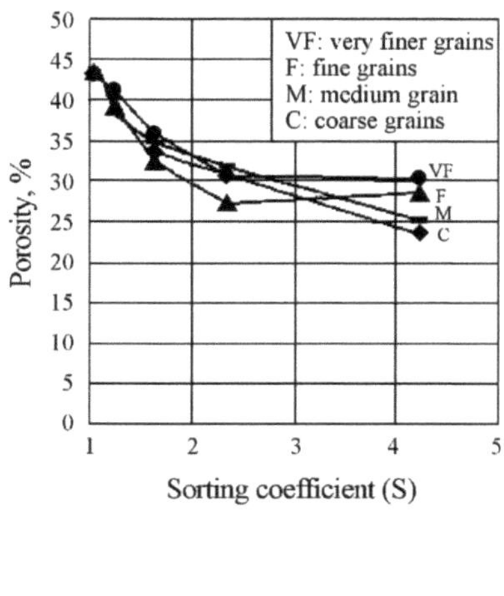

Fig. 2.24 Correlation between porosity and sorting (Yang et al. [34])

smaller the sorting coefficient is, the larger the porosity of the rock is. As sorting coefficient is close to 1, the influence of grain-size on porosity is weakened. That is to say, the porosity of a rock is independent of grain-size only when grains are very uniform (namely, sorting coefficient ≈ 1).

2.2.3.3 Grain Shape

The primary porosity of rocks mainly depends upon the shape (sphericity and roundness), sorting, and packing pattern of sediment particles. Angularity of sediment may increase or decrease porosity, depending if the particles bridge openings or are packed together like pieces of a mosaic [35]. On the whole, porosity often increases with sphericity and roundness of grains, but the maximum porosity often occurs in unconsolidated, subangular, well-sorted sedimentary rocks (Fig. 2.25).

2.2.3.4 Mineral Components of Rocks

Sandstones are commonly composed of five principal ingredients: rock fragments (lithic grains), quartz grains, feldspar grains, matrix, and cement [36]. In sandstones, quartz is the most abundant mineral; feldspar is the next abundant (Table 2.13). Matrix often consists of clay minerals and silt-grade quartz grains. Clay minerals can fill in pores during diagenesis. Cement also precipitates around and between grains during diagenesis [36]. Common cements are argillaceous, calcareous, and siliceous matter. A few of ferruginous matters also exist in clastic rocks.

Owing to the effect of wettability, quartz sandstones are better than feldspar sandstones on the capacity of holding oil. Feldspar often has better wettability than quartz does, which leads to a thicker liquid film on the surface of feldspar particles than on the surface of quartz particles. The liquid films on grain surfaces are usually

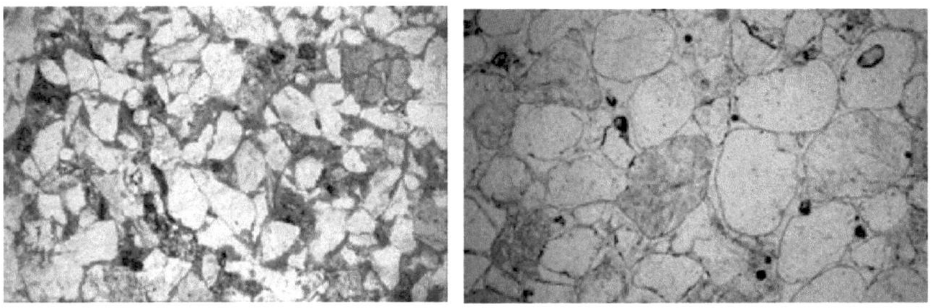

Left is a sandstone with very angular grains. The spaces between the grains are filled with blue resin. Right the grains are well rounded. Their rims have a light covering of hematite (red iron oxide), which gives a pinkish tinge to the stone.

Fig. 2.25 Thin section photographs of two sandstones

Table 2.13 Mineral composition of conventional sandstones

Mineral component	Quartz	Feldspar	Clay minerals	Carbonate minerals
Average content (%)	65	10–15	5	<1

Table 2.14 The effect of argillaceous matter to sandstone porosity (杨胜来 2004)

Content of argillaceous cement (%)	<2	2–5	5–10	10–15	15–20
Porosity (%)	28–34	29–31	25–30	<25	<20

immovable. Therefore, the pore space, where fluids are held, is slightly decreased by the liquid films adsorbed on the surface of particles in feldspar sandstones.

Cements in a rock usually reduce rock porosity because part of the voids is occupied by them (Table 2.14). The major components of calcareous cement are calcite and dolomite. Calcareous cement has greater effect on porosity than argillaceous cement. The porosity of sandstones will be sharply decreased as the amount of calcareous matter is higher than about 3 % in a rock.

2.2.3.5 Buried Depth and Compaction

With the increase in depth of reservoir rocks, geostatic pressure and formation temperature gradually increase, the grain arrangement thus becomes more compact. This change means the inelastic and irreversible locomotion occurring on the grain arrangement, the porosity of the rock is then sharp declined. Figure 2.26 shows the relationship between sandstone porosity and the buried depth of the rock.

As overburden pressure is increasing, grains in rocks become increasingly compact. The partial pressolution occurs at the places where grains contact each other. However, dissolved minerals, such as quartz, may also form new crystals in pore, which leads to a further decrease in porosity. In severe cases, it can cause pores to disappear and a rock to lose its permeability.

The relationship between the porosity and buried depth of carbonate rocks is shown in Fig. 2.27. The porosity of carbonate rocks also decreases remarkably with the decrease in depth. Research shows that carbonate sediments deposit in a quite confined environment, and have close relationship with biological action. After diagenesis, carbonate rocks are susceptible to various physicochemical changes in different environments, such as dissolution under groundwater, recrystallization under appropriate temperature and pressure, etc. As a result, multifarious pore patterns may be formed in carbonate rocks; and primary porosity are easy to change because of the effects of environments.

In a word, the factors affecting primary porosity mostly include sedimentary environment, dolomitization, chemical erosion, and tectonic stress.

Compared with sandstones, carbonate rocks are relatively brittle and thus easy to generate fractures by tectonic stress. The total porosity of carbonate rocks is thus greatly increased. Under the same stress conditions, dolomite may generate more fractures, limestone follows, and marl is the worst.

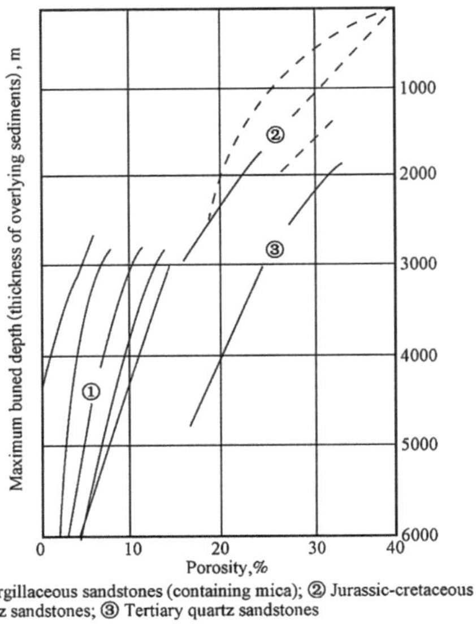

Fig. 2.26 Porosity-depth diagram of sandstone (Mayer-Curr 1978, from [21])

Fig. 2.27 Porosity-depth diagram of carbonate rocks (Schmoker et al. 1982, from [21])

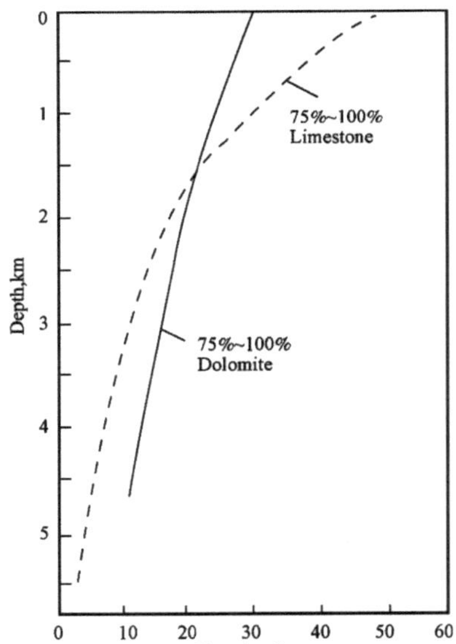

In addition, the solution and transport of dissoluble minerals, such as halite, anhydrite, in sedimentary rocks can also increase the porosity of rocks.

2.2.4 Measurement of Porosity

Rock porosity can be determined with two kinds of techniques: core analysis and well logging. Well logging is an indirect method. It explains rock porosity through the relationship between the electrical characteristics of rocks and porosity. This method involves complex principles, thus not introduced here. The essential method is to measure porosity with a core in the laboratory. The cores are extracted from wells. Generally speaking, the porosity determined with cores is more accurate than that from down-hole tools (e.g., well logging), but suffering from sampling problems [31].

All methods developed for determining porosity in the laboratory are based on the definition of porosity. From the following definition:

$$\phi = \frac{V_p}{V_b} = \left(1 - \frac{V_s}{V_b}\right) \tag{2.29}$$

where ϕ is the porosity of a rock, %; V_p is the total pore volume of the rock, cm^3; V_b is the bulk rock volume of the rock, cm^3; V_s is the total volume of solid particles composing the rock matrix, cm^3.

We see that rock porosity is related with three basic parameters: bulk volume, pore volume, or particle/matrix volume. It can be easily determined according to any two of the three parameters. The techniques used to measure these parameters are simply described as follows.

In these techniques, the sample should be clear and dry. The sample may be either a regular geometric shape (cube, cylinder) or any other shape.

2.2.4.1 Measurement of Bulk Volume V_b

The bulk volume of a sample can be determined by either a vernier caliper or other methods depending on the geometry of the sample is regular or not.

The common methods are: (i) vernier calipers, (ii) Archimedes' method (coating the sample with paraffin; saturating the sample with kerosene), and (iii) fluid displacement (mercury measurement) [37].

Vernier caliper

This method is only suitable for this kind of sample which has a regular geometry (cube or cylinder) and is not chipped or notched. The precision of the common vernier caliper is 0.02 mm (SY/T 5536-1996, China). If the sample is a perfect, smooth cylinder, the length and diameter measured by a caliper can give a very accurate bulk volume of the sample [31]. Note that the results should be the arithmetic average of several measurements but not one measuring. Repeatability and accuracy of measuring mainly depend on the surface texture of the sample.

In the repeat helium expansion determinations of porosity on regular samples, the caliper bulk volume of the samples with smooth surface textures is used, and the measuring precision should fall within ± 0.3 porosity percent regardless of actual porosity [31]. Inaccuracies can arise with samples having very high permeability. High-permeability sandstones are usually friable, and have large pore size. Such rocks do not produce smooth surface texture when cylinder plugs are drilled, and the accurate determination of sample bulk volumes becomes difficult. Therefore, the caliper bulk volume of such samples is less accurate, and other methods for bulk volume should be used.

Archimedes Method—coating the sample with paraffin

It is a gravimetric method ground on Archimedes' principle, and has higher accurate for bulk volume measurement. The procedure of measurement is simply described as follows:

(a) Clean and dry a rock sample.
(b) Weigh the sample in its dry state to give the dry weight, w_1.
(c) Coat the sample with a thin layer of paraffin in melted paraffin to make it waterproof.
(d) Weigh the coated sample in air and give the weight, w_2;
(e) Immerse the coated sample into water, and weigh it in water, then give the weight, w_3.

Clearly, w_3 is less than w_2. The loss of the weight of the sample in water is caused by the buoyancy of water. It is the weight of water displaced by the immersed sample.

The volume of water displaced by the sample is just equivalent to the volume of the immersed sample. So, the bulk volume of the sample can be determined according to the volume of the coated sample and the volume of the paraffin layer on the sample's surface. Namely, it is:

$$V_b = \frac{w_2 - w_3}{\rho_w} - \frac{w_2 - w_1}{\rho_p} \tag{2.30}$$

where V_b is the bulk volume of the sample, cm^3; ρ_w is water density, g/cm^3; ρ_p is paraffin density, g/cm^3.

The method is mainly developed for loose, poorly consolidated, or easily collapsed and crashed rocks. The shape of samples may be irregular. As a result, this method finds wide applicability in oilfields.

Archimedes Method-saturating the sample with kerosene

An alternative method for bulk volume measurement is way based on the sample fully saturated with kerosene. A dried sample is first weighed in air, then fully saturated it with a given-density kerosene. Suspended under a balance, the saturated sample is weighed in air, and then in the kerosene in which the sample is saturated. The weights of the saturated sample in air and in kerosene are denoted by W_{dry} and W_{sus}, respectively.

According to Archimedes' principle, the buoyance acting on the sample in kerosene is equal to the weight of kerosene displaced by the suspended sample. The bulk volume of the sample can then be determined by the following formula:

$$V_b = \frac{W_{\text{dry}} - W_{\text{sus}}}{\rho_o} \tag{2.31}$$

where W_{dry} is the weight of the saturated sample in air, g; W_{sus} is the weight of the saturated sample in kerosene, g; ρ_{o} is the density of kerosene, g/cm^3.

This method allows the bulk volume of any irregular sample to be found accurately, but the sample cannot be collapsed or crashed.

Note that there are few sources of significant error in this method, provided no fluid drains from the plug whilst it is weighed in air [31]. The most difficult thing is determining how much excess fluid should be removed from the surface of the plug, especially for Vuggy limestones. This kind of sample should use special analysis method, namely *whole core (full-diameter core) measurements*.

Fluid Displacement

This method notes the displacement of fluids on a graduated scale when the rock sample is placed in a container containing the fluid. If the fluid automatically enters the pores, the result loses its veracity. Mercury is usually used as a measuring fluid. The fluid used for saturating the sample can also be used as the displacing fluid.

The apparatus for mercury displacement is shown in Fig. 2.28. The sample is first placed in the test chamber, then mercury is pumped into the chamber, which leads to the sample being immersed in mercury. The bulk volume of the sample is thus the difference between chamber volume and mercury volume pumped into the chamber. So, the bulk volume of a sample can be rapidly measured.

Under normal circumstances, mercury cannot penetrate into the voids of conventional rocks due to mercury molecules bigger than rock pores. However, it may

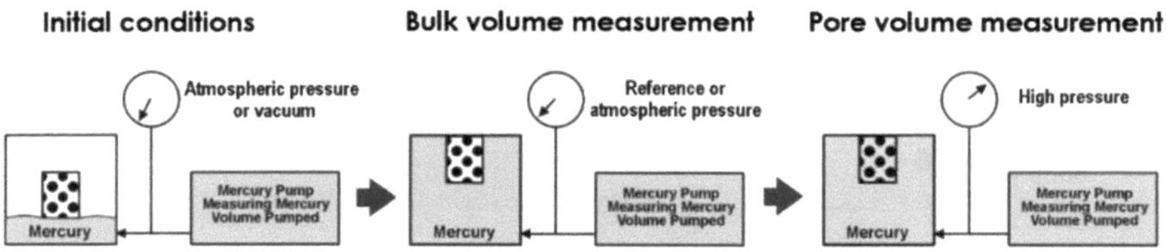

Fig. 2.28 Mercury injection apparatus (Paul [31])

give erroneous results if the sample has large pores or fractures. In addition, there is also a tendency to give a high bulk volume if any air is trapped on the surface of the sample.

Accordingly, this method is valid only if no mercury enters the voids of the sample. This method is appropriate for the measurement of irregular samples. It is worth noting that after measurement, the sample can no longer be used for other experiments and must be safely disposed of.

2.2.4.2 Measurement of Pore Volume V_p

In addition to determining the porosity, pore volume is also a vital rock parameter in various dynamic investigation of petroleum reservoir engineering, such as flow test, displacement test, and the study on microscopic mechanism of enhanced oil recovery (EOR), etc.

In the laboratory, the determination of the pore volume of a rock is chiefly based on the measurement of the volume of fluid contained in the rock. Many methods can be used to measure the pore volume of a rock. The common methods include (i) gas porisimetry, (ii) Liquid saturation method, (iii) buoyancy method, and (iv) mercury injection [37].

Different fluids, such as gas, kerosene, have different molecular size. Therefore, the results measured using different fluids are very different. Because gas can enter very small pores, e.g., subcapillary pores, the pore volume measured using gas is very close to the total pore volume of a sample. Kerosene, however, can only enter bigger pores, e.g., capillary and supercapillary pores, which leads to the result equivalent to the effective pore volume of the sample.

Gas porosimeter

Gas porosimeter is shown in Fig. 2.29. It consists of two chambers connected by tubing equipped with a valve. A clean and dried core sample should be prepared in advance.

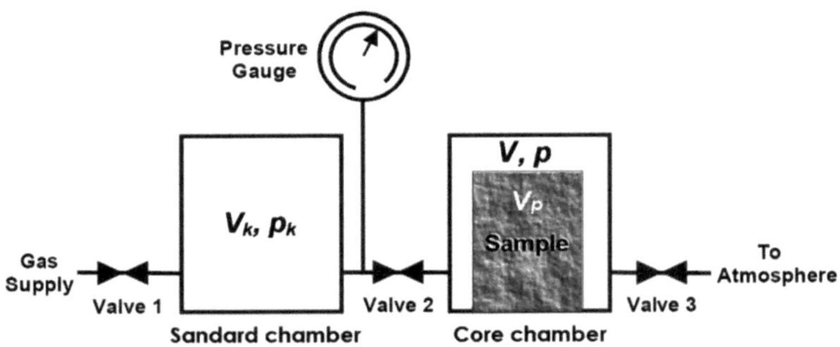

Fig. 2.29 Principle diagram of gas porosimeter

This method relies on the ideal gas law, or rather Boyle's law. The dried sample is sealed in a core chamber of given volume (V) at atmospheric pressure (p_{0030}) (Fig. 2.29). Be careful, no void space between the sample and the wall of the chamber should be remained. This chamber is attached by a valve to another chamber (standard chamber) of given volume (V_k). The standard chamber is filled with gas at a given pressure (p_k). After the valve between the two chambers is slowly opened, the gas of standard chamber expands tardily to the core chamber at isothermal conditions. When the pressure of gas in the two chambers arrives a stable value, e.g., p, the volume of gas expanding to the core chamber is equal to the pore volume (V_p) of the sample. In terms of Boyle's Law, the pressure times the volume for a gas is constant. Thus Boyle's law can be written for the equilibrium system as follows:

$$V_k \cdot p_k = p\left(V_k - V_p\right) \tag{2.32}$$

The pore volume V_p is then determined by the following expression:

$$V_p = \frac{V_k(p_k - p)}{p} \tag{2.33}$$

where V_k is the volume of standard chamber, cm^3; p_k is the pressure of standard chamber before gas expanding, MPa; p is the pressure of the system after gas expanding, MPa; V is the volume of gas expanding to the core chamber, cm^3; V_p is the pore volume of the sample, cm^3.

Any gas can be used, but the commonest is helium. The porosity determined by this method is known as "gas porosity." Despite the fact that the pore volume measured by this method is the connected pore volume of a rock, the value of the porosity calculated by this pore volume is regarded as the total porosity of the rock in respect that gas molecules (e.g., helium) are so small to enter almost all pores.

Consequently this method gives higher porosities than fluid saturation method and mercury injection method do. The method is fast, accurate, insensitive to mineralogy, and leaves the sample available for further petrophysical analysis. It can also be used to irregular samples. Inaccuracies may arise when samples are very low porosity. Low-permeability samples require longer equilibration times to allow helium to diffuse into the narrow pores.

Liquid saturation method [31]

This method measures the effective (connected) pore volume of the rock on because it is based on measuring the volume of fluid saturating the rock. The procedure of this method is as follows:

(a) Clean and dry a rock sample.
(b) Weigh the sample in its dry state to give the dry weight, W_{dry}.
(c) Fully saturate the sample in a wetting fluid. Conventionally toluene, dichloromethane or kerosene are used, but now it is more common to saturate the sample with a synthetic brine which is similar to the formation water in

reservoir conditions, i.e., they contain the same concentrations of major dissolved salts as the formation water.

(d) Weigh the saturated sample after drying any excess fluid from its surface to give its saturated weight, W_{sat}.
(e) Determine the density ρ_{fluid} of the saturating fluid by weighing a given volume of it.

The pore volume is then determined by the following equation:

$$V_p = \frac{W_{sat} - W_{dry}}{\rho_{fluid}} \tag{2.34}$$

where W_{dry} is the weight of the rock in its dry state, g; W_{sat} is the weight of the saturated sample in air, g; ρ_{fluid} is the density of fluid saturating the sample, g/cm^3.

This method is very simple and easy to operate. However, the volatilization of kerosene may result in errors depending on the time of weighing the sample.

The porosity measured by this method is often named "liquid porosity."

Buoyancy Method [31]

This method is somewhat similar to the liquid saturation method, and also measures the effective (connected) pore volume of a rock. The apparatus is shown in Fig. 2.30.

The procedure of this method is as follows:

(a) Clean and dry a rock sample.
(b) Weigh the sample in its dry state to give the dry weight, W_{dry}.
(c) Fully saturate the rock in a wetting fluid as before.

Weigh the saturated sample suspended in a bath of the same fluid with which it was saturated to give its suspended weight, W_{sus}. This is shown in Fig. 2.30. Note

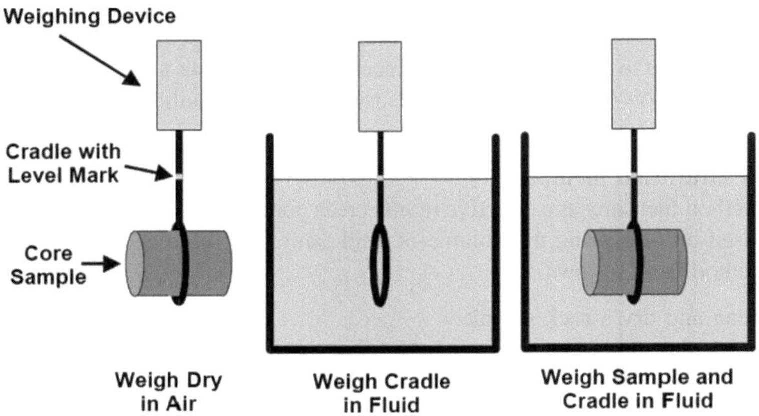

Fig. 2.30 The buoyancy method for porosity determination

that a cradle is needed to suspend the sample in the fluid, and this has a weight W_{cradle} when in the fluid. So the actual measured weight when the sample and cradle are suspended in the fluid is (W_{sus} + W_{cradle}). The effect of the cradle must be taken away, hence the weight of the cradle when suspended in the fluid (up to the same level), W_{cradle}, must be taken account of in the final calculation.

Determine the density ρ_{fluid} of the saturating fluid by weighing a known volume of it.

The pore volume is then obtained by the following expression:

$$V_p = V_{bulk} - V_{matrix} = V_{bulk} - \frac{\left(W_{dry} - (W_{sus} + W_{cradle}) + W_{cradle}\right)}{\rho_{fluid}} \tag{2.35}$$

where V_{bulk} is the bulk volume of the sample, cm^3; V_{matrix} is the volume of solid particles composing the rock matrix, cm^3; W_{dry} is the weight of the rock in its dry state, *g*; W_{sus} is the weight of the saturated sample in the same fluid with which it was saturated, *g*; W_{cradle} is the weight of the cradle suspending the sample in the fluid, *g*; W_{sus} + W_{cradle} means the total weight of the sample and cradle when they are suspended in the fluid, *g*; ρ_{fluid} is the density of fluid saturating the sample, g/cm^3.

Mercury injection [31]
The rock is evacuated and then immersed in mercury. At laboratory pressures, mercury will not enter the pores of most rocks. Mercury displacement can therefore be used to calculate the bulk volume of the rock. However, the pressure on the mercury is raised in a stepwise fashion, which can force the mercury into the pores of the rock (Fig. 2.28). If the pressure is high enough, the mercury will invade all the pores. A measurement of the amount of mercury lost into the rock provides the pore volume directly. The porosity can then be calculated from the bulk volume and the pore volume. Clearly this method also measures the connected porosity. In practice, there is always a small pore volume that is not accessed by the mercury even at the highest pressures. This is pore volume that is in the form of the minutest pores. So the mercury injection method will give a lower porosity than the two methods described above. This is a moderately accurate method. It has the advantage that it can be done on small irregular samples of rocks, and the disadvantage that the sample must be disposed of safely after the test.

Other Techniques
Other techniques include (i) analyzing all evolved fluids (gas + water + oil) and assuming that their volume is equal to the volume of the pore space, (ii) CT scanning, and (iii) NMR techniques [31].

2.2.4.3 Measurement of Matrix Volume

The matrix/solid volume of a rock sample may be measured either by gas porosimeter, or by immersion method.

Gas porosimeter

The principle of measuring the matrix volume of a sample is the same as that of measuring pore volume by gas porosimeter (Fig. 2.29). First, the dry sample is introduced in the core chamber (no core sleeve) of given volume (V). The residual void volume in the core chamber is the chamber volume (V) minus the total solid volume of the sample (V_s). In the same way, open the valve between the two chambers slowly, and let gas expands from core chamber to another at constant temperature. The pressure of the system is p when gas expansion arrives to an equilibrium state. The Boyle's law is then written for the equilibrium system as follows:

$$V_k \cdot p_k = p(V_k + V - V_s)\ V_k \cdot p_k = p(V_k + V - V_s) \tag{2.36}$$

The solid volume V_s can be calculated by the following expression:

$$V_s = V - \frac{V_k(p_k - p)}{p} \tag{2.37}$$

where V_k is the volume of standard chamber, cm^3; p_k is the pressure of standard chamber at initial state, MPa; p is the pressure of the system at equilibrium state, MPa; V is the volume of core chamber, cm^3.

Immersion method

The solid volume V_s can also be determined by the buoyancy effecting on the solid matrix of the sample immersed in a liquid (Fig. 2.31). First of all, weigh the sample in its dry state to give the dry weight, W_{dry}. Then fully saturate the sample in a wetting fluid with a density ρ at measurement temperature. After that, suspend the saturated sample in the same liquid, and weigh the saturated sample in the liquid to give its suspended weight, W_{sus}.

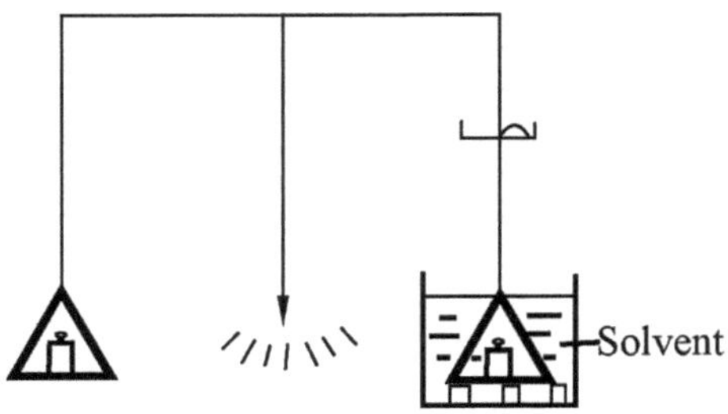

Fig. 2.31 Apparatus of immersion method

The solid volume V_s can be calculated by the following formula:

$$V_s = \frac{W_{dry} - W_{sus}}{\rho} \tag{2.38}$$

where W_{dry} is the weight of the sample in its dry state, g; W_{sus} is the weight of the saturated sample in the same fluid with which it was saturated, g; ρ is the density of fluid saturating the sample, g/cm^3.

Owing to the buoyancy is dependent of the density of liquid, carbon tetrachloride (CCl_4, $\rho_{20°C} = 1.6$) or chlorothene NU ($C_2H_3Cl_3$, $\rho_{20°C} = 1.32$) are commonly used. The former is heavier and much more toxic than the latter.

2.2.4.4 Unconsolidated Core Analysis

Unconsolidated core means the loose or weakly cemented rock cores in which there is no or few cement between the particles in the rocks (Fig. 2.32).

As shallow hydrocarbon reservoirs becomes increasingly common, friable, or unconsolidated core is often encountered. Unconsolidated core gives rise to particular problems in coring, storage, handling, plugging, and analysis [31].

Unconsolidated sand is arguably the most difficult material to work with in core analysis. Its extremely friable nature means that any rough handling damages the pore structure irreversibly, and samples can turn into a pile of mud in your hand. In order to obtain results which accurately reflect reservoir properties, extreme care must be taken during every step of coring, handling, transportation, and preparation. Additionally, laboratory equipment and procedures need to be specially designed to minimize sample disturbance.

The most common way of handling, shipping, storage, and plugging this type of core is in a frozen state [31]. The core is frozen with liquid nitrogen or dry ice as soon as it emerges from the coring barrel. It is then placed in a special core holder for next pretreatment of the core and relevant measurements.

Before analysis, an unconsolidated core is necessary to be especially treated by some encapsulating materials which can undergo washing, sample preparation, and

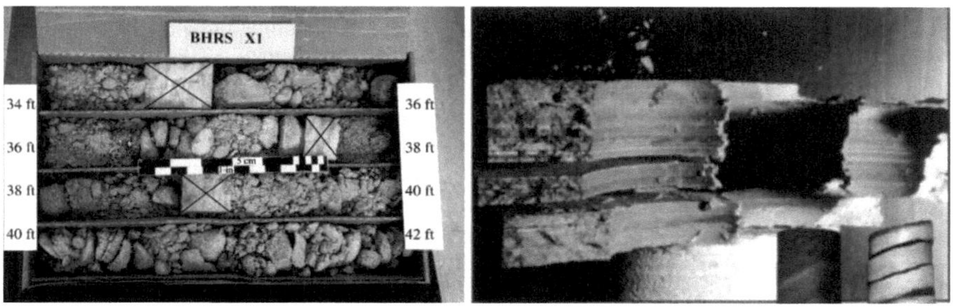

Fig. 2.32 Unconsolidated cores

measurement. Research is being conducted around the world to find materials that can stabilize and protect the unconsolidated sample. Various organic compounds and other materials are available as shown in Table 2.15.

What materials can be selected for protecting unconsolidated cores depends on the proposed core analysis programs. Techniques have been developed to protect friable rocks during all stages of sample preparation. With careful selection and application, these techniques can help provide accurate core analysis data.

In measurements, in order to let the fluids to easily flow through the core and not damage the core, the two ends of the core should be enveloped by screen clothes or other porous materials. The opening size of screen clothes should not only be small enough to prevent the particles in the sample flow away, but also be large enough to keep the moving particles away from blocking the screen opening. Commonly used screens are 200 mesh or 120 mesh.

Generally, two or more layers of screens are suggested to use for unconsolidated cores. If two layers of screens are used, one should be fine openings for preventing the particles flow away, and another is bigger openings for increasing the strength of the cover.

Furthermore, there should not be any chemical reactions between the screen materials and the solvents (e.g., brine) used in the process of washing and testing the cores. Therefore, the screen made of high quality rustless steel is often used.

When an unconsolidated sample is strictly treated by enveloped material and screen clothes, it can be measured with the same methods for conventional rocks.

Table 2.15 Various encapsulating materials for unconsolidated cores (He and Tang [38])

Material	Advantage	Disadvantage
Lead sleeve	Malleable, good adhesion to samples	Reacting with some brines and mercury
Teflon tape	No chemical reaction with other substances	Difficult to accurately determine porosity and permeability
Heat-shrinkable Teflon tube	No chemical reaction with other substances, easy to use	Poor adhesion to rock samples at low confining pressure. Samples are probably distorted due to tube shrinking under heating and pressurizing
Aluminum sleeve	Malleable, good adhesion to samples	At low confining pressure, reacting with some brines may results in poor adhesion to the sample
Epoxy paint or other coating material	Cheap, easy to use	Dissolving in organic solvents, poor mechanical strength, and probably seeping into the sample
Tin coating	Malleable, good adhesion to samples	Reacting with some brines and mercury

2.2.5 Representative Elementary Volume

2.2.5.1 Representative Elementary Volume (REV)

Application of the macroscopic laws in physics to porous media assumes that these media are interiorly continuous. That to say, the values of physical properties (porosity, permeability, saturation) of a porous medium can be defined at each point of the medium as a differentiable function of the point considered. Discontinuity, however, is the fundamental characteristic of a porous medium since at microscopic scale a point is either in the solid or in the void space, if we take the variable "porosity" as an example. The problem of discontinuities is quite common in physics, but what makes porous media so special is that the dimensions of the minimum volumes to be taken into account in order to include the effect of these discontinuities may vary considerably in the same medium, depending on the property considered.

Let O be a mathematical point inside the domain occupied by the porous medium. Consider a volume ΔV_i (say, having the shape of a sphere) much larger than a single pore or grain, for which O is the centroid. For this volume, we may determine the porosity of the medium:

$$\phi_i = \left(\Delta V_p\right)_i / \Delta V_i \tag{2.39}$$

where ΔV_i is the given volume of a medium, cm^3; $(\Delta V_p)_i$ is the volume of void space within ΔV_i, cm^3.

To determine how small ΔV_i should be in order for ϕ_i to represent the porosity in the neighborhood of O, a sequence of values ΔV_i, $i = 1, 2, 3, \ldots$ may be obtained by gradually shrinking the size of ΔV_i, around O as a centroid: $\Delta V_1 > \Delta V_2 > \Delta V_3 \cdots$.

For large values of ΔV_i, the porosity ϕ_i may undergo gradual changes as ΔV_i is reduced, especially when the considered domain is inhomogeneous (e.g., layers of sandstone). Below a certain value of ΔV_i depending on the distance of O from boundaries of inhomogeneity, these changes or fluctuations tend to decay, leaving only small-amplitude fluctuations that are due to the random distribution of pore sizes in the neighborhood of O. However, below a certain value ΔV_0 we suddenly observe large fluctuations in the porosity ϕ_i. This happens as the dimensions of ΔV_i approach those of a single pore. Finally, as $\Delta V_i \rightarrow 0$, converging on the mathematical point O, ϕ_i will become either one or zero, depending on whether O is inside a pore or inside the solid matrix of the medium. Figure 2.33 shows the relationship between ϕ_i and ΔV_i.

The porosity ϕ_i of the medium at point O is defined as the limit of the ratio $(\Delta V_p)_i/\Delta V_i$, as $\Delta V_i \rightarrow \Delta V_0$:

$$\phi(O) = \lim_{\Delta V_i \to \Delta V_0} \phi_i(O) = \lim_{\Delta V_i \to \Delta V_0} \frac{\left[\left(\Delta V_p\right)_i(O)\right]}{\Delta V_i} \tag{2.40}$$

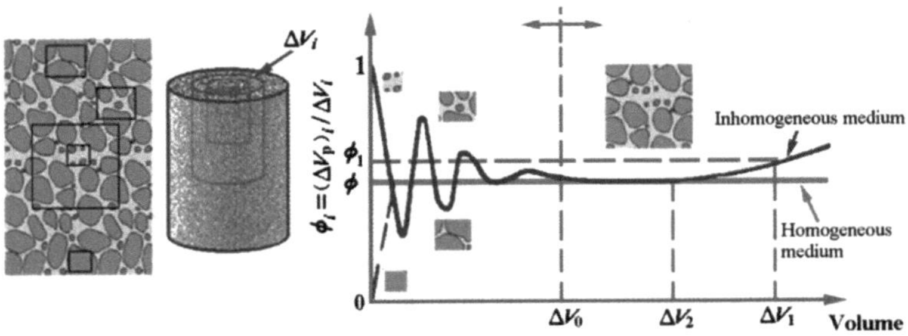

Fig. 2.33 Representative elementary volume (modified after Bear [33])

For values of $\Delta V_i < \Delta V_0$, we must consider the actual presence of pores and solid particles; in this range there is no single value that can represent the porosity at O (Fig. 2.33). The volume ΔV_0 is therefore the representative elementary volume (REV) or the physical (or material) point of the porous medium at the mathematical point O.

The limiting process in Eq. (2.40) is sometimes called the extrapolated limit. Obviously, the limit $\Delta V_i \rightarrow 0$ is meaningless. From the definition of the REV it follows that its dimensions are such that the effect of adding or subtracting one or several pores has no significant influence on the value of ϕ.

We shall assume that both ΔV_0 and ΔV_p vary smoothly in the vicinity of O. Then:

$$\phi(O) = \lim_{O \rightarrow O'} \phi(O') \tag{2.41}$$

The expression means that it is a continuous function of the position of O within the porous medium.

Thus, by introducing the concept of porosity and the definition of REV, we have replaced the actual medium by a fictitious continuum in which we may assign values of any property (whether of the medium or of the fluids filling the void space) to any mathematical point in it.

2.2.5.2 Significance of REV

REV means the smallest rock's volume whose properties can characterize the average physical properties of the reservoir rock.

Based on the concept of REV, we can use small samples to measure the physical properties (e.g., porosity, permeability, and saturation) of conventional rocks. Small core is drilled from full-diameter cores, and also called conventional core or core plug which is generally less than 1 inch (2.54 cm) in diameter (Fig. 2.34). But for these rocks having large-size fractures or dissolved caverns, whole cores

Fig. 2.34 Full diameter core and core plugs cut for laboratory measurements

(full-diameter core) are necessary for the measurement of physical properties of rocks. Full-diameter core is the core taken at the time of well drilling (Fig. 2.34).

2.2.6 Compressibility of Reservoir Rocks

Rocks when buried at reservoir depths are subject to both internal and external stresses (Fig. 2.35). The internal stress results from the pressure of fluids in rock pores. The external stresses are exerted by the weight of the overburden (overlying formations) and any accompanying tectonic stresses. Overburden pressures vary from area to area depending on factors such as depth, nature of the structure, consolidation of the formation, and possibly the geologic age and history of the rocks [39]. Depth of the formation is the most important consideration, and a typical value of overburden pressure is approximately 0.023 MPa per meter of depth [39].

The weight of the overburden simply applies a compressive force to the reservoir. The pressure in the rock pore spaces does not normally approach the overburden pressure. A typical pore pressure, commonly referred to as the reservoir pressure, is approximately 0.01 MPa per meter of depth, assuming that the reservoir is sufficiently consolidated so the overburden pressure is not transmitted to the fluids in the pore spaces [39].

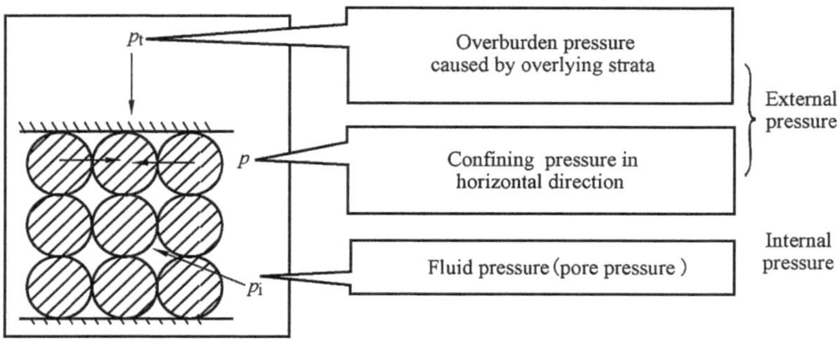

Fig. 2.35 Schematic diagram of stresses in reservoir

Before production, internal pressure (pore pressure) and external pressure (overburden pressure, confining pressure) in reservoirs maintain balance each other. Once the reservoir is disturbed by drilling and consequently production, the pressure balance is broken.

The pressure difference between overburden and internal pore pressure is referred to as the *effective overburden pressure* [39]. During pressure depletion (oil reservoir production), the initial pore pressure decreases and the effective overburden pressure increases since the overburden load remains constant. This increase in pressure difference causes the following effects:

(a) The bulk volume of the reservoir rock is reduced;
(b) Sand grains constituting the rock expand;
(c) Fluids within the pore spaces expand.

The first two volume changes tend to reduce the pore space of the rock. Obviously, both the decrease in pore volume and the expansion of fluids will continuously drive fluids flowing out from pores. Moreover, the output of fluids from the reservoirs will lead to the farther decline in pore pressure and the release of the elastic energies of rock and fluids. The elastic energy of a rock is actually an exhibition of rock compressibility. The elastic energies from the rock and fluids are thus the major drive energies of oil production in an undersaturated oil reservoir, especially in the initial stage of reservoir production.

2.2.6.1 Rock Compressibility

Reservoir rocks can be compressed due to their porosity. Rock compressibility is defined as the reduction in pore volume per unit of rock volume with a unit change in reservoir pressure, expressed as C_f. Mathematically it is given by:

$$C_f = \frac{1}{V_b} \cdot \frac{\Delta V_p}{\Delta p} \tag{2.42}$$

where C_f is the rock compressibility, 1/MPa; V_b is the bulk volume of the rock, m^3; ΔV_p is the change in pore volume, m^3; Δp is the change in reservoir pressure, MPa.

In Europe or the United States, rock compressibility is commonly defined as the fractional change in pore volume of the rock with a unit change in pressure and given by the following equation:

$$C_p = \frac{1}{V_p} \cdot \frac{\Delta V_p}{\Delta p} \tag{2.43}$$

where C_p is the rock compressibility, 1/MPa; V_p is the pore volume of the rock, m^3; ΔV_p is the change in pore volume, m^3; Δp is the change in reservoir pressure, MPa.

Due to the relationship $V_p = \phi V_b$, the compressibility C_f is related to the compressibility C_p by the following expression: $C_f = \phi C_p$, where ϕ is the porosity of the rock, and expressed in fraction.

Obviously, Eq. (2.42) or (2.43) exhibits the reduction in pore volume caused by reservoir pressure drop. It is the compressibility of a rock. The reduction of pore volume will continuously drive oil flowing from the reservoir to a well. Therefore, rock compressibility not only denotes the elastic energy of a rock but also characterizes the ability of the rock driving oil.

Despite the values of rock compressibility typically in the order of $1 - 2 \times 10^{-6}$ 1/MPa, considerable elastic energies can be released from reservoir rocks due to their huge bulk volume.

Example 2.1 Calculate the reduction in the pore volume of a reservoir due to a pressure drop of 2 MPa. The reservoir original pore volume is one million cubic meters with an estimated rock compressibility of 5×10^{-4} MPa^{-1}.

Solution

Applying Eq. (2.42) gives

$$\Delta V_p = C_f \cdot V_b \cdot \Delta p = 5 \times 10^{-4} \times 10^6 \times 2 = 1000\left(m^3\right)$$

We can see that the decrease in pore volume is caused by released elastic energy of the rock is notable. So the elastic energy of a rock is an important energy of drive oil flowing, especially in the early stage of the undersaturated oil reservoir production.

The reduction in the pore volume due to pressure decline can also be expressed in terms of the changes in the reservoir porosity. Equation (2.43) can be rearranged to give:

$$C_p = \frac{1}{V_p/V_b} \cdot \frac{\Delta V_p/V_b}{\Delta p} = \frac{1}{\phi} \cdot \frac{\Delta \phi}{\Delta p} \tag{2.44}$$

It can be obtained from Eq. (2.44):

$$C_p \partial p = \frac{1}{\phi} \cdot \partial \phi \tag{2.45}$$

Integrating the above relation gives:

$$\int_{p_0}^{p} C_p \partial p = \int_{\phi_0}^{\phi} \frac{1}{\phi} \cdot \partial \phi \tag{2.46}$$

Then

$$\phi = \phi_0 e^{-C_p \Delta p} \tag{2.47}$$

where C_p is the compressibility of a rock, 1/MPa; Δp is the change in pressure, MPa; p_0 is original pressure, MPa; p is current pressure, MPa; ϕ_0 is the porosity of a rock at original pressure, %; ϕ is the porosity of the rock at pressure p, %;

Equation (2.47) shows the relationship between rock porosity and reservoir pressure. Rock porosity changes with reservoir pressure. Equation (2.47) can also be used for the correcting porosity values from laboratory to reservoir conditions.

As mentioned above, a decline in reservoir pressure will cause two effects: pore volume decrease and fluids expansion, and both of which will drive fluids in pores flowing from the reservoir to a well. As a result, there are two kinds of elastic energies in reservoirs, one from the rock matrix; another from fluids in rock pores.

When the reservoir is being depleted, the total oil amount displaced by formation elastic energy clearly includes two parts: one is the part displaced by rock matrix expansion, denoted by ΔV_{of}; another is displaced by fluid expansion, expressed as ΔV_{oL}.

Namely,

$$\Delta V_o = \Delta V_{\text{of}} + \Delta V_{\text{oL}} \tag{2.48}$$

where ΔV_o is the total oil amount displaced by formation elastic energy, m^3; ΔV_{of} is the oil amount displaced by rock matrix expansion, m^3; ΔV_o is the oil amount displaced by fluid expansion, m^3.

From Eq. (2.42), the oil amount displaced by rock matrix equals to the reduction in pore volume caused by the expansion of rock grains. It is given by the following expression:

$$\Delta V_{\text{of}} = \Delta V_p = C_f V_b \Delta p \tag{2.49}$$

where ΔV_{of} is the oil amount displaced by rock matrix expanding, m^3; ΔV_p is the change in pore volume of the rock, m^3; C_f is the rock compressibility, 1/MPa; V_b is the bulk volume of the rock, m^3; Δp is the change in reservoir pressure, MPa.

From the definition of rock compressibility C_f, a similar parameter (C_{L}) is introduced to describe the compressibility of fluids. Considering the fluids in rock pores, the oil amount displaced by fluids expansion is calculated as follows:

$$\Delta V_{oL} = \Delta V_L = C_L V_b \Delta p = C_L \phi V_b \Delta p \tag{2.50}$$

where ϕ is rock porosity, f; ΔV_L is the change in fluid volume, m^3; V_p is the pore volume of the rock, m^3; and V_b is the bulk volume of the rock, m^3.

Then substitute Eqs. (2.45) and (2.46) for ΔV_{of} andΔV_{oL} in Eq. (2.52):

$$\begin{aligned} \Delta V_o &= \Delta V_{\text{of}} + \Delta V_{\text{oL}} = C_f V_b \Delta p + C_L \phi V_b \Delta p \\ &= (C_f + C_L \phi) V_b \Delta p \end{aligned} \tag{2.51}$$

Let

$$C_t = C_f + C_L\phi \tag{2.52}$$

Then

$$\Delta V_o = C_t V_b \Delta p \tag{2.53}$$

Here, C_t is named as the total compressibility of the formation, 1/MPa.

2.2.6.2 Formation Compressibility

Total compressibility of the formation is defined as the total change in pore volume and fluid volume in unit rock volume with a unit change in reservoir pressure:

$$C_t = \frac{1}{V_p} \cdot \frac{\Delta V_o}{\Delta p} \tag{2.54}$$

where C_t is the total compressibility of the formation, 1/MPa; ΔV_o is the total oil amount displaced by the formation, which is equivalent to the total change in pore volume and fluid volume, m^3; V_b is the bulk volume of formation, m^3; Δp is the change in reservoir pressure, MPa.

Formation compressibility is the term commonly used to describe the total compressibility of a formation. It also represents the total ability of the formation driving oil by the elastic energies of the formation (i.e., the elastic energies of rock and fluids).

Numerically

$$C_t = C_f + C_L\phi \tag{2.55}$$

In an undersaturated reservoir, the total compressibility includes: the compressibilities of oil, irreducible (connate) water, and rock matrix. Namely:

$$C_t = C_f + (C_o S_o + C_w S_w)\phi \tag{2.56}$$

Contrarily, in a saturated reservoir:

$$C_t = C_f + \left(C_o S_o + C_w S_w + C_g S_g\right)\phi \tag{2.57}$$

where C_o, C_w, and C_g is the compressibilities of oil, connate water, and natural gas, respectively, 1/MPa; S_o, S_w, and S_g are the saturations of oil, connate water, and natural gas, respectively, %; ϕ is the porosity of the rock, *f*.

Note that, in spite of C_L being about two orders of magnitude larger than C_f, the order of the product $C_L\phi$ is almost the same with that of C_f because the porosity is only about 20 % of the rock. The total elastic energy of a rock is roughly equivalent to that of fluids in the rock and cannot be ignored.

2.3 Fluid Saturation of Reservoir Rocks

The saturation of reservoir fluids is another important physical property of reservoir rocks. The knowledge of fluid saturation is much necessary in every phase of reservoir production. For example, it is used for the estimation of initial oil or gas in place, i.e., geological reserves, at discovery stage, and for the identification of reservoir zones where a large quantity of oil is left behind. It is also involved in the evaluation of the enhanced oil recovery process [40].

2.3.1 *Fluid Saturation*

The pores in petroleum reservoirs are always completely saturated with fluids. In general, most rocks are completely saturated with groundwater. Under the right conditions, some of the pores in rocks may be occupied by other liquids, such oil or/and gas. In reservoir pores, actually, there is never an occasion or location where nothing exists (i.e., truly "void space"). In petroleum reservoirs, the pores may be completely filled with the following fluids: (1) oil and its associated impurities in the liquid phase; (2) natural gas and its associated impurities in the vapor phase; (3) water—either connate water or water that flowed or was injected into the reservoir. Because of the difference between the specific gravities of fluids, reservoirs exhibit dominance of a particular fluid saturation at different depths. For example, an oil zone having high oil saturation can be overlain by a gas cap and underlain by formation water.

During the development of oil reservoirs, it is basically concerned that the volumes occupied by oil, gas, and water severally as these volumes represent the amount of oil, gas, and water in reservoirs. The quantity of a liquid in rock pores can be described by the term, *saturation*. The relative amounts of oil, gas, or water when more than one phase is present in the rock can then be characterized by individual phase saturation.

2.3.1.1 Concept of Fluid Saturation

Fluid saturation is defined as the fraction of pore volume occupied by a given fluid in a reservoir rock, often expressed as percentage.

$$S_i = \frac{V_i}{V_p} \times 100\,\% = \frac{V_i}{\phi V_b} \times 100\,\% \tag{2.58}$$

where the subscript i denotes the ith fluid in the reservoir rock, such as oil, gas, or water; V_i is the volume of the ith fluid filled in rock pores, cm^3; V_p is the total volume of rock pores, cm^3, V_b is the bulk volume of the rock, cm^3.

In the pores of oil-bearing rocks, there always remains some water which was there before hydrocarbon being trapped. If oil, gas, and water simultaneously exist in reservoir rocks, the saturation of each fluid is defined by the following expressions:

Oil saturation:

$$S_o = \frac{V_o}{V_p} \times 100\,\%. \tag{2.59}$$

Gas saturation:

$$S_g = \frac{V_g}{V_p} \times 100\,\%. \tag{2.60}$$

Water saturation:

$$S_w = \frac{V_w}{V_p} \times 100\,\%. \tag{2.61}$$

where S_o, S_g, and S_w denote oil saturation, gas saturation, and water saturation severally, %; V_p is the pore volume of a rock, cm^3; V_o, V_g, and V_w represent the oil volume, gas volume, and water volume in the rock, respectively, cm^3.

And at any time during the life of an oil or gas reservoir, the following relationship must hold true in the reservoir. It is the continuity condition.

$$S_o + S_g + S_w = 100\,\% \tag{2.62}$$

For example, a reservoir having an oil saturation of 70 % implies that 70 % of the pore space in the rock is occupied by liquid petroleum. This saturation value represents an aggregate of hydrocarbon components present as liquid. In this case, there are only two fluids, oil, and water, in the reservoir, the remaining 30 % of the pore space is thus occupied by formation water, as expected. If a gas cap exists, overlying the oil zone in the reservoir, gas saturation would dominate in the top section. Note that it is common for oil or gas saturation to be zero, but water saturation is always greater than zero.

In a word, in a hydrocarbon reservoir, the sum of initial fluid saturations, if more than one fluid exists in reservoir, should be 100 % of the pore space.

For example, in an oil reservoir, the following expression is valid, when no free gas is present (undersaturated reservoir):

$$S_{\text{oi}} + S_{\text{wi}} = 100\,\%. \tag{2.63}$$

where S_{oi} is initial oil saturation, %, and S_{wi} is initial water saturation, %.

In a gas reservoir, if there is no liquid condensate in the rock, initial gas saturation can be calculated in a manner similar to Eq. (2.63):

$$S_{gi} + S_{wi} = 100\ \% \tag{2.64}$$

where S_{gi} is initial gas saturation, %.

Producing oil and gas reservoirs usually exhibit initial hydrocarbon saturations in excess of about 70 %. The rest of the pore space is filled with formation water that is not mobile in most instances. This water saturation is often referred to as reducible water or connate water saturation which fills the pores during deposition of the rock.

If oil, gas, and water simultaneously exist in a reservoir, such as the oil reservoir with a gas cap as shown in Fig. 2.36, the saturations of the three phases must add up to 100 %. At some depths, however, the saturation of one fluid could be zero. Descending from the top of the reservoir to the bottom, the saturations of gas, oil, and formation water under initial conditions of the reservoir can be summarized as follows.

In gas cap:

$$S_{gi} + S_{wi} + S_{ogr} = 100\ \% \tag{2.65}$$

where S_{ogr} is the interstitial-oil saturation in the gas cap, %.

In gas-oil transition zone:

$$S_g + S_o + S_{wi} = 100\ \%;\ S_o > S_{ogr} \tag{2.66}$$

In oil zone:

$$S_{oi} + S_{wi} = 100\ \%;\ S_g = 0 \tag{2.67}$$

In oil-water transition zone:

$$S_o + S_w = 100\ \%;\ S_w > S_{wi}\ \text{ and }\ S_o < S_{oi} \tag{2.68}$$

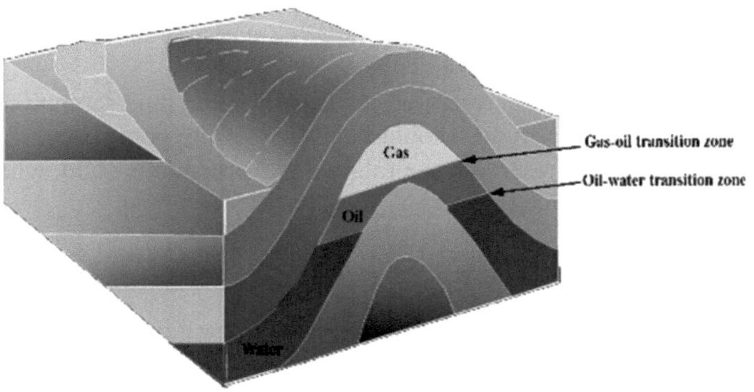

Fig. 2.36 Schematic diagram of oil-gas reservoir

In bottom water zone:

$$S_w = 100\ \%;\ S_o = 0 \tag{2.69}$$

In general, fluid saturations in a reservoir are not expected to vary significantly in lateral directions unless geologic barriers occur. However, they do vary in vertical direction in transition zones as a consequence of capillary effects. Topics on capillary pressure and transition zone are discussed in Chap. 3.

2.3.1.2 Common Fluid Saturations

After production, fluid saturations in a reservoir will alter significantly with time. This phenomenon can be observed due to the appearance of a new phase (such as free gas in oil reservoir or condensate liquid in gas reservoir) [40]. It can also be found following the injection of a driving fluid (such as water or gas) to enhance oil recovery.

The following saturation concepts are commonly used in different stages of reservoir life: *initial saturation, current saturation, residual saturation.*

Initial saturation means the fraction of pore volume occupied by a given fluid (oil, gas or water) before an oil/gas reservoir production. Initial water saturation is also known as irreducible water saturation. Initial oil/gas saturations are vital parameters of determining oil/gas geologic reserves and predicting the performance of reservoir production. However, they are not easy to be determined accurately. Generally, the initial oil/gas saturation is customarily determined according to the value of initial water saturation.

Individual fluid saturations vary with time and location in a producing reservoir. Following oil production, oil saturation decreases gradually, and water saturation and gas saturation increase gradually due to reservoir pressure decrease and water invasion/injection. Thereby, current saturation means the oil, gas, or water saturation during reservoir production, and usually called oil, gas, or water saturation for short. As long as the dissolved gas always remains in solution, the following expression is valid during the life of an oil reservoir:

$$S_o + S_w = 100\ \% \tag{2.70}$$

where S_o and S_w are separately the oil and water saturations at any stage of reservoir production, %.

The values of S_o and S_w depend the location in reservoir and the time of reservoir production.

In addition, several landmark saturations in the life of a reservoir will be briefly described as follows.

Initial water saturation

The common belief is that a reservoir initially saturated by water; then the oil comes from migration, and the water is expelled from the reservoir and replaced by oil that fills the pores in the reservoir rock. Practically, the water saturation will never get to zero percent because some of the water will be trapped in some small pores and pore-throats. During hydrocarbon accumulation in the reservoir, water saturation may be reduced to a very small value, typically 10–40 %, till no more water can escape from the pores of the reservoir.

The water trapped in rock pores and becomes immobile called *residual water*. The saturation of this part of water is called *residual water saturation*. Sometimes it is also referred to as *connate water saturation*, *primary water saturation*, or *irreducible saturation* depending on different viewpoints from different investigators (Table 2.16).

The water saturations listed in Table 2.16 indicate essentially the initial water saturation in a petroleum reservoir before production. In addition, the following symbols for initial water saturation: Swi, Swc, S_{wir} often occur in petroleum literatures [41]. Care must be taken to ensure correct interpretation of the symbol. The following definitions should help.

S_{wir}—irreducible water saturation, being or below which water cannot flow [41].
S_{wc}—connate water saturation. "Connate" implies born, originated with the formation of rocks. It is the saturation of water trapped in the pores during formation of the rock [41]. It may or may not be irreducible.
S_{wi}—initial/original water saturation. It truly means the water saturation on discovery, but it may or may not be irreducible [41].

In most cases, the term *irreducible water saturation* is prevailing in petroleum engineering, and exactly means the initial/original water saturation. In order to avoid confusion, therefore, the initial water saturation and irreducible water saturation are consistently expressed with symbol S_{wi} in the following description.

Table 2.16 Names and meanings of initial water saturation in a petroleum reservoir

Name	Meaning
Primary water saturation	From the viewpoint of sedimentology, it means the saturation of water sealed in pores of sediment during the process of diagenesis
Connate water saturation	
Irreducible water saturation	From the viewpoint of the formation of oil/gas reservoirs, it means the saturation of water remained in pores of rocks during hydrocarbon accumulation
Residual water saturation	
Irreducible saturation	From the viewpoint of fluid displacement, it means the saturation of water which could not be expelled from pores by hydrocarbon during petroleum reservoirs
Critical saturation	

Note that irreducible saturation usually means the saturation below which the fluid cannot flow in the rock. In this case, the reservoir fluid is adhered to the surface of rock particles due to the effect of adsorptive force and interfacial tensions, which will be described in Chap. 3, between fluids and rock particles or between two immiscible fluids in rock pores.

From the origin of initial water in rock pores, as is known to all, connate/original water in reservoirs either surrounds the grains or fills the small pores or dead-end pores for the sake of wettability (see Chap. 3). Hydrocarbons occupy the center of the large pores and cracks in most petroleum reservoir (Fig. 2.37).

Experience shows that there is a good relationship between initial water saturation and the physical properties of sandstones. In general, it is found to vary inversely with the porosity and permeability of rocks with larger pores in basin-wide studies. For example, in coarse-grained sandstones, oolithic and vuggy limestones, initial water saturation is usually relatively lower, while very fine-grained sandstones have a relatively high initial water saturation. In convectional sandstones, initial water saturation is round about 20–30 %, but it may be about 40–50 % in low-permeability reservoirs, and probably lower than 10 % in high-permeability reservoirs.

Based on the values of the porosity and initial water saturation in an oil reservoir, moreover, the original oil in-place (OOIP) of the oil reservoir can be easily estimated. OOIP is the vital basic data for reservoir development design and reservoir production.

Example 2.2 The oil-bearing area of an oil reservoir is 16 km^2, and the net-pay thickness of the reservoir is 10 m. The porosity of the reservoir is 20 %. The initial water saturation is 30 %. The volume factor of in-place oil is 1.2. The relative density of the oil is 0.85. Estimate the original oil in place (oil geological reserves) of the reservoir.

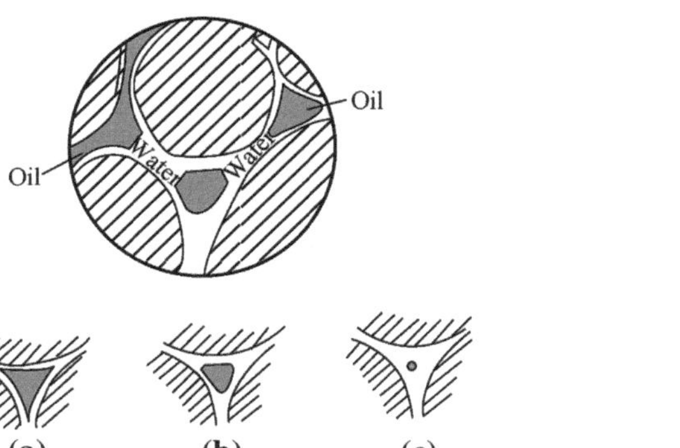

Fig. 2.37 Diagram showing the distribution of oil (nonwetting fluid) in pores of a rock filled with water (wetting fluid) (Monicard [42])

Solution

The volume of original oil in-place in the reservoir is:

$$V_o = Ah\phi(1.S_{\text{wi}}) = 16 \times 10^6 \times 10 \times 20\,\% \times (1{-}30\,\%) = 2.24 \times 10^7\left(\text{m}^3\right)$$

The original oil in-place of the reservoir is then (surface volume or quality):

$$N = V_o/\text{B}_o = 2.24 \times 10^7/1.2 = 1.87 \times 10^7\left(\text{m}^3\right)$$

$$N' = N \times \rho_\text{o} = 1.87 \times 10^7 \times 0.85 = 1.59 \times 10^7(\text{t})$$

Residual oil saturation and remaining oil saturation

At the end of the productive life of a reservoir, the oil that still remains in rock pores is referred to as *residual oil*. The fraction of pore volume occupied by residual oil in rocks is known as *residual oil saturation*, generally expressed as S_{or}.

Residual oil saturation points to a value below which oil is no longer mobile within porous media [40]. Residual oil saturation can be obtained from the core displacement in the laboratory. Knowledge of residual oil saturation is of great interest to reservoir engineers as it can be used to estimate the ultimate recoverable reserves of oil reservoirs. Some studies prefer the term *remaining oil saturation* (ROS) in quantifying the oil left behind following the primary or a subsequent recovery [40]. But the meanings of the two terms are essentially different.

Following the depletion of a typical reservoir based on natural drive mechanisms or secondary drive, a substantial portion of movable oil is usually left behind in the reservoir [40]. The fraction of pore volume occupied by this movable oil left in reservoir rocks is known as *remaining oil saturation*, simply denoted by S_o. The remaining oil saturation encountered in the reservoir could be rather high. Reservoir engineers are interested to know the distribution of remaining oil saturations in a matured field, especially in the zones of bypassed oil for further recovery. The task requires multidisciplinary studies involving petrophysics, geology, geophysics, production, and reservoir simulation.

Note that remaining oil usually refers to the oil remaining in reservoirs and being movable. It generally includes the oil either in unswept dead oil area or in low-permeable layers/areas. The most important is that remaining oil can be recovered by secondary oil production method. Residual oil, however, means the immovable oil left in reservoirs after secondary oil recovery, and existing pores in extremely dispersed state. It is thus the target of tertiary oil recovery.

In the late of secondary oil recovery, an oil reservoir will implement appropriate improved or enhanced oil recovery (IOR or EOR) operations. Following a successful enhanced oil recovery operation in an oil reservoir, the remaining oil saturation is further reduced in the reservoir. For example, the study of a typical oil reservoir throughout its life cycle may lead to the following scenario, as shown in Table 2.17.

It is apparent from Table 2.17 that not all of the movable oil can be produced even by implementing improved or enhanced recovery methods. Sometimes, it is

Table 2.17 Typical changes in oil saturation throughout the reservoir life cycle (Abdus et al. [40])

Average value of oil saturation in the reservoir	Saturation (f)	Source of data
Initial	0.80	Petrophysical studies
Following primary production	0.68	Production history/reservoir studies
Following improved and enhanced recovery	0.55	Production history/reservoir studies
Irreducible/residual	0.22	Petrophysical studies

no longer feasible to produce a reservoir in an economic sense when oil saturation is lower than a limit value. The ultimate recovery from oil reservoirs is found to average around 35 % worldwide [40]. It is the primary goal of reservoir engineers to improve recovery based on technological innovation and new approaches in reservoir management, including detailed studies and intensive monitoring.

In a gas reservoir, similarly, there also exist such terms: residual gas saturation and remaining gas saturation. Not repeat here. However, as gas is much more mobile than oil and water, in contrast to oil reservoirs, the residual gas volume at the end of a gas reservoir life is markedly low [40].

Movable oil saturation [40]

It is important to recognize that only a fraction of the oil in place is ultimately produced in most reservoirs. This poses a challenge to attain better recovery, requiring a thorough understanding of reservoir behavior. This necessitates the estimation of movable oil saturation, which represents the maximum volume of oil that can be moved or produced ultimately from a reservoir. Hence, movable oil saturation is defined as the difference between initial oil saturation and residual oil saturation. Mathematically, it is expressed as:

$$S_{om} = S_{oi} - S_{or} = 1 - S_{wi} - S_{or} \tag{2.71}$$

where S_{om} is movable oil saturation, %; S_{oi} is initial oil saturation, %; S_{wi} is initial water saturation, %; S_{or} is residual oil saturation, %.

Similarly, movable gas saturation in a gas reservoir can be estimated based on the initial and final gas saturation.

Critical gas saturation [29]

Consider an oil reservoir in which no gas evolves out of liquid phase within the reservoir as long as the reservoir produces above the bubblepoint. When the reservoir pressure declines below the bubblepoint, gas evolves out of liquid phase but is not immediately mobile. Following the buildup of free gas saturation to a certain threshold value, referred to as critical gas saturation, the vapor phase begins to flow toward the wellbore. Critical saturation is a term used in conjunction with increasing fluid saturation.

2.3.2 Factors Affecting Fluid Saturation

Fluid saturation distribution is fairly complex in a reservoir, and has great uncertainty. Generally, the distribution of fluids in rock pores depends on many factors related with rock and fluid. For example, it is affect by the physical properties of rocks and fluids, the interactions between the rock and fluids, and the affinity between the rock and fluids, a property known as wettability (see Chap. 3).

The pore structure of a rock and the surface property of rock grains are the most important factors affecting fluid saturation. In general, larger pore-size, better connectivity of pores and smooth pore wall give better permeability and smaller flowing resistance of hydrocarbons in rock pores, which results in higher hydrocarbon saturation and lower irreducible water saturation.

Different hydrocarbon fluids have different compositions, which give different physical properties of fluids, and thus lead to different hydrocarbon saturations in reservoirs. For example, high-viscosity oil, which has larger size molecules and stranger acting forces between fluid molecules and between fluid and rock, is not easy to enter smaller pores. Therefore, there is higher residual water saturation and lower oil saturation in reservoirs in this situation and vice versa.

Besides factors described above, residual oil saturation is also controlled by the following factors: the pore-throat size ratio, pressure difference, development process, and so on. The smaller are the pore-throat size ratio and the specific surface area of a rock, the easier is the displacement of oil in pores, and thus the residual oil saturation is lower. Different wetting types of rocks (oil wetting or water wetting) give rise to different patterns of water displacing oil, which thus leads to different residual oil saturations in different reservoirs. And higher pressure difference between the reservoir and the production wells makes oil easy to flow in pores and thus results in lower residual oil saturations in the reservoir.

2.3.3 Measurement of Fluid Saturation

Like other property parameters of rocks described above, the fluid saturations in various phases of a reservoir life may be determined either by direct measurement of V_o, V_g, and V_w of a cored sample (in this case only residual liquids are measured, if the core has not been preserved carefully), or by the indirect method, e.g., capillary pressure measurement and well logging.

In general, the fluid saturations in a core have been flushed by the mud filtrate and are not representative of the in situ reservoir conditions. Only samples cored with oil-based mud and taken high in the oil column (where the relative permeability to water is ignorable) have a value of water saturation at in situ reservoir conditions, after having performed the correction for the loss of dissolved gas in oil and stress effect. In the case of oil-based mud filtrate invasion, water saturation will remain immobile if Sw is less than 50 % and mild surfactants are used in the mud.

Discussion in this section is only restricted to the directly methods in the laboratory. Capillary pressure method will be introduced in Chap. 3, but well logging techniques is not concerned here.

By means of the definition, fluid saturation is just related with the pore volume and fluid volume contained in a rock. Here is only the introduction of the measurement of fluid volume.

Fluid volume can be determined either by extracting the fluids from a core sample or by heating the core to vaporize fluids from the sample. In determining the fluid volume of a sample, two techniques are commonly employed: vaporizing fluids from the sample, known as *retort distillation*, and leaching fluids from the sample, known as *solvent extraction*.

2.3.3.1 Retort Distillation

Figure 2.38 shows a typical oven of retort distillation which mainly consists of a heating element, a condenser and a graded receiver. The rock sample is placed in a cylindrical metal holder with a screw cap on the top. The top is sealed and the sample holder is placed within a retort oven. The sample, which can be small cylindrical core plug or crushed core, is heated to high temperature to distill the fluids which are then condensed and received. A temperature controller raises either in stages or directly the temperature of the core to a selected level, at which point the water within the core is vaporized and then recovered. The temperature is then increased to 650 °C to vaporize and distill oil from the sample. The vaporized oil and water are subsequently collected in a small graded tube through a condenser to be measured. The volumes of oil and water are read and recorded for future calculations.

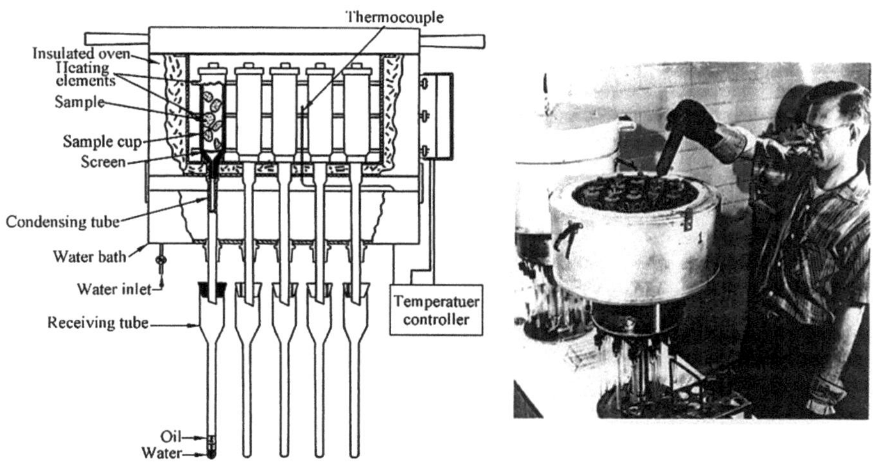

(a) (b)

(a) Diagrammatic illustration (Robert P. Monicard, 1980); (b) Picture of a conventional retort (Core Lab, 1983)

Fig. 2.38 Oven of retort distillation at atmospheric pressure

Then the saturations of oil, water, and gas can be determined as follows:

$$S_o = \frac{V_o}{V_p} \times 100\ \%, \quad S_w = \frac{V_w}{V_p} \times 100\ \% \tag{2.72}$$

$$S_g = 100 - S_o - S_w \tag{2.73}$$

where V_p is the pore volume of the sample, given, cm^3; V_o is the oil volume collected from the sample in the experiment, cm^3; V_w is the water volume collected from the sample in the experiment, cm^3.

The advantage of this method is that it takes a relatively short time, typically less than 24 h, and multiple samples can be run at the same time. The disadvantage is that heating process burns oil to the pore surfaces. When heated to higher temperatures (such as 650 °C), the oil tends to crack and coke. This is known as the *coking effect* and thus results in oil recovery less than the initial amount in the sample [44]. Sometimes, the loss of the oil could be up to 30 %. A correction (Fig. 2.39a) has been empirically developed to overcome this problem. The correction is often based on calibration curves that had been made previously using oil samples similar to the oil from the core samples. If no oil sample is available, a standard correction is generally proved adequate. For example, two typical oil correction curves are shown in Fig. 2.39a.

In addition, another problem in the heating process is the removal of crystallization water, which is the bound water in clays and other hydrates, when temperature rises to 650 ºC or over the temperature. Subsequently, the water recovery is higher than it should be. Figure 2.39b presents an example of water recovery versus time. The first plateau represents the volume of water held in the pores by capillary force. The second plateau is the additional water due to the vaporizing of the crystallized water. In this way, the result can be calibrated for the given samples.

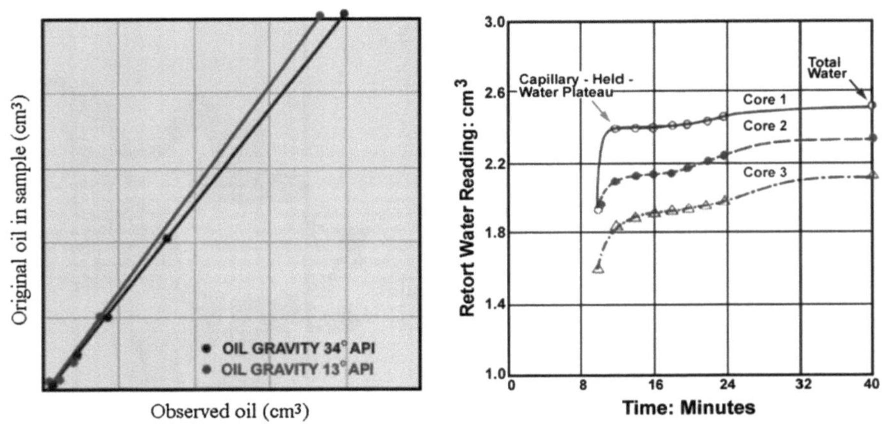

(a) Retort oil correction curves (b) Retort water correction curves

Fig. 2.39 Retort correction curves

A final disadvantage of this method is it destroys the sample; therefore no further measurement can be done with the sample.

Example 2.3 The corrected volumes of oil and water recovered from the retort distillation are 4.32 and 1.91 cm^3, respectively. Prior to this experiment, the bulk volume of the sample containing oil and water was measured to be 34.98 cm^3 and the matrix volume of the sample was 26.34 cm^3. Determine the saturations of this sample.

Solution

The following stepwise procedure is presented.

(a) The pore volume of the sample is: $V_p = V_b - V_s = 34.98 - 26.34 = 8.64$ (cm^3).

(b) Applying Eqs. (2.59) and (2.61), the oil and water saturations are separately:

$$S_o = \frac{V_{\text{oi}}}{V_p} \times 100\,\% = \frac{4.43}{8.64} \times 100\,\% = 50\,\%$$

$$S_w = \frac{V_{\text{wi}}}{V_p} \times 100\,\% = \frac{1.91}{8.64} \times 100\,\% = 22\,\%$$

(c) The gas saturation cannot be measured directly and therefore is determined by the Eq. (2.62):

$$S_g = 100 - S_{\text{oi}} - S_{\text{wi}} = 100 - 50 - 22 = 28\,\%$$

2.3.3.2 Solvent Extraction

Laboratory devices of this method, known as extractors (Dean–Stark type, Fig. 2.40), are closed vessels containing a liquid organic solvent (toluene, pentane, octane, xylene, acetone, carbon disulphide, etc.) that is miscible with the oil but not the water. The sample can be either immersed in the solvent or exposed to dripping condensed vapors.

In Dean–Stark extraction method, the solvent is heated to temperatures close to 110 °C so that the vapor of the solvent to rise through the core and to leach out oil and water from the sample. The water condenses in a cooled tube in the top of the apparatus and is collected in a graduated cylinder (Fig. 2.40). The oil removed from the samples remains in solution in the solvent. The solvent and oil continuously cycle through the extraction process.

The volume of water collected from the sample can be read from the graduated tube. And the weight of water is calculated from the volume of water by the relationship

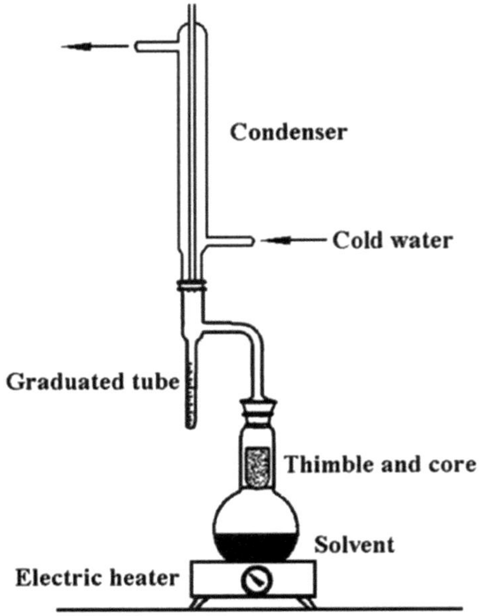

Fig. 2.40 Dean–Stark type extractor

$$W_w = \rho_w V_w \tag{2.74}$$

where ρ_w is water density in g/cm^3; V_w is the volume of water received from the sample, cm^3.

The method provides an accurate measurement result of water content in a sample. The content of oil leached from the sample can then be calculated by the difference between the weight of water recovered and the total weight loss after extraction and drying, namely:

$$W_o = W_{\text{wet}} - W_{\text{dry}} - W_w \tag{2.75}$$

where W_o is the oil amount contained in the sample; W_{wet} is the weight of the sample prior to the extraction (original sample, not cleaned!); W_{dry} is the weight of the weight of the sample after the test, cleaned and dried; W_w is the weight of water collected from the sample. The units of all weights are in gram.

Oil volume is then calculated as:

$$V_o = W_o/\rho_o \tag{2.76}$$

where ρ_o is oil density in g/cm^3; V_o is the oil volume in the sample, cm^3.

The saturations of oil, water, and gas are then determined by Eqs. (2.72) and (2.73).

The advantages of this method are that the measurement results are quit accurate, and the sample can be used for further analysis. The primary disadvantage of this method is the long time it takes to complete the measurement [44]. For example, a complete extraction may take several days to weeks in the case of low API gravity crude or presence of heavy residual hydrocarbon deposit within the core. Low-permeability rock may require longer extraction time, sometimes weeks.

Example 2.4 A saturated sample is extracted in extraction apparatus by heating the solvent. When the reading of graduated tube becomes constant, stop extracting and record the volume of water collected, $V_w = 1.4\ \text{cm}^3$. After cooling, remove the core and dry it. Obtain the weight of the dry sample, $W_{\text{dry}} = 53$ g. Using the saturation method, saturate the sample again with fresh water ($\rho_w = 1.00\ \text{g/cm}^3$) and weigh it. Its weight is 58 g (W_{wet2}). Given that the weight of the saturated sample is 57 g (W_{wet1}); the oil density is 0.88 g/cm^3. Determine the saturation of oil and water.

Solution

The calculation of oil and water saturations is as follows:

(a) The pore volume of the sample:

$$V_p = \frac{V_{\text{wet}2i} - V_{\text{dry}}}{\rho_w} = \frac{58-53}{1.00} = 5\ \text{cm}^3$$

(b) Determine oil weight by Eq. (2.70):

$$W_{\text{o}} = W_{\text{wet1}} - W_{\text{dry}} - W_{\text{w}} = 57 - 53 - 1.4 = 2.6\ \left(\text{cm}^3\right)$$

(c) Oil volume:

$$V_o = \frac{W_o}{\rho_o} = \frac{2.6}{0.88} = 2.96\ \text{cm}^3$$

(d) By Eqs. (2.59) and (2.61), the oil and water saturations are:

$$S_o = \frac{V_{oi}}{V_p} \times 100\ \% = \frac{20.96}{5} \times 100\ \% = 59.2\ \%$$

$$S_w = \frac{V_{\text{wi}}}{V_p} \times 100\ \% = \frac{1.4}{5} \times 100\ \% = 28\ \%$$

(e) The gas saturation is therefore calculated by the Eq. (2.62):

$$S_g = 100 - S_{\text{oi}} - S_{\text{wi}} = 100 - 59.2 - 28 = 12.8\ \%$$

2.3.3.3 Vacuum Distillation

The oil content in whole cores or large core samples is often determined by vacuum distillation. The assembly of a vacuum retort is shown in Fig. 2.41. An additional vacuum pump is added to a retort system for the sake of completely distilling the fluids contained in the center of the larger core. This technique can fully remove oil from the sample without sample destruction.

In the test, the sample is placed within a leak-proof vacuum system, and is heated to higher temperature, usually lower than 450 °C. The fluids contained in the sample are vaporized and collected after being condensed by liquid nitrogen or other coolant, such as mixture of methyl alcohol and dry ice. The fluids distilled from the sample are collected in a calibrated receiving tube which is immersed in a cold bath of alcohol and dry ice at about −75 °C. This prevents the oil from being drawn into the vacuum system. After that, the volumes of oil and water should also be subject to the same corrections as in the atmospheric retort distillation.

2.3.3.4 Error Correction of Fluid Saturation Measurement

A common method of obtaining fluid saturations is the measurements taken on core samples. Unfortunately, the fluid content in the core in the laboratory is altered by two processes: (1) the flushing of mud and mud filtrate into the adjacent formation, and (2) the release of confining pressure as the core is pulled to surface. Besides, during the test, there are also some errors caused by handing, such as washing cores in water, drying cores at high temperatures, or the lack of protective measures for the core.

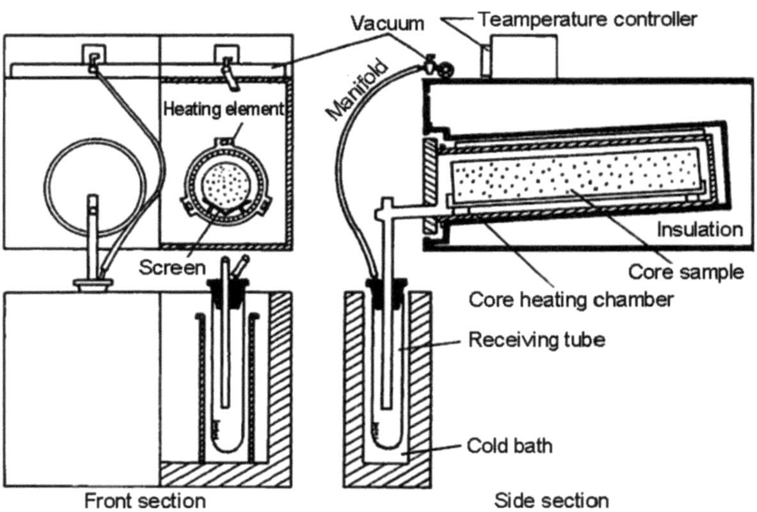

Fig. 2.41 Vacuum distillation (Monicard [42])

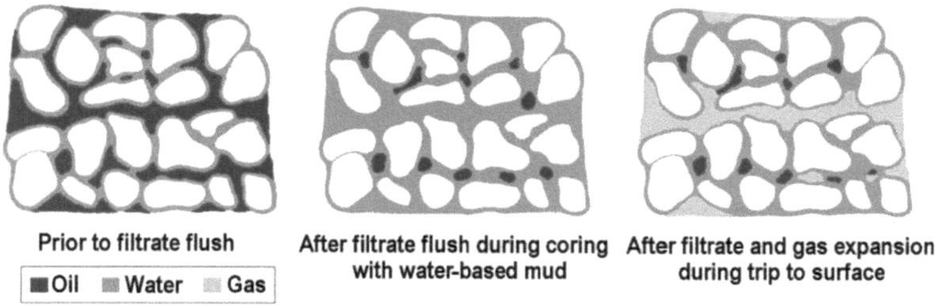

Fig. 2.42 Saturations in sand during coring and recovery (CoreLab [43])

Figure 2.42 illustrates on a microscopic level the invasion process of a water-based mud into an oil-bearing formation. The left diagram is prior to being penetrated by the mud, therefore the saturations present are the connate water and oil. The middle diagram shows the formation after the mud has penetrated and fluid invasion has flushed the original reservoir fluids. Note the increase in water saturation during this time. The right diagram shows the gas expansion as the core is brought to the surface. Invasion by mud filtrate is shown in situ by a reduction in oil saturation and an increase in water saturation. To minimize the invasion problem, it is suggested to use oil-based muds (OBM) as the coring fluid [44].

When a core is brought from in situ to the surface, the fall in pressure and temperature will result the change in oil saturation due to the dissolved gas escaping from oil. An example of saturation changes in the core from in situ to surface conditions is shown in Fig. 2.43. Note the significant decrease in oil saturation due

Fig. 2.43 Example of saturation changes in the core from insitu to surface conditions (Helander [45])

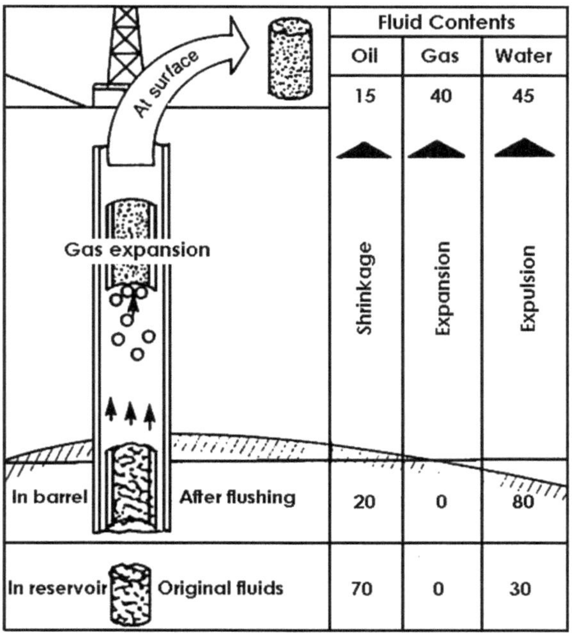

to the invasion process. Also, note the gas expansion as the core is brought to surface, subsequently expelling the fluids in the core. In this illustration, primarily water is expelled. Finally, as the pressure and temperature are reduced, the oil will shrink in volume, therefore also reducing the saturation [44]. In addition, evaporation is inevitable during handling or transporting the saturated sample if the necessary precautions were not taken.

Other research has led to using empirical factors to correct measured core saturations to original conditions (Amyx et al. 1963). The deviation in oil saturation can be corrected in two ways:

Correct the deviation caused by pressure drop using oil volume factor:

$$S_{o1} = S_o^{'} \times B_o \tag{2.77}$$

where B_o is oil volume factor, dimensionless; S'_o is the oil saturation measured in the laboratory, %; S_{o1} is the oil saturation corrected by oil volume factor, %.

Correct the deviation in sampling using empirical coefficient:

$$S_o = S_{o1} \times 1.15 \tag{2.78}$$

where 1.15 is empirical coefficient; S_o is the oil saturation corrected by empirical coefficient, %. It is equivalent to the oil saturation at reservoir conditions.

2.4 Permeability of Reservoir Rocks

In addition to porosity, permeability is also a rock property depending on the pore space of rocks. However, permeability is a measure of the ability of a rock to transmit fluids through the pore network in the rock whereas porosity just characterizes the capacity of the rock holding fluids. In petroleum reservoirs, rock permeability is one of the most valuable characteristics that the reservoir engineer seeks to determine. The oil in a reservoir rock can be extracted through bore holes only if the rock is permeable. A solid of no porosity would be impermeable to a fluid, and it is also possible for a rock of high porosity to be impermeable, namely zero permeability. Such a rock, though saturated with oil will give no production when the reservoir is drilled.

Permeability, as the name implies (ability to permeate), is a measure of how easily a fluid can flow through a rock. It is a critical property in defining the flow capacity of a rock. The study of permeability is perhaps the most important single study in the science of petroleum production. Without a sound knowledge of the permeability of a reservoir, it is impossible to estimate the probable rate of production of oil or the best methods of economic production. Consequentially, rock permeability is an indispensable basic parameter of reservoirs for the oil production and the dynamic investigation of petroleum reservoir engineering.

2.4.1 *Darcy's Law and the Absolute Permeability of Rocks*

Darcy's law, published in 1856, has served as the basis for all modern analysis of fluids flowing through porous mediums. In order to get quantitative information on the permeability of sand layers to water about 150 years ago, Darcy designed a simple experiment of water flowing through a sand-packing column, and hence developed an important empirical expression which was further developed to the vital Darcy law used in the petroleum industry today. At present, Darcy law is the most reasonable approximation for the description of the behavior of a fluid passing through the complicated and tortuous porous medium.

2.4.1.1 Darcy Law

In order to study the behavior of water flowing through sand filter, Henry Darcy, a French hydrology engineer, in 1856, conducted experiments to emulate the flowing of a subsurface aquifer and determine the flow rate of water through the sand filter.

Darcy used the apparatus shown in (Fig. 2.44), where he used a vertical sand pack through which water flowed under the influence of gravity while measuring the fluid pressures at the top and bottom of the pack by the heights of piezometric

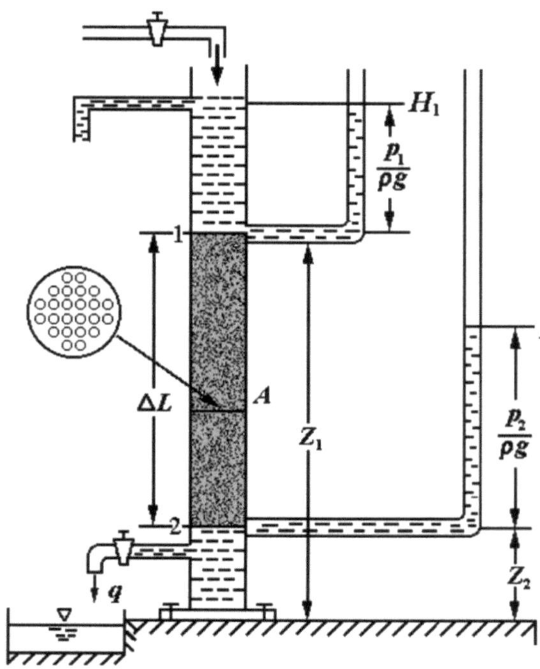

Fig. 2.44 The apparatus of Darcy experiment

tubes. Here the drop in fluid pressures can be calculated from H_1–H_2 providing the density of the fluid is known.

Darcy found that the volumetric flow rate of water through the packed column is a function of (a) the dimension of the porous medium, and (b) the difference in hydraulic head, as given in the following:

$$Q \propto \frac{A(H_1 - H_2)}{L} \tag{2.79}$$

where Q is the volumetric flow rate, cm^3/s; A is the cross-sectional area of flow, cm^2; H_1, H_2 are the hydraulic head at points 1 and 2, respectively, cm water column; and L is the length of the porous medium, cm.

After a proportionality constant was introduced to Eq. (2.79), an empirical expression, called Darcy's law, was generalized by Darcy. Darcy's law states that the volumetric flow rate of a homogeneous fluid through a porous medium is proportional to the pressure gradient and to the cross-sectional area normal to the direction of flow, and inversely proportional to the length of the porous medium and the viscosity of the fluid. It may be stated in several different forms depending on the flow conditions. The following outlines its common forms:

$$Q = K\frac{A\Delta p}{\mu L} \times 10 \tag{2.80}$$

where K is a proportionality constant for the medium in Darcy (D) unit; Q is the volumetric flow rate of fluid, cm^3/s; A is the cross-sectional area of flow, cm^2; L is the length of porous medium, cm; ΔP is the pressure difference between inlet and outlet of the sand pack column, MPa; μ is the viscosity of fluid, mPa s (1cP=1 mPa s); $\Delta P/L$ is the pressure gradient, MPa/cm.

Special attention should be paid to the unit of pressure. It is the unit of "MPa" here (Darcy unit system). But in China's oilfields, "atm" is often used for pressure (field unit system). If the unit of pressure is "atm," Darcy's formula should be expressed as: (1 atm = 0.0981 MPa $\approx$ 0.1 MPa)

$$Q = K\frac{A\Delta p}{\mu L} \tag{2.81}$$

In the formula above, the units of other parameters remain unchanged except for pressure.

The law as stated in its original form involves some limitations originally propounded by Darcy [46]. The experimentally derived form of Darcy's law was limited to homogeneous (single-phase) incompressible fluids. It implies that for the heterogeneous or multiphase fluid systems the law may not be valid. Since its discovery, it has been found valid for any Newtonian fluid. In the petroleum industry Darcy's law is formulated in terms of pressure gradient and generalized for oil and gas flow, which led to the concept of multiphase flow. Darcy's equation may be adjusted to account for unsaturated and multiphase flow. This was accomplished by separating the properties of the rock from that of the fluid by manipulating the proportionality constant K, thereby obtaining a "generalized" law as shown in Eq. (2.80) [47]. By doing so, permeability becomes a property of the porous media, dictated by pore geometry alone. Likewise, despite it was established under the condition of unconsolidated sands, it may as well be found valid for consolidated rocks, such as sandstone, carbonate rock and so on. In addition, Darcy did not envisage the possibility of the fluid interacting with and perhaps hydrating the porous medium. Yet the absence of such effects must be explicitly assumed in the above-stated formulation of the law [46].

In Darcy's law, it is the generalized use of the term "fluid." This includes any liquid—not only water—and all gases not interacting with the porous medium. The distinction between them is expressed only by the differences in their viscosity, which is explicitly introduced in the relationship between the rate of flow and the driving force.

2.4.1.2 Absolute Permeability of Rocks

When Darcy's law is used for describing the flowing of fluids in reservoir rocks, proportionality constant K is known as the *absolute permeability* of the rock in the case of single-phase flow in the rock. The absolute permeability of a rock is a measure of its ability to transmit fluid where the fluid completely saturates the porous medium [40].

In Eq. (2.80), when pressure gradient remains unchanged for a sample in single-phase flow, the relationship between flow rate Q and proportionality constant K is linear, which means that the ability of the rock transmitting fluid increases with the increase in proportionality constant. Therefore, the proportionality constant K, named absolute permeability, characterizes quantitatively the permeable property of the rock and the ability of the rock conducting fluids.

Permeability is specifically expressed in Darcy, but a few reservoir rocks have permeabilities more than 1 Darcy. In geological applications, the Darcy is commonly too large for practical purposes in fields, so the millidarcy (mD) is commonly used, where 1000 mD = 1D, following the conventions of the metric system.

Permeability varies greatly both in lateral and vertical directions in conventional reservoir rocks. Although the reservoir permeability is found to vary from less than 1 mD to several Darcies, commercial oil-bearing reservoirs operate between a few to a few hundred millidarcies in many instances. A rock whose permeability is 5 mD or less is called a tight sandstone or a dense limestone, according to its composition [48]. In general a cutoff of 1 mD is applied to oil-bearing reservoir rocks, below which the rock is not considered as a reservoir rock unless unusual circumstances apply (e.g., it is a fractured reservoir.) [31]. The gradations of permeability for reservoir rocks are shown in Table 2.18 according to the industry standard of China.

Table 2.18 Gradation of reservoir permeability [23]

Clastic rock		Carbonate rock	
Gradation	Permeability (K) (mD)	Gradation	Permeability (K) (mD)
Extremely high	K $\geq$ 2000		
High	500 $\leq$ K < 2000	High	$\geq$ 100
Moderate	50 $\leq$ K < 500	Moderate	10 $\leq$ K < 100
Low	10 $\leq$ K < 50	Low	1 $\leq$ K < 10
Extremely low	1 $\leq$ K < 10	Extremely low	K < 1
Ultra-low	K < 1		

It should be noted that permeability can be much less than 1 mD in unconventional reservoirs such as tight gas and shale gas reservoirs. Since gas requires much less driving force to move in the porous media, certain gas reservoirs may produce economically when the rock permeability is much less. In general, an oil reservoir having has very low permeability (in the low single digits of millidarcies) may not be viewed as a good candidate for substantial production over a long period [40]. One notable exception occurs with a fractured reservoir having low matrix permeability. In this case, fluid flow may occur predominantly through a network of highly conductive microchannels [40].

The explicit meaning of "Darcy" may be stated as follows: a porous medium would have a permeability of 1 Darcy when a fluid having viscosity of 1 cp flows at a rate of 1 cm^3/s under a unit pressure gradient, 1 atm/cm. That is:

$$\begin{aligned} 1D &= \frac{(1\,\text{cm}^3/\text{s})(1/100\,\text{dyn}\cdot\text{s/cm}^2)(1\,\text{cm})}{(1\,\text{cm}^2)(981{,}000\,\text{dyn/cm}^2)} \\ &= 1.02\times10^{-8}\,\text{cm}^2 \approx 10^{-8}\,\text{cm}^2 = 1\ \mu\text{m}^2 \end{aligned}$$

Therefore, the permeability of a rock has the dimension of area (L^2). The area unit (L^2) is physically related to the cross-sectional area of pore-throats in a rock. Relatively large pore-throats of a rock imply relatively large values of L^2 and thus correspond to relatively high values of permeability. That is to say, the permeability of a rock represents the average cross-sectional area of pores in the rock.

Thereby, the greater the average cross-sectional area of rock pores, the higher the rock permeability, the better the permeable property of the rock , and the easier fluids flow through the rock.

2.4.1.3 Validity of Darcy's Law

From Eq. (2.81), Darcy's law reveals a linear relationship between the flow rate q and the pressure difference ΔP for a fluid through a porous medium. That is to say, Darcy's law is a linear law. Darcy's law is valid only if the flow through a porous medium is linear or laminar. Typically, if the curve of Q-ΔP is a straight line

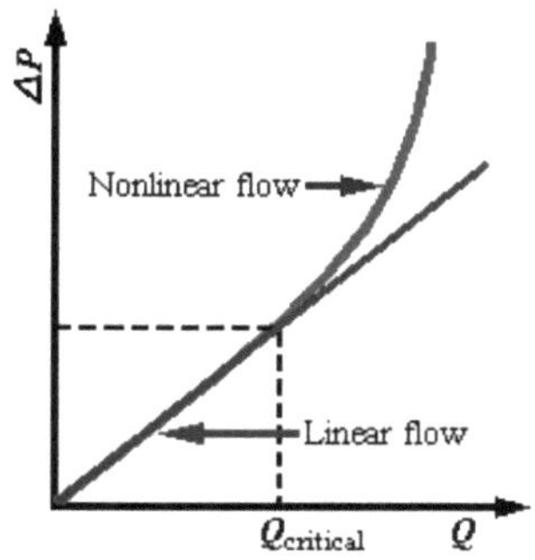

Fig. 2.45 Relationship between flow rate and pressure difference

across coordinate origin, the flow of fluid through porous medium is linear (Fig. 2.45).

However, it has been proved by experiments that the linear of Darcy's law exists only in certain flow regime. Turbulent or nonlinear flow occurs when pressure difference is large enough (Fig. 2.45).

The validity of Darcy's law may be verified by calculating the value of a dimensionless number used in fluid dynamics: *Reynolds number*. In general, the Reynolds number has different definitions depending on the character of flow.

From the experiments on pipe flow, Reynolds number found that the flow is laminar so long as the velocity of flow is less than a lower critical velocity v_c. Based on the analogy between pipe flow and the flow through porous mediums, the flow through rocks may be assumed to depend upon the dimensions of the pore space. F. I. Kotyakhov suggested that the critical velocity of flow in rocks may be determined by Reynolds number. For the flow in rocks, the Reynolds number may be typically defined as:

$$R_e = \frac{v\rho\sqrt{k}}{17.5\mu\phi^{3/2}} \tag{2.82}$$

where R_e is Reynolds number; v is the velocity of fluid through a rock, cm^3/s; μ is fluid viscosity, mPa s; ϕ is rock porosity, %; K is rock permeability, D.

Experience has shown that the critical Reynolds number of average reservoir rocks is about 0.2–0.3. For flow of fluids in a given reservoir rock, if Reynolds number is less than the critical Reynolds number (0.2–0.3), the velocity of the fluid is less than the critical velocity and the flow is laminar (linear).

Example 2.5 The absolute permeability of a core plug is measured with brine. The rock sample is 4 cm long and 3 cm^2 in cross-sectional area. The brine has a viscosity of 1.0 mPa s and is flowing at a constant rate of 0.5 cm^3/s under a 2.0 atm pressure drop. Calculate the absolute permeability of the sample.

Solution
Applying Darcy's equation, i.e., Equation (2.81), gives:

$$K = \frac{Q\mu L}{A\Delta p} = \frac{0.5 \times 1 \times 4}{3 \times 2} = 0.333(\mathrm{D})$$

Example 2.6 Rework the preceding example assuming oil of 2.0 mPa s is used to measure the permeability. Under the same pressure drop, the flow rate is 0.25 cm^3/s.
Solution
Applying Darcy's equation yields:

$$K = \frac{Q\mu L}{A\Delta p} = \frac{0.25 \times 2.0 \times 4}{3 \times 2} = 0.333(\mathrm{D})$$

From the two examples, we can see that rock permeability has nothing to do with the fluids flowing through the rock. Actually, it only depends on the texture of a rock, namely depending on the pore structure of the rock.

2.4.2 Rock Permeability Measurement

Typically, rock permeability is measured on cores in the laboratory by application of Darcy's law under steady-state conditions. During the measurement of rock permeability, the following conditions must be satisfied:

(a) Laminar (viscous) flow of a fluid in the rock;
(b) No reaction between the fluid and the rock;
(c) Only one and incompressible fluid present at 100 % pore space.

In the laboratory, oil, gas or water may be used for the measurement of rock permeability. Except condition (a), actually, other two conditions are not easy to satisfy in laboratory. When oil or water is used for the test, the measurement result may deviate from the real value of rock permeability due to the physical adsorption of oil on the surface of rock particles or the clay swelling caused by passing water in the rock. Consequently, this measurement is generally performed on clean and dried samples; and dry gases (usually air, nitrogen or helium) are often used as measuring fluid. The reasons include:

(a) The state of steady flow can be rapidly reached;
(b) Dried gas does not alter the surface structure of minerals in the rock;
(c) 100 % fluid saturation for one fluid in rock pores is easy to build;
(d) The availability of gases.

The only problem is the gas expansion in rock pores, which results in obvious slippage effect in the measurement.

2.4.3 Gas Permeability and Klinkenberg Effect

2.4.3.1 Equation of Gas Permeability

Rock permeability is often measured on rock cores with permeameter in the laboratory. Figure 2.46 shows the flow chart of a permeameter. Air or nitrogen is usually used as the measuring fluid. Upstream and downstream pressures are measured by manometers on both sides of the core and the flow rate of the gas is measured by means of a calibrated flowmeter at outlet.

Darcy's equation defining permeability is linked to laminar flow in porous media. This laminar flow cannot be achieved in gas flowing. Suppose there is a core sample in core holder. When measuring, let upstream pressure and downstream pressure be p_1 and p_0 , respectively, as shown in Fig. 2.47.

When gas flows through the core at constant temperature, the volumetric flow rate Q of the gas will vary with pressure because of the great compressibility of gas. Therefore, the values of Q at each cross-sectional area are different, which result in no validity of Darcy's law in this flow. Suppose the flow of gas through the core is steady (not change with time), the mass flow rate of gas is then the same at each cross section.

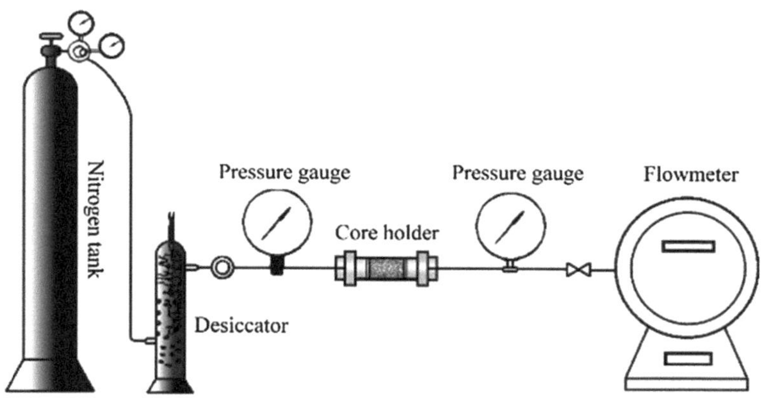

Fig. 2.46 Typically flow chart for a permeameter

Fig. 2.47 Flow of gas in a core

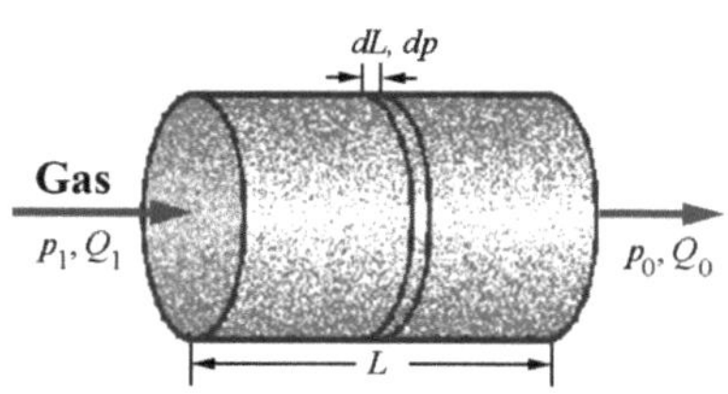

Assume that the gas flowing in a rock follows ideal gas behavior (at low pressures), the following relationships are valid according to Boyle's law:

$$Qp = Q_0 p_0 = \text{const.} \tag{2.83}$$

where Q is the volumetric flow rate of a gas at any slice of the core, cm^3/s; p is the pressure at the slice corresponding to the flow rate Q, MPa; Q_0 is the volumetric flow rate of the gas at pressure p_0, cm^3/s; p_0 may be the pressure of outlet or inlet, MPa.

From Eq. (2.83), we have

$$Q = Q_0 p_0 / p \tag{2.84}$$

Suppose the volumetric flow rate of the gas is constant at each slice of the core, a differential equation of Darcy's law can be applied:

$$Q = -K\frac{A}{\mu}\cdot\frac{\mathrm{d}p}{\mathrm{d}L}\times 10 \tag{2.85}$$

Note that a negative sign appears on the right side of Eq. (2.85) as gas flows from high pressure to low pressure.

Combine Eqs. (2.84) and (2.85), then:

$$K = -\frac{Q_0 p_0 \mu}{A}\cdot\frac{\mathrm{d}L}{p\mathrm{d}p}\times 10^{-1} \tag{2.86}$$

Separate variables and integrate for length L:

$$\int_{p_1}^{p_2} Kp\mathrm{d}p = -\int_0^L \frac{Q_0 p_0 \mu}{10A}\mathrm{d}L \tag{2.87}$$

Then

$$K_a = \frac{2Q_0 p_0 \mu L}{A\left(p_1^2 - p_2^2\right)}\times 10^{-1} \tag{2.88}$$

where K_{a} is the permeability of rock to gas, D; p_1 is upstream pressure, MPa; p_2 is upstream pressure, MPa; p_0 may be downstream pressure or upstream pressure, MPa; Q_0 is the volumetric flow rate of the gas at pressure p_0, cm^3/s.

Equation (2.88) is the formula of gas permeability. It can be seen that the gas permeability is a function of the square difference of pressure, but not a function of pressure difference.

2.4.3.2 Klinkenberg Effect

Klinkenberg (1941) [49] discovered that the gas permeability of a rock is different from the permeability measured using nonreactive liquids at the same conditions. The permeability of a sample to gas is always larger than that to liquid. The difference is considered to be caused by *slippage effect*, a phenomenon well known with respect to gas flow in capillary tubes. Klinkenberg postulated, on the basis of his experiments, that liquids have a zero velocity at the surface of rock particles, while gases exhibit some finite velocity at the grain surface. In other words, the gases exhibited *slippage* at the grain surface. The slippage resulted in a higher flow rate of the gas at a given pressure difference.

This effect is related with the different physical actions of a liquid and a gas flowing in a porous network. At lower pressures, this effect is most noticeable while it gradually becomes less noticeable as pressure increases.

Darcy's low describes the flow of viscous fluids, e.g., conventional liquids, in porous media. When a viscous liquid flows in a capillary tube, the velocities of the liquid in the tube gradually decrease from the center to the wall of the tube. The velocity profile of the liquid in tube is conic shape. Namely, the velocity of the liquid at the wall of the tube is zero, and the velocity at the center of the tube is the maximum (Fig. 2.48).

According to the kinetic theory of gas, however, gas molecules are considered as tiny spheres with diameters of approximately 1/10,000 μ. At atmospheric pressure, gas molecules move extremely fast (near the velocity of sound) and stochastically collide between gas molecules. The mean free path of a gas is inversely proportional to gas pressure. It may be very large for low-pressure gases.

When flowing in a rock, at pressures much higher than atmospheric pressure, most of gas molecules collide each other instead of with the walls of pores because of the large number of molecules in the unit volume of the rock. The internal friction of gas flow in the pores is chiefly ascribed to the collisions between gas molecules.

On the other hand, when the pressure is low enough, the distance between gas molecules is comparable to the pore sizes. In this situation, the internal friction causing by collisions of gas molecules tends to disappear and the flow is clearly affected by the collisions between gas molecules and the wall of pores molecular flow then occurs and friction factor is no longer meaningful in this flow. Because there are collisions instead of adsorptions between gas molecules and the wall of pores, gas molecules is definitely mobile rather than immovable on the internal

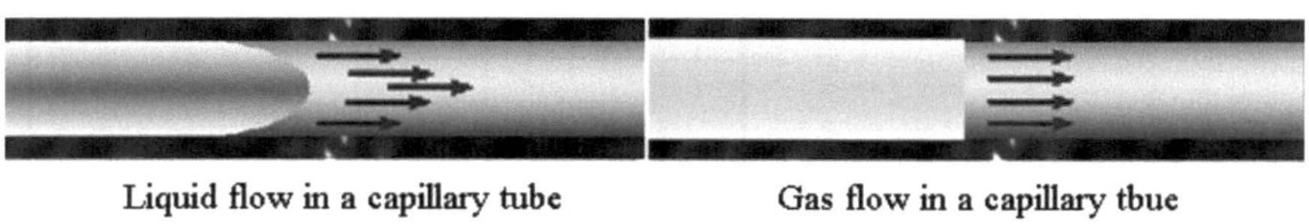

Fig. 2.48 Velocity profiles for a viscous fluid and a gas in capillary tubes

surfaces of pores. The velocity of gas molecules on the internal surfaces of pores is absolutely not zero. Comparing with the flow of a liquid in capillary tube, gas molecules flowing in a tube seem to slip from the surfaces of pores. Thus, whenever the mean free path of gas molecules approaches the dimensions of pores, gas molecules are always in motion on the pore surfaces and contribute an additional flux which results in an increase in the permeability of a rock. This phenomenon is called *slippage phenomenon (effect)*, or *Klinkenberg effect*.

Klinkenberg has reported variations in the permeability of a rock when different gases are used to measure the rock (Fig. 2.49). Figure 2.49 shows that gas permeability is a linear function of the reciprocal of average pressures for permeability measurement; and each gas has different linear relationships. Moreover, the lighter the gas molecules, the greater the slope of the straight lines. The results shown in Fig. 2.49 indicate that gas permeability increases with the decrease in the average pressure and the molecular weight of gas. The gas permeability of a roc is not a constant.

However, when extrapolating the lines in Fig. 2.49 to infinite average pressure, all lines extend to the same point at the permeability axis. The value of the point was designated K_∞, and called equivalent liquid permeability or Klinkenberg permeability.

Based on the experimental results, Klinkenberg related the apparent permeability K_a measured for gas at an average pressure $\bar{p}$ to the true permeability K_∞, by an equation with the form:

$$K_a = K_\infty\left(1 + \frac{b}{\bar{p}}\right) \quad (2.89)$$

where K_a is gas permeability, *D*, K_∞ is named equivalent liquid permeability, Klinkenberg permeability, or absolute permeability, *D*; $\bar{p}$ is average pressure, and defined as the arithmetic mean of upstream flowing pressure and downstream flowing pressure, namely: $\bar{p} = (p_1 + p_2)/2$, MPa; *b* is a constant for a given gas in a given medium, called *slip coefficient*.

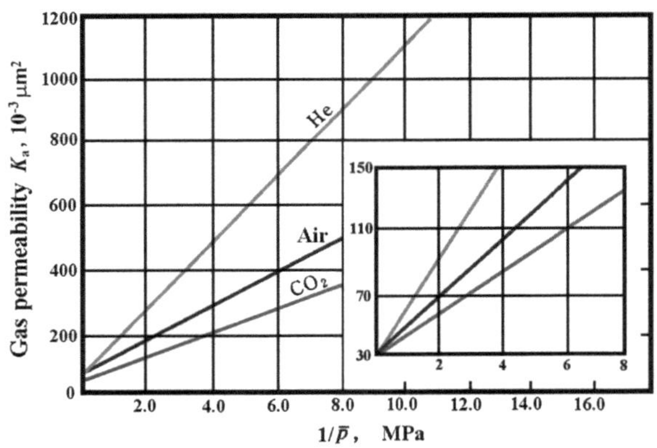

Fig. 2.49 The permeabilities to gases at different average pressures (Klinkenberg [49])

Slip coefficient b depends upon the mean free path of gas molecules, the average pressure $\bar{p}$ and the pore size of a porous medium, and can be expressed as:

$$b = \frac{4C\lambda\bar{p}}{r} \tag{2.90}$$

where $C \approx 1$; r is the average pore radius of the porous medium; λ is the mean free path of gas molecules at average pressure $\bar{p}$; λ is related with the size of the gas molecule and the density of gas molecules by the following expression:

$$\lambda = \frac{1}{\sqrt{2}\pi d^2 n} \tag{2.91}$$

where d is the diameter of the gas molecule; n is the density of gas molecules at average pressure $\bar{p}$.

Equations (2.89)–(2.91) indicate that gas permeability is directly dependent on the size of gas molecules, the pore size of a rock, and the average pressure for measurement. The smaller the size of gas molecules and pore radius, the lower the average pressure, and the larger the gas permeability K_a; and the slippage effect is more remarkable. Therefore, slippage effect is an essential reason causing the departure of gas permeability from the rock permeability. Many phenomena described above can be explained from the slippage point of view.

It needs to be emphasized that Klinkenberg effect (slippage effect) is physically significant in any situation where the mean free path of gas molecules in a porous medium approaches the pore dimension, i.e., the effect can be ignored when molecular collisions with the pore wall rather than between gas molecules are significant. Gas permeability is then enhanced by "slip flow." Therefore, It is not hard to see that slippage effect is greater in fine-grained, lower permeability rocks than in higher permeability rocks (Fig. 2.50). Generally, the correction for slippage effect is necessary for rocks which absolute permeability is lower than 10 mD.

The most reliable way to correct Klinkenberg effect is the graphic method based on the experimental data. It is procedure is as follows:

(a) For a given sample, measure several sets of permeabilities to a given gas;
(b) Plot each datum of gas permeability versus the reciprocal of average pressure ($1/\bar{p}$) in a cartesian coordinates system. Then draw the straight line fitting these data.
(c) The absolute permeability is obtained by extrapolating the line to the point of $1/\bar{p} = 0$ on the gas permeability axis. The intercept (K_∞) of K_a axis is the absolute permeability of the sample. Namely, $K = K_\infty$.

Example 2.7 The following data (Table 2.19) were obtained during permeability measurement in the laboratory. Given that the length of a sandstone core is 1.30 cm; its area is 1.36 cm^2; the viscosity of the gas is 0.0183 mPa s.

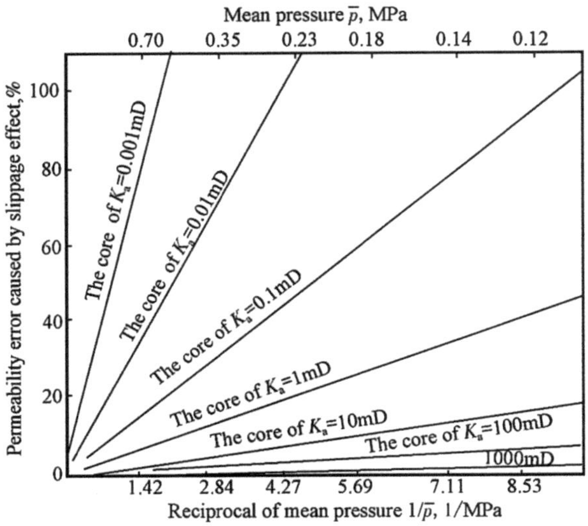

Fig. 2.50 Correction chart for Klinkenberg effect (CoreLab [43])

Table 2.19 The results of gas permeability measurement for a sandstone core

No.	Flow rate, Q (cm^3/s)	Upstream pressure, p_1(MPa)	Reciprocal of average pressure, $1/\bar{p}$ (1/MPa)	Gas permeability, $K_a(D)$
1	0.405	0.113	9.4	0.512
2	1.790	0.150	8.0	0.501
3	3.456	0.186	7.0	0.492
4	6.179	0.233	6.0	0.488

The downstream pressure for each test is 0.1 MPa. Determine the absolute permeability of the core.

Solution

Using Eq. (2.88), the gas permeability for each measurement is calculated as follows:

$$K_{a1} = \frac{2Q_0 p_0 \mu L}{A\left(p_1^2 - p_0^2\right)} \times 10^{-1}$$
$$= \frac{2 \times 0.405 \times 0.1 \times 0.0183 \times 1.30}{1.36(0.113^2 - 0.1^2)} = 0.512(D)$$

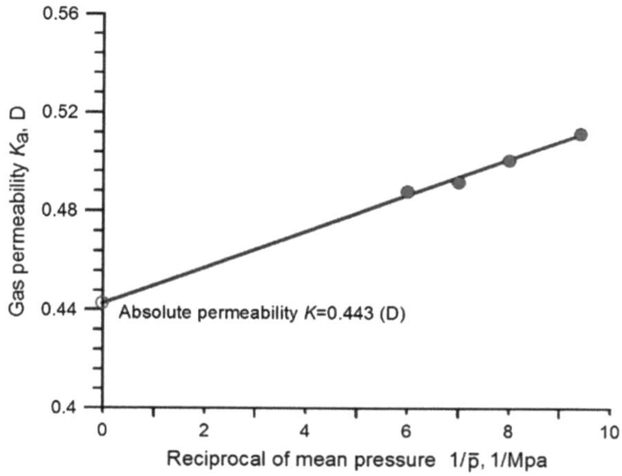

Fig. 2.51 The correction for the Klinkenberg effect using experimental data of gas permeability

Similarly,

$$K_{a2} = 0.501(D);$$
$$K_{a3} = 0.492(D);$$
$$K_{a4} = 0.488(D)$$

The gas permeabilities, listed in Table 2.19, are then plotted versus $1/\bar{p}$ as shown Fig. 2.51; and a best-fit line drown through these points. The absolute permeability of the core is read at the intercept as 0.443 D.

2.4.4 *Measurement of Rock Permeability*

2.4.4.1 Type of Core Analysis

In the laboratory, there are two categories of core analysis measurement. They are conventional core analysis and special core analysis.

Conventional core analysis [42]
In conventional core analysis, two cases can be considered: partial analysis and complete analysis

Partial analysis only includes porosity and air permeability measurements. This kind of analysis is only for exposed and not fresh or extracted samples.

In this case, air permeability and porosity are customarily measured using a perfectly geometrical sample, which is cut as either a cylinder or a cube. Before measurements, the sample should be carefully washed with solvent. Then the analysis for the sample includes the following measurements:

(a) Air permeability either with a variable head permeameter or with a constant head permeameter;

(b) Horizontal and vertical permeabilities with a cubic sample; if not, two cylindrical samples must be collected.
(c) Porosity either by immersion (see Sect. 2.2.4) which immediately gives the true mass densities or by gas expansion or by some other methods, but the first probably gives the best results.

Complete analysis includes the measurements of all physical properties of a rock, namely, the porosity, air permeability, and water, gas, and oil saturation measurements. This complete analysis can only be carried out for fresh or well-preserved samples. In this case, the method which seems to give the best results for fluid saturation is the so-called fluid summation method.

For finishing these analyses, a fresh core is divided into three parts, each for one testing. Using fresh samples, complete analysis should include the following measurements:

(a) A cube (for porosity, horizontal, and vertical permeability) or a cylinder (for porosity, horizontal permeability) is collected from one part.
(b) One part (about 40 g) is used for determination of gas saturation.
(c) One part selected as far as possible from the core center is coarsely ground: approximately 125 g. It is used for determination of oil and water saturation through the atmospheric pressure distillation method.

Special core analysis [42]
Special core analysis includes the following contents:

(a) Capillary pressure: i) by the restored state method; ii) by mercury injection, evaporation, or centrifuging.
(b) Liquid permeability.
(c) Relative gas/oil permeability.
(d) Relative water/oil permeability.
(e) Water flood susceptibility test.
(f) Formation factor and resistivity ratio.
(g) Other measurements: wettability, etc.

These tests are necessary for the study of reservoir problems and make it possible to predict field behavior in time and to estimate reserves. They are usually made in specialized laboratories.

2.4.4.2 Measurement of Rock Permeability

In the laboratory, several techniques can be used for permeability measurements of cores, depending on sample dimensions and shape, degree of consolidation, type of the fluid used (gas, water or oil), ranges of confining and fluid pressure applied, and permeability range of the core.

For consolidated rocks, samples are generally collected in geometric shapes:

(a) Either full cylinders with the following approximate dimensions: diameter = 2.3–4.0 cm and height = 2.5–5.0 cm.
(b) Or cubes with 2 cm sides.
(c) Or hollow cylinders for radial circular flow measurements.

When the state of consolidation is insufficient, it is necessary to make a small assembly or to coat:

(a) For example, to coat a fragment with wax in order to measure permeability in side wall core analysis or for unconsolidated sand.
(b) Coating with bakelite or plastic ("plasticore" by means of a thermosetting plastic).

A special technique is necessary for measuring the permeability of cores from fissured limestone deposits. These measurements are made on whole cores.

In a word, the laboratory measurement of rock permeability can usually be divided into: conventional small core analysis and whole core analysis.

Measurement of small core
In general, a cube (K_h, and K_v) or a cylinder (K_H) sample is collected, washed with a combination of a solvent extraction apparatus and a centrifuge and then dried in an appropriate manner. Cylinder sample may be cut either parallel or perpendicular to the bedding plane of the rock, depending on the direction of flow of interest.

In the laboratory, gas permeability is measured by passing a gas of known viscosity through a core sample of measured dimensions and then measuring flow rate and pressure drop. Then the gas permeability values will be corrected for the Klinkenberg effect by means of laboratory data so as to obtain absolute permeability of the rock. Two types of instruments are usually used in the laboratory:

(a) Constant head permeameter, core laboratories type.
(b) Variable head permeameter, IFP type.

Measured values using constant head equipment range from a low of 0.1 mD to 20 D. Data accuracy declines at high and low permeability values and is within ±0.5 % of true value otherwise.

Constant head permeameter (Core Laboratories type) This equipment is designed for plug or whole core permeability measurements. This experiment may be used for single or multiphase, compressible fluid or liquid measurements and can also be used under reservoir pressure and temperature.

Figure 2.52 shows a diagram of a constant head permeameter. A clean, dry sample is placed in a holder; it must fit snugly and allow no air to bypass along the sides of the sample. Air is usually used as gas flow. Upstream and downstream pressures are measured by manometers on both sides of the core and air flow is measured by means of a calibrated outlet. Air permeability can then be calculated using Eq. (2.88).

Hassler core holder may be used with this instrument. The Hassler system is an improvement of the rubber plug system whose tightness is limited at certain pressures. The core is placed in a flexible rubber tube (Fig. 2.53). The Hassler cell has these advantages:

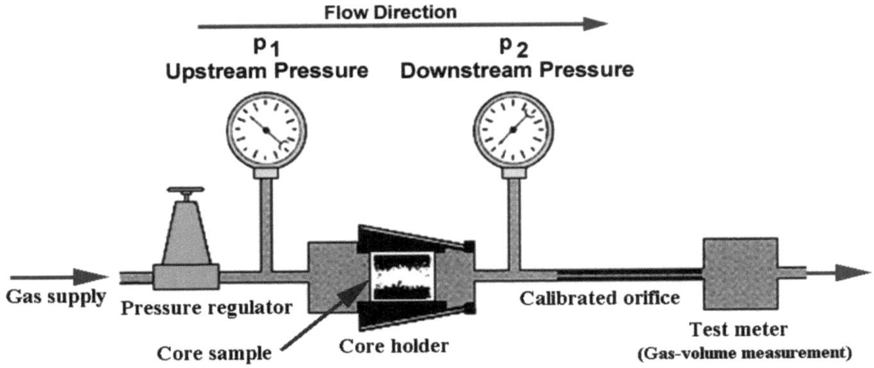

Fig. 2.52 Schematic diagram of a constant head permeameter

Fig. 2.53 Hassler type core holder (Torsæter and Abtahi [50])

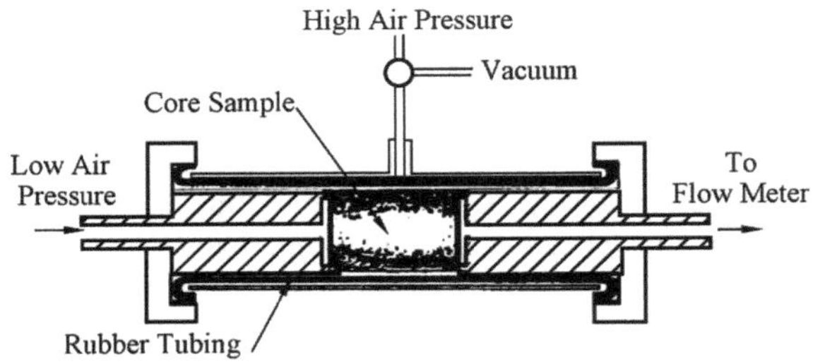

(a) Excellent tightness.
(b) Can be used for samples of different sizes.
(c) Much higher pressure or Δp can be used.

It can be used for measuring relative permeability.

Variable head permeameter (IFP type)
At present, the method of "flow-tube air permeability" is widely used in China. Figure 2.54 shows the apparatus of the method, which is also called variable head permeameter. It consists of the following parts [42]

(a) Core holder with a rubber stopper.
(b) A glass tube for flow rate measurements. This tube consists of three different sections with diameter gradations from the top down of 1–10–50 mm so that there are three ranges of sensitivity (for $1 < k < 2000$ mD) [42].
(c) A constant level water tank.
(d) A rubber bulb for bringing water up to desired point (or for suction under partial constant vacuum).
(e) A valve for cutting off the suction bulb from the inner space.

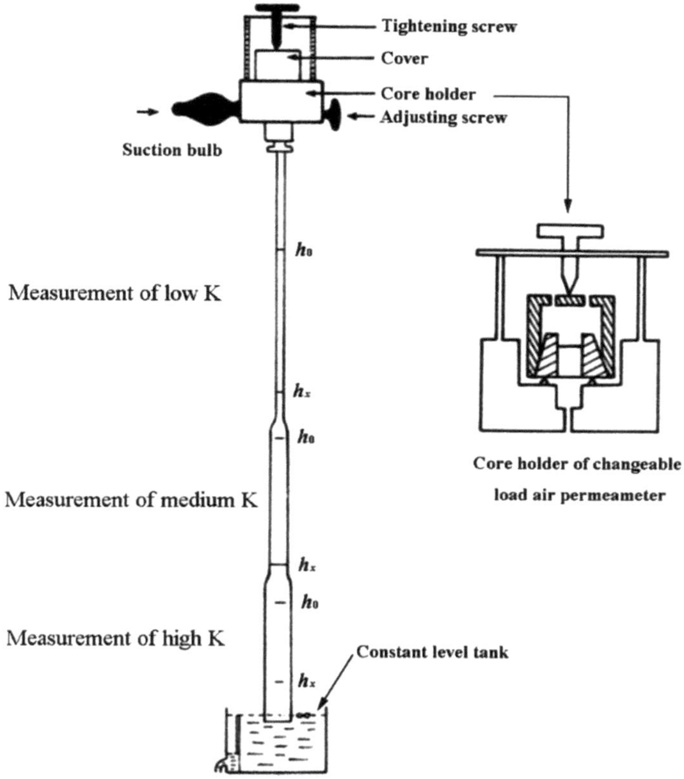

Fig. 2.54 Variable head permeameter, IFP type (Torsæter and Abtahi [50])

In the test, the top of the core keeps connected with the atmosphere. When air is pumped out by suction bulb, the water level in tube is quickly brought to a desired position h_0 of the tube. When pumping is stopped, the water level in tube falls under hydrostatic head; and the pressure difference is thus built between the two ends of the core due to the negative pressure in tube. Then outside air enters into the tube through the core.

The speed of water level falling down depends on the permeability of the core. That is to say, if the permeability of the core is great, the water level falls fast. Conversely, it falls slowly. Accordingly, when the time of water level falling from the position h_0 to the position h_x in the tube is measured, the permeability of the core can be given by an equation with the following form:

$$K_a = \frac{B\mu L}{TA} \times 10^6 \qquad (2.92)$$

where K_a is air permeability, mD; μ is air viscosity, mPa s; L is sample length, cm; A is the cross-sectional area of the sample, cm^2; T is the time of water level falling from h_0 to h_x in the tube, s; B is a constant depending on the geometry and units of the apparatus, cm^3/atm.

Constant B can be determined either by measurement or by calculating as follows:

$$B = C\left\{-2F[\ln(Cp_0 - h_x) - \ln(Cp_0 - h_0)] - \frac{V^* - 1033.6p_0F + h_0F}{Cp_0}\left[\ln\frac{Cp_0 - h_x}{h_x} - \ln\frac{Cp_0 - h_0}{h_0}\right]\right\} \quad (2.93)$$

where F is the cross-sectional area of flow tube, cm^2; p_0 is atmospheric pressure; atm; h_0 is the height between the water level in the tank and the position h_0 in the tube, cm; h_x is the height between the water level in the tank and the position h_x in the tube, cm; C is a constant, C = 2067.2; V^* is the void volume over h_0 in the tube, cm^3.

The flow tube consists of three different sections with different inside diameter, which corresponds to different values of ($h_0 - h_x$). Thereby, different sections of the tube may be selected to measure depending on the permeability of the core.

Measurement of whole core

Whole core, also called full-diameter core, measurements are normally restricted to carbonates and formations containing vugs and fractures. These measurements can be made on sandstones, but care must be taken to ensure that the invasion of drilling fluid solids does not reduce horizontal permeabilities. This type of reduction can happen in sandstones as well as in chalky limestones. In the latter, ground limestone powder that has been produced by the temperature and by pressure conditions encountered while coring is sometimes plastered on the core, causing permeability reduction. This powder must be removed prior to whole core horizontal permeability measurements, as it forms a "skin" that will result in erroneously low permeability values. This material can sometimes be cut away easily when the core is wet, otherwise sand blasting or rasping of the dry core may be necessary to remove the plaster coat.

The plugs in this case were cut from the center of the full-diameter core and were not subjected to damage. The data indicate that permeability reduction was greatest for the high-permeability section, but all full-diameter values were reduced.

Measurement of vertical permeability

The permeability which is measured by fluid flowing vertically through the bedding planes of a sample is known as vertical permeability of the sample.

Vertical permeability on a full-diameter sample is easily measured using a Hassler core holder, as illustrated in Fig. 2.55a. The sample is placed in the apparatus and the rubber tubing is then collapsed around the core. With the introduction of high pressure air to the holder, the tubing provides a seal along the sides

Fig. 2.55 Horizontal and vertical permeability measurement apparatus for full-diameter cores

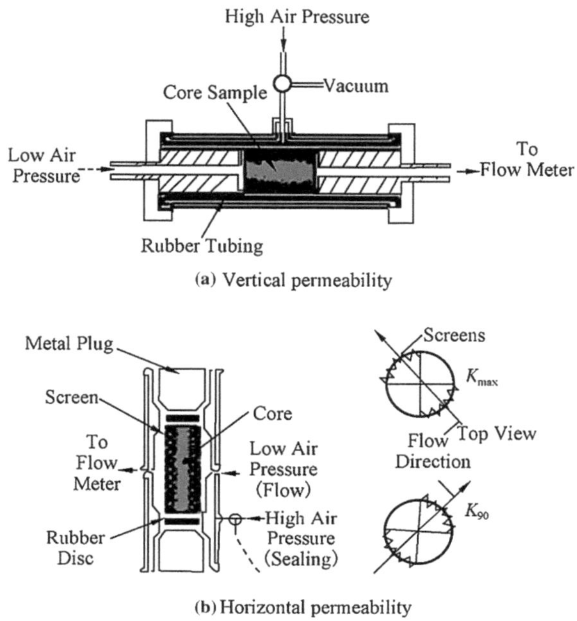

of the sample. Low-pressure air is introduced on the upstream end of the core and the measurements needed to compute permeability are easily made. After upstream and downstream pressures and flow rates being measured, gas permeability can be determined by Eq. (2.88).

Measurement of horizontal permeability

Permeability measured with gas flowing parallel to the bedding planes of a sample is called horizontal permeability of the sample.

Measuring the horizontal permeability of a whole core is more complex. Screens are placed on opposite sides of the full-diameter core (Fig. 2.55b) and the sample is then moved into the core holder. Non-permeable rubber disks are placed on each end of the sample and the rubber tubing is again for collapsing around the core. Low pressure air, introduced into the center of the holder, passes through the rubber boot, intersects with the screen, and flows vertically in the screen. The air then flows through the full-diameter sample along its full height and emerges on the opposite side, where the screen again allows free flow of the air to the exit port. In this test, the flow length is actually a function of the core diameter. The cross-sectional area of flow is a function of the length of the core sample and of the core diameter. The screens are selected to cover designated outer segments of the full-diameter sample. In most cases, the circumference of the core is divided into four equal quadrants and the screens occupy two opposite quadrants.

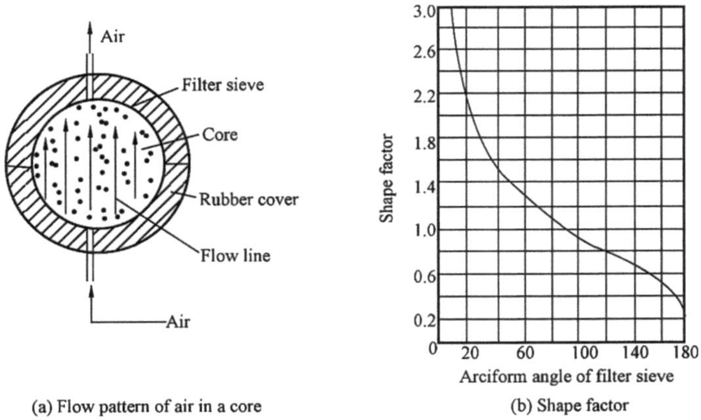

Fig. 2.56 Illustration of horizontal flow in core and shape factor chart (He [21])

In this measurement, a shape factor is introduced to gas permeability equation as follows:

$$K_{\mathrm{ha}} = E \cdot \frac{2Q_0 p_0 \mu}{L(p_1^2 - p_2^2)} \times 10^{-1} \tag{2.94}$$

where K_{ha} is the horizontal gas permeability of the sample, D; L is core length, cm; p_1 is upstream pressure, MPa; p_0 is downstream pressure or upstream pressure, MPa; p_2 is downstream pressure, MPa; Q_0 is the volumetric flow rate of gas under p_0, cm^3/s; E is shape factor; μ is gas viscosity, mPa s.

Shape factor generally depends on the diameter of the sample and the arciform angle of filter sieve which is used for assigning gas to the lateral surface of the sample, e.g., $E = 1$ if the area of filter sieve equals to the 1/4 lateral area of the sample (Fig. 2.56b).

It needs to be emphasized that at least twice measurements are necessary for horizontal permeability of a full-diameter sample. The second measurement is made at right angles to the first. When fractures are visible in the core, the first measurement is made parallel to these fractures. The maximum horizontal permeability is recorded as $k_{\max}$ and the other at 90° to the direction of this flow is labeled k_{90} (Fig. 2.55b).

Measurement of radial permeability

The permeability which is measured under the condition of gas flowing radially in a cylindrical sample is called radial permeability of the sample.

The apparatus for circular radial flow test is shown in Fig. 2.57. According to Darcy's law, gas radial permeability can be determined by the following expression:

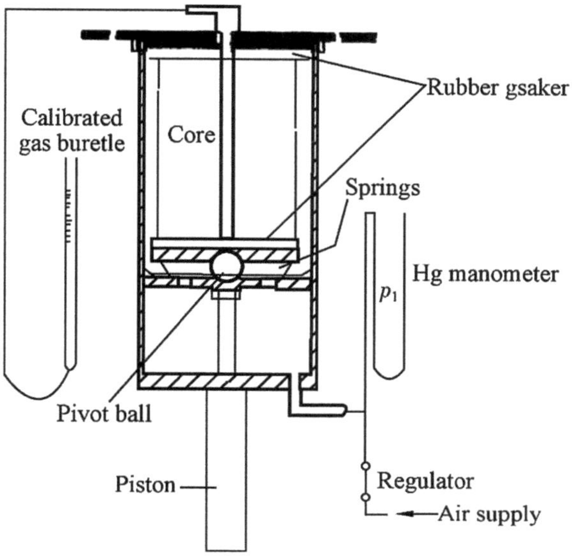

Fig. 2.57 Full-diameter radial permeameter (Monicard [42])

$$K_{ra} = \frac{Q_0 p_0 \mu \ln\left(\frac{d_e}{d_w}\right)}{\pi h \left(p_1^2 - p_2^2\right)} \times 10^{-1} \tag{2.95}$$

where K_{ra} is gas radial permeability of sample, D; d_e is sample diameter, cm; d_w is the diameter of the inner pore in the sample, cm; h is the height of the sample, cm; p_1 is upstream pressure, MPa; p_0 is downstream pressure or upstream pressure, MPa; p_2 is downstream pressure, MPa; Q_0 is the volumetric flow rate of gas under p_0, cm^3/s; μ is gas viscosity, mPa s.

The point of the test is that the small inner pore must be drilled in midmost of the cylinder core so as to keep absolutely same radius of flowing in all directions.

2.4.5 Factors of Affecting Rock Permeability

Permeability is a measure of the ability of a rock to transmit fluids. It is a critical property to define the flow capacity of a rock. From the sedimentary point of view, numerous factors affect the magnitude and/or direction of rock permeability, including various factors in the process of sedimentary and diagenesis, such as rock texture, sedimentary structure, overburden pressure, and formation temperature and so on. The effects of these factors on rock permeability are much more complicated than that on rock porosity.

2.4.5.1 Sedimentation and Tectonism

Rock texture

Since the pore structure and porosity of a rock are basically determined by roc skeleton texture, it controls the permeability of the rock as well. From the rock texture point of view, following factors are mainly involved: grain-size, sorting, grain arrangement, the component, and content of cementing, and cementation type, etc.

Experimental evidence indicates that rock permeability is direct proportional to the square of grain diameter ($K \propto cd^2$, where c is a characteristic of the rock properties) at certain sorting. Therefore as grain-size increases, an increase in permeability occurs. Figure 2.58 illustrates the significant effect of grain-size and sorting on rock permeability. It is not as dramatic as grain-size; however, the illustration does show that an increase in sorting (better or well sorted) will improve the permeability. This is why in gravel pack operations the selection of the gravel is important, both from a size and sorting viewpoint.

Particle shape and texture influence permeability to a great extent. Elongated or irregular particles create flow paths more tortuous than the spherical particles (Fig. 2.59). Also, particles with a rough surface texture provide more frictional resistance to flow than smooth textured particles. Thus, both elongated shape and rough texture reduce the rate of fluid flowing through the rock, i.e., reduce rock permeability.

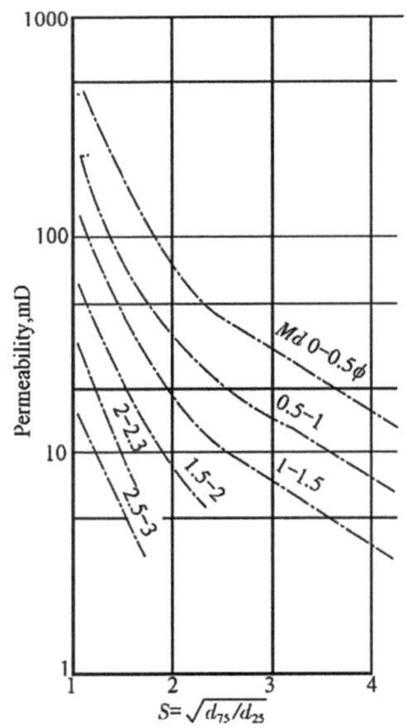

Fig. 2.58 Effect of grain-size and sorting on permeability (Bear and Weyl [51])

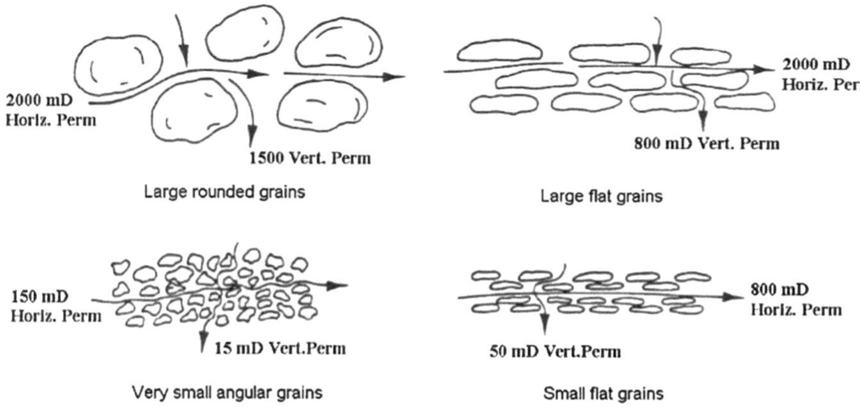

Fig. 2.59 Textural parameters and permeability (Link 1982)

As shown in Fig. 2.59, the shape and packing of grains also affect the directivity of permeability. Notice in these examples, the more angular the grains or the flatter the grain shape, a more pronounced anisotropy develops.

Reservoirs with directional permeability are called *anisotropic* [12]. Anisotropy greatly affects fluid flow characteristics of the rock. The difference in permeability measured parallel and vertical to the bedding plane is the consequence of the origin of the sediment, because grains settle in the water with their longest and flattest sides in a horizontal position. Subsequent compaction of the sediment increases the ordering of the sand grains so that they generally lie in the same direction [12].

Platy minerals such as muscovite, and shale laminations, act as barriers to vertical permeability. In this case, the ratio of vertical permeability (i.e., perpendicular to the bedding plane) to horizontal permeability (i.e., along the bedding plane) (K_h/K_v) generally ranges from 1.5 to 3 and may exceed 10 for some reservoir rocks. Sometimes, however, K_v is higher than K_h due to fractures or vertical jointing and vertical solution channels. Joints act as barriers to horizontal permeability only if they are filled with clay or other minerals.

The importance of the clay minerals as a determinant of permeability is often related not only to their abundance but also to their mineralogy and composition of the pore fluids. The clay minerals, which coat the grain surfaces, expand and/or become dislodged due to changes in the chemistry of the pore fluids or mud filtrate invasion, will considerably reduce the permeability.

In sandstone rocks, fracturing is not an important cause of the secondary permeability, except where sandstones are interbedded with shales, limestones, and dolomites. In carbonates, the solution of minerals by percolating surface and subsurface acidic waters as they pass along the primary pores, fissures, fractures, and bedding planes, increases the permeability of the reservoir rock.

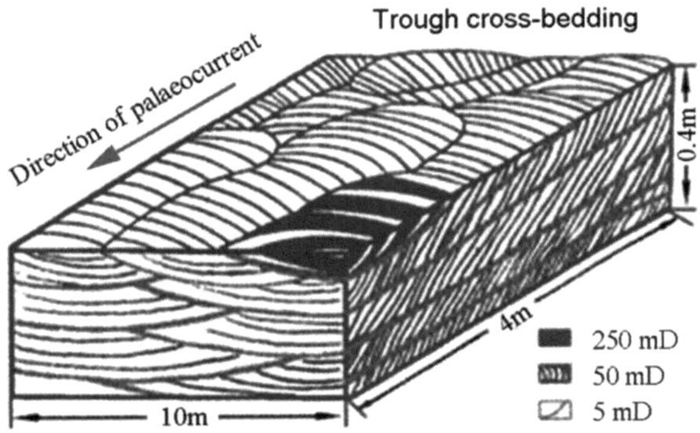

Fig. 2.60 Permeability directivity in beddings

Sedimentary structure

Permeability is a vector property, and as such, is greatly affected by directional heterogeneity within a rock. The commonest cause of such heterogeneities is bedding. It is a general rule that the vertical permeability within a reservoir is lower than that in the bedding plane (horizontal permeability). In most cases, the vertical permeability is often about a third of that in the horizontal direction. It should be noted that some of the difference between the vertical and horizontal permeabilities results from differences in the way the local stress fields in the vertical and horizontal directions compact pores and close microcracks [31].

In addition, along different directions (such as perpendicular or parallel to the direction of palaeocurrent), there are different permeabilities in a rock. Permeabilities in different directions sometimes may be a difference of several orders of magnitude (Fig. 2.60).

In a word, the skeleton texture of rocks determines the magnitude of rock permeability while sedimentary structure controls the directivity of rock permeability.

Fracture

Both in clastic and carbonate rocks, there are fractures. Among the factors affecting rock permeability, fracture always plays an important role.

In most strata, fracture often has little contribution to total pore space; but it has significant effect on rock permeability because the isolated pores in rocks can be connected with each other by fractures, which thus increase the ability of rock transmitting fluids.

In conventional rocks, the formation and distribution of fractures mainly depend on the lithology and position of reservoir rocks in structure. In general, fractures may be formed in the position of structure with larger curvature. The more brittle the rock , more fractures occur in the rock.

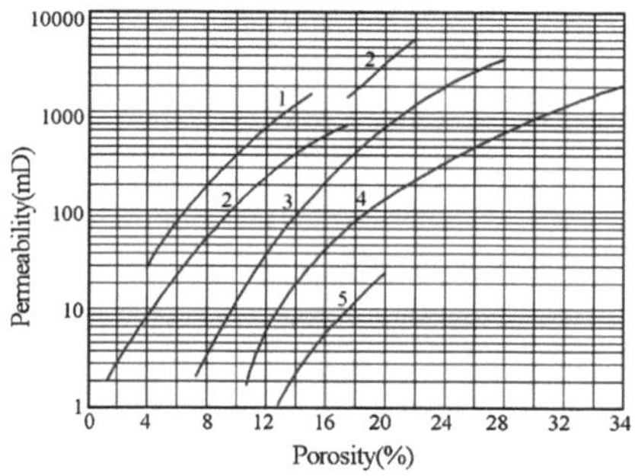

1-Coarse and very coarse grains; 2-Coarse and medium grains; 3-Fine grains; 4-Argillaceous particles; 5- Clay particles

Fig. 2.61 Permeability-porosity relationship (Chilingarian and Wolf 1964, form [21])

2.4.5.2 Pore Structure

Permeability is highly dependent on the pore structure (pore size, tortuosity, connectivity, porosity, etc.) of a rock. The most popular correlation expressing permeability as a function of porosity and pore size is derived by Kozeny–Carman:

$$K = \frac{\phi r^2}{8\tau^2} \tag{2.96}$$

where K is rock permeability, D; ϕ is rock porosity, f; r is average pore radius, cm; τ is the tortuosity of rock pores, dimensionless.

Obviously, rock permeability increases with the porosity and the pore size of a rock, and decrease with the increase in the tortuosity of rock pores. Figure 2.61 shows a relationship of permeability versus porosity obtained from rock samples or porous media. In addition, experience has shown that rock permeability increases as well with the increase in the connectivity of pores and the decrease in the aspect ratio.

In a word, the larger the pore sizes, the better the pore structure, the higher the rock permeability.

2.4.5.3 Diagenesis

Geostatic pressure

The action of geostatic pressure in rocks leads to a decrease in pore space as a result of rock particles becoming more closely packed. Permeability is thus sensitive to pressures that compact the rock. This compaction can occur in any direction not just

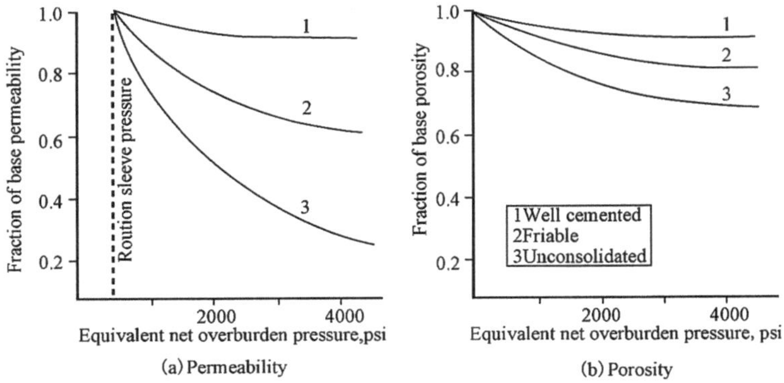

Fig. 2.62 Comparison of the effect of overburden pressure on permeability and porosity (Paul [31])

vertically. However, vertical compaction is usually the most important. Figure 2.62a shows the effect of increasing the hydrostatic confining pressure on the permeability of a rock. As shown on the figure, well-consolidated rock has a high modulus of rigidity and does not reduce in volume significantly under confining pressures. The change in slope for the second sample could be due to crushing of the sand grains or fracture closure point. Permeability measurements are normally made on samples of this type with only an adequate confining pressure to prevent bypass of the flowing fluid around the sample. Unconsolidated or poorly consolidated rock undergoes much greater reduction of permeability under confining pressure.

Formation temperature

Research shows that a complementary relationship occurs on the effects of overburden pressure and formation temperature to rock permeability.

Figure 2.62 shows the effect of overburden pressure on permeability compared to the effect upon porosity. It can be seen that overburden pressure affects permeability much more than porosity. This is because permeability is very sensitive to the tortuosity of pore paths in which fluid flows, and such changes are associated with very small changes to the rock porosity. Overburden pressure compacts the rock pressing the grains together. The size of the pores reduces little, but the pore-throats that control the passage of fluids between the pores undergo much greater closure, affecting the permeability to a greater extent.

H.A. Byplakof et al. found that the effect of pressure on permeability will decrease with the increase in temperature, especially under lower pressure. This is because the rock matrix and the fluids in rock pores expand when temperature increases, which exactly counterworks the compaction of the rock. Necessarily, the decrease in permeability caused by pressure increase is weakened.

It is worth noting that rock permeability is not fixed for a given reservoir. Some human factors, such as the physical and chemical reactions caused by poor compatibility between rock and outside working fluid, as well as the inappropriate treatment and unreasonable recovery rate, will lead to the change in rock permeability.

Cementation
Whatever in the early or in the late diagenetic stage, both the deposit of cement materials and cementation will decrease the pore space of a rock and increase the aspect ratio of pore system. These changes finally lead to a decrease in rock permeability.

2.4.6 *Estimation of Rock Permeability*

Theoretically, a reservoir rock may be equivalent to an ideal porous medium which consists of capillary tubes (Fig. 2.63). The estimation of rock permeability here is based on the principle of equivalent filtrational resistance between real rock and the ideal porous medium.

Consider a rock of cross-sectional area A and length L as an ideal porous medium which is made up of straight capillary tubes in a parallel, with the spaces between the tubes sealed by a cementing material (Fig. 2.62b). If all capillary tubes are the same radius r (cm) and length L (cm), the total flow rate Q (cm^3/s) of a fluid through the bundle of tubes, according to Poiseuille's equation, is:

$$Q = \frac{nA\pi r^4 \Delta p}{8\mu L\tau} \times 10 \tag{2.97}$$

where r is the radius of capillary tubes, cm; τ is the tortuosity of rock pores, dimensionless; n is the number of capillary tubes in unit cross-sectional area of the ideal medium; Δp is the pressure loss over length L, expressed in MPa. A is the

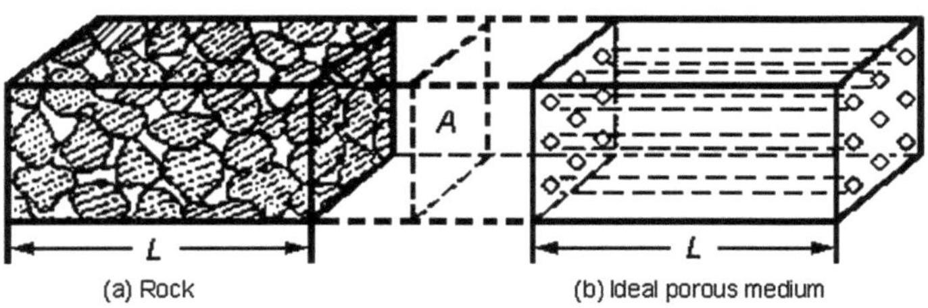

Fig. 2.63 Schematic diagram of real rock and ideal porous medium

cross-sectional area, including cemented zones, of this bundle of capillary tubes, namely the ideal medium, cm^2. μ is fluid viscosity, mPa s.

On the other hand, Darcy's law can be used to characterize the flow of the same fluid through the rock. The flow rate is thus:

$$Q = K\frac{A\Delta p}{\mu L} \times 10 \tag{2.98}$$

where K is rock permeability, D. Other parameters are the same as above.

Combine Eqs. (2.97) and (2.98), and solving for K gives:

$$K = \frac{n\pi r^4}{8\tau} \tag{2.99}$$

By definition, the porosity of the ideal porous medium is:

$$\phi = \frac{V_p}{V_b} = \frac{nA\pi r^2 L}{AL} \tag{2.100}$$

Combine Eqs. (2.99) and (2.100), an estimation of rock permeability is obtain as follows:

$$K = \frac{\phi r^2}{8\tau^2} \tag{2.101}$$

where K is rock permeability, D; ϕ is rock porosity, f; τ is the tortuosity of rock pores, dimensionless; r is the average radius of rock pores, cm.

If $\tau = 1$, $K = 0.125\phi r^2$;
If $\tau = 1.2$, $K = 0.087\phi r^2$;
If $\tau = 1.4$, $K = 0.064\phi r^2$.

Besides, rock permeability can also be determined or estimated by other methods, such as well logging, well tests, production data, or other correlations based on rock permeability. These methods refer to the relevant books available.

2.4.7 *Permeability of Fractured Rocks**

In principle, the permeability established in the case of a conventional reservoir remains valid in the case of a fractured reservoir. But in the presence of two systems (matrix and fractures), permeability has to be redefined in relation to matrix ("matrix" permeability), to fractures ("fracture" permeability) and to the fracture-matrix system ("fracture-matrix" permeability or total system permeability).

This redefinition may create some confusion in relation to a fractured reservoir and fracture permeability, which may be interpreted either as "single-fracture permeability" or as "fracture network permeability," or sometimes as entire "fracture-bulk volume permeability." The resulting expression of permeability is, therefore, examined in more detail. Therefore, some expressions of permeability involving fracture will be examined and discussed in detail.

2.4.7.1 Permeability of Pure Fracture System

Single-fracture case

The feature of fractures must be first described for defining fracture permeability. Generally, the density n and width b of fractures are often used to depict the fractures of a rock.

The density n of fractures is the ratio of the total length of fractures in flowing cross section to the area of the flowing cross section:

$$n = \frac{L}{A} \tag{2.102}$$

where A is the flowing cross-sectional area, cm^2; L is the total length of fractures in the flowing cross section, cm.

Then, fracture porosity can be denoted by the ratio of fracture area to rock area:

$$\phi_f = \frac{Lb}{A} = nb \tag{2.103}$$

where ϕ_f is fracture porosity, f; b is the average width of fractures, cm.

In single-fracture case, the flow cross section is represented only by the fracture void areas (excluding the surrounding matrix area). In a simplified case of a block, where the fracture is parallel to the flow direction (Fig. 2.64, fracture ① is parallel to the horizontal flow direction), the flow rate through the fracture is given by,

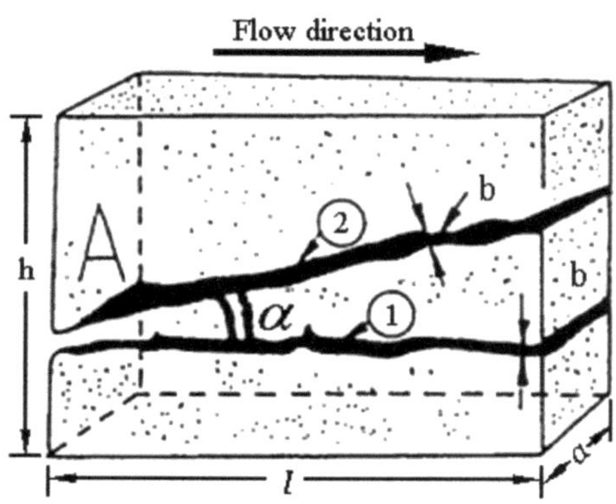

Fig. 2.64 Matrix block containing two fractures: fracture 1 ($\alpha = 0$); fracture 2 ($\alpha > 0$)

$$q_f = ab\frac{b^2}{12\mu} \cdot \frac{\Delta p}{l} \quad (2.104)$$

where a is the width of the block, cm; ab is the effective "real flow cross-section" of a single fracture as shown in Fig. 2.64: $A = ab$, cm^2; l is the length of the block, cm; p is pressure drop between two ends of the block, atm.

If the single fraction forms an angle with the flow direction (Fig. 2.64, fracture ②), the real cross-section (ab) will remain unchanged, but the fracture will be projected on the flow direction:

$$q_f = ab\frac{b^2 \cos^2 \alpha}{12\mu} \cdot \frac{\Delta p}{l} \quad (2.105)$$

On the other hand, based on the Darcy concept, if limited to the entire cross-flow section. $A = ab$, the rate is expressed by:

$$q = A\frac{K_f}{\mu} \cdot \frac{\Delta p}{l} = ab\frac{K_f}{\mu} \cdot \frac{\Delta p}{l} \quad (2.106)$$

where K_f is the permeability of single fracture, cm^2.

The further comparison of Eqs. (2.105) with (2.106) will lead to,

$$K_f = \frac{b^2}{12}\cos^2 \alpha \quad (2.107)$$

For a fracture system having n fractures of similar orientation, namely the fracture system is formed by parallel groups of fractures; the fracture permeability is expressed by:

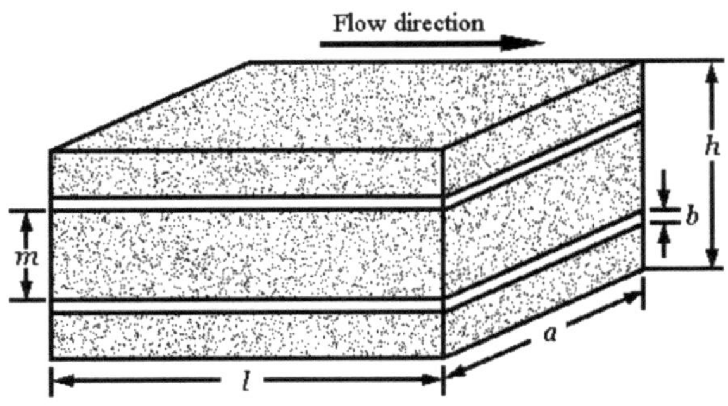

Fig. 2.65 Multi-fracture layer. (Fractures are parallel to the bedding planes)

$$K_f = \frac{\cos^2 \alpha}{12} \sum_i^n b_i^2 \tag{2.108}$$

where α is the included angle of fractures with the flow direction, °; b_i is the average width of ith fracture, cm; K_f is fracture permeability, cm^2.

Multi-fracture case

If, instead of a single fracture, the flow is examined through a fracture system formed by several parallel fractures (n) as shown in Fig. 2.65, separated by matrix of height "m," then the flowing equation (similar to the case of single fracture) will give:

$$q = n \times ab \frac{b^2}{12\mu} \cdot \frac{\Delta p}{l} = ah \frac{K_f}{12\mu} \cdot \frac{\Delta p}{l} \tag{2.109}$$

Thus:

$$K_f = \frac{b^2}{12} \cdot \frac{nb}{h} \tag{2.110}$$

In terms of the definition, the fracture porosity based on Fig. 2.65 is:

$$\phi_f = \frac{nabl}{ahl} = \frac{nb}{h} \tag{2.111}$$

Substituting ϕ_f for nb/h in Eq. (2.110), then:

$$K_f = \frac{b^2 \phi_f}{12} \tag{2.112}$$

where ϕ_f is fracture porosity, f; b is the average width of fractures, cm; K_f is fracture permeability, cm^2.

2.4.7.2 Permeability of Fracture-Matrix System

The permeability of a fracture-matrix system may be represented by the simple addition of the permeabilities of matrix K_m and fractures K_f:

$$K_t = K_m + K_f \tag{2.113}$$

where K_t is the (total) permeability of a rock, cm^2; K_m is the matrix permeability of the rock, cm^2; K_f is the fracture permeability of the rock, cm^2.

If Eq. (2.113) refers to the block described in Fig. 2.64 where the specific permeabilities of matrix and fracture were discussed, it is evident that the total permeability will depend on flow direction. Any change in direction of flow will change the value of K_f, since K_f depends on the relationship between fracture and flow directions.

If the fracture network of the rock consists of n parallel fractures which form an angle (α) with the flow direction, the total permeability is expressed by combining Eq. (2.108) as:

$$K_t = K_m + \sum_{i}^{n} K_{\mathrm{fi}} \cos a \tag{2.114}$$

where K_{fi} is the fracture permeability of ith fracture, expressed in cm^2, and determined by Eq. (2.112).

2.5 Average of Rock Parameters

Property parameters, e.g., porosity, permeability, and so on, of reservoir rocks are the vital foundation for the evaluation of reservoir properties, the estimation of geologic reserve and the performance design of hydrocarbon reservoirs. But the original rock parameters, obtained directly or indirectly from wells, are usually an aggregate of scattered data in a wide range of scale. Because smaller scale heterogeneities always exist, the values of rock properties from core analysis or well logging must be averaged to represent the characteristic of the entire reservoir or individual beddings for studies of petroleum engineering. The commonly used averaging techniques are presented below.

2.5.1 Arithmetic Mean

Arithmetic mean is the simplest method; and it can be expressed as follows:

$$\bar{\phi} = \frac{\sum_{i=1}^{n} \phi_i}{n}, \quad \bar{K} = \frac{\sum_{i=1}^{n} K_i}{n} \tag{2.115}$$

where $\bar{\phi}$ is the average porosity of given reservoir unit, %; ϕ_{i} is the porosity of each sampling point, %; $\bar{K}$ is the average permeability of the rock unit, mD; K_{i} is the permeability of each sampling point, mD; n is the number of sampling points.

This method is mainly used to determine the average values of rock parameters from well logging for individual beds.

2.5.2 Weighted Average

The averaging technique is often used to determine the average values of parameters of reservoirs. It also includes the following two kinds of methods.

Generally weighted

This averaging can be calculated either by thicknesses of layers or by areas controlled by sampling points, or by volumes represented by sampling points. These methods, taking permeability for an example, are briefly depicted as follows.

Thickness-weighted:

$$\bar{K} = \frac{\sum_{i=1}^{n} K_i h_i}{\sum_{i=1}^{n} h_i} \tag{2.116}$$

where h_i is the thickness of each layer or bed, m; K_i is the permeability of each layer or bed, mD. n is the number of data.

This method is often used to determine the average values of reservoir properties in the direction vertical to the bedding.

Area-weighted:

$$\bar{K} = \frac{\sum_{i=1}^{n} K_i A_i}{\sum_{i=1}^{n} A_i} \tag{2.117}$$

where A_i is the influenced area of each sampling point (e.g., each well), m^2; K_i is the mean permeability in area A_i, mD; n is the number of sampling points.

Table 2.20 Permeability data from a core analysis report

Depth (m)	Permeability (mD)
3998–4002	150
4003–4007	110
4007–4009	170
4010–4015	130

Table 2.21 The length (h) of cores and the product of the length and permeability of each core

Thickness of bedding, h (m)	Permeability, K (mD)	$K_i h_i$
4	150	600
4	110	440
2	170	340
5	130	650
$\sum h_i = 15$		$\sum K_i h_i = 2030$

Table 2.22 Permeability data from a core analysis report

Sample	Length of cores, h (m)	Permeability (mD)
1	0.2	80
2	0.1	20
3	0.5	50
4	0.1	30
5	0.3	70

Table 2.23 The product of the length and permeability of each core

Sample	Length of cores, h (m)	Permeability, K (mD)	$h_i \ln(K_i)$
1	0.2	80	0.88
2	0.1	20	0.30
3	0.5	50	1.96
4	0.1	30	0.34
5	0.3	70	1.27
–	$\sum h_i = 1.2$	–	$\sum h_i \ln(K_i) = 3.96$

This method is often used to determine the average values of reservoir parameters in the direction parallel to the bedding.

Volume-weighted:

$$\bar{K} = \frac{\sum_{i=1}^{n} K_i A_i h_i}{\sum_{i=1}^{n} A_i h_i} \tag{2.118}$$

where A_i is the influenced area of each sampling point, m^2; h_i is the reservoir thickness corresponding to the area A_i, m; K_i is the mean permeability of each sampling point, mD.

This method is mainly used in dynamic research of petroleum reservoir engineering to determine the average values of parameters in an oil group.

Example 2.8 Permeability data from a core analysis report are listed in Table 2.20. Calculate the average permeability of the reservoir from which the core is sampled.

Solution

From Table 2.20, the length of each core can be determined by the depth range of coring. The results are listed in the first column of Table 1.21. The product of the length and permeability of each core are then calculated and listed in the last column of Table 2.21.

Then, the average permeability of the reservoir is:

$$\bar{K} = \frac{\sum_{i=1}^{n} K_i h_i}{\sum_{i=1}^{n} h_i} = \frac{2030}{15} = 135.3\ (\text{mD})$$

Specially weighted

It is rare to encounter a homogeneous reservoir in actual practice because heterogeneity is common in petroleum reservoirs. All physical properties of rocks vary in petroleum reservoir, merely different variations for different properties. Generally, rock permeability is more variable than porosity, and more difficult to determine. Yet a good knowledge of permeability distribution in a reservoir is critical to the prediction of reservoir depletion by any recovery process. Therefore, it is necessary to find an average permeability to represent the flow characteristics of the entire reservoir or individual unit. The proper way of averaging permeability depends on how permeabilities were distributed as the rock was deposited.

Following special permeability-averaging techniques are commonly used to determine an appropriate average permeability to represent an equivalent homogeneous system.

Logarithmic average permeability:

$$\log \bar{K} = (\log K_1 + \log K_2 + \cdots + \log K_n)/n \tag{2.119}$$

Reciprocal average permeability:

$$\frac{n}{\bar{K}} = \frac{1}{K_1} + \frac{1}{K_2} + \cdots + \frac{1}{K_n} \tag{2.120}$$

Harmonic average permeability:

$$\bar{K} = \frac{\sum_{i=1}^{n} h_i}{\sum_{i=1}^{n} h_i / K_i} \tag{2.121}$$

Equation (2.121) is basically used for calculating the average value of vertical permeability of oil layer.

Geometric average permeability:

Experiments have illustrated that the most probable permeability behavior of a heterogeneous medium (formation), made up of n randomly distributed samples (regions) of differing uniform permeabilities, $K1$ to Kn, approaches that of a uniform system having a permeability that is equal to the geometric average. The geometric average is defined mathematically by the following relationship:

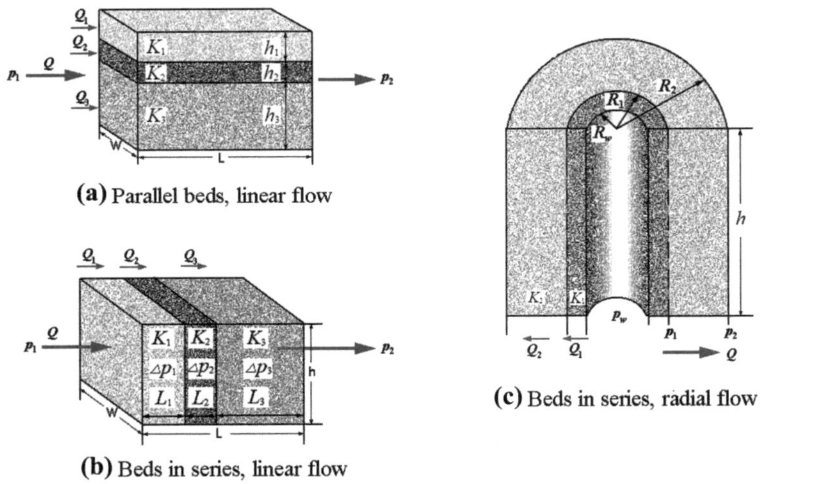

Q_i:flow rate; P_i:pressure;ΔP_i:pressure difference; K_i:permeability; h_i:thickness of a bed; L_i:width of the rock; R_i: radius of the bed.

Fig. 2.66 Types of composite layers

$$\bar{K} = \exp\left[\frac{\sum_{i=1}^{n} h_i \ln(K_i)}{\sum_{i=1}^{n} h_i}\right] \tag{2.122}$$

where K_i is the permeability of core sample i, mD; h_i is the thickness of core sample i, m; n is the total number of samples.

If the thicknesses (h_i) of all core samples are the same, Eq. (2.122) can be simplified as follows:

$$\bar{K} = \sqrt[n]{K_1 \times K_2 \times \cdots \times K_n} \tag{2.123}$$

Example 2.9 Given the following core data (Table 2.22), calculate the geometric average permeability of the cores.

Solution

According to the data of Table 2.22, the product of the length and permeability of each core can be calculated. The results are listed in the last column of Table 2.23.

Then the geometric average permeability is:

$$\bar{K} = \exp\left[\frac{\sum_{i=1}^{n} h_i \ln(K_i)}{\sum_{i=1}^{n} h_i}\right] = \exp\left(\frac{3.96}{1.2}\right) = 52.22(\text{mD})$$

2.5.3 Average Based on Flow Process

Permeability variations can occur laterally or vertically in a reservoir as well as in the vicinity of a wellbore. Since permeability is a measure of the ability of a rock transmitting fluids, the most accurate permeability behavior of a heterogeneous rock, which is made up of randomly distributed regions of differing permeabilities, should be determined by the flow process of fluid through the rock.

The detailed analysis is extremely complex. However, it is possible to analyze two simple systems of different permeabilities that occur within core analysis and petroleum reservoirs. These are (a) flow through beds in series, and (b) flow through beds in parallel (Fig. 2.66).

2.5.3.1 Parallel Beds—Linear Flow

Linear flow through parallel beds without vertical communication is show in Fig. 2.66a. Each bed has different permeability. The pressures at the inlet (p_1) and outlet (p_2) of the complete unit will be the same for all beds, but each bed i will transport a different fraction Q_i of the total flow rate Q. Thus:

$$Q = \sum_{i} Q_i \tag{2.124}$$

Fig. 2.67 Parallel beds, radial flow

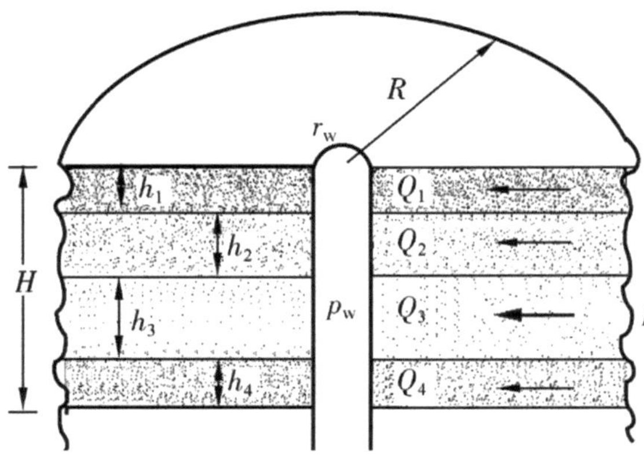

Suppose bed i has a length L, a width W, a net thickness h_i, and a uniform permeability K_i. Noting that the total thickness of the complete unit is the sum of each net thickness ($H = \Sigma h_i$). Applying Darcy's law for the linear flow of a fluid with viscosity μ gives:

$$\frac{\bar{K}Hw(P_1 - P_2)}{\mu L} = \sum_i \frac{K_i h_i w(P_1 - P_2)}{\mu L} \tag{2.125}$$

where $\bar{K}$ is the average permeability of the complete unit.

After canceling common terms, and substitute h_i for H, we obtain following expression:

$$\bar{K}\sum_i h_i = \sum_i K_i h_i \tag{2.126}$$

Solving for $\bar{K}$ gives:

$$\bar{K} = \frac{\sum_i K_i h_i}{\sum_i h_i} \tag{2.127}$$

Apparently, the average permeability for linear flow through parallel beds of different permeabilities equals the thickness-weighted average of the individual permeability. If each bed has the same thickness, the average permeability is then the arithmetic average of the each permeability.

2.5.3.2 Parallel Beds—Radial Flow

This is the case of several parallel layers in which fluid simultaneously flows toward the well (Fig. 2.67). Each layer supplies a flow rate Q_i, determined by Darcy's law. The total flow rate through all layers is $Q = \sum Q_i$.

In the case of radial flow through a parallel combination of beds, we have:

$$Q_i = \frac{2\pi K_i h_i (P_i - P_w)}{\mu \ln \frac{R}{r_w}} \tag{2.128}$$

$$Q = \frac{2\pi \bar{K} H (P_i - P_w)}{\mu \ln \frac{R}{r_w}} \tag{2.129}$$

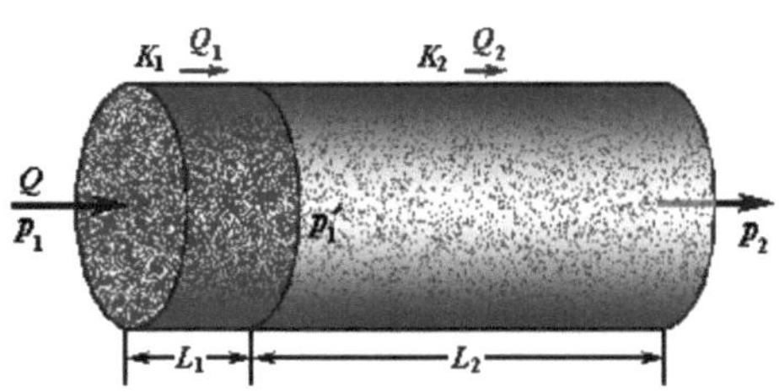

Fig. 2.68 Illustration of contaminated core

where $\bar{K}$ is the average permeability of the parallel combination unit, D; K_i is the permeability of the ith bed, mD; H is the total thickness of the parallel beds, cm; P_i is the terminal pressure of the system, atm; P_w is bottom hole pressure, atm; R is the radius of formation edge, cm; rw is well radius, cm; Q_i is the flow rate supplied by each layer, cm^3/s; Q is the total flow rate through the parallel beds, cm^3/s.

Substituting Q_i for Q in Eq. (2.128), we obtain:

$$\frac{2\pi \bar{K} H (P_i - P_w)}{\mu \ln \frac{R}{r_w}} = \frac{2\pi (P_i - P_w) \sum_i K_i h_i}{\mu \ln \frac{R}{r_w}} \tag{2.130}$$

Simplify Eq. (2.130), then:

$$\bar{K} = \frac{\sum_i K_i h_i}{\sum_i h_i} \tag{2.131}$$

It is clear that the average permeability for radial flow in parallel beds is the same relationship as the linear flow, namely the thickness-weighted average permeability.

2.5.3.3 Beds in Series—Linear Flow

Consider a core contaminated by drilling fluid. Generally, a thin mud cake is formed at one end of the core after the drilling fluid experiment. Mud cake and the clean part of the core constitute a series combination of beds with different permeabilities (Fig. 2.68). Determine the average permeability of the core contaminated.

Suppose:

- mud cake: thickness = L_1, cm; permeability = K_1, D;
- clean part of the core: thickness = L_2, cm; permeability = K_2, D;
- pressures at the two ends of the core: p_1 and p_2, atm;
- pressure at the cross section of the mud cake contacted with the clean part of core is p_1', atm;
- total flow rate of fluid passing through the core is Q, cm^3/s;
- fluid viscosity is μ, mPa s
- flow rate through mud cake is Q_1, cm^3/s; flow rate through the unpolluted part of the core is Q_2, cm^3/s.

Under the condition of continuous steady flow, the flow rate of the single-phase fluid is constant, namely $Q = Q_1 = Q_2$; and the total pressure drop Δp is equal to the sum of the pressure drops across each bed:

$$\Delta p = \Delta p_1 + \Delta p_2 \tag{2.132}$$

Applying Darcy's formula for the mud cake and the clean part of the core separately:

For the mud cake:

$$Q_1 = -K_1 \frac{A(p' - p_1)}{\mu L_1} \tag{2.133}$$

Then:

$$\Delta p_1 = (p' - p_1) = -\frac{Q_1 \mu L_1}{AK_1} \tag{2.134}$$

For the clean part of the core:

$$Q_2 = -K_2 \frac{A(p_2 - p')}{\mu L_2} \tag{2.135}$$

Then:

$$\Delta p_2 = (p_2 - p') = -\frac{Q_2 \mu L_2}{AK_2} \tag{2.136}$$

For the whole core:

$$Q = -\bar{K} \frac{A(p_2 - p_1)}{\mu L} \tag{2.137}$$

where $\bar{K}$ = the average permeability of the core.

Then:

$$\Delta p = (p_2 - p_1) = -\frac{Q \mu L}{A\bar{K}} \tag{2.138}$$

Substituting for the pressure drops in Eq. (2.132) by Eqs. (2.134), (2.136) and (2.138):

$$\frac{Q\mu L}{A\bar{K}} = \frac{Q_1 \mu L_1}{AK_1} + \frac{Q_2 \mu L_2}{AK_2} \tag{2.139}$$

Simplify Eq. (2.139):

$$\frac{L}{\bar{K}} = \frac{L_1}{K_1} + \frac{L_2}{K_2} \tag{2.140}$$

Then:

$$\bar{K} = \frac{L_1 + L_2}{L_1/K_1 + L_2/K_2} \tag{2.141}$$

This is the average permeability of linear flow through beds in series. Compare it with Eq. (2.121), we see that Eq. (2.141) is exactly the expression of the harmonic mean of permeability.

2.5.3.4 Beds in Series—Radial Flow

Permeability in the neighborhood of the well changes with the effects of well completion depending on the method and treatment. Consider the formation near a wall. Part of it is damaged by drilling fluid, as shown in Fig. 2.66c. The formation thus turns into a combination of two zones in series. The average permeability of the damaged formation needs to determine.

As the flow of single-phase through formation is radial, the combination of polluted zone and unpolluted formation is equivalent to a series combination of layers.

Suppose:

- Polluted zone: radius = R_1, cm; permeability = K_1, mD; edge pressure of polluted zone = p_1, atm;
- Unpolluted formation: permeability = K_2, mD; supply radius of oil layer = R_2, cm, terminal pressure = p_2, atm;
- Radius of wellbore is R_w, cm; bottom hole pressure is p_w, atm;
- Thickness of the formation is h, cm;
- Total flow rate through the formation is Q, cm^3/s;
- Viscosity of fluid is μ, mPa s;
- Flow rate through polluted zone = Q_1, cm^3/s; flow rate through unpolluted formation = Q_2, cm^3/s;

Under the condition of continuous steady-state flow, the rate of redial flow is constant, namely $Q = Q_1 = Q_2$; and the total pressure drop Δp is equal to the sum of the pressure drops across each zone:

$$\Delta p = \Delta p_1 + \Delta p_2 \tag{2.142}$$

Substituting for the pressure drop by applying redial Darcy's equation gives:

$$\frac{Q\mu \ln \frac{R_2}{R_w}}{2\pi \bar{K} h} = \frac{Q_1 \mu \ln \frac{R_1}{R_w}}{2\pi K_1 h} + \frac{Q_2 \mu \ln \frac{R_2}{R_1}}{2\pi K_2 h} \tag{2.143}$$

So that,

$$\bar{K} = \frac{\ln \frac{R_2}{R_w}}{\frac{1}{K_1} \ln \frac{R_1}{R_w} + \frac{1}{K_2} \ln \frac{R_2}{R_1}} \tag{2.144}$$

The calculation can be extended to a larger number of rings. And the calculation can also be utilized for estimating the effects of mud invasion, acidizing, or fracturing.

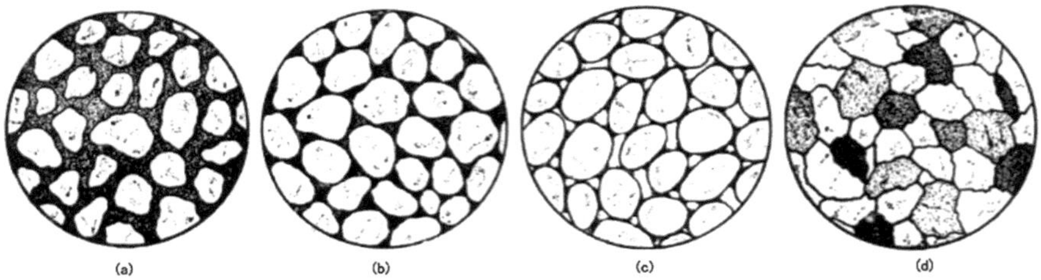

(a) Basal cementation (b) Porous cementation (c) Contact cementation (d) Inlaid cementation

Fig. 2.69 Illustration of cementation type for clastic rocks (Zhu [4])

2.6 Sensitivity of Reservoir Rocks

In reservoir rocks, there are always some minerals which are sensitive to some factors (such as invading fluids, flow rate, and so on) and thus cause a decrease in reservoir permeability in the process of oil/gas production. This phenomenon is known as *formation/reservoir damage* or *reservoir sensitivity*. Formation damage is an undesirable operational and economic problem that can occur during the various phases of an oil/gas reservoir life time, including well drilling, hydraulic fracturing, oil/gas production, workover operations, etc. Formation damage may be caused by various processes, including chemical, physical, biological, and thermal interactions of formation and fluids, and deformation of formation under stress and fluid shear. In order to avoid, control, and remedy formation damage, the sensitivity of reservoir rocks needs to well understood. And various properly experimental techniques are necessary to evaluate the sensitivity of petroleum reservoirs.

In sandstones, rock sensitivity is mainly caused by cementing agents in rocks. The sensitivity of reservoir rocks is thus dependent on the components, content, and distribution of cementing agents in sandstones.

In this section, the cementing agent and cementation type of sandstone reservoirs are first briefly reviewed. After that, the mineralogy and sensitivity of reservoir rocks, and the method of evaluating reservoir sensitivity are presented.

2.6.1 Cementing Agent and Cementation Type

2.6.1.1 Cementing Agent

Cementing agents, or cements for short, are minerals which are chemically precipitated from pore fluids around and between grains in clastic rocks. The cement plays a part in welding and hardening clastic sediments during lithification. It forms an integral and important part of the rock. The formation of cement agents affects the porosity and permeability of the rock, and always make reservoir quality worse.

Many minerals may become cements. The most common in petroleum reservoirs is clay minerals, but calcite, other carbonates iron oxides, and so on also undergo the process.

In clastic rocks, cements generally are crystalline and noncrystallized authigenic minerals, and the content of cements is generally less than 50 %. In sandstones, and the most common cements—in decreasing order of frequency—are calcite, dolomite, quartz, siderite, anhydrite, muscovite, kaolinite, chlorite, and hematite, and so on. In total, more than 40 minerals in rocks are known as cementing agent. They mainly include argillaceous, siliceous, calcareous, ferruginous, sulfate cements, etc. Perceptibly, different cement agents give rise to different sensitivities and result in different damage to petroleum reservoirs.

2.6.1.2 Cementation Type

In clastic rocks, the distribution and contact pattern of cement agents with grains is known as *cementation type*. The cementation type of a rock primarily depends on the component, amount, and formation conditions of cementing agents. Generally, the following four cementation types may be found in rocks: basal, porous, contact, and inlaid cementations (Fig. 2.69).

Basal cementation

There is a lot of fillings between clastic particles. Particles without contacting with each other, float in fillings in such cementation (Fig. 2.69). This type of cementation generally represents the rapid sedimentation of high-density fluids. Basal cementation is also called progenetic cementation because it is formed in the period of sedimentation. In this kind of rock, the cementation is very strong owing to the large amount of cements; and the major void space is the micropores occurring in cementing agents. As a result, the quality of the rock with basal cementation is poor, and this kind of rock can hardly hold oil or gas.

Porous cementation

It occurs in the rocks with grain-supporting structure. Clastic particles constitute the bracket-like structure, and most of particles are of point contact (Fig. 2.69b). In porous cementation, cements are relatively less, and only fill heterogeneously in the pores between particles. The strength of porous cementation is weaker than that of basal cementation. Therefore, the quality of rocks with porous cementation is better than that of rocks with basal cementation.

Table 2.24 Physical properties of rocks with different cementation types

Physical property	Contact cementation	Porous cementation	Basal cementation
Porosity (%)	23–30	18–28	8–17
Permeability (mD)	50–1000	1–150	<1

Contact cementation

In this kind of cementation, clastic particles are of point contact or long contact (Fig. 2.69c); and cements are much fewer, usually less than 5 %. Cement deposits the places where particles contact each other. Cements are mostly authigenic mineral or weathered detritus; and argillaceous cements are very common. Relatively, the quality of rocks with contact cementation is the best. They can hold oil and gas very well.

Inlaid cementation

In the period of diagenesis, clastic particles of sandy sediments can contact more tightly due to the effect of pressure solution in evidence. The contact formed between particles evolves into long contact and concavo-convex contact, even into sutured contact (Fig. 2.69d). As a result, inlaid cementation is formed between particles. In this kind of cementation, clastic particles sometimes cannot be clearly distinguished from siliceous cements; and it looks like no cements between particles. Therefore, this kind of cementation is also known as no-cement cementation. The rock with inlaid cementation has no capacity of holding oil and gas.

It is obvious that the porosity and permeability of rocks are greatly influenced by the cementation type of rocks. Table 2.24 lists the physical properties of rocks which are from the formations of tertiary system in North China depression.

2.6.2 Sensitive Minerals in Cements

The sensitivity of reservoir rocks is basically induced by the sensitive minerals in cement agents of rocks. Sensitive minerals means the minerals which are sensitive to water, oil, salt, or other matters invading into oil beds, or sensitive to the velocity of fluid flowing through the bed during the process of oil production; and the permeability of oil beds is thus decreased by the sensitivity of minerals.

In petroleum reservoirs, sensitive minerals usually include the following groups: water-sensitive minerals, salinity-sensitive minerals, acid-sensitive minerals, alkali-sensitive minerals, and minerals sensitive to flow velocity.

Water-sensitive minerals

In petroleum reservoirs, water-sensitive minerals mean the minerals which are sensitive to aqueous solutions with lower salinity than that of formation water. On contacting low-salinity water, some physical and chemical changes will occur in water-sensitive minerals, including hydration swelling, dispersion, desquamation, and migration. Thereby, once water sensitivity occurs in a reservoir, the rock pores and throats will be severely blocked, and the permeability of the reservoir will be greatly declined.

Water-sensitive minerals generally include various clay minerals, such as montmorillonite, illite/smectite mixed-layer minerals, chlorite/smectite mixed-layer minerals, hydromuscovite, etc.

Salinity-sensitive minerals

In petroleum reservoirs, if the salinity of formation water is changed by invading fluids, the following changes will occur in some minerals in the reservoir: hydrating and swelling, dehydrating and shrinking, and even cracking and breaking off. These changes finally lead to a great deal of fines occurring in pores. The fines then migrate with fluids flowing in pores, and could pile up at narrow throats. The reservoir permeability could be terribly damaged at the end. The phenomenon is called *salinity sensitivity*. These minerals are known as salinity-sensitive minerals.

Salinity sensitivity minerals usually include gypsum, barite, calcite, celestite, halite, magnetite, etc.

Acid-sensitive minerals

Acid-sensitive minerals denote the minerals which react with acid invading in reservoirs and then generate chemical precipitation; and in the process of chemical reaction, fine particles could be released from the rock matrix. In petroleum production, acidizing (pumping acidizing fluids into wells) is the most essential well stimulation. In well acidizing, the acid sensitivity of minerals can cause rock pore-throats to be seriously plugged by the precipitates formed in reaction and the fines released from the rock matrix. As a result, the permeability of reservoir rocks can be notably declined.

Acid-sensitive minerals generally include hydrochloric acid-sensitive minerals and hydrofluoric acid-sensitive minerals. Common hydrochloric acid-sensitive minerals are iron-containing minerals, such as Fe-rich chlorite, ferroan calcite, ankerite, hematite, siderite, pyrite, glauconite, hydrobiotite, etc. These minerals react with hydrochloric acid and release ferrous ions (Fe^{2+}). In oxygen-rich solutions, Fe^{2+} can be converted to the precipitate $Fe(OH)_3$ (Fe^{3+}). And under acidic conditions (5–6 pH), Fe^{2+} becomes the colloidal precipitate. These precipitate will block the pores and finally cause serious reservoir damage. For example, the chemical reactions of Fe-rich chlorite with hydrochloric acid are:

$$3FeO \cdot Al_2O_3 \cdot 2SiO_2 \cdot 3H_2O + 6HCl = 3FeCl_2 + 2Al(OH)_3 + 2SiO_2 \cdot 3H_2O$$

$$2FeCl_2 + 3H_2O + 3O = 2Fe(OH)_3\downarrow + 2Cl_2\uparrow$$

In order to prevent the damage of Fe^{2+} and $Fe(OH)_3$ to reservoirs, addition agents, such as iron chelating agent, oxygen-cleaner, and so on, can be added into acidizing liquids in well acidizing.

Hydrofluoric acid-sensitive minerals are Ca/Mg-containing carbonate minerals, mainly including limestone ($CaCO_3$), dolomite ($CaMg\ (CO_3)_2$), anorthite, scolecite zeolite, and so on. Reacting with hydrofluoric acid, these minerals produce indissoluble fluoride which can plug pores and finally cause reservoir severe damage.

In addition, part of silicate minerals, such as monox, quartz, feldspar, etc., are also sensitive to hydrofluoric acid. Under certain conditions, they can react with hydrofluoric acid and produce the precipitates of fluorosilicate and fluorin-aluminate and silicone gel. The petroleum reservoirs are then damaged by these precipitates

Alkali-sensitive minerals
Alkali-sensitive minerals are the minerals which can react with the high-pH fluids invading in the reservoir and then disperse, desquamate, or generate new silicate precipitates and silicon gelinite. As a result, rock pores are plugged by these fine particles and precipitate; and the permeability of reservoir rocks is markedly decreased

Alkali-sensitive minerals include feldspar, albite, chalcedony, microcrystalline quartz, hydromuscovite, and opal.

Minerals sensitive to flow velocity
In petroleum reservoirs, the high-speed flow of fluids can cause fines of minerals to fall off the rock matrix and to flow with the fluid. The rock pores could thus be blocked by migrating fines, which then leads to a decline in rock permeability. This sensitivity is caused by the flow velocity of fluids flowing in reservoir rocks. So, the minerals sensitive to the flow velocity of fluids in rocks are referred to as velocity-sensitive minerals.

Velocity-sensitive minerals are usually the minerals whose particle size is less than 37 μm. Velocity-sensitive minerals may be clay minerals or non-clay minerals, such as kaolinite, microcrystalline quartz, microcrystalline feldspar, etc.

2.6.2.1 Clays and Clay Mineralogy

Petroleum reservoirs contain more or less clay minerals. Clays are one of the most important groups of minerals that destroy the permeability of reservoir rocks because some of them are sensitive to various invading fluids during oil/gas production. Clays are a very complex family of minerals.

Clay minerals
Clays are generally platy-shaped or fibrous-shaped minerals that present in sedimentary rocks as packs of crystals. Clay minerals are very small in size. It is normally considered to be less than 2 μm in clay size according to standard particle size classifications. They can only see with aid of electron microscope.

Table 2.25 Description of the authigenic clay minerals (Faruk [52])

Mineral	Molecular formula	Morphology
Kaolinite	$Al[4Si_4O_{10}](OH)_8$	Stacked plate or sheets
Chlorite	$(Mg, Al, Fe)_{12}[(Si, A1)_8O_{20}](OH)_{16}$	Plates, honeycomb, cabbagehead rosette or fan
Mite	$(K_{1\text{-}1.5},Al_4[Si_{7\text{-}6.5},Al_{1\text{-}1.5}O_2](OH)_4$	Irregular with elongated spines or granules
Smectite	$(1/2Ca, NA)_{0.7}(Al, Mg, Fe)_4[(Si, Al)_8\ O_{20}]{\cdot}nH_20$	Irregular, wavy, wrinkled sheets, webby or honeycomb
Mixed Layer	Illite-smectite	Ribbons substantiated by filamentous morphology
	Chlorite-smectite	

From the composition, clays are hydrous aluminum silicate minerals, sometimes with variable amounts of iron, magnesium, alkali metals, alkaline earths, and other cations. They include two categories: *crystalline layered silicate* (*phyllosilicates*) and *noncrystalline silicate*. Most of crystalline clay minerals have a scaly or platy shape, and are layered structure. A few clay minerals have a nemaline or baculine shape, and are chain-layered structure. Contacting with water, they become soft and plastic. Many clay minerals have the ability of strong adsorption and ion exchange.

In sandstone reservoirs, the most common clays are the phyllosilicates, such as kaolinite, montmorillonite, illite, and chlorite; and noncrystalline silicates are rare. Typical clay minerals are shown and described in Table 2.25.

In sandstones, various clay minerals can be found. Based on the features of crystal structure, caly minerals can be classified on the whole into three groups: (1) kaolinite group, (2) smectite (or montmorillonite) and illite group, and (3) chlorite group. Mixed-layer clay minerals can be formed from several of the three basic groups.

Kaolin group generally includes the minerals: kaolinite, dickite, halloysite, and nacrite [53]. Smectite group includes dioctahedral smectites, such as montmorillonite and nontronite, and trioctahedral smectites, for example saponite [52]. Illite group includes the clay-micas. Illite is the only common mineral [52]. Chlorite group includes a wide variety of similar minerals with considerable chemical variation [52].

By nature, the sensitivity of clay minerals, or the damage of clay to petroleum reservoirs, depends on the type and the amount of the exchangeable cations, such as K^+, Na^+, Ca^{2+}, in clays and the type of layered structures of clay minerals.

Structure of clay minerals

Research shows that different crystal structures of clays often lead to different behaviors of swelling and fine migration of clay minerals. Therefore, the knowledge of clay structure is necessary for a good understanding of mineral sensitivity.

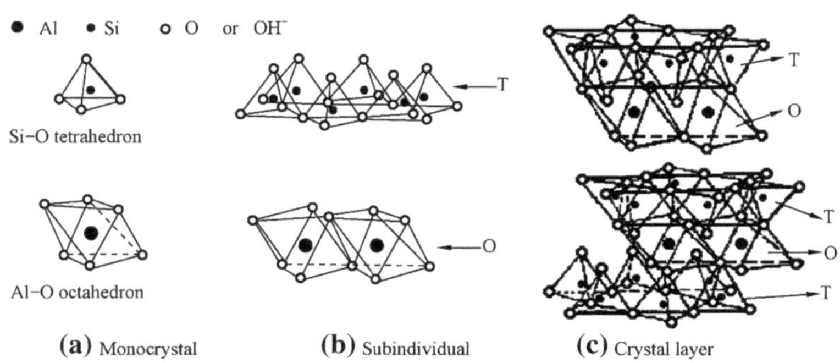

Fig. 2.70 Schematic diagram of the structure of clay minerals (何更生 1994)

The primary crystal units of clay minerals are the monocrystals of Si-O tetrahedron and the monocrystal of Al-O octahedron (Fig. 2.69a). Each tetrahedral monocrystal consists of 4 oxygen ions (O^{2-}) on its vertex and 1 silicon ion (Si^{4+}) in its center. In the same way, each octahedral monocrystal consists of 6 oxygen ions (O^{2-}) or hydroxide ions (OH^{-1}) on its vertex and 1 aluminum ion (Al^{3+}) or magnesium ion (Mg^{2+}) inside it.

Clay minerals are characterized by overlapped two-dimensional *sheets* of corner sharing Si-O tetrahedron and AlO_4 octahedron, namely:

$$\textbf{Monocrystal} \overset{\text{Horizontal extension}}{\rightarrow} \textbf{Subindividual} \overset{\text{Vertical overlap}}{\rightarrow} \textbf{crystal} \overset{\text{Extension and overlap}}{\rightarrow} \textbf{Clay mineral}$$

The sheet of Si-O tetrahedron is formed by monocrystals of Si-O tetrahedron extending horizontally. Each tetrahedron shares three of its vertex oxygen ions with other tetrahedra forming a hexagonal array in two-dimensions. The fourth vertex is not shared with another tetrahedron and all of the tetrahedra "point" (fourth vertex) in the same direction; i.e., all of the unshared vertices are on the same side of the sheet (Fig. 2.70b). A sheet of Si-O tetrahedron can be expressed in $n(Si_4O_{10})$, and is called subindividual of Si-O tetrahedron for short. Generally, it is expressed in symbol *T* (Fig. 2.70b).

As such, the sheet of Al-O octahedron is formed by the lateral extension of Al-O monocrystals. Each octahedron, inclining to the same direction, shares two of its vertex oxygen ions (O^{2-}or hydroxide ions (OH^{-1}) to form octahedral sheet (Fig. 2.70b). The unshared vertex on the surface of connected side also forms part of one side of the octahedral sheet but an additional oxygen ion is located above the gap in the tetrahedral sheet at the center of the six tetrahedra. A sheet of Al–O octahedron is also called subindividual for short, and expressed in symbol *O* (Fig. 2.70b). Subindividual of Al–O octahedron can be expressed in n $[Al_4(OH)_{12}]$ if aluminum ions (Al^{3+}) are in the center of octahedrons, or in n $[Mg_6(OH)_{12}]$ if magnesium ions (Mg^{2+}) are in the center of octahedrons.

After that, clay minerals may be formed by overlapped tetrahedral sheets and octahedral sheets. According to the way that tetrahedral and octahedral sheets are packed into *crystal layers*, clay minerals can be categorized into the following groups:

1. Structure of TO (1:1 clay)
2. Structure of TOT (2:1 clay)
3. Structure of TOT·O (2:1 + 1 clay)

A 1:1 clay consists of one tetrahedral sheet and one octahedral sheet in each crystal layer (Fig. 2.71), and Kaolinite group is such structure. A 2:1 clay consists of an octahedral sheet sandwiched between two tetrahedral sheets (Fig. 2.72), and examples are smectite, illite, and attapulgite. Smectite/illite group is such structure. A 2:1 + 1 clay consists of an octahedral sheet sandwiched between two tetrahedral sheets, and an external octahedral sheet. That is to say, a 2:1 + 1 clay consists of one TOT and one T crystal layers (Fig. 2.73). Chlorite group is such structure.

Fig. 2.71 Kaolinite structure

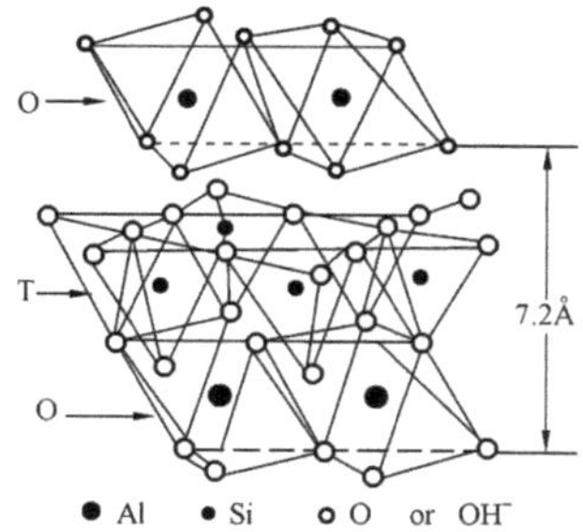

Fig. 2.72 Smectite structure

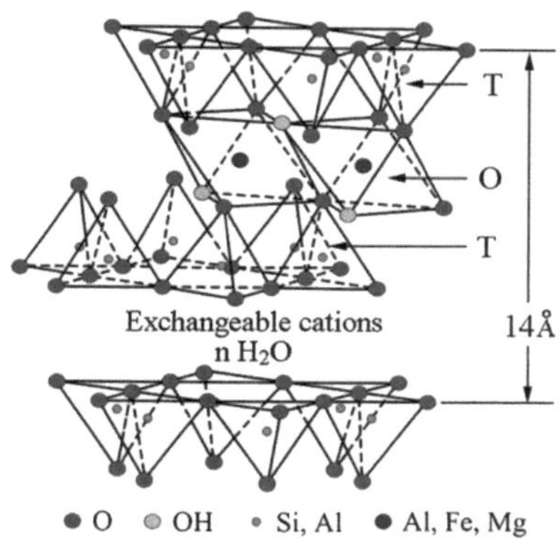

Physicochemical properties of clay minerals.

Kaolinite group

As shown in Fig. 2.71, kaolinite is a layered silicate mineral, with the 1:1 structure (TO). Kaolinite crystal is a triclinic system. In the two sides (up and down side) of a crystal layer, there are different elements. One side of a TO layer is composed of oxygen atoms, while another side contains hydroxide ions and oxygen atoms. When TO crystal layers are packed together, hydrogen bonds will be formed by the oxygen atoms of one crystal layer linked to the hydrogen atoms of another crystal layer in Kaolinite structure. The hydrogen bonds formed in Kaolinite leads to a tighter joint and closer spaces between crystal layers than that between crystal layers of Montmorillonite, and thus water can hardly enter in crystal layers. In addition, the capacity of exchange cation (K^+) is relatively low in kaolinite crystal. Therefore, kaolinite is a relatively stable, nonswelling clay, and not easy to hydrate; but it easily disperses and migrates under the action of external force.

In spite of a few hydrogen bonds, in fact, van der waal's force is the major combination force between crystal layers of kaolinite. The link between crystal layers is relatively weak, and the hardness of kaolinite is thus lower. Cleavages are common in kaolinite. Accordingly, kaolinite can hardly undergo the action of mechanical force. Under the effect of high-speed fluids, cleavages will split into scaly fines, and these fines will then migrate with the flow of fluids in rock pores. During the migration of fine particles, these fines will block the pores or throats of the rock, and then lead to a significant decline in rock permeability.

Smectite/illite group

Smectite group has a 2:1 structure (TOT). A 2:1 clay consists of one octahedral sheet (T) sandwiched between two tetrahedral sheets (O) (Fig. 2.72). A notable feature of smectite is that the high valence cations ($A1^{3+}$, Si^{4+}) in crystal may be replaced by low valence cations (Mg^{2+}, Ca^{2+}, Na^{1+}, etc.) in solution. For example, $A1^{3+}$ is replaced by Mg^{2+}, and Si^{4+} is replaced by Na^{1+}. The results of cation exchange are: (1) additional negative charges on clay due to a loss of positive charges. Consequently, the negatively charged clay will adsorb further the

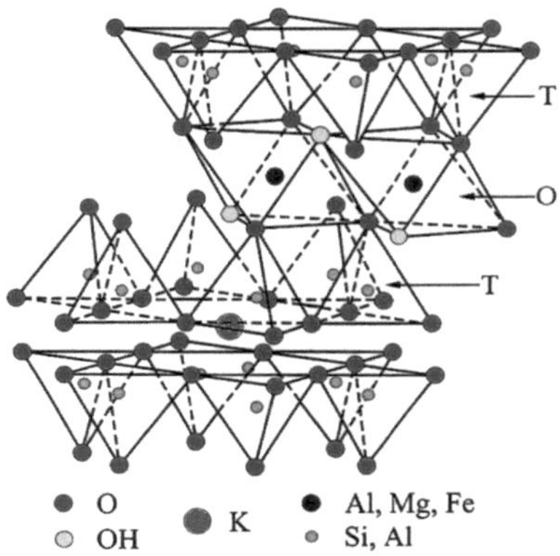

Fig. 2.73 Illite structure

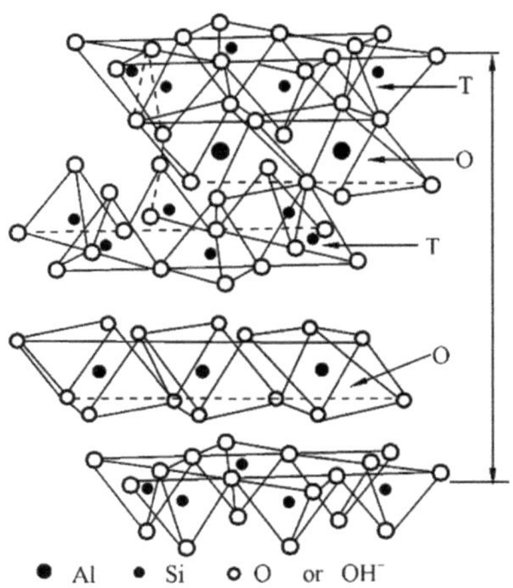

Fig. 2.74 Chlorite structure

exchangeable cations (Mg^{2+}, Ca^{2+}, Na^{1+}, etc.) between crystal layers or on the surface of crystal to balance additional negative charges; (2) unbalanced charges providing a favorable condition for next ion exchange. As a result, smectite has a large cation exchange capacity, and thus is easily swelling and dispersion.

Between the crystal layers of smectite, there is only van der waal's force, which leads to a weak link and wide space between the crystal layers (Fig. 2.72). Thereby, water molecules or other polar molecules are easily adsorbed by the cations into the crystal interlayers. When exposed to low ionic-strength aqueous solutions, the interlayer cations in smectite will adsorb a lot of water molecules to form thick envelopes of water films over the clay (Fig. 2.72). This process will cause a visible increase in the space between crystal layers and thus lead to clay swelling. Usually, the volume of a smectite clay can be increased 13–20 times by its swelling. Exposed to low-salinity aqueous solutions, as a result, the hydration of smectite will cause crystal layers to expand and crack. Under the effect of fluid flow, moreover, a smectite clay will split along the cleavages into thin and curved squamae. These squamae will further coagulate to glomeration and migrate with fluid flowing in rock pores.

However, the cation exchange capacity of illite is much lower than that of smectite in despite of the similarity in their structure (Fig. 2.73). Moreover, the occurrence of potassium ions (K^+) between crystal layers of illite also lead to a strong binding force between crystal layers (Fig. 2.73), which thus make it very hard for water molecules to enter into the space between crystal layers of illite. Therefore, illite is a nonswelling clay.

Chlorite group

As shown in Fig. 2.74, chlorite is a 2:1 + 1 clay (TOT·O), namely, an "O" sheet is sandwiched between two "TOT" units. In this structure, a lot of hydrogen bonds occurs excepting the van der waal's force between crystal layers, which leads to a very strong binding force between crystal layers. Chlorite is also a nonswelling clay.

However, iron cations, such as Fe^{2+} or Fe^{3+}, are common in chlorite. For example, chamosite is an iron-rich mineral. It is often associated with pyrite and siderite, and may become limonite after weathering. The most important is that iron-rich chlorite will react with acid and generate precipitates under acidic conditions. These precipitates fills in rock pores and finally cause severe damage to reservoirs. Iron-rich chlorite is an acid-sensitive mineral.

Unstable mechanism of clay minerals

Once contacting with water, clay minerals will swell, disperse or flocculate in water. "Swelling" means the case where clay adsorbs water and the bulk volume of the clay increase greatly. "Dispersion" is the case where clay particles separate and suspend in a liquid. "Flocculation" denote the phenomenon that the clay particles in a liquid aggregate to form clumps or "flocs." All the phenomena are the result from clay instability, but different unstable mechanisms responsible for different phenomena.

Swelling

Exposed to aqueous solution, clay minerals will adsorb water molecules and swell. The phenomenon is also known as *clay hydration*. Based on the hydration mechanism, clay hydration can be classified to two phases.

Surface hydration (crystalline swelling)

Surface hydration occurs when clays are exposed to concentrated brine or aqueous solutions containing large amount of divalent or multivalent cations. Surface hydration is caused by the hydration of interlayer cations. This leads to weak clay swelling and less formation damage.

In the process, the adsorption energy of water on the surface of clays is the primary action force. The amount and the thickness of water adsorbed on the clay basically depend on the hydration energy of cations and the charge density on the surface of clay. Research shows that when the exchangeable cations of montmorillonite are Ca^{2+}, Mg^{2+}, and H^{+}, the interlayer attractive force of clay increases, the hydrated film is thus thinner, and the arrangement of water molecules on the clay is directional and regular; whereas when the exchangeable cations are Na^{+}, the interlayer attraction of clay decreases, the hydrated film becomes thick, and the arrangement of water molecules on the clay is not regular. Therefore, sodium montmorillonite swells more than calcium montmorillonite. Generally, the order of clay swelling capacity is as follows: montmorillonite > mixed-layer clays contained swelling layer > illite > kaolinite.

The affection of clay composition to the capacity of surface hydration can be explained from the cation exchange point of view. Water-sensitive 2:1 phyllosilicate clay minerals have permanent layer charge because of the isomorphous substitution in either the octahedral sheet (typically from the substitution of low charge species such as Mg^{2+}, Fe^{2+}, or Mn^{2+} for Al^{3+}) or the tetrahedral sheet (where Al^{3+} or occasionally Fe^{3+} substitutes for Si^{4+}) [54]. For example, it is common for smectites to have both tetrahedral charge and octahedral charge. These isomorphous substitutions lead to net negative charges on the crystal layers which must be satisfied by the presence of charge-balancing cations somewhere else in the structure. Therefore, when clays contact with aqueous solution, the crystal layers will be forced to split due to the net negative charges on the crustal layers (repulsive force), and water molecules are easy to enter into the interlayers following the cations adsorbed by the clay from water to balance the negative charges on the interlayers. The tendency of the cation exchange of clays depends on *Law of Mass Action* and ionic valence of cations between layers. The exchange of cations is directly determined by the adsorption energy of exchangeable cations between layers. As a general rule, the higher the valence of cations, the stronger the attractive force of cations on clays. As a result, the higher the valence of cations in clays, the more difficult clay hydration The weaker the repulsive force between clay particles, the more difficult it is for the clays to disperse.

The order of replaceability of the common cations in clays from most to least easy cations is $Li^{+} > Na^{+} > K^{+} > Rb^{+} > Cs^{2+} > Mg^{2+} > Ca^{2+} > Sr^{2+} > Ba^{2+} > H^{+}$ [55]. With the exception of potassium in illites, the firmness with which cations are held in the clay structure increases with the valence of the cation. Hence, the

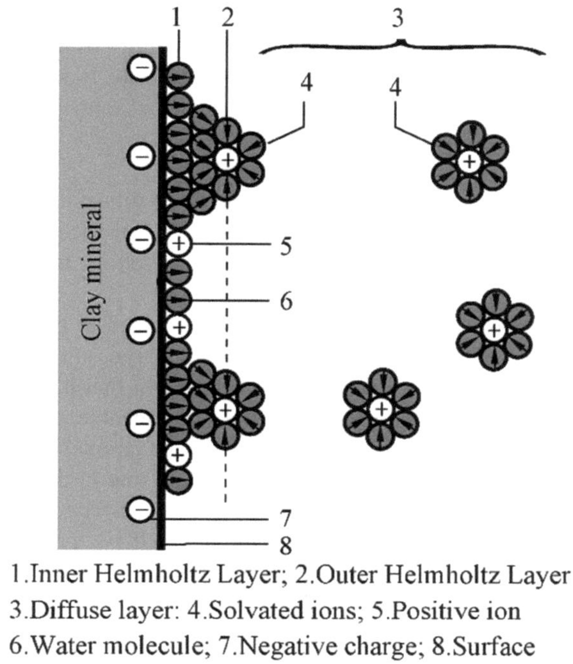

Fig. 2.75 Model of double electrode layer (from Wikipedia)

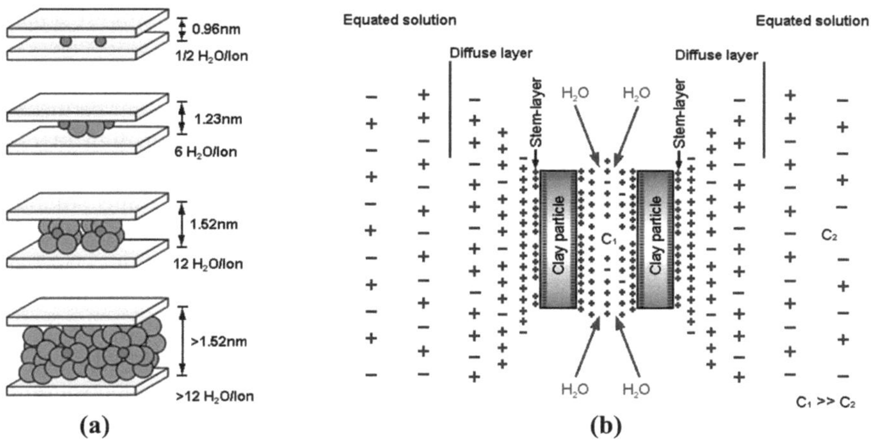

(a) Crystalline swelling results in an increase in the space between interlayers of swellable clay particles by the hydration of cations; (b) Expansion of neighbour clay particles by osmotic pressure

Fig. 2.76 Crystalline and osmotic swelling of clay minerals (Joerg et al. 2011 [57])

difference of dilatability between sodium montmorillonite and calcium montmorillonite can be explained by the higher adsorption energy of calcium ions (Ca^{2+}). The replaceability of calcium ions form montmorillonite is smaller than that of sodions, the repulsive force and dilatability between Ca^{2+}-clay particles are smaller than that between Na^{+}-clay particles.

Osmotic swelling

Osmotic swelling occurs when clays are exposed to dilute solutions or solutions containing large quantities of sodions (Na^{+}). It is caused by the electric double layer formed on the surface of clay minerals. Osmotic swelling makes stronger clay swelling and more formation damage.

In aqueous solution, exchangeable cations are attracted to the clay particles by the negative electric field arising from the negative charge on the particles. Hence, the electric field acts as a semi-permeable membrane in that it will allow water to enter the double layer but will not allow the exchangeable cations to leave the double layer [56]. Thus, when the total ion (cations and anions) concentration in the double layer between the clay particles is higher than that in the aqueous solution (pore fluid), the water in fluid diffuses into the double layer to dilute its ion concentration. This phenomenon creates an osmotic repulsive pressure between the clay particles. The osmotic pressure is primarily related with the difference of the ion concentrations between the clay double layer and surrounding fluid. Because of the osmotic pressure, the distance between clay particles increases greatly, and then the clay swelling occurs.

When the salinity of an invading fluid is lower than that of formation water, the water in invading fluid is attracted to the surface of the clay due to higher ion concentration on the clay surface than that in the invading fluid. As a result, oriented water film is formed on clay surface, and the repulsive force of double layers is thus increased (Fig. 2.75). As the repulsive interaction of double layers on clay surface, clay particles push each other away, which then leads to continuous clay swelling. Consequently, the osmotic equilibrium of the semi-permeable membrane on clay particles is a key factor to affect the hydration film of clay swelling. In other words, osmotic hydration is the major factor of clay swelling (Fig. 2.76). For example, when montmorillonite is exposed to low-salinity solution, surface hydration, occurring firstly, makes its crystal layers swell and doubles its volume; the following osmotic swelling continues the hydration of the clay, and leads its volume to increase by many times as well. Montmorillonite is the most easily expansive clay minerals.

In a word, clay swelling is related with many factors; and the following factors are the key to control the swelling of clays: the type of clays, the composition of exchangeable cations in clays, the ion composition, and concentration of aqueous solution.

Flocculation and dispersion

Flocculation and dispersion are also the important features of clay-water system. When clay particles clot in aqueous solutions, it is known as the flocculation or

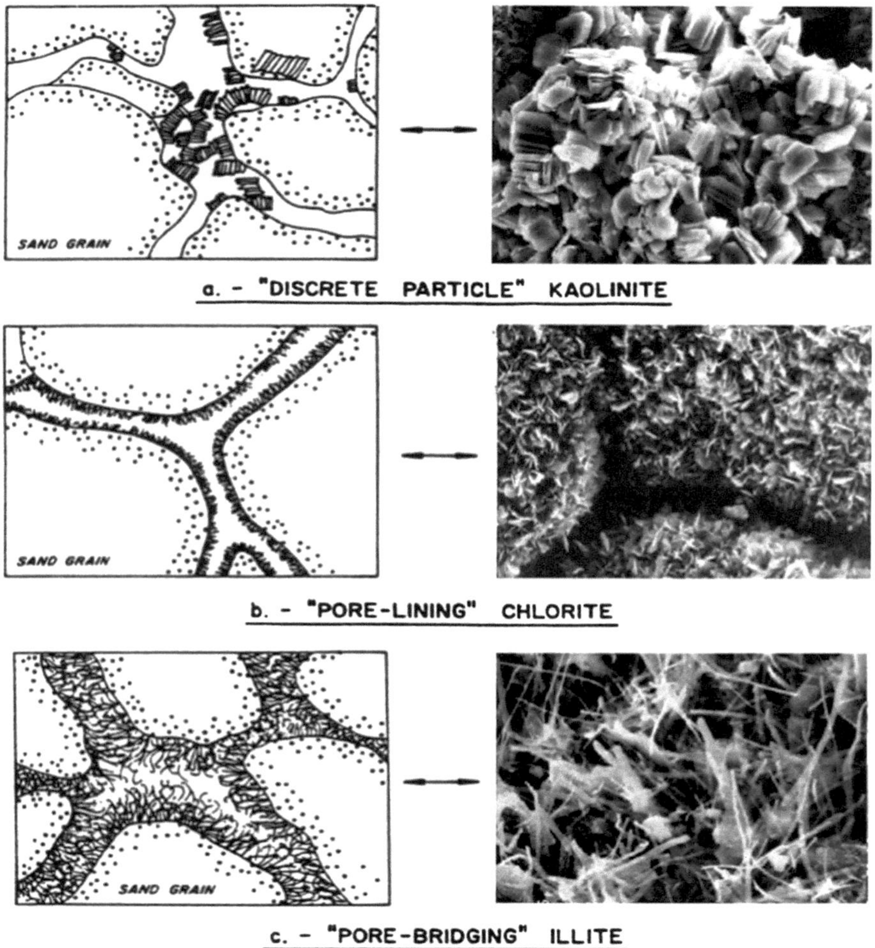

Fig. 2.77 Occurrences of clay minerals in sandstone

aggregation of clays. When clays split and disperse in aqueous solutions, it is known as the dispersion of clays. The flocculation and dispersion of clays can be explained by the theory of double layer of counterions (Fig. 2.75).

In electrolyte solutions, negatively charged anions and positively charged cations coexist in the solution by reason of ionization. When in contact with an electrolyte solution, the negatively charged clays will attract the cations in the solution to keep the charge balance of clays; and this process will lead to an electrical double layer formed on the surface of clays (Figs. 2.75 and 2.76).

Primarily, the negative clay attracts part of cations to form a firmly attached layer around the surface of the clay; the layer of cations attracted on the clay is known as the Stern layer. When other cations in the solution are further attracted by the negative clay, they will be repelled by the Stern layer as well as the cations that are trying to approach the clay. The attraction-repulsion dynamic process results in a diffuse layer of oppositely charged ions (counterions) on the clays. The concentration of counterions is higher near the surface of the clay; and it gradually decreases with distance to the clay until the charge equilibrium has been reached in the solution.

When counterions are far away from negative clay particles, clay particles are scattered in the solution due to the repulsion of negative charges on the surface of the clay particles. However, when counterions are near the clay particles in the solution, clay particles attract each other owing to the effect of charge balance and flocculation, in this way, occurs in the solution.

Investigations have shown that Ca^{2+}, Mg^{2+}, or H^+ in solutions is favorable for clay flocculation, but Na^+ makes clay disperse. Moreover, the increase in electrolyte concentration is also favorable for clay stabilization and flocculation.

In the original conditions, clay minerals in petroleum reservoirs are stable on account of high salinity of formation water. However, the low salinity of invading fluids or Na^+ in invading fluids will cause clay minerals to disperse and migrate, and finally result in serious formation damage.

Occurrence of clay minerals

The occurrence of clay minerals has great effects on the flowing of fluids in petroleum reservoirs. The occurrences and distribution of clay minerals in reservoir rocks primarily depend on the source of sediment, sedimentary environment, and hydrodynamic conditions. Based on the shape of clay mineral aggregate in rock pores and the the effects of clays on physical properties of reservoir rocks, the occurrences of clay minerals may be divided into three types, as shown in Fig. 2.77.

Patch type (Discrete particle type)

"Patch type," as shown in Fig. 2.77a, means the occurrence of clay minerals, like "disperse patches," packs in the rock pores. Pores thus become locally narrow and micropores between the clay particles are well developed. In this kind of occurrence, kaolinite, a few of needle-shaped mica and montmorillonite, etc., are very common. As a result, fine migration easily arises because of the effect of high-speed fluids.

Table 2.26 Physical properties of rocks with different occurrences of clay minerals

Rock sample	Occurrence	Grain-size (μm)	Sorting coefficient	Mean pore size	Porosity (%)	Permeability (mD)
1	Patch type	346	1.23	242	14.1–24.9	150–1173
2	Bridging type	150	1.36	0.9	8.45–19.1	0.09–0.31

Fig. 2.78 SEM photo of montmorillonite and quartz grains in sandstone

Fig. 2.79 SEM photo of kaolinite

Pore-lining type

In this occurrence, clay minerals, like the lining of pores, arrange parallel to the grain surface with the grains wrapped partially or fully (Fig. 2.77b). This type of occurrence belongs mainly to montmorillonite, illite, chlorite, and so on. The fluid suffers a weak resistance when flowing through them. Generally, fine migration is rare, but hydration and swelling are very common in this type of occurrence. The pore-throats are thus narrowed. If micropores are developed well, the damage of water blocking may even be caused.

Pore-bridging type

The hairy or fibrous clays, such as chlorite, illite (hydromica), easily form bridges in pores between grains (Fig. 2.77c). However, the "bridges" in pores are easily broken off by the shock of fluids through the pores. As a result, fine migration occurs along the flow of fluids. In such type of occurrence, micropores are easily developed. When fluids pass through the micropores, the flow is very circuitous. In this way, the resistance of fluid flowing through the rocks is much increased. The permeability of the rock to fluid is thus remarkably decreased. The quality of the petroleum reservoir thus becomes very poor.

In addition, the kaolinite, shell-type smectite, illite-smectite interlayer minerals, and chlorite-smettite interlayer minerals sometimes overlap mutually at pore-throats, and could narrow the pore-throats due to a high potential of hydration, swelling, and water blocking.

Fig. 2.80 SEM photos of illite

In most cases, there is not only one kind of occurrence of clay minerals in sandstone reservoirs. It is very common that several occurrences coexist in reservoirs. It should be noted that the occurrence of clay minerals usually has greatly influence on the physical properties of reservoir rocks (Table 2.26).

Major reservoir problems caused by common clay minerals in petroleum reservoirs

Montmorillonite

The crystal of montmorillonite is quite fine. A sheet of montmorillonite crystal is usually in the shape of curly blade. Its aggregate is petal-like or honeycomb (Fig. 2.78). The most notable reservoir problem induced by these minerals is seriously water sensitivity, especially when sodium montmorillonite occurs in the reservoir. If exposed to water, sodium montmorillonite may expand by 5–9 times volume. As a result, reservoir permeability will be significantly declined.

Fig. 2.81 SEM photos of chlorite and apatite

Kaolinite

Kaolinite is the most common clay mineral in sandstone reservoirs. It often fills in pores in page-like, worm-like shapes (Fig. 2.79).

The potential impact of kaolinite on reservoirs is as follows:

(a) Kaolinite fills intergranular pores, and thus turns original intergranular pores into tiny intercrystalline pores. Tiny intercrystalline pores contribute little to the permeability of rocks. Rock permeability is thus visibly decreased.
(b) The adhesion of kaolinite aggregate on the surface of rock particles is weak. Under the shear stress of fluids through the rock, kaolinite aggregate easily crashes or falls off from the surface of rock particles, and move with the fluid in rock pores. Consequently, the pores or throats of the rock could be severely blocked.

Illite

Illite has the most complex appearance. But its single crystal is tiny. From the genesis point of view, Illite can be, by its shape, divided into two categories: irregular flakes illite and hairy, fibrous or ribbons illite (Fig. 2.80). The former generally distributes on the surface of rock particles, and thus affects rock permeability by reason of narrowing the pores which contribute to the flow of fluids through the rock. The latter, however, has much more complex effects on the pore system of rocks. It may bridge in pores and then led to many tiny pores. The pore space of the rock becomes more complicated. The permeability of the rock is thus destroyed. In addition, when contacting with fresh water or affecting by the shear stress of fluids, fibrous illite may disperse or break to pieces and migrate with the fluid. The pores could be plugged by the fragment of illite. Rock permeability is greatly declined.

The cation exchange capacity of illite is smaller than that of smectite, but higher than that of kaolinite. Illite has scarcely water sensitivity. Due to the huge specific

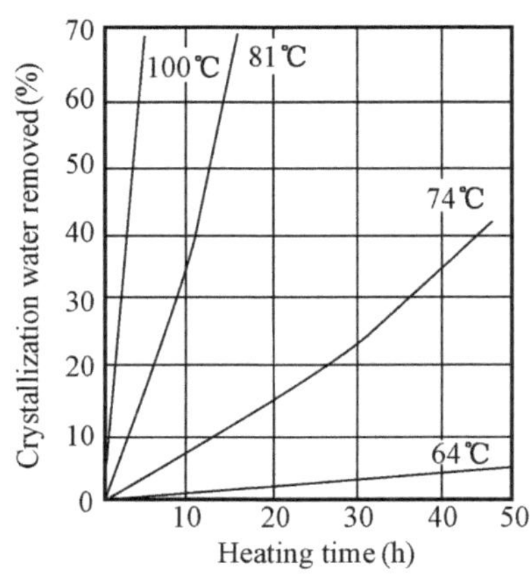

Fig. 2.82 Dehydration curves of gypsum at different temperatures (He [21])

surface area increased by illite, however, a rock may has a high irreducible water saturation.

Chlorite
Chlorite often covers the surface of rock particles in the shape of willowleaf, or fills in pores in the shape of pompon-like aggregates (Fig. 2.81). It is bridge-type growing in pores. In petroleum reservoirs, common chlorite is iron-rich minerals. When contacting with chlorhydric acid, it releases iron ions and then generate colloidal precipitate of ferric hydroxide. The colloidal precipitates will plug pore-throats and throats, and finally result in serious reservoir damage.

2.6.2.2 Sulfate Cement

The sulfate minerals in reservoir rocks mainly include gypsum ($CaSO_4{\cdot}nH_2O$) and anhydrite ($CaSO_4$). A remarkable characteristic of sulfate minerals is dehydration at high temperature. It affects the accurate determination of the water saturation of reservoir rocks. If the amount of sulfate in a reservoir rock is more than 5 %, it will significantly affect the veracity of the irreducible water saturation measured in the laboratory.

When heated to a certain temperature (e.g., 64 °C), crystallization water in gypsum will vaporize slowly out from the gypsum. If the temperature is over 80–100 °C, crystallization water vaporizes quickly. With the increase in temperature, the speed of water vaporization doubles. Figure 2.82 shows gypsum dehydration curves at different temperatures.

In core analysis, gypsum dehydration should be taken into account. It may result in a visible experimental error. For example, in the measurement of the water saturation of a core, if the core is extracted with toluene (boiling point 105 °C), the result (water saturation) is on the high side. The sum of the saturation of fluids containing in the core, sometimes, is even higher than 100 %.

In order to avoid the error caused by gypsum dehydration at high temperature, two methods can be used in core analysis: (a) washing core at low temperature using centrifugal machine; (b) using the azeotropic liquid mixed by chloroform and methanol in the proportion of 13:87. The boiling point of the azeotropic liquid, 53.5 ºC, is lower enough than the temperature of gypsum dehydration. The azeotropic liquid does not cause the dehydration of gypsum.

2.6.2.3 Carbonate Cement

Carbonate minerals are one of the major cements of clastic rocks. Carbonate minerals include limestone ($CaCO_3$), dolomite ($CaMg(CO_3)_2$), ferroan calcite, ankerite, siderite ($FeCO_3$), Na_2CO_3, K_2CO_3, and so on. In sandstones, calcite ($CaCO_3$) and dolomite are the most common. In clastic reservoirs, the content of carbonate cement varies widely, generally ranging from 9.5 to 21 %. Occasionally, some rocks, e.g., quartz sandstones, contain a little carbonate minerals.

An important feature of carbonate minerals is that they can react with acid. When contacting with acid, carbonate minerals may improve rock permeability or cause a damage to the rock depending on the type of the acid. When reacting with hydrochloric acid, most of carbonate minerals, except for ferroan minerals, can be dissolved and no chemical precipitates formed. In this situation, chemical reaction causes no damage to reservoir rocks. The carbonate minerals which can react with acid but no precipitates generating are called *acidic minerals*. On the contrary, when mud acid (a mixture of hydrofluoric acid and hydrochloric acid) is used in well stimulation, calcareous minerals can react with the hydrofluoric acid in mud acid and produce precipitates. The precipitates will then block in pore-throats and finally cause serious formation damage. In this case, these carbonate minerals are known as *acid-sensitive minerals*.

To hydrochloric acid, the common acidic minerals are carbonate minerals, including calcite ($CaCO_3$), dolomite ($CaMg(CO_3)_2$), sodium salt (Na_2CO_3), hoevellite (K_2CO_3), and so on. But these minerals are sensitive to fluorhydric acid.

Based on the features that acidic minerals can dissolve in acid without precipitates formed, the permeability of petroleum reservoirs can be improved by pumping acidic fluids into the reservoir. The method that uses acidic fluids and improves rock permeability is called *acidizing*. It is one of the essential well stimulations in petroleum production. With this method, the output of oil/gas wells or the injectability of water wells can be greatly improved.

In oilfields, well acidizing basically depends on the type and amount of acidic minerals in reservoir rocks. The most common acidizing fluid used in sandstone reservoirs is mud acid (a mixture of hydrofluoric acid and hydrochloric acid) whereas hydrochloric acid is the major acidizing fluid used in carbonate reservoirs. In conventional sandstones, however, hydrochloric acid stimulation can be used only when the carbonate content in the rock is more than 3 %.

A typical chemical reaction of carbonate minerals with hydrochloric acid is as follows:

$$CaCO_3 + 2HCl = CaCl_2 + H_2O + CO_2 \uparrow$$

Based on the equation, the amount of carbonate in a rock can be determined. The amount of carbonate can be calculated according to the volume of CO_2 released from the reaction above. If using the amount of $CaCO_3$ represents the total amount of all carbonate in the rock, the content of carbonate in the rock can be calculated by the following expression:

$$C = \frac{V\rho}{4.4W} \times 100\,\% \tag{2.145}$$

Table 2.27 Sensitivity minerals in reservoirs and their sensitivity (Shen [58])

Sensitive minerals	Potential problems	Sensitivity degree	Conditions of sensitivity	Method of controlling sensitivity
Montmorillonite	Water sensitivity	Highest	Fresh water	Fluids with high-salinity, anti-swelling agent
	Velocity sensitivity	Middling	Fresh water, high flow velocity	Acidizing treatment
	Acid sensitivity	Middling	Acidizing treatment	Inhibiting agent of acid
	Alkali sensitivity	Middling	Chemistry flooding	Inhibiting agent
Illites	Velocity sensitivity	Lowest	High flow velocity	Low flow velocity
	Plug micropore	Middling	Fresh water	Fluids with high salinity, anti-swelling agent
	K_2SiF_6 precipitates	Lowest	Hydrofluoric acid	Inhibiting agent of acid
Kaolinite	Velocity sensitivity	Middling	High flow velocity, high PH, high transition pressure	Low flow velocity, stabilizing agent of particles, low transition pressure
	$Al(OH)_3$ precipitates	Middling	Acidizing treatment	Inhibiting agent of acid
	Alkali sensitivity	Middling	Chemistry flooding	Inhibiting agent
Chlorite	$Fe(OH)_3$ precipitates	Highest	Rich-oxygen, after acidizing	oxygen scavenger
	MgF_2 precipitates	Middling	Hydrofluoric acid at high PH	Inhibiting agent of acid
	Velocity sensitivity	Lowest	High flow velocity, high pH	Low flow velocity
Carbonate	CaF_2 or MgF_2 precipitates	Middling	Hydrofluoric acid	Before acidizing, wash with hydrochloric acid, inhibiting agent of acid
Ankerite	$Fe(OH)_3$ settling	Middling	High pH, rich-oxygen	Inhibiting agent of acid, oxygen scavenger
Pyrite	Sulfide settling	Middling	Ca^{2+}, Sr^{2+}, Ba^{2+} in fluid	anti-scaling agent
Siderite	$Fe(OH)_3$ precipitates	Lowest	High PH, rich-oxygen	Inhibiting agent of acid, oxygen scavenger
Mixed-layer clay	Velocity sensitivity	Middling	High flow velocity	Low flow velocity
	Water sensitivity	Middling	Fresh water	Fluid with high-salinity, anti-swelling agent
	Acid sensitivity	Lowest	Acidizing treatment	Inhibiting agent of acid
Feldspar	Feldspar precipitates	Lowest	Hydrofluoric acid	Inhibiting agent of acid
Quartz particles uncemented	Velocity sensitivity	Middling	High flow velocity, high transition pressure	Low flow velocity, low transition pressure

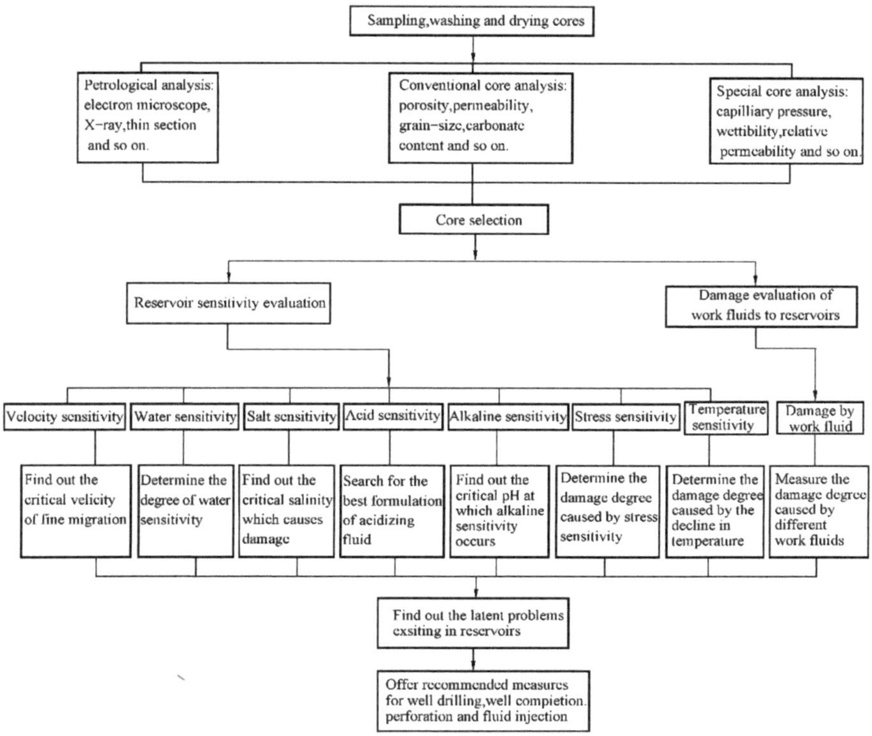

Fig. 2.83 Experimental flowchart of laboratory evaluation for formation damage (何更生 1994)

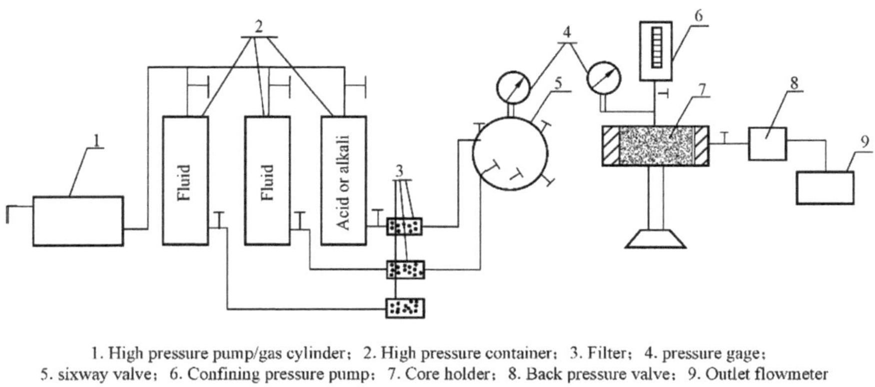

Fig. 2.84 Schematic illustration of core flow test (SYT5358-2010 China)

where C means the amount of $CaCO_3$, %; V is the volume of CO_2 released from the reaction between the rock and HCl, cm^3; ρ is the density of CO_2 gas in excremental temperature and pressure, g/cm^3; W is the total weight of the core sample, g.

In a word, there are a lot of sensitive minerals in petroleum reservoirs. Different minerals cause different reservoir problems and different potential damages to reservoirs (Table 2.27). Therefore, it is quite necessary to use special techniques to evaluate reservoir sensitivity in petroleum production.

2.6.3 Assessment Methods of Reservoir Sensitivity

In general, petroleum reservoir may have several kinds of sensitivity, typically including water sensitivity, salinity sensitivity, acid sensitivity, alkali sensitivity, and flow velocity sensitivity. The evaluation on reservoir sensitivity chiefly focuses on the effects of physical-chemical reactions between invading fluids and the rock on the permeability of the rock. Systematical evaluation for reservoir sensitivity includes a series of experimental analysis (Fig. 2.83), e.g., petrology analysis, conventional core analysis, special core analysis, core flow testing, and so on. The measurement of reservoir sensitivity is primarily based on the core flow testing. A flow chart of core flow testing is shown in Fig. 2.84. It is a typical procedure of the evaluation on reservoir sensitivity. The testing procedures of common reservoir sensitivity are briefly described below.

2.6.3.1 Evaluation on Flow Velocity Sensitivity

In formations, unconsolidated clays and particles (diameter <37 μm) are totally called "*fines.*" Usually, these fines are not tightly fixed on the surface of grains.

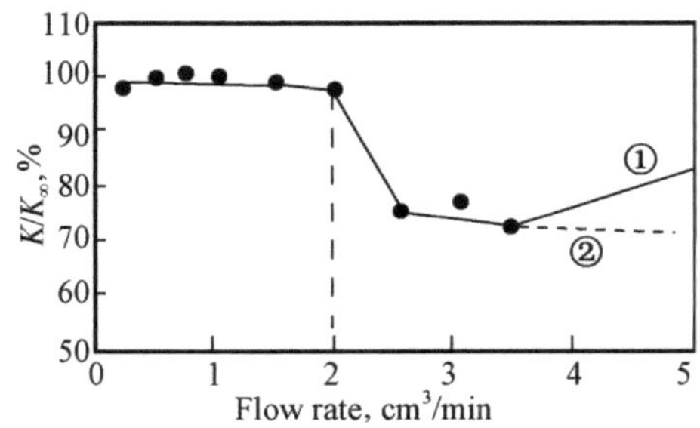

① fine sflow out; ② no fine flows out.

Fig. 2.85 Experimental curve of flow velocity sensitivity (Shen [58])

They can migrate with the fluids flowing through the pores and easily pile up at narrow throats to form "*bridge plug.*" A lot of "*bridge plugs*" finally result in a remarkably decrease in rock permeability, and the reservoir is thus damaged.

Experiments indicate that fine migration generally increases with the flow velocity of fluids in a rock, but different rocks exhibit different sensitive degrees to flow velocity. Some responds weak, which is regarded as *insensitivity to velocity*. On the contrary, if rock permeability decreases considerably with the increasing velocity, the rock is sensitive to velocity. When the flow velocity of fluids causes a remarkable decline in rock permeability, the velocity is known as *critical velocity*, symboled v_c. If a rock is sensitive to the flow velocity, the critical velocity or critical flow rate (Q_c) should be determined because it is an important basis for pertoleum engineers to determine the reasonable production of oil wells or the reasonable injection rate of water wells.

Generally, critical velocity can be determined from the experiment of flow velocity sensitivity. The experiment of measuring flow velocity sensitivity aims at finding the relationship between the alteration of rock permeability and the flow velocity of the fluid passing through the rock.

Experimental procedure

(a) Preparation of the test fluid. According to the industry standard, a test fluid should not have any physical and chemical reaction with the rock. In general, formation water or kerosene is customarily used. In addition, the test fluid should be first filtrated through the filter membrane with openings of 0.22 μm diameter.

(b) Setting of confining pressure. Put a saturated core sample in the core holder. Measure gas permeability first. Then keep the same flow direction of fluid injection with the gas flow when gas permeability is measured. Let the confining pressure of the core slowly increase to 2 MPa. During the test, confining pressure should always be kept 1.5–2.0 MPa higher than the inlet pressure of the core.

(c) Design of flow rate. Following series of flow rates: 0.1, 0.25, 0.5, 0.75, 1.0, 1.5, 2.0, 3.0, 4.0, 5.0, 6.0 cm^3/min, is often used to inject formation water to the core in the order of increasing flow rate and then measure the permeability of the core under steady flow for each flow rate. Note that the pressure gradient in each test should be less than 2 MPa/m. Besides, initial flow rate and its intervals should be chosen according to the value of gas permeability of the core.

(d) Plot the curve of permeability (or permeability/equivalent liquid permeability) to flow rate, as shown in Fig. 2.85.

When injection velocity is higher than the critical velocity, namely injection flow rate higher than the critical flow rate, this curve shows a sharp decline in permeability, such as the flow rate of 2.5 cm^3/min shown in Fig. 2.85. It indicates that the rock is sensitive to this flow velocity (rate). After that, the interval of flow rate may be properly increased. If the critical rate cannot be found in whole test, end the test

Table 2.28 Graduation of flow velocity sensitivity (SYT5358-2010 China)

Damage rate of permeability (%)	Damage degree of permeability
$D_v \leq 5$	No damage
$5 < D_v \leq 30$	Weak
$30 < D_v \leq 50$	Middle to weak
$50 < D_v \leq 70$	Middle to strong
$D_v > 70$	Strong

when flow rate is over 6.0 cm^3/min. For low-permeability tight rock core, the test should be ended when flow rate is lower than 6.0 cm^3/min but pressure gradient has been over 2 MPa/cm.

It is noteworthy that the fines migrating with the fluids may either form "bridge plug" in narrow pores or throats or be swilled out from the outlet of the core by the fluid. When flow rate is enough higher, part of fines may be rinsed out of the core. The permeability of the rock sample may increase gently at the late period of the test (Fig. 2.85).

Evaluation of flow velocity sensitivity

The rate of permeability change

The rate of permeability change caused by velocity sensitivity is defined as [59]:

$$D_{\mathrm{vn}} = \frac{|K_n - K_i|}{K_i} \times 100 \quad \% \tag{2.146}$$

where D_{vn} is the rate of change in permeability corresponding to each flow velocity, %. It is also called the damage rate of permeability; K_{i} is the initial permeability measured at the minimum flow rate, mD; K_{n} is the permeability corresponding to each flow rate before critical velocity, mD.

Critical velocity

According to experimental results, plot the curve of $K_{\mathrm{n}}/K_{\mathrm{i}}$ to flow rate. The critical velocity can then be determined based on the curve and the value of D_{vn}. From Fig. 2.85, we can see that with the increase in flow rate (flow velocity), the sample permeability could occur obvious decline at a certain flow rate. If the decline in permeability exceeds 20 %, it marks the occurrence of critical velocity.

Table 2.29 Graduation of water sensitivity (SYT5358-2010 China)

Damage rate of water sensitivity (%)	Damage degree of water sensitivity
$D_w \leq 5$	No damage
$5 < D_w \leq 30$	Weak
$30 < D_w \leq 50$	Middle to close to weak
$50 < D_w \leq 70$	Middle to close to strong
$70 \leq D_w < 90$	Strong
$D_w > 90$	Extremely strong

The last flow velocity (flow rate) before 20 % decline in permeability is known as *critical velocity*, as the flow rate 2 cm^3/min shown in Fig. 2.85.

Damage rate of permeability

Damage rate of permeability is determined by the following expression [59]:

$$D_v = \max(D_{v1}, D_{v2}, D_{v3}, \ldots, D_{vn}) \tag{2.147}$$

where D_v is the damage rate of permeability, %; D_{v1}, D_{v2}, ..., D_{vn} are the rates of permeability change corresponding to different flow velocities (flow rates), respectively, %.

Based on the damage rate of permeability, the damage degree of flow velocity to permeability can then be evaluated. According to the industry standard, the graduation of permeability damage is listed in Table 2.28.

2.6.3.2 Evaluation on Water Sensitivity

Under the original conditions of petroleum reservoirs, clay minerals in rocks are steady due to the higher salinity of formation water. During the process of well drilling, well completion, and water flooding, low-salinity fluids will invade into the reservoir. On contacting low-salinity invading fluids, water-sensitive minerals will swell, disperse, and even migrate with fluids. The reservoirs will be seriously damaged. The purpose of water sensitivity evaluation is to find out the degree of permeability damage caused by water sensitivity of minerals. In general, *water sensitivity index* can be used to determine the degree of permeability damage caused by water sensitivity. Water sensitivity index is defined as follows [59]:

$$D_w = \frac{|K_i - K_w|}{K_i} \times 100\ \% \tag{2.148}$$

where D_w is water sensitivity index, also called damage rate of water sensitivity, %; K_i is the permeability measured by the initial fluid. Initial fluid may be formation water or standard brine. The salinity of a standard brine should equal to that of the formation water which is from the reservoir, mD; K_w is the permeability measured by distilled water, mD.

According to the industry standard, the gradation of water sensitivity is shown in Table 2.29.

2.6.3.3 Evaluation on Salinity Sensitivity

Clay minerals in reservoir rocks could swell, shrink, cracking when obvious change occurs in the salinity of formation water. If the salinity of formation water is lowered by the invading fluid, some clay minerals in the rock will swell and disperse; whereas some of them will shrink and crack when the salinity of formation

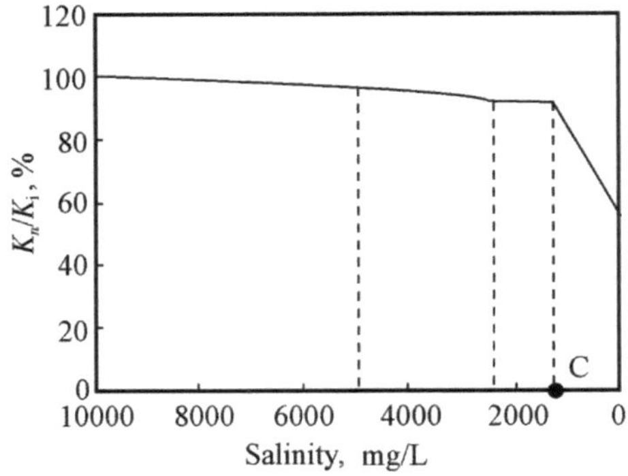

Fig. 2.86 Experimental curve of salinity sensitivity evaluation (沈平平 1995)

water is increased by invading fluids. In both cases above, clay minerals will release a lot of fines, which finally results in fine migration and formation damage.

Therefore, it is very necessary to evaluate the salinity sensitivity of a reservoir rock in petroleum production. The aim of evaluation on salinity sensitivity is to find out the *critical salinity* or *critical salt concentration* at which the rock permeability declines visibly. This test is called *salinity sensitivity evaluation experiment*.

Experimental procedure

Because the decline in rock permeability may be caused by either low-salinity fluids or high-salinity fluids, the salinity sensitivity test usually includes two parts: lowering-salinity test and increasing-salinity test.

Lowering-salinity test

Lowering-salinity test should be based on the result of water sensitivity test. If the rate of permeability damage measured by distilled water in water sensitivity test is lower than 20 %, the lowering-salinity test is not necessary; whereas it is necessary if the rate of permeability damage is higher than 20 %.

The salinity interval of the test fluid: the salinity interval is generally determined in term of the permeability measured by the test fluid and the distilled water in water sensitivity test. If the damage rate of permeability at one interval of salinity is higher than 20 %, the interval of salinity should be shortened. Lowering-salinity test should include at least four tests using different salinity fluids.

The flow rate of the test should be determined according to the results of flow velocity sensitivity test. In lowering-salinity test, the initial fluid, either being formation water or standard brine, should be first used to measure the rock permeability. After that, the series of lower salinity fluids are then used in turn to test.

Increasing-salinity test

This test is necessary only for the following two cases: the salinity of invading fluids is higher than that of formation water; or the salinity sensitivity evaluation for special requirement.

The salinity interval of the test fluid: it is determined according to the salinity difference between the invading fluid and formation water. Generally, at least three fluids of different salinities should be used for one test; but a large salinity difference between the invading fluid and formation water suggests a denser salinity interval of test fluids.

In the same way, the flow rate of the test should be determined according to the results of flow velocity sensitivity test. The initial fluid, either being formation water or standard brine, should be first used to measure the rock permeability. Then, the series of higher salinity fluids are in turn used to test.

In each testing, flow rate should be constant. When the volume of the fluid injecting into the core is about 10–15 times PV (pore volume) of the core, stop injecting fluid. Then under the conditions of testing, keep the fluid inside the core at least 12 h so as to let the minerals in the rock contact and react adequately with the injected fluid. Afterwards, using the same salinity fluid and the same flow rate displaces the core, measure the permeability of the core.

Critical salinity

The rate of permeability change caused by the salinity of injected fluid is calculated by [59]:

$$D_{sn} = \frac{|K_i - K_n|}{K_i} \times 100\ \% \tag{2.149}$$

where D_{sn} is the rate of permeability change corresponding to *n*th salinity of the injected fluid, %; K_{i} is the permeability measured using the initial fluid, mD; K_{n} is the permeability corresponding to *n*-th salinity of the injected fluid, mD.

Plot the experimental curve of permeability ratio (K_n/K_i) to the salinity, as shown in Fig. 2.86. It should be plotted in terms of the order of salinity in the experiment whatever in lowering-salinity test or increasing-salinity test.

Based on the experimental curve and the data of D_{sn}, the critical velocity can be easily determined. From Fig. 2.86, it can be seen that a visible decline in permeability can occur at a certain salinity. The last sanility before 20 % decline in permeability is known as *critical sanility*, as the point *C* in Fig. 2.86.

2.6.3.4 Evaluation on Acid Sensitivity

Acid sensitivity means the phenomenon that the minerals in reservoir rocks react with acidizing fluids invading into the rocks, and then cause a decline in rock permeability due to the occurrence of precipitate and fines released from the reaction between minerals and acidizing fluids.

Acid sensitivity is one of the most primary factors inducing formation damage. In the acid stimulation of production wells, the acidizing formulation should be carefully selected according to the mineral composition of reservoir rocks. An appropriate acidizing formulation or stimulation can correctly improve the quality of the reservoir around the well. On the contrary, the reservoir is probably damaged again by the acidizing, and the secondary damage will lead to a further decrease, but not increase in the oil-gas output of production wells. Even worst, no output from the well.

Table 2.30 Gradation of acid sensitivity (SYT5358-2010 China)

Damage rate of acid sensitivity (%)	Damage degree of permeability
$D_{ac} \leq 5$	No
$5 < D_{ac} \leq 30$	Weak
$30 < D_{ac} \leq 50$	Middle to close to weak
$50 < D_{ac} \leq 70$	Middle to close to strong
$D_{ac} > 70$	Strong

Acid sensitivity evaluation aims at the optimization of acidizing formulations so as to maximize the efficiency of the acid stimulation.

For conventional acidizing experiments, the experimental fluid is customarily the fluid of 15 %HCl or 12 %HCl + 3 %HF. In general, 15 %HCl is used in carbonate reservoirs, whereas 12 %HCl + 3 %HF is used in conventional sandstones.

Experimental procedure

(a) First of all, measure the permeability of the sample using the KCl brine whose salinity should equal to the salinity of formation water. Note that the direction of fluid flow in the sample should be the direction of fluids flow in reservoirs when measuring the permeability of the sample.
(b) Pump acidizing fluid into the sample from the inverse direction of permeability measurement. If the sample is a sandstone, the amount of 0.5–1.0 PV acidizing fluid is enough, whereas a carbonate sample usually needs the amount of 1.0–1.5 PV 15 % HCl.
(c) Stop pumping the acidizing fluid. Stimulate the process of shut in and wait for the reaction between the acidizing fluid and the sample. Usually, half an hour is enough for the reaction in carbonate samples, but a sandstone sample needs at least one hour for the reaction.
(d) Pump the KCl brine into the sample again along primary direction (the direction of permeability measurement), and displace fully the acidizing fluid out of the sample. The salinity of the KCl brine is the same with the formation water.
(e) Measure the permeability of the acidized sample. Calculate the variation in permeability of the sample, and evaluate the acid sensitivity of the sample.

Evaluation of acid sensitivity

The damage rate of acid sensitivity is defined as follows [59]:

$$D_{\text{ac}} = \frac{K_i - K_{\text{acd}}}{K_i} \times 100 \quad \% \tag{2.150}$$

where D_{ac} is the damage rate of acid sensitivity, %; K_i is the permeability of the sample before acidizing treatment, mD; K_{acd} is the permeability of the sample after acidizing treatment, mD.

Plot the experimental curve of the permeability ratio (K_{acd}/K_i) to the cumulative injection volume of the acidizing fluid. According to the industry standard, the gradation of acid sensitivity is shown in Table 2.30.

2.6.3.5 Evaluation on Alkaline Sensitivity

Formation water is generally neuter or weak alkaline, pH value ranging 4–9. In petroleum production, some fluids invading into the reservoir could be alkaline, e.g., drilling fluids, alkaline flooding fluids in EOR. Their pH values are usually higher, about 8–12. When invading into petroleum reservoirs, alkaline fluids can lead to a series of physical and chemical reactions due to the increase in pH value of formation water.

When invading into petroleum reservoirs, high-PH fluids can react with formation water and lead to precipitates of divalent cations and emulsible substance. They can also react with paraffin and asphaltene in crude oil and generate flocculation and deposition, called *organic fouling*, in the reservoir around the well hole. All of these reaction products could plug pores and cause a remarkable decline in reservoir permeability.

In addition, clay minerals can dissolve partially in alkali solutions. After the cleavage of clays being dissolved, the structure of clay minerals is damaged; and a lot of fine particles are thus released from clay minerals. Fine migration occurs. As a result, rock pores could be blacked by migrating fines and reservoir permeability could be visible decreased. Generally, the order of the solubility of clay minerals in alkali solutions is: $NaOH > KOH > Na_2CO_3 > NaHCO_3$.

The purpose of alkaline sensitivity evaluation is to find out the critical pH value at which the alkaline sensitivity of reservoir rocks begins to occur. The results of this test lay the important foundations for the design of all kinds of work fluids in petroleum production.

In an ordinary way, the experimental fluid is the specified KCl solution which has the same salinity with the formation water. A 8 % (mass fraction) KCl solution is often used if the data of formation water is not available.

Experimental procedure

(a) The preparation of alkaline solutions: the pH value of the first KCl solution should be 7.0. Then gradually increase the pH value of KCl solutions by 1–1.5 pH value till the value 13.0.
(b) Measure the initial permeability of the sample with the KCl solution whose salinity equals to the salinity of formation water.
(c) According to the increasing order of pH, inject in turn the alkaline solutions into the sample. In each injection, stop injecting solution when the volume of injected solution is about 10–15 times PV of the sample.
(d) Wait for at least 12 h till clay minerals reacts completely with the alkali solution.
(e) Using the same KCl solution displaces the core again, and measure the permeability of the sample.

Table 2.31 Gradation of alkaline sensitivity (SYT5358-2010 China)

Damage rate of alkaline sensitivity	Damage degree of permeability
$D_{al} \leq 5$	No
$5 < D_{al} \leq 30$	Weak
$30 < D_{al} \leq 50$	Middle to close to weak
$50 < D_{al} \leq 70$	Middle to close to strong
$D_{al} > 70$	Strong

Evaluation of alkaline sensitivity

The rate of permeability change caused by the alkaline solutions injected into the sample is defined as follows [59]:

$$D_{a\ln} = \frac{K_i - K_n}{K_i} \times 100\ \% \tag{2.151}$$

where D_{aln} is the rate of permeability change caused by the alkaline solutions injected into the sample, %; K_i is the sample permeability measured by the first KCl solution, mD; K_n is the sample permeability after displaced by n-th alkaline solution, mD.

Plot the experimental curve of the permeability ratio (K_{aln}/K_i) to the pH value of alkali solutions. The critical pH value can be determined according to the value of D_{aln} and the shape of the curve. With the increase in pH values of alkali solutions, the damage rate of permeability could exceed 20 % at a certain pH value. The last pH value before 20 % damage rate of permeability is known as *critical pH value.*

The rate of permeability damage caused by alkaline sensitivity can be determined by the following expression [59]:

$$D_{al} = \max(D_{al1}, D_{al2}, D_{al3},\ \ldots, D_{aln}) \tag{2.152}$$

where D_{al} is the damage rate of permeability, %; D_{al1}, D_{al2}, ..., D_{aln} are the rate of permeability change corresponding to each alkaline solution, respectively, %.

According to the industry standard, the gradation of alkaline sensitivity is shown in Table 2.31.

Exercise

Define the following terms and discuss their significance in reservoir production

(1) Grain-size composition	(2) Nonuniform coefficient
(3) Sorting coefficient	(4) Specific surface area
(5) Effective porosity	(6) Rock compressibility
(7) Total compressibility of formation	(8) Irreducible water saturation
(9) Initial gas saturation	(10) Critical gas saturation
(11) Rock absolute permeability	(12) Equivalent liquid permeability

(continued)

(continued)

(13) Darcy's law	(14) Slippage effect
(15) Clay mineral	(16) Velocity sensitivity
(17) Water sensitivity	(18) Salinity sensitivity
(19) Acid sensitivity	(20) Alkaline sensitivity

Question

1. What are the main methods for grain-size analysis? What are the principles of main methods for grain-size analysis?
2. How to express grain-size distribution? What statistical parameters may be used to evaluate grain-size distribution?
3. What are the features of grain-size distribution?
4. The cumulative grain-size distribution curves of two rock samples are shown in Fig. 2.87. According to the curves, plot schematically the grain-size distribution curves of the two samples, and analyze qualitatively the grain-size distribution features of the two samples.
5. What are the methods for the measurement of specific surface area of a rock?
6. Why does the flow rate of water from Mariotte bottle equals the flow rate of air through the core when the water column of piezometer was kept unchanged in the experiment of specific surface area measurement by gas permeability?
7. Can the specific surface area of a core be measured by the method of gas permeability when the core is saturated by a liquid? Why?
8. What is the significance of the specific surface area of a rock?
9. How to express the distribution of pore size? How to classify the pores of a rock according to the pore size of the rock?
10. What is the range of rock porosity in petroleum reservoirs? What is the relationship between the porosities of ϕ_a, ϕ_e, and ϕ_{ff}?

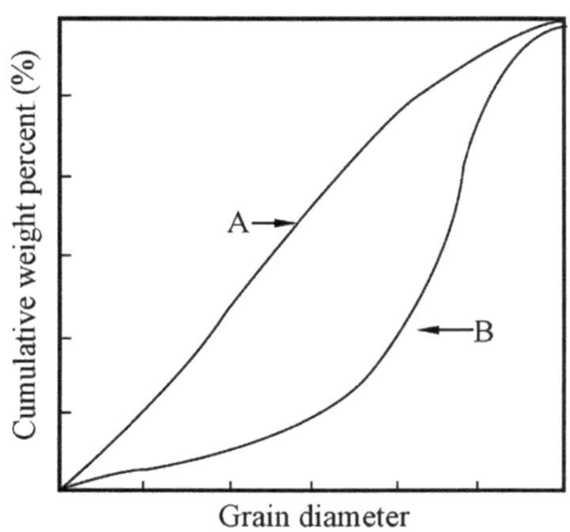

Fig. 2.87 The cumulative grain-size distribution curves of two samples

11. What are the methods for the measurement of rock porosity? Describe the basic principle of porosity measurement in the laboratory.
12. What are the major factors of affecting rock porosity?
13. What is the physical meaning of 1 Darcy?
14. What is the commonly used unit of rock permeability? Give the conversion of different units.
15. What conditions must be satisfied in the measurement of rock permeability using Darcy's equation?
16. What factors does the rock permeability mainly depend on? Is the rock permeability dependent on the properties of the fluid flowing through the rock?
17. What is the conventional range of rock permeability in petroleum reservoir?
18. Is the rock absolute permeability equal to the liquid permeability or the gas permeability of the rock? Which is higher between any two of them? Why?
19. How to determine the absolute permeability of a rock from the data of gas permeability measured in an experiment?
20. Is it true or false, the following expression: the higher the porosity of a rock, the higher the permeability of the rock? Why?
21. What are the major factors of affecting fluid saturation?
22. How many methods are there for fluid saturation measurement? What method is more appropriate for the fluid saturation measurement of the rock containing crystallization water?
23. What are the main components of cement agents? What is the cementation type of rocks? How many cementation types are there in general in rocks?
24. What are the commonly sensitive minerals in cement agents? What are the features of each sensitive minerals?
25. What is the purpose of reservoir sensitivity evaluation?
26. What are the major reasons, in your opinion, causing formation damage after learning the knowledge of the physical properties of reservoir rocks?
27. What experiments does the reservoir sensitivity evaluation include?
28. What is the principle of permeability measurement in the laboratory? Describe the procedure of data processing of permeability measurement. Plot the experimental flow chart, and mark the name of each instrument in flow chart.
29. Why should the carbonate content in rocks be measured?

Derivation

1. From what you have learned, educe the following expression of the total compressibility of an oil reservoir:

$$C_t = C_f + \phi(S_o C_o + S_w C_w)$$

where C_t is the total compressibility of formation; C_f is rock compressibility; C_o, C_w are oil and water isothermal compressibility, respectively; S_o, S_{wi} are oil and reducible saturation, separately.

2. Educe the following equation from the definition of specific surface areas:

$$S = \phi S_p = (1 - \phi) S_S$$

where S, S_P, and S_s denote separately the specific surface areas defined by view volume, pore volume, and matrix volume of a rock. ϕ is rock porosity.
3. Derive the following estimating equation of rock specific surface area from grain-size distribution:

$$S = C\frac{6(1-\phi)}{100}\sum_{i=1}^{n} G_i/d_i$$

where S is the specific surface area defined by the bulk volume of a rock; ϕ is rock porosity; G_i% is the weight percent of grains with average diameter d_i; d_i is the average diameter of ith fraction grains.
4. Derive the following equation from Darcy's law, and explain the meaning of each symbol in the equation:

$$K_a = \frac{2Q_o P_o \mu L}{A(P_1^2 - P_2^2)}$$

Calculation

1. The porosity and specific surface area, measured in experiments, of a sandstone are 23 % and 950 cm^2/cm^3, severally. Estimate the permeability of the sandstone if pore tortuosity $\tau = 1$ and 1.4 separately.
2. The results of grain-size analysis for two rock samples are shown in Table 2.32. Assume the shape coefficient of rock particles $C = 1.3$.
   Answer the following question:
   (a) Plot the grain-size distribution curves and cumulative grain-size distribution curves of the two samples.
   (b) Calculate the nonuniform coefficients and sorting coefficients of the two samples.
   (c) Calculate the specific surface areas of the samples.
   (d) What conclusions can you drawn out from the calculations above?
3. The weight of a dry sample is 32.0038 g. After saturated by kerosene, the weight of the sample in kerosene is 22.2946 g. The weight of the saturated

Table 2.32 Results of grain-size analysis for samples

No. of samples	Porosity (%)	Opening size of sieves (mm)					
		1–0.5	0.5–0.25	0.25–0.1	0.1–0.05	0.05–0.01	<0.01
1	27.0	5.05	53.35	24.7	7.95	5.75	3.05
2	28.3	0	2.95	76.15	10.5	5.4	4.7

sample in air is 33.8973 g. Determine the pore volume, porosity, and view density of the sample. Given that the density of kerosene is 0.8045 g/cm^3.

4. The weight of a core containing oil and water is 8.1169 g. After the core is extracted, 0.3 cm^3 water is separated from the core. The weight of the dried core is 7.2221 g. After saturated by kerosene, the weight of the core in kerosene is 5.7561 g while it is 8.0535 g in air. Calculate the water saturation, oil saturation, and porosity of the core. Given that water density is 1 g/cm^3; oil density is 0.8760 g/cm^3; and the density of kerosene is 0.8 g/cm^3.
5. The weight of a sample contained oil and water is 6.5540 g. After extracted and dried, its weight is 6.0370 g. The volume of water separated from the sample is 0.3 cm^3. The porosity of the sample, measured by saturated kerosene, is 0.25. Suppose the view density of the sample is 2.0 g/cm^3; oil density is 0.8750 g/cm^3; and water density is 1 g/cm^3. Determine the oil, gas, and water saturations of the sample.
6. The length of a core is 10 cm. Its cross-sectional area is 2 cm^2. Under the pressure difference of 0.15 MPa, let the oil, whose viscosity is 2.5 mPa s, flow through the core. The flow rate of the oil is 0.0080 cm^3/s. If the core is completely saturated by oil, calculate the absolute permeability of the core. If it is completely saturated by the brine whose viscosity is 0.75 mPa s, determine the flow rate of the brine under the pressure difference of 2.5 atm.
7. A sample, 2.77 cm length and 3.35 cm diameter, is completely saturated by the brine. The viscosity of the brine is 1 mPa s. The brine flows through the sample under the pressure difference of 251 cm mercury column. The downstream pressure is 1 atm. The flow rate of the brine is 0.02 cm^3/s. Calculate the liquid permeability of the sample when the brine flows through the sample. If the sample is completely saturated by gas with 0.0175 mPa s viscosity and let the gas flow through the sample, the flow rate of the gas is 0.12 cm^3/s. Output pressure is 1 atm. Determine the gas permeability of the sample. Compare the gas permeability to that measured by brine. What conclusions can you draw from the comparison? Why?
8. Given that the reducible water saturation of an undersaturated oil reservoir is 24 %. The isothermal compressibilities of oil and water are 70×10^{-4} 1/MPa and 4.5×10^{-4} 1/MPa, respectively. The rock compressibility is 1.4×10^{-4} 1/MPa. The average porosity of the reservoir is 27 %. Calculate the total compressibility of the reservoir. If the oil-bearing volume of the reservoir is 1500 m^3, initial reservoir pressure is 27 MPa, and the saturation pressure of the oil is 20 MPa, estimate the recoverable reserves of the oil reservoir.
9. A laterally heterogeneous rock, consisting of three beds, is shown in Fig. 2.88. The porosities and permeabilities of three beds are listed in Table 2.33. Calculate:

   (a) Arithmetic means of the porosity and permeability of the rock
   (b) Average porosity of the rock based on equal storing capacity.
   (c) Average permeability of the rock based on the equivalent filtrational resistance.

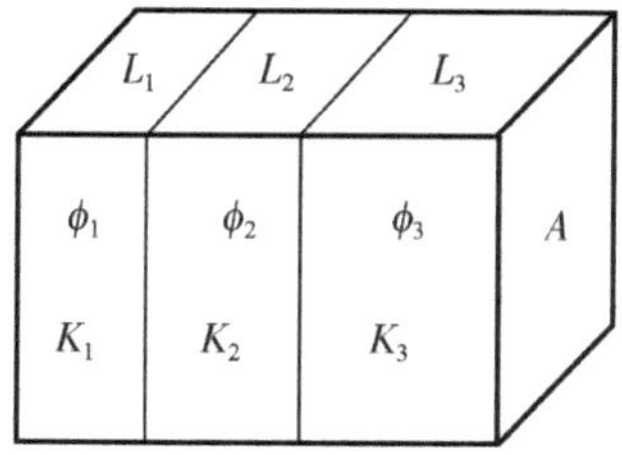

Fig. 2.88 Illustration of the heterogeneous rock

Table 2.33 Porosities and permeabilities of the beds in a heterogeneous rock

Beds	Thickness (cm)	Porosity (%)	Permeability (mD)
1	3	18	160
2	14	22	320
3	11	24	350

Table 2.34 The porosities and permeabilities of the cores from a well

No.	Coring depth (m)	Thickness of cores (m)	Porosity (%)	Permeability (mD)
1	2280.22–2280.96	0.76	12.5	800
2	2280.98–2286.22	5.24	3.82	250
3	2286.22–2289.20	2.98	28.1	210
4	2289.20–2290.00	0.8	29.5	200
5	2290.00–2290.85	0.85	16.8	150

10. The porosities and permeabilities of the cores from a well are shown in Table 2.34. Assume that the cross-sectional area of each core is the same, and porosity and permeability are homogeneous in the direction parallel to the bedding planes.
    Evaluate:
    (a) Arithmetic means of the porosity and permeability of the whole coring section.
    (b) Weighted averages of the porosity and permeability of the whole section.
    (c) Equivalent porosity of whole section according to equal storing capacity.
    (d) Equivalent permeability of whole section based on the principle of equivalent filtrational resistance.
11. The average pore radius of a reservoir rock, from pore-size distribution curve, is 1×10^{-4} cm. The porosity of the rock is 10 %. Estimate the permeability of the rock. The pore tortuosity of the rock is given: $\tau = 1$.
12. A reservoir contains clay minerals which include illite/smectite mixed-layer, chlorite/smectite mixed-layer minerals, illite and dilatable chlorite, etc. Due to the potential damage, water sensitivity analysis is

Table 2.35 Results of water sensitivity experiments

No. of cores	K_L (mD)	K_W (mD)
1	251.0	191.0
2	100.0	55.5
3	3.79	1.01

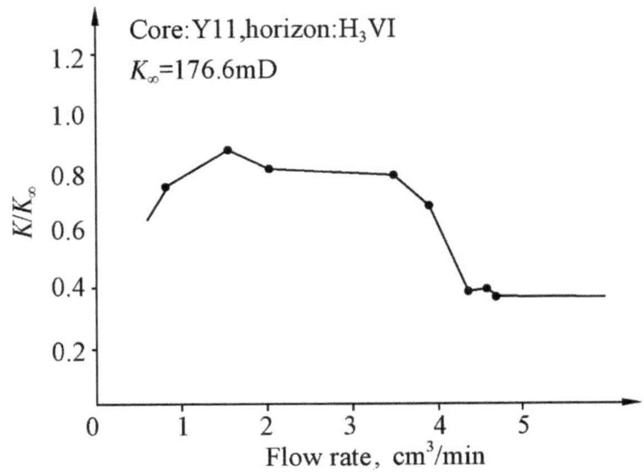

Fig. 2.89 Experimental result of velocity sensitivity of a core

carried out with three cores of the reservoir. The analysis results are shown in Table 2.35. K_L is the permeability measured using formation water while K_w is measured using pure water. Calculate water sensitivity index of each core and evaluate the water sensitivity degree for each core.

13. The velocity sensitivity experiment of a core from a formation is down with kerosene. The K/K_∞—Q curve, platted according to the experimental result, is shown in Fig. 2.89.

Determine:

1. The critical velocity of the core from the curve in Fig. 2.89.
2. What conclusion can you draw out from the experimental result in Fig. 2.89.

References

1. Daniel N, Lapedes (1978) McGraw-Hill encyclopedia of the geological sciences. Graw-Hill
2. Cecil GL(1949) Principles of petroleum geology. In: The century earth science series. Appleton-Century-Crofts, Inc., New York
3. Knut B (2010) Petroleum geoscience: from sedimentary environments to rock physics. Springer, London
4. Zhu XM (2008) Sedimentary petrography. Petroleum Industry Press, Beijing
5. http://www.yzrmjx.net/NewsDetail.asp?id=23

6. http://www.azom.com/article.aspx?ArticleID=1417
7. Paolo N, Kornelia S (2006) Nucleic acids and proteins in soil. Springer, Berlin
8. Ruben DK, Nikolaĭ VP, Billy LE (2000) Coastal processes in tideless seas. ASCE Publications
9. Sam B Jr (2009) Petrology of sedimentary rocks, 2nd edn. Cambridge university press
10. Folk RL, Ward WC (1957) Brazos river bar: a study in the significance of grain size parameters. J Sediment Petrol 27:3–26
11. Folk RL (1966) A review of grain-size parameters. Sediment 6:73–93
12. Djebbar T, Erle CD (2012) Petrophysics: theory and practice of measuring reservoir rock and fluid transport properties. Gulf Professional Publishing, Waltham
13. Oleg DN, Stanislav N (2009) Handbook of non-ferrous metal powders: technologies and applications. Elsevier, Oxford
14. Carman PC (1938) The determination of the specific surface of powders. J Soc Chem Ind Trans Commun Lond 57(225)
15. Evangelos T, Arun SM (2011) Modern drying technology, experimental techniques. Wiley-VCH, Darmstadt
16. Brunauer S, Emmett PH et al (1938) Adsorption of gases in multimolecular layers. J Am Chem Soc 60:309–319
17. Jorden JR, Campbell FL(1984) Well logging I—rock properties, borehole environment, mud and temperature logging SPE of AIME
18. Richard CS (2000) Applied sedimentology. Academic, San Diego
19. Gary N (2009) Sedimentology and stratigraphy. Wiley-Blackwell, Oxford
20. Erik F (2004) Microfacies of carbonate rocks: analysis, interpretation and application. Springer
21. He GS (1994) Reservoir physics. Petroleum Industry Press, Beijing
22. Jacob B (1988) Dynamics of fluids in porous media. Courier Dover Publications
23. SY/T 6285-2011 (2011) Evaluating methods of oil and gas reservoirs (China)
24. Arville IL, Frederick FB (2001) Geology of petroleum. AAPG Foundation, 2001-6-1
25. Collins RE (1961) Flow of fluids through porous materials. Reinhold Publishing Corporation, New York
26. Society of Petroleum Engineers (2004) SPE reservoir evaluation & engineering
27. Richard A, Schatzinger, John FJ (1999) Reservoir characterization: recent advances. AAPG (71)
28. Ahr WM (2008) Geology of carbonate reservoirs: the identification, description and characterization of hydrocarbon reservoirs in carbonate rocks. Wiley, New Jersey
29. Epstein N (1989) On tortuosity and the tortuosity factor in flow and diffusion through porous media. Chem Eng Sci 44:777–779
30. Christopher H, William DH (2002) Water transport in brick, stone and concrete. Spon Press, London
31. Paul G (2001) Petrophysics MSc Course Notes.http://www2.ggl.ulaval.ca/personnel/paglover/CD%20Contents/GGL-66565%20Petrophysics%20English/Chapter%202.PDF; http://www2.ggl.ulaval.ca/personnel/paglover/CD%20Contents/GGL-66565%20Petrophysics%20English/Chapter%203.PDF
32. Carlson MR (2003) Practical reservoir simulation: using, assessing, and developing results. Penn Well Books, Tulsa
33. Bear J (1972) Dynamics of fluids in porous media. Dover Publications, New York
34. Yang SL et al (2004) Reservoir physics. Petroleum Industry Press, Beijing
35. Michael K (2006) Aquifer test data: analysis and evaluation. Water Resources Publication
36. Tucker ME (2011) Sedimentary rocks in the field: a practical guide. Wiley-Blackwell, Oxford
37. SY/T 5336-2006 (2006) Practices for core analysis (China)
38. He GS, Tang H (2011) Reservoir physics. Petroleum Industry Press, Beijing
39. Tarek A (2010) Reservoir engineering handbook. Gulf Professional Publishing, Burlington
40. Abdus S, Ghulam MI, James LB (2007) Practical enhanced reservoir engineering: assisted with simulation software. Penn Well Books, Tulsa

41. Pierre D (2007) Essentials of reservoir engineering. Editions Technip, Paris
42. Monicard RP (1980) Properties of reservoir rocks: core analysis. Editions Tecnip, Paris
43. Corelab (1983) Fundamentals of core analysis. Houston, TX
44. Engler T (2010) Saturation. http://infohost.nmt.edu/~petro/faculty/Engler524/PET524-3a-saturation
45. Helander DP (1983) Fundamentals of formation evaluation. OGCI Publications, Tulsa
46. Morris M (1949) Physical principles of oil production. McGraw-Hill Book Co
47. Henry D, Henry PG (2003) Darcy and other pioneers in hydraulics: contributions in celebration of the 200th birthday of Henry Philibert Gaspard Darcy. Pa. ASCE Publications, Philadelphia
48. Nalco Chemical Company (1979) The NALCO water handbook. McGraw-Hill
49. Klinkenberg LJ (1941) The permeability of porous media to liquids and gases. Drilling and production practice. American Petroleum Inst, pp 200–213
50. Torsæter O, Abtahi M (2003) Experimental reservoir engineering laboratory work book. http://faculty.ksu.edu.sa/shokir/PGE463/Textbook%20and%20References/Exerimental%20Reservoir%20Engineering%20Laboratory%20Workbook.pdf
51. Beard DC, Weyl PK (1973) Influence of texture on porosity and permeability of unconsolidated sand. AAPG Bulletin J 57(2):348–369
52. Faruk C (2011) Reservoir formation damage. Gulf Professional Publishing, Burlington
53. Amethyst G (2007) The clay mineral group
54. Collins DR, Catlow CA (1992) Computer simulation of structures and cohesive properties of micas. Am Mineral 77:1172–1181. http://www.swac.umn.edu/classes/soil2125/doc/s11ch1.htm
55. Carroll D (1959) Ion exchange in clays and other minerals. Geol Soc Am Bull 70:749–779
56. Ladd CC (1960) Mechanisms of swelling by compacted clay. In: Highway research board bulletin, national research council, vol 245. Washington, DC, pp 10–26
57. Joerg R, Tobias B, Reiner D, Siegfried S (2011) Moisture expansion as a deterioration factor for sandstone used in buildings. Environ Earth Sci J63:1545–1564
58. Shen P (1995) The experimental technique of reservoir physics. Petroleum Industry Press, Beijing
59. SY/T 5358-2010 (2010) Formation damage evolution by flow test (China)

Chapter 3
The Physical Properties of Reservoir Fluids

Xuetao Hu

Reservoir fluids refer to those fluids held in reservoir rocks under the conditions of high temperature and high pressure. Generally, reservoir fluids fall into three broad categories: (i) aqueous solutions with dissolved salts, (ii) liquid hydrocarbons, and (iii) gases (hydrocarbon and nonhydrocarbon). In all cases their compositions depend upon their source, history, and present thermodynamic conditions. Their distribution within a given reservoir depends upon the thermodynamic conditions of the reservoir as well as the petrophysical properties of the rocks and the physical and chemical properties of the fluids themselves.

However, the properties of reservoir fluids are constantly changing due to a continuous decline in the temperature and pressure of reservoir fluids when these fluids are brought from reservoirs to surface. Moreover, these changes in the properties of fluids will affect further the output of oil and gas. Therefore, the knowledge of the properties of reservoir fluids and their change is necessary for one to determine and optimize the process parameters of oil and gas production. This chapter will briefly introduce the properties of these reservoir fluids and the property change in oil and gas production.

X. Hu (✉)
School of Oil & Natural Gas Engineering, Southwest Petroleum University, Chengdu, Sichuan, China
e-mail: huxt@swpu.edu.cn

X. Hu et al. (eds.), *Physics of Petroleum Reservoirs*, Springer Geophysics,
DOI 10.1007/978-3-662-55026-7_3

3.1 Introduction of Petroleum

3.1.1 Concepts of Petroleum

3.1.1.1 What Is Petroleum?

At the 11th World Petroleum Congress in 1983, the definition of petroleum is finally determined by three organizations: API, SPE, and AAPG.

Petroleum is a mixture which is mainly composed of hydrocarbon compounds and exists naturally in gas, liquid, and solid in nature. It includes crude oil and natural gas. This is a broad definition of petroleum. Narrowly, petroleum is a flammable liquid or solid mineral mainly composed of a variety of hydrocarbons. Crude oil is the unprocessed liquid petroleum which is recovered from oil wells.

The following definitions come from English dictionaries and encyclopedia.

Petroleum is a thick, flammable, yellow-to-black mixture of gaseous, liquid, and solid hydrocarbons that occurs naturally beneath the Earth's surface, can be separated into fractions including natural gas, gasoline, naphtha, kerosene, fuel and lubricating oils, paraffin wax, and asphalt and is used as raw material for a wide variety of derivative products—*Form the American Heritage® Dictionary of the English Language, Fourth Edition copyright ©2000 by Houghton Mifflin Company. Updated in 2009. Published by Houghton Mifflin Company.*

Petroleum is a dark-colored thick flammable crude oil occurring in sedimentary rocks around the Persian Gulf, in parts of North and South America, and below the North Sea, consisting mainly of hydrocarbons. Fractional distillation separates the crude oil into petrol, paraffin, diesel oil, lubricating oil, etc. Fuel oil, paraffin wax, asphalt, and carbon black are extracted from the residue—*Form Collins English Dictionary—Complete and Unabridged © HarperCollins Publishers, 2003.*

Petroleum [L. petroleum, from Greek: petra (rock) + Latin: oleum (oil)] or crude oil is a naturally occurring, flammable liquid consisting of a complex mixture of hydrocarbons of various molecular weights and other liquid organic compounds, that are found in geologic formations beneath the Earth's surface. Petroleum is recovered mostly through oil drilling. It is refined and separated, most easily by boiling point, into a large number of consumer products, from gasoline and kerosene to asphalt and chemical reagents used to make plastics and pharmaceuticals—*From Wikipedia, the online free encyclopedia.*

3.1.1.2 Concept of Natural Gas

In a narrow sense, natural gas is a combustible gas that is buried in deep ground and mainly composed of hydrocarbon compounds. Besides, there are also some definitions of natural gas from English dictionaries listing as follows:

Natural gas is a mixture of hydrocarbon gases that occurs with petroleum deposits, principally methane together with varying quantities of ethane, propane, butane, and other gases, and is used as a fuel and in the manufacture of organic compounds—*Form The American Heritage® Dictionary of the English Language, Fourth Edition copyright ©2000 by Houghton Mifflin Company. Updated in 2009. Published by Houghton Mifflin Company.*

Natural gas is a gaseous mixture consisting mainly of methane trapped below ground; used extensively as a fuel—*Collins English Dictionary—Complete and Unabridged © HarperCollins Publishers, 2003.*

Broadly speaking, natural gas refers to all gases naturally generated in nature. In fact, not all natural gases from reservoirs are hydrocarbon gases. Some are exactly the nonhydrocarbon gases of high purity, such as N_2, CO_2, etc. For example, the content of H_2S in a natural gas, which is from the fourth member of Shahejie and Kongdian formations in Paleogene in Zhao Lanzhuang structure of Jizhong Depression in North China, is up to 92 %. In Sha touping gas field of Sanshui Basin in Guangdong, China, and the content of CO_2 is up to 99.75 %.

3.1.2 Petroleum Compositions

3.1.2.1 Elements Naturally Occurring in Petroleum

From the chemical element point of view, there are only five kinds of basic elements naturally existing in petroleum compounds. They are C, H, O, N, and S elements. These elements combined with other microelement constitute the various hydrocarbon and nonhydrocarbon compounds of petroleum.

The carbon and hydrogen contents in some crude oils from different places in the world are listed in Table 3.1. It can be noticed that the variation in the contents of carbon and hydrogen stays in a narrow range despite these crude oils came from different places and have different properties. Among these oils, the carbon content only ranges between 83 and 87 % (mass percentage), and the hydrogen content also ranges from 11 to 14 % (mass percentage). Take both elements into consideration, the total contents vary in the range of 95–99 % in general.

The sulfur content of most crude oils from China's oil fields is very low. For example, the crude oils from Daqing and Dagang oil fields contain only 0.12 % sulfur. Even if the oil from Gudao oil field, which contains much more sulfur, still does not fall into the category of high-sulfur oil around the world (Table 3.2).

Conversely, the crude oils from China's oil fields are on the high side of the nitrogen content. The statistical results show that there are only three oil samples whose nitrogen content is higher than 0.3 % in 210 oil samples from oversea oil fields; but most of crude oils from China contain nitrogen higher than 0.3 % [1].

In addition, crude oil also contains some microelements of metal and nonmetal, such as Cu, V, Ni, Mg, Cl, B, P, etc. Among these elements, vanadium and nickel are the most important. The two elements may be 50–70 % of the total microelement in crude oil.

Table 3.1 Carbon and hydrogen composition of crude oils [1]

Production place of crude oil	Element content (mass percentage, %)			H/C atomic ratio
	C	H	C + H	
Daqing, China (crude)	85.7	13.3	99.0	1.86
Shengli, China (mixed oil)	86.1	12.2	98.3	1.68
Kelamayi, China (crude)	86.1	13.3	99.4	1.85
Dagang, China (mixed oil)	85.7	13.4	99.1	1.88
Canada	83.4	10.4	93.8	1.60
Mexico	84.2	11.4	95.6	1.62
Iran	85.4	12.8	98.2	1.80
Colombia	83.6	11.9	95.5	1.67
Roumania	87.2	11.3	98.5	1.56
Russia	83.9	12.3	96.2	1.76

Table 3.2 Element content of sulfur and nitrogen of crude oils [1]

Production place of crude oil	Element content (mass percentage, %)	
	S	N
Daqing, China	0.12	0.13
Shengli, China	0.80	0.41
Gudao, China	1.8–2.0	0.50
Dagang, China	0.12	0.23
Renqiu, China	0.30	0.38
Jianghan, China	1.83	0.30
The highest content in oils around the world	5.50	0.77
The lowest content in oils around the world	0.02	0.02

3.1.2.2 Compounds Naturally Occurring in Petroleum

From the results of component analysis, it can be seen that petroleum consists of two essential kinds of compounds: hydrocarbon and nonhydrocarbon. Hydrocarbon compounds chiefly include paraffin hydrocarbons, naphthenic hydrocarbons, and aromatic hydrocarbons (arene). Among hydrocarbons, paraffin hydrocarbons are the predominant in petroleum components. The nonhydrocarbon includes gum and asphaltene which are composed of O, N, S compounds. In general, the heavy components of crude oil are mainly constituted by nonhydrocarbon compounds.

Natural gas is basically composed of the gasiform light hydrocarbons of C_1–C_4; whereas crude oil mainly consists of the liquid hydrocarbons of C_5–C_{16}. The paraffin hydrocarbons, with carbon atoms higher than C_{17}, are usually solid or semisolid. They are the major components of paraffine and asphaltene.

Hydrocarbon Compounds

The common hydrocarbon compounds present in crude oil are listed as follows:

Paraffin hydrocarbons, also called saturated hydrocarbons
Paraffin hydrocarbons have the general formula of C_nH_{2n+2} and are chemically stable. This series includes both the straight-chain hydrocarbons, such as butane, and the branched-chain hydrocarbons, such as isobutene. They are found in natural gas, in liquid oil and in solid petroleum waxes. The values of n range from 1 in methane gas (CH_4) to 70 in the solid heptacontane ($C_{70}H_{142}$). The number of possible isomers increases rapidly with the carbon number of hydrocarbons. In general, methane, ethane (C_2H_5), propane (C_3H_8), and butane (C_4H_{10}) are gases at ambient temperature and atmosphere pressures, while the paraffins from pentane (C_5H_{12}) to pentadecanes ($C_{15}H_{32}$) are liquids, and the higher members of hydrocarbon are solids. These paraffins are found in almost all crudes, particularly in the volatile fractions.

Naphthenic hydrocarbons
These compounds have a ring or a cyclic structure without double bonds. They are similar in properties to paraffins and chemically stable. They have the same molecular formulae with the olefin series (C_nH_{2n}), if they have the same carbon numbers. Olefin series are unsaturated open-chain hydrocarbons, only occurring in cracked products. An example of naphthene is cyclopentane. Cyclopentanes are the majority of naphthenic-base crudes, typical examples of which are most mid-continent crudes [2]. The naphthenes are a kind of saturated hydrocarbon. Saturated hydrocarbons form a very important proportion of many crudes. They particularly occur in large quantities of the Russian Caucasian oils.

Aromatic hydrocarbons
These compounds contain double bonds and are chemically active. In particular they are easily oxidized. Their general formula is C_nH_{2n-6}. Aromatic hydrocarbons are important but usually account for less than 10 % of the crude oil volume. Aromatic hydrocarbons are found mostly in asphaltic- and mixed-base crudes. Truly aromatic-base crudes are rare.

A crude oil composed mostly of saturated hydrocarbons is termed as "*paraffin-base*" crude [2]. Crudes containing dominant proportions of naphthenic hydrocarbons leave a residuum of complex asphalts on distillation, and hence are termed as "*asphalt-base*" or "*naphthenic-base*" crude [2]. Crudes between the above-mentioned are called *mixed-base crudes*. Generally, paraffin-base crude oils are easy to burn because they have more paraffin hydrocarbons compared with other crudes; and are the most valuable since a large percentage of high-grade lubricating oils can be yielded from them.

Therefore, crude oil is generally referred to as being either "paraffin base", when it contains large proportions of paraffins, or "naphthenic" or "asphalt base" when the naphthenes are the dominant group.

At present, the crudes produced from China's oil fields are chiefly low-sulfur and paraffin-base crude, for example, the crude oils from Daqing oil field often fall into this category. Daqing's crudes are generally low-sulfur, high-wax, and high pour point, and can produce high-quality kerosene, diesel oil, solvent oil, lubricating oil, and paraffin-grade wax. However, the crude oils from Shengli oil field have high-resin content (29 %) and specific gravity (0.91 or so). It is usually sulfur-bearing, mixed-base crudes. Its gasoline fraction has good lead susceptibility. Shengli crudes, rich in paraffin and aromatic hydrocarbons, are thus a good raw material of reforming.

Nonhydrocarbon Compounds

Nitrogen, oxygen, and sulfur elements commonly exist in crude oils in the form of nonhydrocarbon compounds, which are composed of sulfur, oxygen, or/and nitrogen atoms and have very complex ring structure. Nonhydrocarbon compounds constitute the majority of macromolecular compounds in petroleum. Nonhydrocarbon compounds comprise a class of compounds, generally called resin and asphaltene. Asphaltene mainly includes phenols, fatty acids, ketones, esters, and porphyrins. Resin chiefly includes pyridines, quinolines, carbazoles, sulfoxides, and amides. Asphaltene is a kind of macromolecule dispersed in oil in colloidal forms, whereas resin is a kind of amorphous solid dissolved truly in crudes.

In resin-asphaltene, there are polar molecules containing N, O, and S elements. These polar molecules are of higher interfacial activity, and thus have great influence on many properties of crude oil, such as color, density, viscosity, interfacial tension, and so on.

It should be emphasized that there is a wide range of petrochemistry in each petroliferous area. Even in one oil field, considerable changes may occur in the crudes produced from zone to zone.

3.1.2.3 Contents of Paraffin, Resin, and Asphaltene in Crude Oil

Wax Content

Wax content or paraffin content refers to the percentage of paraffin wax and ozocerite in a crude oil at room temperature and pressure. Paraffin wax is a white or light yellow solid, composed of higher members of the paraffin alkane series (C_{16}–C_{35}). Its molecular weight is in the range of 300–450, and melting point is in the range of 37–76 °C. Ozocerite, more complex structure, is a high-boiling point crystalline hydrocarbon of 36–55 carbon atoms. Its molecular weight is about 500–730, and melting point is about 60–90 °C. Paraffin wax and ozocerite dissolve in crude oil underground in the form of colloid. When pressure and temperature decrease, they will separate out from oil.

Resin Content

Resin content means the mass percentage of resins contained in a crude oil, usually ranges from 5 to 20 %.

Resins (or gum) are a mixture of polycyclic aromatic hydrocarbons which have huge molecular weights, e.g., about 300–1000, and contain oxygen, nitrogen, sulfur, etc., elements in crude oils. Resins usually exist in crude oil either in the form of viscous liquid or in semisolid. Resins are easy to dissolve in some organic solvents, such as petroleum ether, lubricating oil, gasoline, chloroform, and so on.

Asphalt Content

Asphaltene is a black solid mixture, with huge molecular weight (generally higher than 1000) and aggregate structure. Asphaltene in petroleum does not dissolve either in chloroform or in carbon bisulfide, but easily dissolve in benzene. Asphalt content is lower in crude oils, usually less than 1 %. As a rule, the increase in asphalt content of crude oils leads to obvious decrease in oil quality.

Despite both are nonhydrocarbon mixture, resin and asphaltene are very different in their composition. Usually, they are different in their molecular weights. However, the more notable differences between them are: (a) resin/gum is a viscous liquid while asphaltene is an amorphous, brittle solid; (b) resin is dissoluble in alkanes of low carbon number while asphaltene is not.

The color of crude oils is various. It may be brown, black-brown, light tan, yellow, light straw, or water white, depending on the contents of lighter hydrocarbon, heavier hydrocarbon, and resin-asphaltene in the oil. In general, the more heavier the hydrocarbon and nonhydrocarbon is in the oil, the darker the color of the oil is, and the higher the density and viscosity of the oil are. Higher contents of nonhydrocarbon in oils make undoubtedly the quality of oils worse.

3.1.2.4 Composition of Natural Gases

Natural gas exists in nature under pressure in reservoir rocks in the Earth's crust, either dissolved in heavier hydrocarbons and water or by itself. It is produced from the reservoir similarly to or in conjunction with crude oil [3].

Natural gas consists principally of gaseous hydrocarbons. The predominant component of a natural gas is methane (CH_4), which typically ranges from 60 to 90 %, followed by decreasing amount of ethane (C_2H_6), propane (C_3H_8), butane (C_4H_{10}), and heavier hydrocarbons which may be present as gases or vapors. The gaseous impurities are chiefly nitrogen, carbon dioxide, and hydrogen sulfide (H_2S), and traces of hydrogen, oxygen, and carbon monoxide (CO) may be present.

The composition of natural gases can vary widely. Tables 3.3 and 3.4 show the typical compositions of natural gases before they are refined. Theses natural gases came from great oil fields and gas fields in China. From Tables 3.3 and 3.4, it can

Table 3.3 The composition of associated gases from great oil fields in China (volume, %)

Oil field	CH_4	C_2H_6	C_3H_8	nC_4H_{10}	iC_4H_{10}	nC_5H_{12}	iC_5H_{12}	C_6	C_7^+	CO_2	N_2	H_2S
Daqing	85.88	3.34	4.54	0.67	1.99	0.35	0.81	0.36	0.16	0.9	1.0	–
Liaohe	87.53	6.2	2.74	0.62	1.22	0.36	0.3	0.21	0.46	0.03	0.33	–
Zhongyuan	75.3	10.17	6.18	1.45	2.6	0.98	0.75	0.65	0.52	0.34	0.43	0.0003
Shengli	87.75	3.78	3.74	0.81	3.31	0.82	0.65	0.06	0.03	0.53	0.02	–
Qiuling inTuha	67.61	13.51	10.69	3.06	2.55	0.68	0.56	0.16	0.09	0.40	0.65	–

Table 3.4 The composition of natural gases from great gas fields in China (volume, %)

Gas field	CH_4	C_2H_6	C_3H_8	nC_4H_{10}	iC_4H_{10}	nC_5H_{12}	iC_5H_{12}	C_6	C_7^+	CO_2	N_2	H_2S
Changqing (Ma 5 member)	95.6	0.6	0.08	0.02	0.01	0.01	0.03	–	–	3.02	0.04	0.0264
Tainan	99.2	–	0.02	–	–	–	–	–	–	–	0.79	–
Pinghu (condensate gas)	77.76	9.74	3.85	1.14	1.19	0.27	0.44	0.34	2.61	1.39	1.27	–
Kekeya (condensate gas)	82.69	8.14	2.47	0.38	0.84	0.15	0.32	0.2	0.14	0.26	4.44	–
South in Sichuan (Weiyuan)	86.47	0.11	–	–	–	–	–	–	–	4.437	8.10	0.879
Chongqing (Wolonghe)	64.91	0.35	0.09	0.047	0.046	0.017	0.016			1.65	0.69	31.95

be seen that natural gases, especially the associated gases, contain a small proportion of $C_5{}^+$ hydrocarbons. When separated, this fraction is light gasoline.

Sometimes, natural gas is considered "dry" when it is almost pure methane, having had most of the other commonly associated hydrocarbons removed. When other hydrocarbons are present, the natural gas is "wet" [3].

The composition of natural gases lines on the field and reservoir from which it is extracted. Since the composition of natural gas is never constant, a standard test method, by which the composition of natural gas can be determined and thus prepared for use, is necessary. The composition of a natural gas can commonly be determined by a gas chromatograph. The result may be expressed either in mole fraction, in volume fraction, or in mass fraction.

Mole fraction, the most commonly used, is defined as

$$y_i = \frac{n_i}{\sum_{i=1}^{N} n_i}, \quad \sum_{i=1}^{N} y_i = 1, \tag{3.1}$$

where n_i is the mole number of ith component of a natural gas; y_i is the mole fraction of ith component, f; N is the component number of the natural gas.

Volume fraction is expressed as

$$y_i = \frac{V_i}{\sum_{i=1}^{N} V_i}, \quad \sum_{i=1}^{N} y_i = 1, \tag{3.2}$$

where V_i is the volume of ith component of a natural gas, m^3; y_i is the volume fraction of ith component, f; N is the component number of the natural gas.

Under low pressure, such as atmospheric pressure, natural gas can be regarded as ideal gas. Thereby, in terms of the state equation for ideal gas, the volume fraction of a natural gas is just equal to its mole fraction under the conditions of ambient temperature and atmospheric pressure.

Mass fraction can be calculated by the following expression:

$$G_i = \frac{m_i}{\sum_{i=1}^{N} m_i}, \quad \sum_{i=1}^{N} G_i = 1, \tag{3.3}$$

where m_i is the mass of any component i of a natural gas, kg; G_i is the mass fraction of component i, f; N is the number of components of the natural gas.

According to the definition of mole number, the mass fraction of a gas can be converted to the mole fraction by the following equation:

$$y_i = \frac{G_i/M_i}{\sum_{i=1}^{N} G_i/M_i}, \tag{3.4}$$

Table 3.5 Mass fraction and mole fractions of a natural gas

Component	Mass fraction, G_i	Molecular weight, M_i	Mole number, n_i	Mole fraction, y_i
Methane	0.71	16.0	0.044	0.85
Ethane	0.14	30.1	0.005	0.09
Propane	0.09	44.1	0.002	0.04
Butane	0.06	58.1	0.001	0.02
Sum	1.00		0.052	1.00

where M_i is the molecular weight of ith component of a natural gas; G_i is the mass fraction of component i, f.

Example 3.1 Determine the mole fraction of a natural gas from its mass fraction given in Table 3.5.

Solution

According the mass fraction of each component in this gas, given in Table 3.5, we have the mole number of each component by the following expression. For example, the number of methane in this gas is

$$n_1 = \frac{G_1}{M_1} = \frac{0.71}{16.0} = 0.0444$$

In the same way, we can determine the mole number of other components. The results are shown in Table 3.5.

Based on the n_i of the gas determined, the mole fraction of the gas can be calculated using Eq. (3.1). For example, the mole fraction of methane in this gas is

$$y_1 = \frac{n_1}{\sum_{i=1}^{N} n_i} = \frac{0.044}{0.044 + 0.005 + 0.002 + 0.001 + 0.052} = 0.85$$

In the same way, we can determine the mole fraction of other components. The results are shown in Table 3.5.

3.1.3 Classification of Crude Oils and Natural Gases

3.1.3.1 Classification of Crude Oil at the Surface

Commercial value of petroleum liquid can be estimated quickly through the measurement of the following physical characteristics: relative density (specific gravity), gasoline and kerosene content, sulfur content, asphalt content, pour point, cloud point, and so on [4]. Therefore, crude oils at the surface are often classified by its physical properties. The common classifications are listed as follows:

Classification by the Relative Density of Dead Oil [1]

Light oil: relative density <0.855, °API > 34;
Conventional oil: relative density: 0.855–0.934, °API: 34-20;
Heavy oil: relative density >0.934, °API < 20.

Classification by the Content of Resin-Asphaltene

In crude oil, resin-asphaltene can form colloidal structure, which has great influence on the flowability and other physical properties of crude oil, and thus leads to high-viscosity crude oil.

Less-resin crude: the content of resin-asphaltene in crude oil less than 8 %;
Resin crude: the content of resin-asphaltene in crude oil ranges from 8 to 25 %;
High-resin crude: the content of resin-asphaltene in crude oil higher than 25 %.

In general, the crude oils produced from most oil fields in China are "less-resin crude" or "resin crude".

Classification by Wax Content

The paraffin content in crude oil often affects its freezing point. Generally, the higher the paraffin content is, the higher the freezing point. High freezing point will cause a lot of troubles during the production and the gathering–transportation of crude oil.

Less-wax oil: the wax content in oil less than 1 %;
Waxy crude oil: the wax content in oil ranges from 1 to 2 %;
High-wax oil: the wax content in oil higher than 2 %.

The wax content of crude oil from oil fields in China changes in wide range. Most crude oils are "high-wax oil" while some crude oils are "less-wax oil".

Classification by Sulfur Content

The sulfur contained in crude oil is harmful. It erodes steel products and is not favorable for oil refining. In addition, the sulfur dioxide generated after oil burning will pollute atmosphere. So, some countries clearly prescribe that crude oil products cannot be sold until the sulfur contained in oil has been removed.

Less-sulfur crude: the sulfur content in oil less than 0.5 %;
Sour crude oil: the sulfur content in oil is between 0.5 and 2 %;
High-sulfur crude: the sulfur content in oil higher than 2.0 %.

Most crude oils produced from oil fields in China are "less-sulfur" crudes.

Classification Based on Oil Composition [5]

This classification is mainly based on the composition of crude oils, namely on the amount of paraffins, naphthenes, and aromatic compounds along with nonhydrocarbon compounds (i.e., resins and aspbaltcncs) in oils.

Paraffin-base crude oil: They contain normal and isoalkanes and less than 1 % S. The asphaltene and resin contents are below 10 %. Percentage of aromatic hydrocarbons is low. They yield gasoline of low octane number and oil of high cetane value. They give high-grade lubricating oils on distillation and leave behind paraffin wax as residue. The oil tends to have high API and pour point and is green in color. Paraffin-base crude oils represent 2 % of the world's oil supply [6].

Naphthene-base crude oil: They contain cycloalkanes (naphthenes). Asphaltic matter is present in quite large proportion. Paraffinic wax is very little or absent. These crude oils yield good quality gasolines and asphalts/bitumens. Asphalt-base crude oil is dominated by the naphthenic hydrocarbon compounds. The oil is black in color and tends to have low °API and a low pour point. Asphalt-base crude oil represents about 15 % of the world's oil supply [6].

Paraffin–naphthene-base crude oils/mixed-base crude oil: They have both paraffinic and naphthenic hydrocarbons together with a certain proportion of aromatic hydrocarbons. On distillation, they yield residue containing both paraffin (waxes) and asphaltic matter, their density and viscosity are more than paraffinic class.

3.1.3.2 Classifications of Formation Oil

Classification by Viscosity

Viscosity is one of the most important properties of in-place oil. Oil viscosity controls the off-take potential of oil wells, and affects the ultimate recovery of oil reservoirs.

Low-viscosity crude: the viscosity of oil in reservoirs lower than 5 mPa s;
Medium-viscosity crude: the viscosity of oil in reservoirs between 5 and 20 mPa s;
High-viscosity crude: the viscosity of oil in reservoirs between 20 and 50 mPa s;
Viscous crude oil: the viscosity of oil in reservoirs higher than 50 mPa s; its relative density is higher than 0.920.

Special Classification

Condensate oil: it means the oil produced from the gas reservoirs, in which liquid oil will condense when reservoir pressure is lower than a special pressure named dew point pressure. This kind of gas reservoir is known as condensate gas reservoirs. The relative density of condensate oils is about 0.75, lower than that of black oils.

Volatile oil: it contains a relatively large fraction of lighter and intermediate components which vaporize easily [7]. They behave liquid-like at reservoir conditions. The reservoir temperature of a volatile oil is near the critical temperature [8]. In production, the liquid oil is easily volatile and shrinking when it is brought to the surface. Generally, its gas–oil ratio is in the range of 210–1200 m^3/m^3. The relative density of the oil is less than 0.825, and its volume factor is higher than 1.72.

High pour point oil: the crude oil whose pour point is higher than 40 °C.

Pour point means the lowest temperature, expressed as a multiple of 5 °F, at which the liquid is observed to flow when cooled under prescribed conditions. Cloud point is the temperature at which paraffin wax begins to solidify and is identified by the onset of turbidity as the temperature is lowered. Both tests qualitatively measure the paraffin content of the liquid [4].

3.1.3.3 Classifications for Natural Gas

Natural gas is often found associated with oil in all oil-producing fields of the world, whereas it is also found in many gas fields in the complete absence of oil. So it may be broadly classified as *associated* when it occurs with oil, and as *nonassociated* when it occurs alone. Distinctly, associated gas coexists with oil in an oil reservoir. But it may be present in the form of solution gas in the oil, or in the form of gas-cap gas, laying above the oil in the reservoir.

Sometimes, natural gas produced from a gas reservoir may contain small amount of heavier hydrocarbons that are separated as a liquid called *condensate oil*. A natural gas containing condensate is said to be *wet* or *rich* gas depending on the amount of heavier hydrocarbons in the gas (Table 3.6). Conversely, a gas is called *dry* or *poor* gas if no liquid hydrocarbon condenses from the gas when it is extracted to the surface (Table 3.6). If the amount of acidic components (e.g., carbon dioxide, hydrogen sulfide) in gas is higher than 1 %, the gas is called *sour*; otherwise called *sweet*.

The common types of natural gases are listed in Table 3.6.

3.1.4 Classifications for Hydrocarbon Reservoirs

Briefly, a petroleum reservoir is a porous and permeable underground formation containing an individual and separate natural accumulation of producible hydrocarbons (oil and/or gas), which is confined by impermeable rock or water barriers and is characterized by a single natural pressure system [9].

Table 3.6 Classification of natural gases

Criterion		Type
Occurrence		Nonassociated gas
		Associated gas
Content of heavier hydrocarbon in natural gases (SC, cm^3/m^3)	$C_3{}^+ < 94$	Poor gas
	$C_3{}^+ > 94$	Rich gas
	$C_5{}^+ < 13.5$	Dry gas
	$C_5{}^+ > 13.5$	Wet gas
Content of sulfur and CO_2 (SC, g/m^3)	<1	Sweet gas
	>1	Sour gas

Naturally occurring hydrocarbons found in petroleum reservoirs are mixtures of organic compounds that exhibit multiphase behavior over wide ranges of pressures and temperatures [10]. These hydrocarbon accumulations may occur in the gaseous state, the liquid state, the solid state, or in combinations of gas, liquid, and solid. These differences in phase behavior of liquids, coupled with the physical properties of reservoir rocks, result in diverse types of hydrocarbon reservoirs with complex behavior [10].

Consequently, there are many classifications for hydrocarbon reservoirs proposed by petroleum geologists and petroleum engineers. Ordinarily, the types of hydrocarbon reservoirs may be classified in terms of the following factors: (i) properties (geologic features) of petroleum reservoirs; (ii) properties and distribution of reservoir fluids; (iii) natural drive types; and so on. We will here focus on the classification based on the properties of hydrocarbon fluids because it is a basement of understanding the complex phase behavior of hydrocarbon systems in reservoir production. Based on the properties of hydrocarbon mixtures, hydrocarbon reservoirs can be broadly classified into five categories (Table 3.7) [11].

Table 3.7 Types of hydrocarbon reservoirs

Type	Feature of hydrocarbon mixtures	Relative γ	Feature underground	Case
Gas reservoir	C_1: dominant C_{2+}: a few		Gas	Dry gas reservoirs Wet gas reservoirs
Condensate gas reservoir	C_1: major C_{7+}: a few	~0.74	Gas underground. Part of gas will condense to oil at lower pressure	Condensate gas reservoirs
Critical hydrocarbon reservoir	$>C_8$	0.7–0.8	No clear interface between oil and gas	Volatile oil reservoirs
Oil reservoir	C_5–C_{16}: main	0.8–0.94	Liquid	Light oil reservoirs Black oil reservoirs
Heavy oil reservoir		0.94–1	Liquid	Viscous crude reservoirs

3.1.4.1 Gas Reservoir

The dominant component in natural gases is methane (CH_4). Others are mainly paraffinic hydrocarbons, such as ethane, propane, and butanes. Natural gas may also contain a small proportion of $C_5{}^+$ hydrocarbons which is liquid fraction when separated at the surface.

Gas reservoirs may produce dry gas or wet gas depending on the content of light hydrocarbon ($C_5{}^+$ hydrocarbons) in the natural gas. Gas reservoirs are thus classified into dry gas reservoir and wet gas reservoir. Their chief characters are described below.

Dry Gas Reservoir

A dry gas reservoir is found initially with hydrocarbons in the gas phase alone, and without liquid hydrocarbons accompanied in the reservoir. During the production of such reservoirs, no liquid hydrocarbons are separated from the gas whether at surface conditions or in the reservoir. Besides conventional gas reservoirs, dry gas reservoirs also include unconventional gas resources, such as coalbed methane (trapped in underground coal deposits), fractured shales containing gas, among others [11].

Wet Gas Reservoir

A wet reservoir contains all hydrocarbons in the gas phase, as in a dry gas reservoir. During the process of pressure decline upon production, the gas remaining in the reservoir would be entirely in a single phase, without any condensation in the subsurface reservoir. However, a small portion of the gas produced through the wells condenses to liquids owing to the decline in pressure and temperature at surface conditions. So, a wet gas reservoir will produce small amounts of liquid oil. This phenomenon is due to the presence of some heavier hydrocarbons in the reservoir gas that condense under surface conditions [11].

3.1.4.2 Condensate Gas Reservoir (Gas Condensate Reservoir)

Condensate gas contains more light hydrocarbons ($C_7{}^+$) than wet gas. All components including heptane (C_{7+}) in condensate gas are initially in the gas phase in the reservoir. Upon depletion of the reservoir, however, a portion of the gas dominated by heavier hydrocarbons may condense to liquids which will deposit in rock pores. This occurs as reservoir pressure declines below the *dew point* of the hydrocarbon fluid. Condensation could be significant near the wellbore due to the relatively large drop in pressure [11]. The phenomenon is referred to as *retrograde*

Table 3.8 Classification of condensate gas reservoirs

Type	Content of condensate oil (g/m³)	Case
Extremely high condensate-content gas reservoir	>600	Kekeya reservoir (N1X4), China Mideng and Wenxi1 reservoirs, China Pinghu4 gas reservoir, China
High condensate-content gas reservoir	250–600	Tazhong1, Yakela gas reservoirs (∈-O), China
Moderate condensate-content gas reservoir	100–250	Qiudong, Wen 1 gas fields in Tuha, China Su1, Su4, Jingzhou 20-2 gas reservoirs in Bohaiwan basin, China
Low condensate-content gas reservoir	50–100	Suinan gas field, Zhongba gas field (T2) in Sichuan, China

condensation since pure substances do not condense, but evaporate under declining pressure. The prevailing reservoir temperature is below the cricondentherm defined as the limit temperature above which a fluid can only exist as gas. So, in condensate gas reservoirs, liquid hydrocarbons will be condensed out from the gas whatever at surface conditions or in reservoirs as long as the pressure is lower than the dew point pressure of the system. Typical characteristics of condensate gas reservoirs are described in the following sections of this chapter.

The liquid hydrocarbons condensed from condensate gas are called condensate oil. Condensate oil is a light-color, low-density, and high-quality crude oil. The relative density (at surface conditions) of condensate oil often ranges between 0.72 and 0.80. According to the standard of petroleum industry in China (GB/T 26979-2011 China) [12], gas condensate reservoirs may be classified into following types by the content of condensate oil in the reservoir (Table 3.8).

3.1.4.3 Oil Reservoir

Hydrocarbons under initial reservoir temperature and pressure are predominatingly liquid hydrocarbons which are totally known as *crude oil*. Crude oil in oil reservoirs is often referred to as light, intermediate, or heavy, which leads to the following classification of oil reservoirs [11].

Volatile Oil Reservoir

Hydrocarbons are initially found in the liquid phase in oil reservoirs. However, an oil reservoir may be discovered with an overlying gas cap when the initial reservoir pressure is lower than the *bubble point* of the crude oil. When the reservoir pressure declines, relatively light-hydrocarbon components can evolve out of the liquid phase in large quantities. Volatile crude is typically light to dark amber color. The API gravity of typical volatile oils is relatively high, 38° or above (relative density 0.83 or lower than 0.83). Crude oil with an API gravity of 45° or greater (relative density 0.80 or lower than 0.80) is referred to as *near-critical oil*, implying the location of the system near the critical point on the *p–T* phase diagram and the abundance of highly volatile hydrocarbons [11]. This kind of reservoir is called near-critical oil reservoir. This kind of reservoir has been discovered in Jilin, Zhongyuan and western region of China, North Sea of United Kingdom, and eastern region of United States.

The definition of bubble point pressure and API gravity is introduced in later section of this chapter.

Black Oil Reservoir

The term "*black oil*" is frequently cited in reservoir simulation studies as various volatile components are collectively considered to be single gas phase due to the relatively low volatility of crude encountered in many reservoirs [11]. It is used here to categorize oil reservoirs that fall in between the volatile and heavy oil reservoirs in terms of the specific gravity of crude. The gravity range of typical crude would be 38°–22.3°API (relative density: 0.83–0.92) (approximately). Abundance of intermediate hydrocarbons is the typical characteristic of black oils. The color of black oils is variable, from green to black.

Heavy Oil Reservoir

Heavy and extra-heavy oil reservoirs exhibit a dominance of heavier hydrocarbons in the composition of crude. According to the standard made at 11th World Petroleum Conference in London, 1983, viscous crude (heavy oil) means the oil having a relative density of 0.934–1.00 under standard conditions at surface. The viscosity of degassed heavy oils at reservoir temperature is about 100 mPa s or greater, even exceeding 10,000 mPa s sometimes. At present, according to API classification, reservoirs with crude's gravity less than 22.3° (relative density higher than 0.92) are considered to be *heavy oil reservoirs* [11].

Heavy oil composition exhibits low hydrogen/carbon ratios and high asphaltene content (Table 3.9). Ultraheavy oil, such as bitumen, tar in oil sands and others, having API gravity in the teens or less (relative density >1) is referred to as *an unconventional resource* [11]. It cannot be produced by conventional methods. A new era is emerging in the petroleum industry in which substantial recovery is expected from unconventional resources in the future based on innovative technology [11].

Table 3.9 Typical compositions and characteristics of reservoir fluids (after [13])

Component	Dry gas	Wet gas	Gas condensate	Near-critical oil	Volatile oil	Black oil
Typical fluid composition in mole (%)						
Hydrocarbons						
Methane	86.12	92.46	73.19	69.44	5&77	34.62
Ethane	5.91	3.18	7.8	7.88	7.57	4.11
Propane	358	1.01	355	4.26	4.09	1.01
i-Butane	1.72	0.28	0.71	089	0.91	0.76
n-Butane		0.24	1.45	2.14	2.09	0.49
i-Pentane	0.5	0.13	0.64	0.91	0.77	0.43
n-Pentane		0.08	0.68	1.13	1.15	0.21
Hexanes		0.14	1.09	1.46	1.75	1.61
Heptanes and higher		0.82	&21	10.04	21.76	56.4
Nonhydrocarbons						
CO_2	0.1	1.41	2.37	1.3	0.93	0.02
N2	2.07	0.25	0.31	0.56	0.21	0.34
Total:	100	100	100	100	100	100
Reservoir fluid characteristics						
Heptane$^+$ ($C_7{}^+$), mol wt		130	184	219	228	274
GOR[a] (scf/stb)	–	105,000	5450	3650	1490	300
Oil gravity[a] (°API)	–	57	49	45	38	24
Relative density				0.80	0.83	0.91
Usual color of liquid phase		Translucent to clear	Straw	Light amber	Light to dark amber	Green to black

[a]Values are provided as a general guideline

3.1.4.4 Tar Sand [9]

Tar sands, also variously called *oil sands* or *bituminous sands*, are loose-to-consolidated sandstone or a porous carbonate rock, impregnated with bitumen [14]. Bitumen is a high boiling asphaltic material with an extremely high viscosity that is immobile under reservoir conditions and vastly different to conventional petroleum [14]. On an international note, the bitumen in tar sand deposits represents a potentially large supply of energy. However, many of the reserves are available only with some difficulty and that optional refinery scenarios will be necessary for conversion of these materials to liquid products because of the substantial differences in character between conventional petroleum and tar sand bitumen (Table 3.10).

Table 3.10 Comparison of the properties of Tar Sand Bitumen (Athabasca) with the properties of conventional crude oil (James and Speight 2011)

Property		Bitumen (Athabasca)	Crude oil
Specific gravity		1.01–1.03	0.85–0.90
API gravity		5.8–8.6	25–35
Viscosity (mPa s)	38 °C/100 °F	750,000	<200
	100 °C/212 °F	11,300	
Pour point (°F)		>50	ca. −20
Elemental analysis (wt%)	Carbon	83	86
	Hydrogen	10.6	13.5
	Nitrogen	0.5	0.2
	Oxygen	0.9	<0.5
	Sulfur	4.9	2
	Ash	0.8	0
	Nickel (ppm)	250	<10.0
	Vanadium (ppm)	100	<10.0
Composition (wt%)	Asphaltenes (pentane)	17	<10.0
	Resins	34	<20.0
	Aromatics	34	>30.0
	Saturates	15	>30.0
Carbon residue (Conradson, wt%)		14	<10.0

Because of the diversity of available information and the continuing attempts to delineate the various world tar sand deposits, it is virtually impossible to present accurate numbers that reflect the extent of the reserves in terms of the barrel unit. Indeed, investigations into the extent of many of the world's deposits are continuing at such a rate that the numbers vary from 1 year to the next. Accordingly, the data quoted here must be recognized as approximate with the potential of being quite different at the time of publication.

The bitumen in tar sand deposits is estimated to be at least, 1.7 trillion barrels (1.7×10^{12} barrels) in the Canadian Athabasca tar sand deposits and 1.8 trillion barrels (1.8×10^{12} barrels) in the Venezuelan Orinoco tar sand deposits, compared to 1.75 trillion barrels (1.3×10^{12} barrels) of conventional oil worldwide, most of it in Saudi Arabia and other Middle Eastern countries. Eighty-one percent of the world's known recoverable bitumen is in the Athabasca tar sands of Alberta, Canada. In addition, bitumen reserves under active development are also included in the official estimate of crude oil reserves for Canada.

In spite of the high estimations of the reserves of bitumen, the two conditions of vital concern for the economic development of tar sand deposits are the concentration of the resource, or the percent bitumen saturation, and its accessibility, usually measured by the overburden thickness. Recovery methods are based either on mining combined with some further processing or operation on the oil sands in situ. The mining methods are applicable to shallow deposits, characterized by an overburden ratio (i.e., overburden depth to thickness of tar sand deposit). For example, indications are that for the Athabasca deposit, no more than 15 % of the in-place deposit is available within current concepts of the economics and technology of open-pit mining; this 10 % portion may be considered as the proven reserves of bitumen in the deposit.

In a word, these categories of reservoirs are based on the relative dominance of light, intermediate, and heavy petroleum fractions, as shown in Table 3.9. From this table, it can be seen that different hydrocarbon reservoirs have different components and different fluid characteristics. As a result, different hydrocarbon systems exhibit different behavior in the process of production. Liquid- and gas-phase behavior exhibited during production by the preceding categories of reservoirs will be discussed as follows.

3.2 Phase Behavior of Hydrocarbon Systems

A substance is solid, liquid, or gas depending basically on the pressure and temperature acting on the substance. As a rule, a decrease in temperature may turn vapor into water, and water into ice. Similarly, a decrease in temperature may alter the phase or state of hydrocarbon mixture. This can happen by not only changing temperature but also pressure. The change in phase states of hydrocarbon mixtures is known as *phase behavior* [15]. What fluids, e.g., oil or oil and gas, will be produced from a reservoir depends on the phase state of hydrocarbon fluids in reservoirs. The phase behavior of hydrocarbon fluids is thus very important in reservoir simulation, reserves evaluation, and so on.

In the process of hydrocarbon accumulation, hydrocarbons may enter a reservoir in the form of gaseous or liquid, which results in complex phase behavior of hydrocarbon systems. The phase behavior of hydrocarbon systems can be characterized on a diagram called phase diagram. A phase diagram of hydrocarbon mixtures shows how the mixtures behave at different conditions. One of the main tasks of petroleum engineers is to investigate the behavior and characteristics of hydrocarbon reservoirs, and thus to determine the course of future development and production that would maximize the profit [10].

The objective of this section is to review the phase behavior of hydrocarbon fluids and illustrate the applications of phase diagram in determining the types of reservoirs and the native hydrocarbon systems.

3.2.1 Basic Knowledge Used in Phase Behavior Investigation

Phase and phase behavior are part of phase equilibrium theory in physical chemistry. Oil–gas systems are typical multicomponent systems, and may also be multiphase systems under specified conditions. The separation and dissolution between oil and gas fall into the category of phase equilibrium in multiphase systems. Before coming to recognize the phase behavior of hydrocarbon mixtures, one should first understand the related terminology in phase equilibrium theory.

3.2.1.1 Terminology

System: it means the researched part of a substance, also known as *substance system*.

Phase: in the physical sciences, a "*phase*" is a region of space (a thermodynamic system), throughout which all physical properties of a material are essentially uniform [16]. Examples of physical properties include density, viscosity, index of refraction, and chemical composition. In short, a phase is a region of material that is chemically uniform, physically distinct, and (often) mechanically separable (from Wikipedia). For instance, the most common is water vapor, liquid water, and ice. There are definite boundaries between water vapor–liquid water, water vapor–ice, and liquid water–ice; thus the vapor, liquid, and solid of water are said to make up a three-phase system. The typical features for phases are as follows:

(a) A distinct boundary or interface occurs between different phases;
(b) Across the interface, the properties of phase are completely different;
(c) A phase can be continuous or discrete in a system;
(d) Multiple components can coexist in one phase.

It should be noted that a distinction is made by some authors between gas and vapor since vapor is defined as any substance in the gaseous state, but a gas is the state of substances under ordinary ambient conditions. In dealing with hydrocarbon mixtures, it is convenient to think of gas and vapor as being synonymous; thus the two words are used interchangeably throughout this chapter.

Phase state: it is a state of systems that are qualitatively different in its characteristics from other of the same system.

Phase transition: it means the change in states of a system owing to the change in the attribute of the system. It is an equilibrium transformation from one phase to another. Common examples are sublimation (solid to vapor), boiling or vaporization (liquid to vapor), and freezing (liquid to solid) [17].

Phase equilibrium: it means a state at which a substance system appears to be at rest; that is to say, no change occurs in phase number or states of the system.

Component: a component means the same kinds of molecules in a system. It is a compound or a substance. For example, methane, ethane, carbon dioxide, and nitrogen in a hydrocarbon mixture are called a component separately.

Pseudocomponent: it indicates a sort of hypothetical component which may include several single components that are similar in their physical properties, i.e., molecular weight, boiling point, and so on. So, a pseudocomponent is not one component, but a group of components; and is characterized by the average

Table 3.11 Compositions and properties of typical hydrocarbon systems [1]

Component or property	Black oil	Light oil	Condensate oil	Dry gas
C_1 (%)	48.83	64.36	87.01	95.85
C_2 (%)	2.75	7.52	4.39	2.67
C_3 (%)	1.93	4.47	2.29	0.34
C_4 (%)	1.60	4.12	1.08	0.52
C_5 (%)	1.15	2.97	0.83	0.08
C_6 (%)	2.59	1.38	0.60	0.12
C_7^+ (%)	41.15	15.13	3.8	0.42
Gas–oil ratio (m^3/m^3)	111	356	3240	18,690
Liquid density underground (g/m^3)	0.8534	0.7792	0.7358	0.7599
Liquid color	Green	Saffron	Light straw	Colorless

properties (average molecular weight, average boiling point, average density, etc.) of the group. For example, common pseudocomponents found that the investigation of petroleum reservoir engineering include $C_{2\text{–}6}$ and C_7^+. Pseudocomponent is primarily used for simplifying the calculation of petroleum reservoir engineering.

Composition: it is broadly defined as a series of amounts in which each represents the proportion of a component in a mixture. In short, it is a proportional constitute of components in a system (mixture) Table 3.11 shows the compositions of typical hydrocarbon mixtures.

3.2.1.2 Phase Diagram

From the viewpoint of classical thermodynamics, an equilibrium state of a system can be described by all attributes of the system, such as temperature, pressure, components, and so on. That is to say, if the attributes of a system is constant, the equilibrium state of the system is unchanged; vice versa.

A phase diagram is a type of chart used to show the conditions at which thermodynamically distinct phases can occur at equilibrium (from Wikipedia). That is to say, phase diagrams exhibit the relationship between the phase states and the attributes (e.g., T, P, n_i, and so on.) of a system. Phase diagram is known as *state diagram of phase equilibrium,* or *state diagram, simply.*

Figure 3.1 is a typical phase diagram of a pure substance. From this diagram, it can be seen that a phase diagram may be a three-dimensional (3D) surface showing the relationship of P–V–T, or 2D lines representing the relationship between any two attributes of pressure (P), specific volume (V), and temperature (T). In addition, a phase diagram may be constituted by three components of a system, named ternary phase diagram (Figs. 3.2, 3.3, and 3.4).

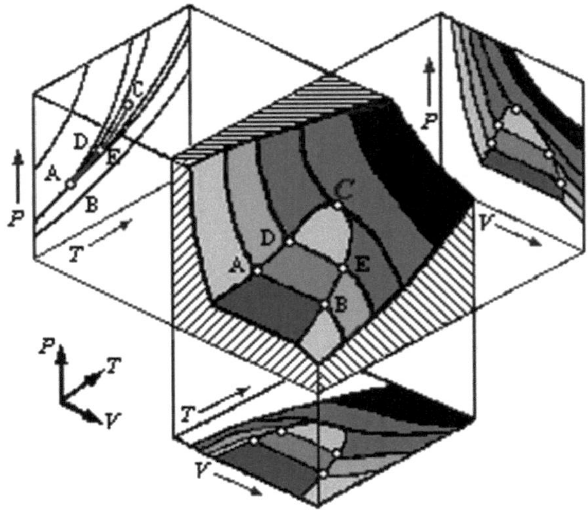

Fig. 3.1 Typical phase diagrams for a pure substance showing p–V–T surface and its projections [18]

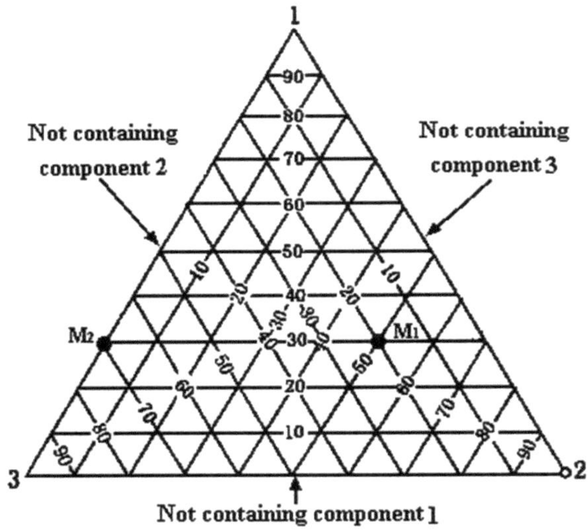

Fig. 3.2 Coordinate system for ternary phase diagram

3D Phase Diagram

According to the attributes of a substance/mixture, we can plot its equilibrium-phase behavior in three-dimensional coordinate system with pressure (p), specific volume (V), and temperature (T) as coordinates (Fig. 3.1). This diagram is called 3D phase diagram or P–V–T diagram. A 3D diagram represents adequately the effects of all attributes to the phase behavior of a system. It shows the phase behavior of a system with a 3D surface, and is thus strongly stereoscopic.

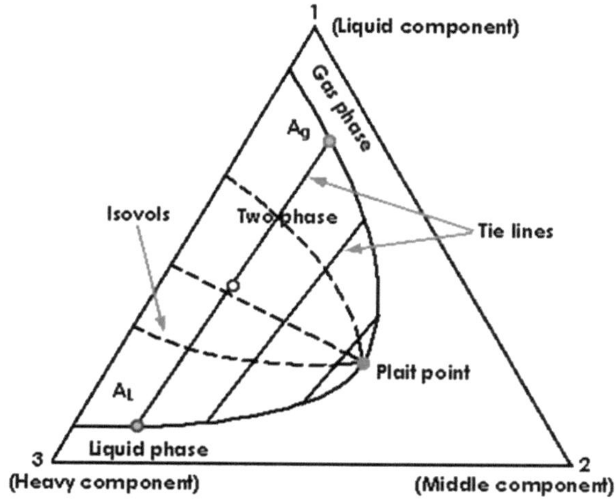

Fig. 3.3 Schematic of a ternary phase diagram for hydrocarbon systems

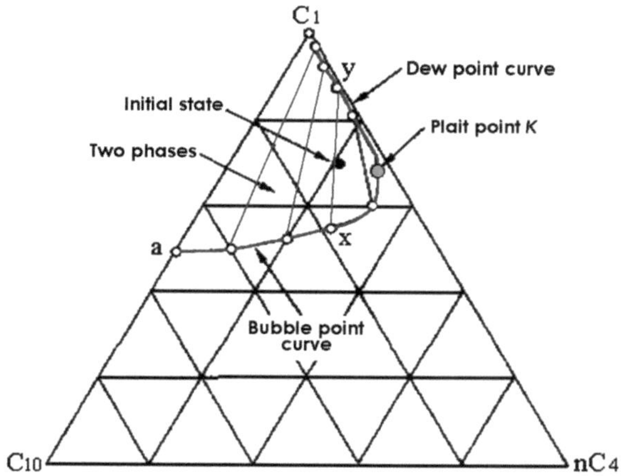

Fig. 3.4 Ternary phase diagram of a mixture consisting of methane, normal butane, and decane [19]

Based on 3D phase diagram, one can have an all-sided understanding of the phase behavior of a system (Fig. 3.1). However, it is difficult to plot a 3D diagram and apply it in practice.

2D Phase Diagram

Orthogonal projections of a *p–V–T* diagram onto the *p–V*, *V–T*, and *p–T* planes give two-dimensional plots (2D diagrams) suited for characterizing the phase behavior in the combinations of any two of the three variables (*p*, *V*, and *T*). Therefore, a 2D phase diagram is essentially a special case of 3D phase diagram when one of the three attributes is constant. Namely, under the conditions of constant temperature, constant pressure, or constant specific volume, a 2D diagram is appropriate for highlighting the phase behavior of a system. So, 2D phase diagram actually includes *pressure–temperature diagram* (*p–T* diagram), *pressure–specific volume diagram* (*p–V* diagram), and *temperature–specific volume diagram* (*T–V* diagram) (Fig. 3.1). Among those diagrams, *p–T* diagram (*pT* diagram) is the most prevalent in the investigation of the phase behavior of hydrocarbon systems (see Fig. 3.8).

Ternary Phase Diagram

A *p–T* diagram is convenient for phase behavior investigation of a system with fixed composition. However, in the process of immiscible flooding in oil reservoirs, displacing fluid is continually injected into the reservoir; and the composition of the hydrocarbon system in the reservoir is thus altered by injecting fluid. In this case, a ternary phase diagram, shown in Fig. 3.3, is more useful for the study on the phase behavior of the system.

In a ternary phase diagram, every apex of triangle denotes a component (Fig. 3.2). At the three apes of triangle, we have 100 % 1, 100 % 2, and 100 % 3 component, respectively. The usual convention is to plot the lightest component at the top (e.g., component 1) and the heaviest component at the lower left (component 3) [4]. Each side of the triangle represents all possible binary combinations of the three components, namely two-component mixtures. For instance, the left side of the triangle represents all possible mixtures of the light and heavy components. On any of the three sides, the fraction of the third component (corresponding to the side) is zero (0 %).

All possible mixtures composed of the three components (1, 2, and 3 components) fall in the diagram of equilateral triangle. That is to say, points within the triangle denote three-component mixtures. Composition is usually plotted according to mole fraction or mole percent of each component. In Fig. 3.2, point M_1 represents a mixture of 20 % component 3, 50 % component 2, and 30 % component 1. Similarly, point M_2 means the mixture consisting of two components: 70 % component 1 and 30 % component 3. For a single ternary diagram, both pressure and temperature are constant; only composition changes.

Figure 3.3 is a schematic of a ternary phase diagram for hydrocarbon systems. This diagram is for a single temperature and a single pressure. Within the triangle, the solid curve denotes the *envelope curve* (*binodal curve*), which is a boundary between the two phase and the single phase. At and inside the envelope curve, two phases exist; otherwise, one phase. The dew point line lies along the top of the saturation envelope, and the bubble point line lies along the bottom. The envelope curve is constituted by the *bubble point curve* and the *dew point curve*, both of which meet at the *plait point* (Fig. 3.4). The plait point denotes the state at which the liquid phase and vapor phase have the same composition (resembles the *critical point* that we will introduce below). In the two-phase region, the straight lines, connecting the bubble point to the dew point, referred to as *tie lines* must be determined experimentally and given on the diagram. Equilibrium tie lines are straight but not horizontal. On a tie line, any mixture (i.e., any point) has completely identical liquid-phase and gas-phase compositions. The composition of any point on a tie line can be given by the coordinate of the point.

Point A represents a mixture of components 1, 2, and 3 which exhibits equilibrium gas and liquid at a given temperature and pressure. Points A_g and A_L represent separately the compositions of gas and liquid phases at the equilibrium state. The quantity of gas, in fraction of total moles of overall mixture, is represented by the length of line A–A_L divided by the length of line A_L–A_g. The quantity of liquid in terms of fraction of total moles of overall mixture is represented by the length of line A–A_g divided by the length of line A_L–A_g. Figure 3.4 shows a typical phase diagram of three-component mixtures of methane, normal butane, and decane.

Ternary phase diagram is very useful in miscible displacement. It will be discussed again in Sect. 4.4 of Chap. 4.

3.2.2 Phase Behavior of Hydrocarbon Systems

Crude oil is a mixture of various hydrocarbons. Its phase behavior is strongly controlled by composition, which leads to the behavior of a hydrocarbon mixture much different from a pure hydrocarbon. For the better understanding of hydrocarbon mixtures, a careful analysis of the phase behavior of a pure substance will be first recalled. After that, the behavior of systems containing two or more independent components will be discussed in detail.

3.2.2.1 Single-Component System

To understand the phase behavior of hydrocarbons, let us first look at a simplistic case: a single-component system. A pure substance is a single-component system; and is the simplest substance system. Its phase diagram is also the simplest. Figure 3.5 shows the p–T phase diagram of a typical single-component system: ethane.

This diagram shows whether ethane exists as a liquid, a vapor (gas), or a combination of both, at different pressures and temperatures. The line GC shown in Fig. 3.5 is known as the vapor pressure curve or boiling point curve [15]. The curve denotes the temperature and pressure conditions at which a liquid substance coexists with its gaseous. Figure 3.5 shows that the vapor pressure of a substance increases with the increase in temperature because of more molecules of the substance escaping from liquid phase to gas phase at higher temperature. The vapor pressure curve divides the major region of the diagram into two areas. One is the area where ethane is a liquid; and another is area in which ethane is gas. Liquid ethane area is above the vapor pressure curve; oppositely, gas area below the curve.

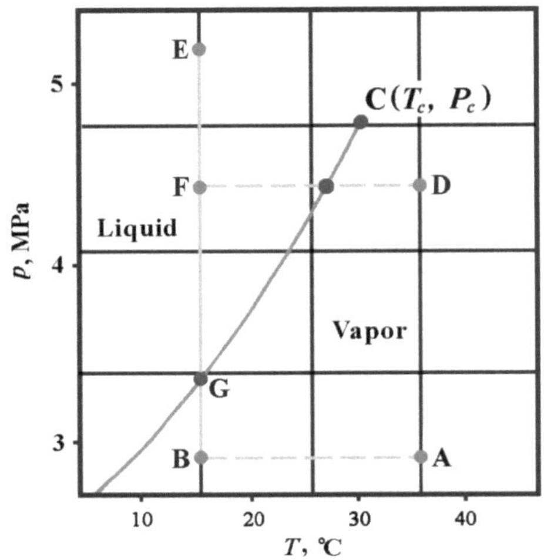

Fig. 3.5 Pressure–temperature diagram of ethane [18]

Only when the temperature–pressure conditions are exactly on the curve, the gas and liquid of ethane equilibrium are coexistent.

The vapor pressure curve terminates at point *C* which is called *critical point*. The temperature and pressure denoted by this point are called the *critical temperature*, expressed by symbol T_c and the *critical pressure*, expressed by P_c. For a pure substance, the critical temperature is defined as the temperature above which a gas cannot be liquefied regardless of the pressure upon the system. Similarly, the critical pressure is the pressure above which vapor and liquid cannot coexist in equilibrium regardless of the temperature. For single-component substances, the liquid and vapor phase are indistinguishable at the critical point. Note that the definitions of critical temperature and pressure are invalid for the systems which have more than one component.

In Fig. 3.5, each temperature–pressure point represents a phase equilibrium state of ethane (phase state). At different temperatures and/or pressures, ethane is in different phase states. As long as any change in the temperature or/and pressure occurs in the system, the phase state of the system is altered. If the change in temperature or/and pressure is across the vapor pressure line (phase boundary), the phase and phase number of the system will be changed: from one single phase to two phases coexisting, and to another single phase, shown as line FB or line FD in Fig. 3.5.

The pressure–volume relation of ethane is shown in Fig. 3.6. The area within the dashed line represents conditions under which gas and liquid phases coexist in equilibrium; the right area outside the dotted line represents the vapor of ethane while the left area outside the line represents the liquid of ethane. Point *C* again designates the critical pressure and temperature (709 psia and 90 °F, respectively). The curve *AC* defines what are generally called bubble point or saturated liquid

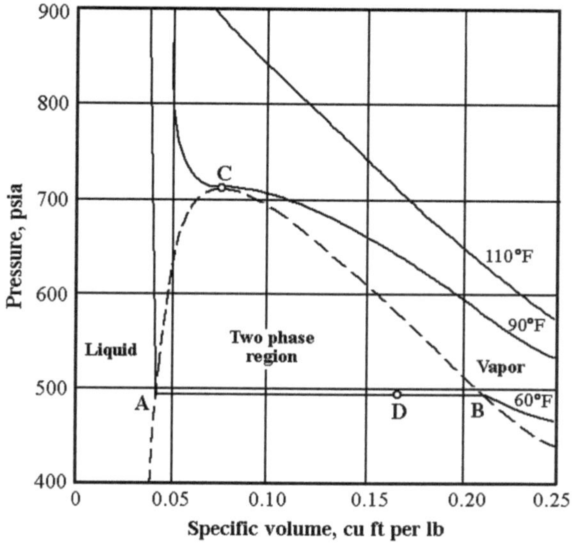

Fig. 3.6 Pressure–volume diagram of ethane showing three isotherms [18]

conditions, and line *CB* designates the properties of ethane at dew point or saturated vapor conditions [20].

To illustrate the use of Figs. 3.5 and 3.6, let us assume that one pound (0.454 kg) of ethane is contained in 0.25 cu ft (0.0071 m^3) of space at 60 °F (15.6 °C) [15]. At the specified temperature and volume conditions, ethane will be in vapor form and exert a pressure of 465 psia (3.21 MPa). This is shown in Fig. 3.6. When the ethane vapor is compressed at a constant temperature of 60 °F, liquid will condense first at 495 psia (3.41 MPa). Point *B*, the state at which the system is entirely in the vapor form except for an infinitesimal amount of equilibrium liquid, is, by definition, the dew point. Further decreases in the volume of the system will result in the formation of larger and larger proportions of liquid in the system, until at point *A* all materials will be in the liquid condition, except for an infinitesimal amount of equilibrium vapor (bubble point). Additional decrease in volume will result in rapid increase in pressure because of the low compressibility of liquid ethane. Since the gas is highly compressible, the slope of the isotherm is much less steep in the gas region than in the liquid region. During the process of moving between points *A* and *B*, the specific volumes of the two phases will remain constant at 0.0423 cu ft per lb (0.00264 m^3/kg) of liquid and 0.2105 cu ft per lb (0.0131 m^3/kg) of vapor.

The line marked 110 °F in Fig. 3.6 illustrates the same process at a temperature above the critical temperature of the substance. The line shows that there is simply an expansion of the substance and that no abrupt phase change occurs. The behavior of other pure substances is similar to that outlined for ethane.

Vapor pressure or equilibrium vapor pressure is the pressure of a vapor in thermodynamic equilibrium with its condensed phases in a closed container. It is

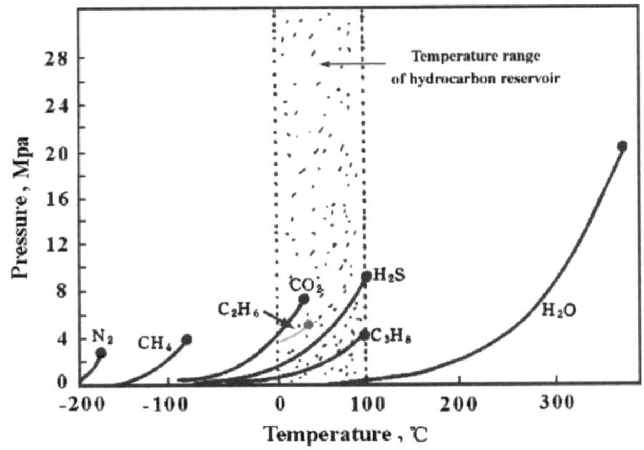

Fig. 3.7 Vapor pressure lines of pure substances [18]

the thermodynamic properties of pure substances. Different pure substances have different thermodynamic properties. Their vapor pressure curves are thus different. Figure 3.7 shows the vapor pressure curves of common fluids found in petroleum reservoirs.

In Fig. 3.7, one can observe that the position of a vapor pressure curve in *p*–*T* diagram is closely related to the nature of a pure substance, such as molecular size, active force between molecules, and so on. As a result, the larger the molecular size is, the stronger the active forces between molecules are, and the more difficult it is for liquid molecules to escape into the vapor phase; and then, the higher the temperatures at given vapor pressures are. In short, the lighter the component is, the lower the temperature range of the vapor pressure curve of the component is (Fig. 3.7).

3.2.2.2 Binary System

Next, let us consider a binary system. A binary system is a mixture. It means there are two independent components in a system. It is also known as a two-component system.

Normally, petroleum engineers do not work with two-component systems in practice; mixtures consisting of many components are usually encountered [4]. But this system is instructive for us to observe the differences of phase behavior between binary systems and single-component systems.

Compared with single-component system, a binary system is just one more component. However, its phase diagram is more complex than a single-component system. Consider a binary mixture, consisting of a light component I and a heavy component II. The typical *p*–*T* phase diagram of the binary mixture is shown in Fig. 3.8. The vapor pressure curves of the two components are also shown in Fig. 3.8.

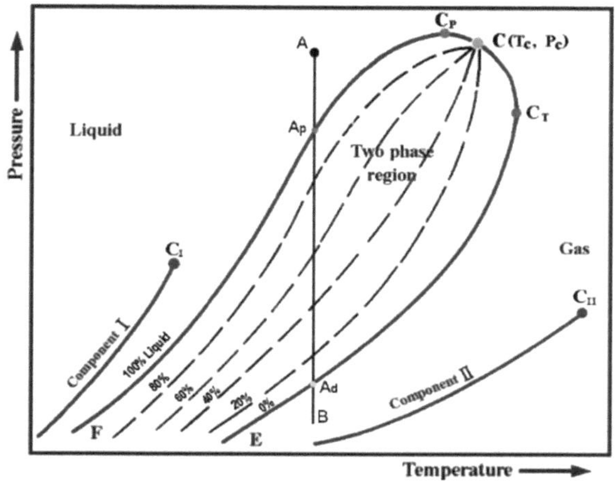

Fig. 3.8 Typical phase diagram of a binary mixture and the vapor pressure lines of the two pure components which constitute the binary mixture

In the diagram of the binary mixture, there is an envelope instead of the two isolated vapor pressure curves. Within the envelope, there is a broad two-phase region, called *saturation envelope, phase envelope*, or *two-phase region* [20]. That is to say, on the envelope and in this region, liquid phase and gas phase of the mixture coexist in equilibrium (Fig. 3.8). On the contrary, outside the envelope, only one phase (gas or liquid) can exist. The envelope is thus called *phase boundary*.

In the *p–T* diagram of the mixture, the envelope consists of two sections which join at the critical point (*C* point). One is located on the left of critical point, called the bubble point curve; another is located on the right of critical point, called the dew point curve.

To understand the concepts of bubble point and dew point, consider a constant temperature expansion illustrated in Fig. 3.8 by line *AB*. At the pressure of point *A*, the mixture is liquid. As pressure is decreased, the liquid expands until the pressure reaches the point $A_{p.}$ Researches have shown that at point A_p, a few molecules are able to escape from the liquid and form a small gas bubble. This point is thus called *bobble point*. The pressure at this point is called *bubble point pressure*, expressed as p_b. Then, the bubble point curve is the *locus of the temperature–pressure points* at which the first gas bubble formed in passing from the liquid to the two-phase region. In short, a bubble point means a temperature–pressure condition at which the first gas bubble appears in a liquid system.

Let the pressure of the system decline unceasingly. When pressure is lower than bubble point pressure, more and more gas appears in the liquid system. Finally, almost all of the liquid evaporates into gas, and only a minute amount of liquid remains in this system at point A_d. This point is called *dew point*. The pressure at this point is known as *dew point pressure*, p_d. Further reduction in pressure to point *B* simply causes an expansion of the gas. Conversely, when reaching the dew point from single gas phase, such as from point *B* to point A_d, the first sign of the second phase will be a small liquid drop formed in gas phase. Therefore, a dew point denotes the temperature and pressure conditions at which the first drop of liquid condensed from a gas system. A dew point curve is the *locus of the temperature–pressure points* at which the first droplet of liquid formed in passing from the gas to the two-phase region.

The dashed lines within the envelope represent constant liquid volume measured as percent of total volume. These lines are known as *isovols* or *quality lines.* Exactly, an isovol is the *loci of points of equal liquid volume percent* within the two-phase region. The numbers on isovols indicate the volume percent of the liquid phase. On each isovol, the liquid volume of each point is equal completely even if the temperature and pressure vary from point to point. For example, the left first dashed line denoted by 80 % in Fig. 3.8 represents that the liquid volume of the system keeps 80 % at every temperature and pressure on the line.

Point C in Fig. 3.8 is the critical point of the binary mixture. The pressure at critical point is known as critical pressure, and the temperature at this point is called critical temperature.

From Fig. 3.8, we can see that liquid and gas can coexist at temperatures and pressures above the critical point in a binary mixture. The definition of critical point as applied to a pure substance does not apply to a mixture. Simply, critical point is the point at which the bubble point line and the dew point line join. Accurately, critical point of a mixture consisting two or more components is the point at which the difference between the nature of the gas and liquid disappears, and no interface occurs between liquid and gas in the system. This definition can also be applied to pure compounds, but is usually neglected due to that simple definition above.

In addition, outside the envelope and near critical point, there is a small region in which it is hard to tell where the phase is a liquid or a gas, and the term of "*supercritical fluid*" is often used to describe the state of fluid in the region.

Therefore, the critical point of a mixture is no longer the point of the highest temperature and highest pressure at which two phases coexist. Distinctly, in the diagram of a binary mixture, the *highest pressure* at which gas and liquid coexist occurs at point C_p, called "*cricondenbar*". It is the *top pressure* on the envelope. In the same way, the highest temperature at which two phases coexist is found at point C_T, called "*cricondentherm*". It is the *top temperature* on the envelope. For a binary mixture, the cricondenbar is the pressure above which no gas can exist regardless of the temperature; and the cricondentherm is the temperature above which liquid cannot be formed regardless of the pressure attained.

Figure 3.9 shows the envelopes of methane and ethane mixtures with different compositions, and the vapor pressure curves of pure methane and pure ethane. From the diagram, the following features can be observed:

(a) The envelope of each mixture lies between the vapor pressure curves of the two pure components. If one component dominates the mixture, such as the mixture containing 97.50 % methane or 95.0 % ethane, the two-phase region of the mixture is narrow and very near to the vapor pressure curve of the dominating component in the mixture. The two-phase region of a mixture is the widest only

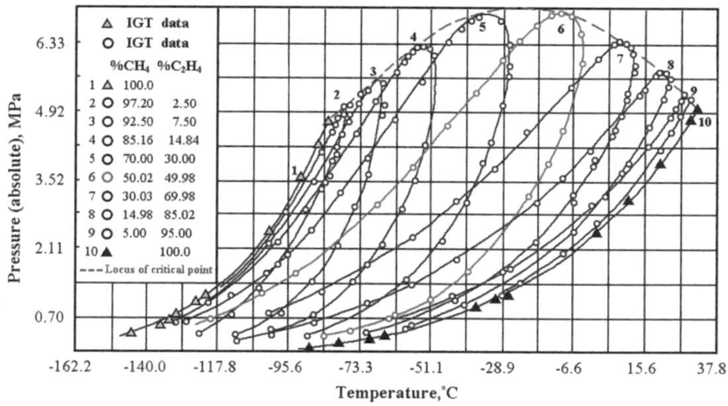

Fig. 3.9 Phase diagrams of mixtures consisting of methane and ethane [21]

when the amount of each component is in about half in the mixture (e.g., the mixture of 50.02 % methane and 49.98 % ethane).

(b) The dashed curve in Fig. 3.9 identifies the locus of the critical points of methane–ethane mixtures. The critical pressure and critical temperature of a binary mixture is determined by the composition of the mixture. The critical temperature of a binary mixture lies between the critical temperatures of the two pure substances in the mixture. The critical pressure of a mixture is usually higher than that of pure substances unless the dominating component has much lower critical pressure than another component. The critical point of a binary mixture mainly depends on the thermodynamic properties of the dominating component of the mixture. For example, the critical point of the mixture containing 97.50 % methane is near the methane (left side of the diagram). The critical temperature of this mixture lies between the critical temperatures of the two pure components. The critical pressure of this mixture is higher than that of methane, and is almost equal to the critical pressure of ethane. As the amount of ethane in the mixture increases, the critical point of the mixture moves toward the right side, until it is near the critical point of ethane (i.e., the mixture containing 95.0 % ethane.).

The *critical loci* of the binary systems composed of normal paraffin hydrocarbons are shown in Fig. 3.10. Obviously, the critical pressures of mixtures are higher than the critical pressures of the components of the mixtures. In fact, a larger difference between molecular sizes of the two components in a mixture causes a higher critical pressure of the mixture [4].

At present, a commonly used method to enhanced oil recovery is carbon dioxide miscible flooding. In the study of CO_2 miscible flooding, hydrocarbons in the reservoir can be regarded as a pseudocomponent; then the system of CO_2 and hydrocarbons is a binary system. Therefore, the optimal proportion of CO_2 injected into the reservoir can be determined according to the phase behavior of series of binary systems (Fig. 3.11).

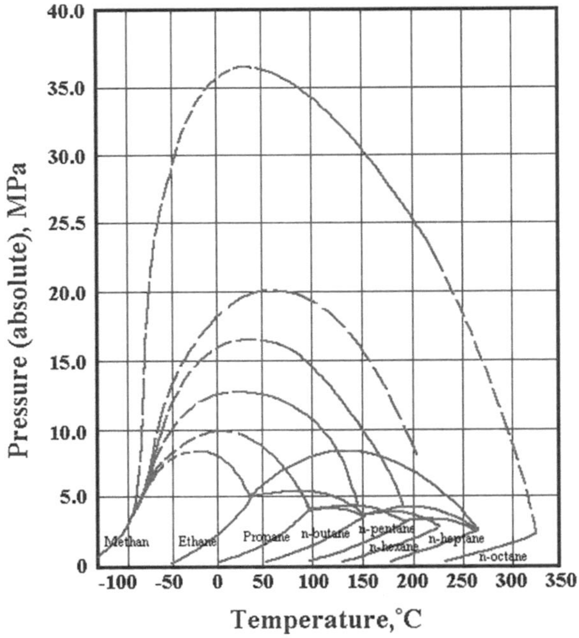

Fig. 3.10 Critical loci of binary *n*-paraffin mixtures [22]

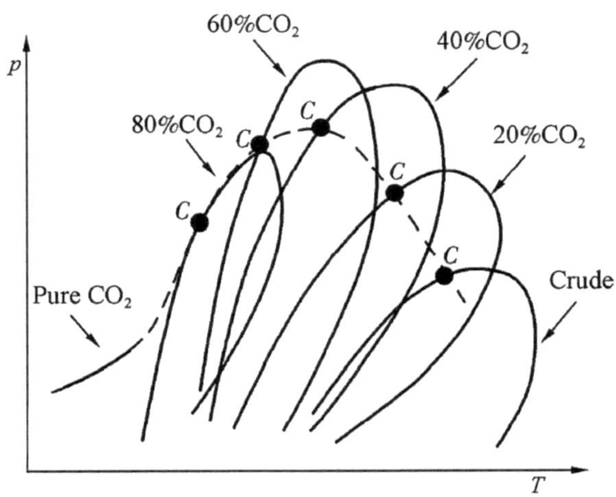

Fig. 3.11 Effect of CO_2 on phase diagram of oil system [18]

3.2.2.3 Hydrocarbon Systems in Petroleum Reservoirs

Hydrocarbon mixture occurring in petroleum reservoir is a truly complicated multicomponent system. Figure 3.12 shows the schematic p–T phase diagram of a typical hydrocarbon mixture. It is very similar with the diagram of a binary system

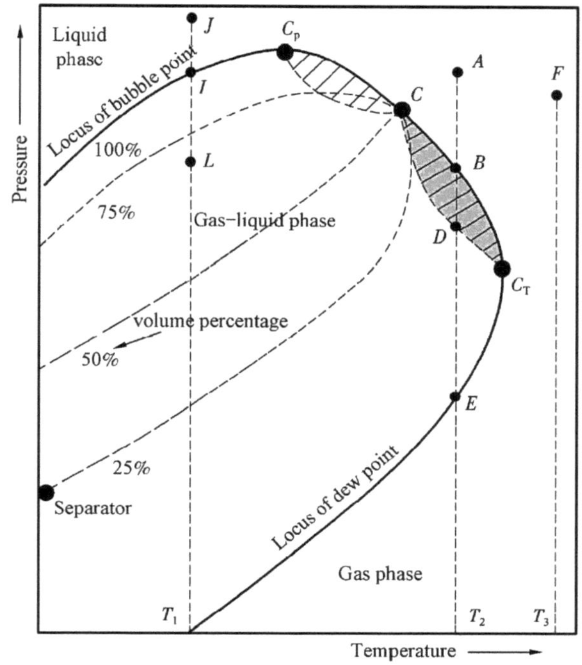

Fig. 3.12 Typical *p–T* phase diagram of hydrocarbon mixtures

(Fig. 3.8). The phase behavior of a multicomponent hydrocarbon mixture is thus the same as that of a binary mixture.

However, there is a slight difference between the two-phase diagram of binary systems and multicomponent mixtures. In the diagram of a multicomponent mixture, there are two special areas, marked in shadow (Fig. 3.12). They are named as *retrograde condensation region*. The shadow area at the left of critical point is called *isopiestic retrograde region* whereas the area lying at the right of critical point is known as *isothermal retrograde region*. In the two regions, the phase behavior may be retrograde, depending on the change in temperature or pressure of a mixture.

First, remember that for a pure substance a decline in pressure causes a phase transition from liquid to gas on the vapor pressure curve. Likewise, in the case of a binary mixture a decline in pressure also causes a phase transition from liquid to gas on the envelope. An example is process *AB* in Fig. 3.8.

To see what happens in retrograde regions, consider an isothermal decrease in pressure illustrated by line *ADE* in Fig. 3.13. Point *A* represents the initial state of a gas mixture. When pressure decreases to point *B*, the dew point of the system, the liquid begins to condense from the gas system. To the position indicated by point B_1, the system is 10 % liquid by volume and 90 % gas. When pressure declines further, the amount of liquid in the system increases continuously till point D. *A* decrease in pressure caused not only a phase transition from gas to liquid, but also

Fig. 3.13 Phase behavior of isothermal retrograde region

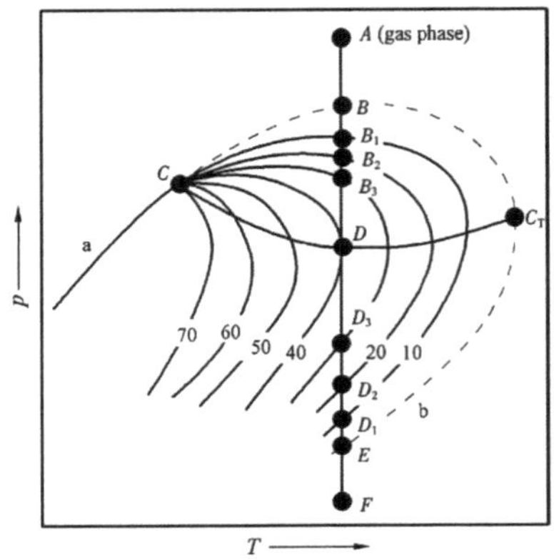

an increase in liquid amount in the system. This is exactly the reverse of the behavior one would expect, hence named *retrograde condensation*. In detail, the retrograde condensation means the reverse behavior or phenomenon of liquid phase increasing with the decrease in pressure. Accurately, the phase behavior from *B* to *D* is also termed as *isothermal retrograde condensation* because it is an isothermal process. Conversely, the reevaporization of a liquid as a result of pressure increase is called *retrograde vaporization*.

When pressure decreases from point *D* to *F*, however, the amount of liquid decreases till it evaporates to gas phase completely. Obviously, the phase behavior from *D* to *E* is normal evaporation. It tells us that retrograde condensation only occurs in limited regions.

Note that the path of pressure depletion passed through the dew point curve twice, the names of *upper dew point* and *lower dew point* are suggested to avoid confusion. The upper dew point sometimes is called the *retrograde dew point*.

The region of retrograde condensation can occur only at the temperatures between the critical temperature and the cricondentherm. A similar retrograde situation may occur during the process of change in temperature at constant pressure, and only occurs at the pressures between the critical pressure and the cricondenbar. This reverse behavior is called *isopiestic retrograde condensation*. In respect of that the development of conventional oil reservoirs nearly follows an isothermal path, the isopiestic retrograde is less important than isothermal retrograde.

How to explain the phenomenon of retrograde condensation? One of the viewpoints, at present, is based on the molecular kinematics. As is known to all, with the decrease in pressure, the distances between molecules increase and the attractive force between molecules is weaken. At point *A* in Fig. 3.13, the system is in gas phase. Along the isothermal path *AF*, when pressure is at the point *B*, the attractive forces between gas molecules become weak and are not big enough to sustain some heavier hydrocarbon molecules to disperse in gas phase; these heavier hydrocarbon molecules thus condense from the gas phase. With the further decrease in pressure, the attractive force between molecules become weaker and weaker, and more and more liquid condense from gas phase. When pressure reaches point *D*, almost all of higher hydrocarbon molecules have condensed from gas phase, and the amount of condensate in system reaches the maximum.

It should be noted that isothermal retrograde condensation occurs only in a special region near the critical point inside the envelope. Out of the range, it does not happen.

3.2.2.4 Application of Phase Diagram

According to the data of PVT experiment for a hydrocarbon mixture, a *p*–*T* phase diagram can be plotted. If the initial temperature and pressure (reservoir condition) of the mixture are given, the type of the hydrocarbon reservoir can be easily determined form the phase diagram. Figure 3.14 shows the schematic illustration of the initial position of various hydrocarbon mixtures in a *p*–*T* phase diagram. Different hydrocarbon mixtures have different initial locations with respect to the envelope in the diagram; and different shapes of envelopes.

In Fig. 3.12, a mixture at point *J* would contain only liquid, i.e., crude oil. If the initial condition (temperature and pressure) of a hydrocarbon mixture is at point *J* shown in Fig. 3.12, the reservoir should be an oil reservoir. As the pressure of the oil reservoir (point *J*) is higher than the bubble point pressure of the system, the crude oil in reservoir is not saturated by dissolved gas in the oil, namely, it is undersaturated. This kind of hydrocarbon reservoir is thus named as *undersaturated reservoirs*.

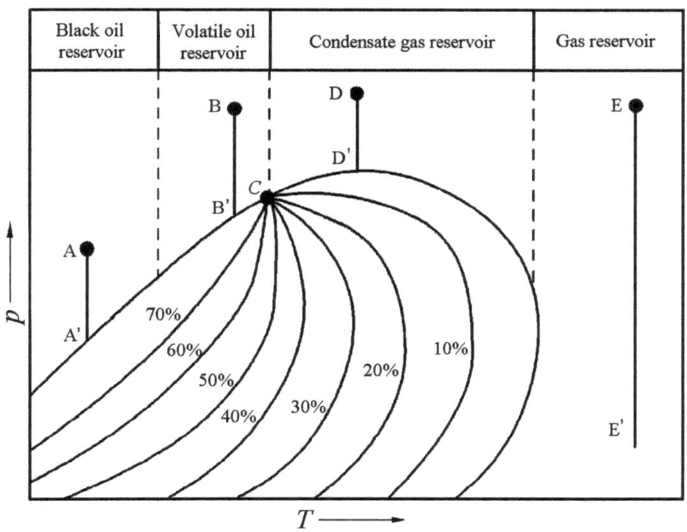

Fig. 3.14 Schematic illustration of the initial position of various hydrocarbon systems in *PT* phase diagram [1]

Reservoir engineers frequently refer to oil reservoirs as either *saturated* or *undersaturated.* In undersaturated oil reservoirs, initial reservoir pressure is higher than its bubble point pressure (i.e., above bubble point curve), and no free gas in reservoir at discovery. However, in saturated oil reservoirs, initial reservoir pressure is lower than its bubble point pressure (i.e., below bubble point curve). Saturation or not is a very important property of oil reservoir in relation to the development strategy of the reservoir. Therefore, in the p–T diagram, a system that exists above the bubble point curve is regarded as undersaturated as no free gas in the reservoir; whereas a system that lies on the bubble point curve or below it is considered as saturated, and may contains free gas in its reservoir. The gas in gas caps is called *free gas*, whilst the gas dissolved in the oil is called *dissolved* or *solution gas*.

If the undersaturated oil reservoir is developed, the reservoir pressure will gradually decline along the dashed line JT_1 in Fig. 3.12. When reservoir pressure reaches point I (bubble point), dissolved gas starts to evolve from the liquid phase. Oil–gas two phases begin to coexist in the reservoir. Further decline in pressure causes more and more gas evolving out of the liquid. Finally, a secondary gas cap may be formed in the reservoir, just like the mixture at point L.

If the initial conditions (temperature and pressure) of a mixture are at point I (bubble point), this mixture is said to be "*saturated*" due to liquid and gas coexistence. The word *saturated* here indicates that the oil contains dissolved gas as much as it can hold. Therefore, the pressure at point I (bubble point) is also called the *saturation pressure* of the system.

Point L would also represent an oil reservoir that contains both oil and free gas at initial state. Such a reservoir is called *saturated reservoir with a gas cap.* In initial conditions, oil–gas two phases coexist in saturated reservoirs. Further production will lead to further liberation of gas from oil. Consequently, this kind of oil reservoir is also known as *degassed reservoir* or *oversaturated reservoir.*

A reservoir denoted by point F in Fig. 3.12 would be a pure gas reservoir because the system is entirely gas phase at initial conditions and in the process of isothermal decrease in pressure. The path of pressure decline does not pass through the envelope, and enters into the two-phase region of the system.

Point A in Fig. 3.12 represents a special system, which is referred to as *condensate gas reservoir.* Its initial pressure is higher than critical pressure, and initial temperature is between critical temperature and cricondentherm. That is to say, a condensate gas reservoir lies affirmatively above or in the region of isothermal retrograde. It is said to be a condensate gas reservoir because a large amount of liquid (condensate oil) will condense from gas during the reservoir production. It is the vital difference between a condensate gas reservoir and a conventional gas reservoir.

When pressure is lower than the upper dew point pressure of a condensate gas reservoir, the retrograde condensation occurs in reservoirs, namely, the heavier components in gas phase condense from gas phase. In this situation, the most valuable components (heavier hydrocarbons) in gas phase will be lost in the reservoirs. The major reasons are

(a) The amount of liquid is so small that the liquid cannot flow together with the gas in the reservoir. Thus, the condensate liquid cannot be brought to the surface.
(b) The condensate liquid will be absorbed on the surface of rock particles, or remains at the corners of pores.
(c) Theoretically, the liquid phase may completely reevaporate to gas phase when pressure is lower than the lower dew point of the system, but it hardly happens because of the first two reasons. With the continuous production, the composition of gas remaining in the reservoir changes ceaselessly. With the decline in reservoir pressure, the gas remaining in the reservoir becomes heavier and heavier. As a result, the envelope of the system alters. It moves to lower right, and is closer to the temperature axis. The change in envelope will also prevent the liquid from reevaporating.
(d) The deposition of condensate liquid in rock pores will increase the resistance of gas phase flowing in pores, even results in off production of gas wells owing of "*hydrocarbon lock*".

In order to reduce the deposition of condensate liquid in reservoirs, i.e., to avoid retrograde condensation occurring in reservoirs, the method of cyclic gas injection should apply to develop further the condensate gas reservoir when pressure declines around upper dew point because cyclic gas injection can effectively maintain the reservoir pressure during production. Actually, many condensate gas reservoirs are found to associate with crude oil, namely, condensate gas in the top (gas cap) and oil in the bottom of the reservoir. In such cases, the reservoir pressure is exactly the dew point pressure of the system. A detailed scheme of cyclic gas injection is necessary to determine how to develop effectively the reservoir.

3.2.3 Phase Diagrams for Typical Hydrocarbon Mixtures

As described above, hydrocarbon reservoirs are usually classified into five systems: dry gas, wet gas, condensate gas, volatile oil, and black oil. Each of them has been defined above because each requires different approaches by reservoir engineers and production engineers [4]. The typical components of production from them are shown in Table 3.9. The schematic envelopes of these systems are shown in Fig. 3.15. The features of each system can be better understood in terms of its envelope. Clearly, with the increase in the content of heavier hydrocarbons in a system, the envelope of the system becomes wider and moves to lower right in the p–T diagram.

From dry gas to black oil, as we all know, the contents of heavy hydrocarbons (C_7^+) in these systems gradually increase. The properties of these systems, such as color, density, viscosity, and so on, also increase by degrees. The changes in their envelopes thus show the obvious regularity (Figs. 3.16, 3.17, 3.18, 3.19, and 3.20). A simple analysis of the phase behavior of these systems is presented as follows.

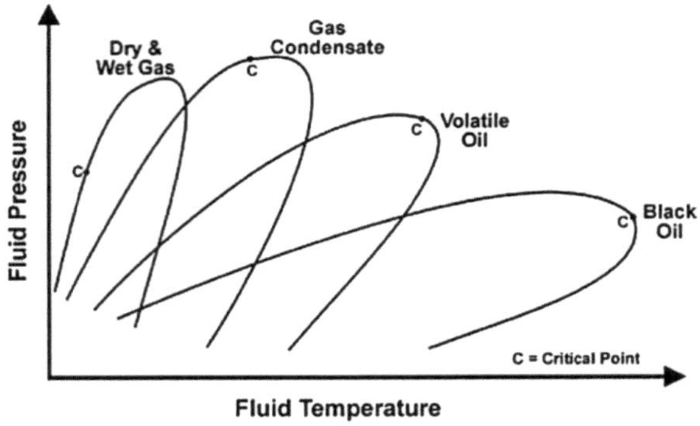

Fig. 3.15 Envelopes for different types of hydrocarbon systems [23]

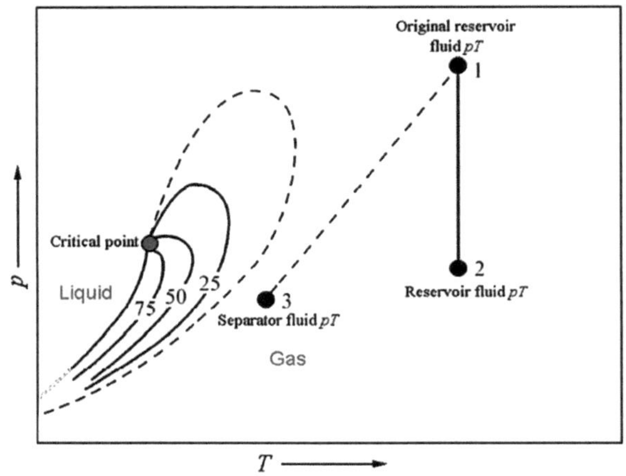

Fig. 3.16 Phase diagrams of a dry gas [18]

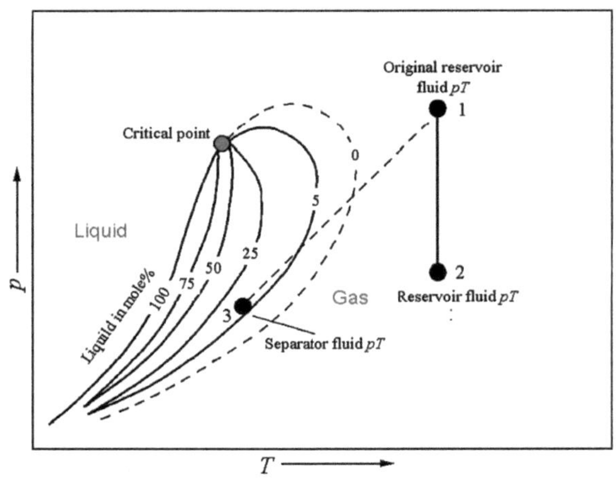

Fig. 3.17 Phase diagrams of a wet gas [18]

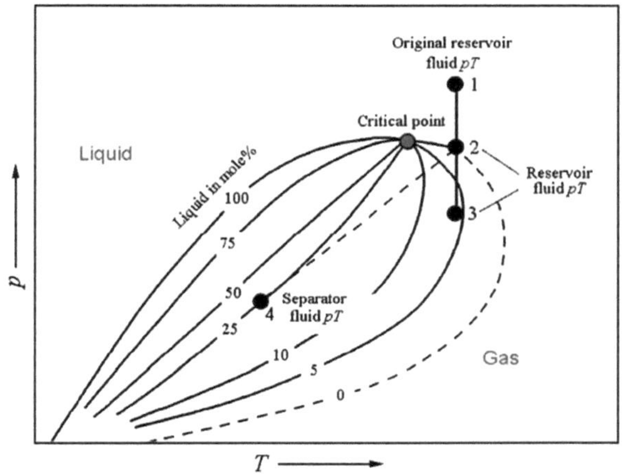

Fig. 3.18 Phase diagrams of a gas condensate [18]

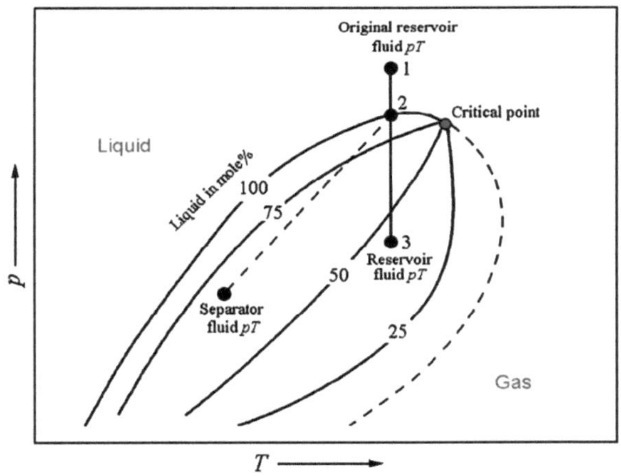

Fig. 3.19 Phase diagrams of a typical volatile oil [18]

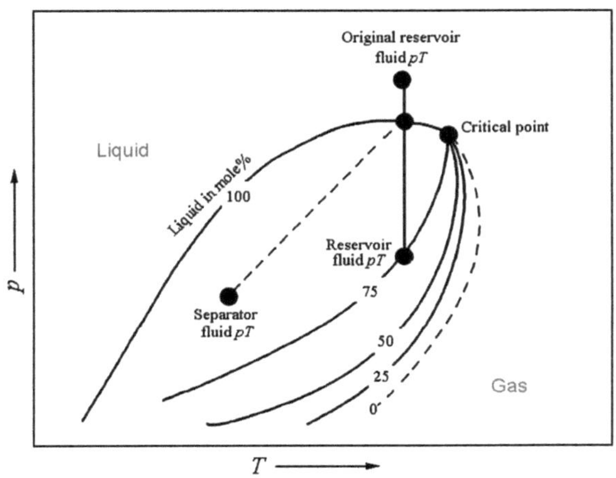

Fig. 3.20 Phase diagrams of a typical black oil [18]

3.2.3.1 Dry Gas Reservoirs

The remarkable features of a dry gas are the predominance of methane and the lack of C_5^+ components in the system (Table 3.9). The content of methane in dry gases may be up to 70–98 %.

A typical dry gas reservoir is shown in Fig. 3.16. The two-phase region is very narrow, and the critical point lies on the lower left of the envelope. The isovols are crowded together near the bubble point curve. Point 1 represents the initial temperature and pressure of a dry gas in reservoir. Point 2 is a state of the dry gas in reservoir during production. Point 3 denotes the temperature and pressure in separator at the surface.

Obviously, the initial temperature of a dry gas reservoir is much higher than the cricondentherm of the gas. During production, reservoir pressure which depletes follow the path of line 1–2. The dashed line 1–3 approximatively simulates the change from reservoir conditions to separator conditions when the gas is brought to surface. Apparently, during production, the path of line 1–3 does not pass through the envelope, which indicates that no liquid condenses as the gas travels from the reservoir to the surface (separator conditions). Therefore, this system is called "*dry gas*".

Theoretically, a dry gas produces no liquid at surface conditions and the gas–oil ratio of a dry gas tends to infinity. However, the definition of a dry gas is of a little arbitrariness. In fact, there is no strict delimitation between a dry gas and a wet gas. Usually, a natural gas, which can produce a little liquid at the surface and has a gas–oil ratio greater than 18,000 m^3/m^3, is considered as a dry gas.

3.2.3.2 Wet Gas Reservoirs

A natural gas is said to be wet if it contains an appreciable natural gasoline content as determined by standard tests. The presence of small amounts of C_{5+} components and the continuing predominance of methane in the gas mixture (see Table 3.9) are the features of wet gas reservoirs.

A typical wet gas reservoir is shown in Fig. 3.17. The two-phase region of a wet gas is wider than that of a dry gas. The location of critical point on the envelope is higher than that of a dry gas. During production, the reservoir conditions (line 12) are always outside the envelope, as shown in Fig. 3.17. Namely, in the reservoir, hydrocarbon mixture is always gas phase throughout the production. However, the separator condition is just located in the two-phase region.

The reservoir temperature of a wet gas reservoir is just above the cricondentherm. During production, the fluids are reduced in temperature and pressure. The temperature–pressure path followed during production just penetrates the envelope (line 13), resulting in the production of gas at the surface with a small associated liquid phase. The liquid is a light oil which is transparent, light-color like water, and the relative density of the oil is often less than 0.78.

The word "*wet*" does not denote the gas is wet but means the gas contains a few heavier hydrocarbons. These heavier hydrocarbons will condense to liquid at surface conditions.

A wet gas usually contains slightly more moderate-size hydrocarbons than a dry gas. The gas–oil ratio (GOR) has fallen as some liquid is being produced. Surface GOR of a wet gas reservoir is often less than 18,000 m^3/m^3. Note also the small specific gravity for $C_7{}^+$ components (0.750), indicating that the majority of the $C_7{}^+$ fraction is made up of the lighter $C_7{}^+$ hydrocarbons [23]. The gas produced from a wet reservoir is sometimes separated and liquefied to the butane and propane.

3.2.3.3 Condensate Gas Reservoirs

A condensate gas is characterized by the increase in the $C_7{}^+$ components and the continuing predominance of methane in reservoir mixture (see Table 3.9).

A typical condensate gas reservoir is shown in Fig. 3.18. The reservoir temperature is such that it falls between the temperature of the critical point and the cricondentherm. The envelope begins to lean to temperature axis; and the range of two-phase region becomes wider than dry gas or wet gas reservoirs. The isovols symmetrically distribute within the envelope.

Due to the reservoir temperature being between the critical temperature and the cricondentherm, the production path then has a complex history. Initially, the hydrocarbon mixture in reservoir is in gas phase, and expands as the pressure drops. This occurs until the dew point curve is reached, where a few amounts of liquids are condensed from the gas phase. If the pressure reduces further, the amount of liquids condensed from the gas phase increases. The process is called isothermal retrograde condensation. Thus, in condensate gas reservoirs, the oil produced at the surface results from a gas existing in the reservoir.

A condensate gas mixture may contain more heavier hydrocarbons than a wet gas mixture, and more lighter hydrocarbons than a crude oil mixture. So, the fluid PT in separator lies in the two-phase region and far from dew point curve. Much more liquid (about 25 %) will be produced from a gas condensate mixture at the surface than from a wet gas mixture (Fig. 3.18).

The GOR of a gas condensate reservoir has decreased significantly, and may be lower than 12,600 m^3/m^3. However, GOR tends to increase with the decline in pressure because some of heavy components condense to liquid and thus are lost in reservoir.

The specific gravity of the $C_7{}^+$ components is increasing, indicating that greater fractions of denser hydrocarbons are present in the $C_7{}^+$ fraction [23]. The condensate oil is usually water white or slightly colored and diaphanous. Its relative density is about 0.74.

3.2.3.4 Volatile Oil Reservoirs

Volatile oils contain relatively fewer heavy molecules and more intermediates (defined as ethane through hexanes) than black oils. Volatile oils also have been called *high-shrinkage crude oils* and *near-critical oil* [4]. Table 3.9 shows that the fraction of gases in a volatile oil reservoir is reduced, and the fraction of denser liquid hydrocarbons is increased, compared with the previously discussed reservoir types.

The phase diagram for a typical volatile oil is shown in Fig. 3.19. The phase envelope of a volatile oil is relatively wider than that of a gas condensate, with a higher critical temperature due to its larger concentration of heavy hydrocarbon. The phase diagram of a volatile oil is somewhat different from the black oil phase diagram (Fig. 3.20); and the temperature range covered by the phase envelope is somewhat smaller, but of more interest is the position of the critical point. The critical point is up to right side of the envelope. The reservoir temperature is lower than critical temperature, and the fluid is therefore a liquid in the reservoir. The reservoir temperature is and near the critical temperature, hence, volatile oils are referred to as near-critical oils. The isovols are not evenly spaced but are slightly shifted upward and towed the bubble point curve (Fig. 3.19).

The vertical line 1–3 shows the path taken by the constant temperature reduction in pressure during production. Notice that a small reduction in pressure below the bubble point (point 2) causes the release of a large amount of gas in the reservoir. An isovol with a much lower percent liquid crosses the separator conditions (Fig. 3.19). Hence the name "*volatile oil*" is used. The oil produced at point 2 in Fig. 3.19 will shrink by more than one-half often three-quarters, on the trip to the stock tank. Namely, only about 50–25 % liquid remains in surface separator. It is observably less than the amount of liquid separated from a heavy oil mixture. Therefore, such kind of oil is also called high-shrinkage crude oil. Volatile oils should be produced through three or more stages of surface separation to minimize this shrinkage [4].

The dividing line between black oils and volatile oils is somewhat arbitrary. Volatile oils are identified as having initial producing gas–oil ratios between 360 and 580 m^3/m^3 [4]. The producing gas–oil ratio increases as production proceeds and reservoir pressure falls below the bubble point pressure of the oil. The stock tank oil gravity is usually 40°API (relative density 0.82) or higher (less than 0.82) and increases during production as reservoir pressure falls below the bubble point [4]. The stock tank oil is deeply colored (usually brown, orange, or sometimes green).

3.2.3.5 Black Oil Reservoirs

Black oils, or ordinary oils, are the most common type of oil reservoirs. The name does not reflect the color, but to distinguish it from the volatile oil. Black oils consist of a wide variety of chemical species including large, heavy, nonvolatile molecules. They generally include more than about 20 mol% heptanes and heavier components (see Table 3.19).

A phase diagram of the typical black oil is given in Fig. 3.20. The phase diagram predictably covers a wider temperature range than that of a volatile oil. The reservoir temperature is lower than critical temperature. The critical point is well up the slope of the phase envelope. Because black oils contain a lower fraction of volatile components, a much larger pressure drop below the bubble point is required before significant volumes of gas are released from oil. This is reflected by the position of the isovols in the phase diagram, where the isovols are grouped near the dew point curve.

The vertical line 1–3 indicates the reduction in pressure at constant temperature that occurs in the reservoir during production. When reservoir pressure lies anywhere along line 1–2, the oil reservoir is said to be *undersaturated*. The word *undersaturated* is used in this sense to indicate that the oil could dissolve more gas if more gas was present. If the reservoir pressure is at point 2, the oil is at its bubble point and is said to be *saturated*. The oil contains as much dissolved gas as it can hold. A reduction in pressure will release gas to form a free gas phase in the reservoir.

When a reservoir pressure declines along line 2–3, additional gas evolves from the oil as it moves from the reservoir to the surface. This causes some shrinkage of the oil. But the shrinkage is low due to isovols crowding near the dew point curve. Namely, a large drop in pressure below the bubble point causes the release of a small amount of gas in the reservoir. For example, when reservoir pressure reaches point 3, the reservoir mixture just shrink by one-quarters (25 %). Beside, separator conditions lie well within the phase envelope, also indicating that a relatively large amount of liquid arrives at the surface.

Black oils are also known as *low-shrinkage oils* because they liberate relatively small amount of gas either in the reservoir or the separators, leaving relatively larger amount of stabilized oil compared to volatile oils (high-shrinkage oils). The kinds of oil contain relatively more of heavier hydrocarbons than the light oils (volatile oils).

Black oils are characterized as initial producing gas–oil ratios (GOR) of 360 m^3/m^3 or less [4]. The GOR of heavy oils is even less than 90 m^3/m^3. Producing gas–oil ratios of black oils increase during production when reservoir pressure is lower than bubble point pressure of the oil.

The stock tank oil usually has a gravity below 30°API (relative density 0.876) [4]. The stock tank oil is very dark, indicating the presence of heavy hydrocarbons, often black, sometimes with a greenish cast, or brown.

3.3 Properties of Natural Gas

Natural gas is a naturally occurring mixture of low-molecular-weight hydrocarbons and nonhydrocarbon gases found in porous formation underground. It is nearly everywhere considered to be the gaseous phase of petroleum. However, the properties of natural gas are considerably different from liquid petroleum because there is greater distance between the molecules in natural gas than in crude oil. For example, pressure has much greater effect on the density of a natural gas than on oil's density. The notable features of natural gases are much easily compressible and flowable.

The following are natural gas' physical properties that are often encountered by petroleum engineer.

3.3.1 Molecular Weight and Density of Natural Gas

3.3.1.1 Molecular Weight of Natural Gas

Natural gas is a mixture of hydrocarbon and nonhydrocarbon gases. There is no fixed molecular formula for any natural gas. Therefore, it is impossible to calculate the molecular weight of a natural gas in terms of its molecular formula like a pure substance.

Molecular weight is a definition for a specific molecule but not for a mixture of different molecular species. In order to determine the average mass of all molecules in a gaseous mixture, the term, *apparent molecular weight*, is used.

The molecular weight of a natural gas is defined as the mass of one mole (22.4 L) natural gas at standard conditions (0 °C, 760 mm mercury column). Obviously, it is a hypothetical molecular weight, and just represents the average molecular weight of all molecules in a natural gas. As a result, it is an apparent molecular weight of a natural gas. For short, it is still called the *molecular weight* of a natural gas.

Numerically, the molecular weight of a natural gas is estimated in terms of the composition of a gas. Gas composition measured in the laboratory is customarily reported in mole fraction. Suppose y_i is the mole fraction of the ith component in a natural gas. The apparent molecular weight of the gas can be estimated using a mixing rule such as

$$M = \sum_{i=1}^{N} y_i M_i, \tag{3.5}$$

where M is the apparent molecular weight of a natural gas; y_i is the mole fraction of component i in the gas, f; N is number of components; M_i is the molecular weight of

Table 3.12 Mole compositions of a natural gas

Components	Mole fraction, y_i	Molecular weight, M_i	y_iM_i
Methane	0.85	16.0	13.60
Ethane	0.09	30.1	7.71
Propane	0.04	44.1	1.76
Butane	0.02	58.1	1.16
Sum	1.00		M = 19.23

component i in the gas. The molecular weights of pure components (M_i) are available in textbooks on organic chemistry or petroleum fluids.

Generally, the apparent molecular weights of natural gases vary with their compositions. For instance, the molecular weights of natural gases from dry gas reservoirs usually range from 16.82 to 17.98.

Example 3.2 Calculate the molecular weight of a natural gas according to its mole composition given in Table 3.12.

Solution

According to the mole composition of the gas, the values of y_iM_i for each component are first calculated, which are listed in the last column of Table 3.12. The molecular weight of the gas is then determined by Eq. (3.5):

$$M = \sum_{i=1}^{N} y_i M_i = 13.6 + 7.71 + 1.76 + 1.16 = 19.23$$

3.3.1.2 Density of Natural Gas

The density of a natural gas is defined as the mass per unit volume of natural gas:

$$\rho_g = \frac{m}{V}, \tag{3.6}$$

where ρ_g is the density of a natural gas, g/cm^3 or kg/m^3; m is the mass of the natural gas, g or kg; V is the volume occupied by the natural gas, cm^3 or m^3.

Considering the density of a natural gas is related to the temperature and pressure, it is often measured at *standard conditions* in the laboratory. It should be noted that the standard conditions here mean the temperature of 20 °C and the pressure of 0.1 MPa (NIST standard). This is a standard in engineering, which is also called *normal temperature and pressure* (from Wikipedia, the free encyclopedia).

3.3.1.3 Relative Density of Natural Gas

Relative density of a natural gas is defined as the ratio of the density of the natural gas to the density of dry air, both determined at standard conditions (20 °C and 0.1 MPa).

$$\gamma_g = \frac{\rho_g}{\rho_a}, \tag{3.7}$$

where γ_g is the relative density of a natural gas, dimensionless; ρ_g is the density of the natural gas, g/cm^3; ρ_a is the density of dry air, g/cm^3.

Note that, in Europe and the United States, gas *specific gravity* (the same symbol, γ_g) is widely used in petroleum industry to characterize natural gases. It is defined as the ratio of the density of gas to the density of air, both measured at standard conditions, which are different from those in China, and represented by T_{sc} = 60 °F (15.6 °C) and p_{sc} = 14.7 psia (0.1 MPa). One should note that there is little difference between the values of specific gravity and relative density of a gas.

As the molecular weight of dry air is usually taken as a constant, 28.97 (about 79 % nitrogen and 21 % oxygen), Eq. (3.7) can be written as

$$\gamma_g = \frac{M}{28.96} \tag{3.8}$$

So, the apparent molecular weight of a natural gas can be easily estimated by Eq. (3.8) if the relative density of the natural gas is given. In a word, this parameter is widely used in petroleum engineering owing to ease to obtain.

In general, the relative densities of natural gases vary between 0.55 and 0.9. It may be bigger than one if there are enough heavier hydrocarbons or nonhydrocarbons in the natural gas.

3.3.2 Equation of State and Z-factor Chart for Natural Gas

All fluids follow the physical laws that define their state under given physical conditions. These laws are mathematically represented as equations, which are consequently known as *equations of state* (EOS) [24]. This kind of equation is essentially correlated with *p*, *V*, and *T* for any fluid, and can be expressed, on a mole basis, as

$$F(p, V, T) = 0 \tag{3.9}$$

Therefore, an equation of state for natural gases is an analytical expression relating the pressure *p*, to the temperature *T* and the volume *V* of the gas.

3.3.2.1 Ideal Gas Equation

The equation of state for an ideal gas is the relationship of p, V and T obeyed by the ideal gas. The ideal gas is a hypothetical gas, but it serves as a useful tool to explain the more complex real gas behavior. Moreover, the equation for the ideal gas is also the basis of developing other equations for real gases.

As is known to all, an ideal gas has the following properties:

(a) The volume occupied by gas molecules is negligible compared with the volume occupied by the gas;
(b) There is no attractive or repulsive forces among the molecules or between the gas molecules and the wall of the container except during collisions;
(c) All collisions between the gas molecules are purely elastic, implying no internal energy loss on collision.

Then, the equation of state for an ideal gas is often written as

$$pV = nRT, \tag{3.10}$$

where p is the pressure of the gas, MPa; V *is* the volume of the gas, m^3; n is the mole number of the gas, Kmol; R is the universal gas constant, $R = 0.008314$ MPa m^3/(Kmol K); T is the absolute temperature of the gas, K.

From Eq. (3.10), we can see that the PVT behavior of a gas described by the equation above is independent of the sort of gas. If the PVT behavior of a gas obeys Eq. (3.10) at any p–T conditions, the gas is an ideal gas. However, real gases are not ideal gases because the molecules of real gases have sizes and there are acting forces between the molecules of real gases. Therefore, any real gas does not absolutely obey the equation of state for the ideal gas. However, the equation of the ideal gas is a good approximation of the behavior of many gases under many conditions, such as at high temperatures and low pressures (usually less than 0.4 MPa), although it has several limitations. Moreover, it gives us a starting point for developing the equation of state that can adequately describe the behavior of real gases at high pressure.

At low pressures, most real gases almost behave like an ideal gas. For simplifying engineering calculation, a natural gas may be considered as an ideal gas under low pressures (less than 0.4 MPa). In reservoirs (high pressure), however, a natural gas behaves far from an ideal gas does.

3.3.2.2 Equation of State for Natural Gases

After ideal gas equation, hundreds of equations for real gases were proposed. The customarily used equation of state in petroleum engineering is the so-called *compressibility equation of state* or *real gas equation* [4]. It is developed by inserting a correction factor into the ideal gas equation because most real gases do not behave drastically different from that ideal gas does. In most cases, the deviation from ideal behavior of real gases can be modified by a correction factor; and the results can satisfy the accuracy in engineering application.

The compressibility equation of state for real gases is developed as

$$pV = ZnRT, \tag{3.11}$$

where the correction factor Z is called *compressibility factor*, *deviation factor*, or *Z-factor*, dimensionless. p is gas pressure, MPa; V is the volume of a gas, m^3; n is the mole number of the gas, Kmol; R is gas constant, $R = 0.008314$ MPa m^3/(Kmol K); T is the absolute temperature of the gas, K.

Equation (3.11) can be applied to any gas that does not undergo a phase transition at any temperature and pressure, thus it is also called *real gas equation*. It should be noted that the Z-factor of a gas is not a constant but varies with the attributes (T, p, and composition, for gas mixtures) of the gas. As a result, the Z-factor of a gas should be first determined when Eq. (3.11) is applied to the gas.

Combining Eq. (3.10) with Eq. (3.11), we see that the Z-factor of the ratio of the molar volume of a gas to the molar volume of an ideal gas at the same temperature and pressure as

$$Z = \frac{V_{\text{actual}}}{V_{\text{ideal}}}, \tag{3.12}$$

where Z is the compressibility factor of a gas, dimensionless; V_{actual} is the volume of real gas at given temperature and pressure, m^3; V_{ideal} is the volume of ideal gas at the same temperature and pressure as the real gas, m^3.

If a compressibility factor is less than 1, it means that the volume of a real gas is less than that of ideal gas under given conditions. In this case, the Z-factor indicates that the real gas is easier to compress than ideal gas at the conditions due to the attractive forces between the molecules of the real gas. Contrarily, a compressibility factor, bigger than one, indicates that the real gas is more difficult to compress than ideal gas at the same conditions due to the repulsion forces between the molecules of real gas. When Z is unit, the volume of real gas is exactly equal to that of ideal gas. It represents that the real gas has the same compressibility with ideal gas at the same conditions. From here, we see that Z-factor describes the integrated effects of the intermolecular forces and the molecular size of real gas on the behavior of the gas, and characterizes the compressibility difference between actual gas and ideal gas. In a word, Z-factor is a useful thermodynamic property for modifying the ideal gas law to account for the real gas behavior [24].

On the other hand, Z-factor can be considered as a correction factor added in ideal gas equation. That is to say, the deviation from ideal behavior can be reasonably modified using the factor. From this viewpoint, Z-factor represents a deviation from ideal behavior of real gases. Therefore, Z-factor is also called "*deviation factor*".

Actually, the deviation from ideal behavior of a real gas varies with temperatures and pressures; and different real gases have different deviations even at the same conditions. Therefore, the Z-factor for real gases has been experimentally correlated as a function of pressure, temperature, and composition of gas mixtures. For any real gas, however, Z-factor should be determined by experiments. Figure 3.21 shows the correlation of methane's Z-factor as a function of its attributes.

Figure 3.21 shows that the shape of Z-factor curves is consistent with our knowledge of the behavior of gases. As we all know, at lower pressures, gas molecules are relatively far apart, thus a real gas can sufficiently behave like an ideal gas because both intermolecular attractions and molecular size are ignorable. In Figs. 3.21, 3.22, and 3.23, it can be found that Z-factors are approaching the value of 1.0 at very low pressures, namely real gases can indeed behave like an ideal gas at very low pressures.

At moderate pressures, the distance between the molecules of real gases is reduced. The intermolecular attractions then cause the gas to be compressed more easily. Namely, in this case, the volume of the real gas is less than that of an ideal gas behaving. Therefore, the Z-factor is less than 1.0.

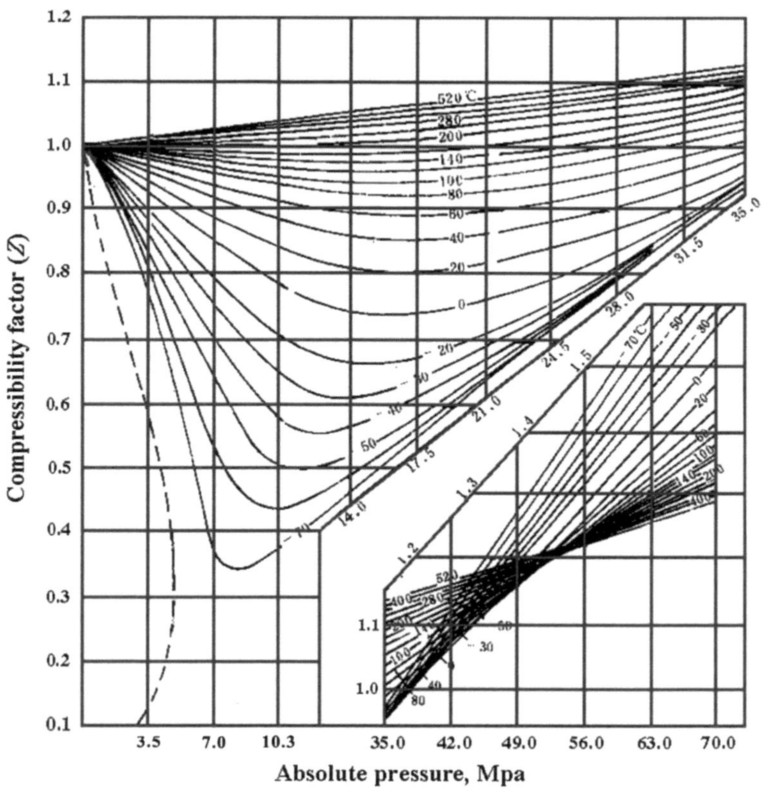

Fig. 3.21 Compressibility factor for methane [25]

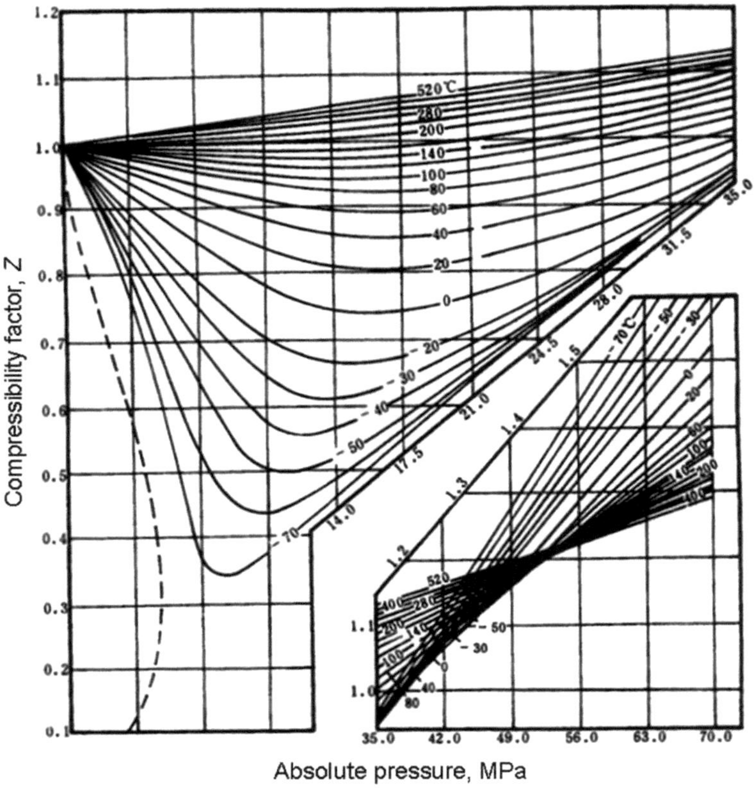

Fig. 3.22 Compressibility factor for ethane [25]

At higher pressures, the molecules are so close that the intermolecular repulsions cannot be ignored. In this case, intermolecular repulsions and molecular size make it not easy to compress the gas. So, the volume of the gas is greater than that of an ideal gas behaving. The *Z*-factor is thus greater than 1.0.

In general, the *Z*-factor for specific gases can be read from *Z*-factor charts that plot *Z* as a function of pressure at constant temperature. For example, Figs. 3.21 and 3.22 are two *Z*-factor charts for pure hydrocarbon gases.

Natural gas is a mixture of many pure gases. Its *Z*-factors can never be read from the *Z*-factor charts for any pure gase. Therefore, a generalized *Z*-factor chart based on the law of corresponding states is developed for mixed gases.

Example 3.3 Calculate the mass of methane in a container with volume of 2 m^3 at 7.0 MPa and 30 °C.

Solution

Because the gas pressure is higher than 4.0 MPa, the methane cannot be considered as an ideal gas. According to Eq. (3.11), the mass of the methane is

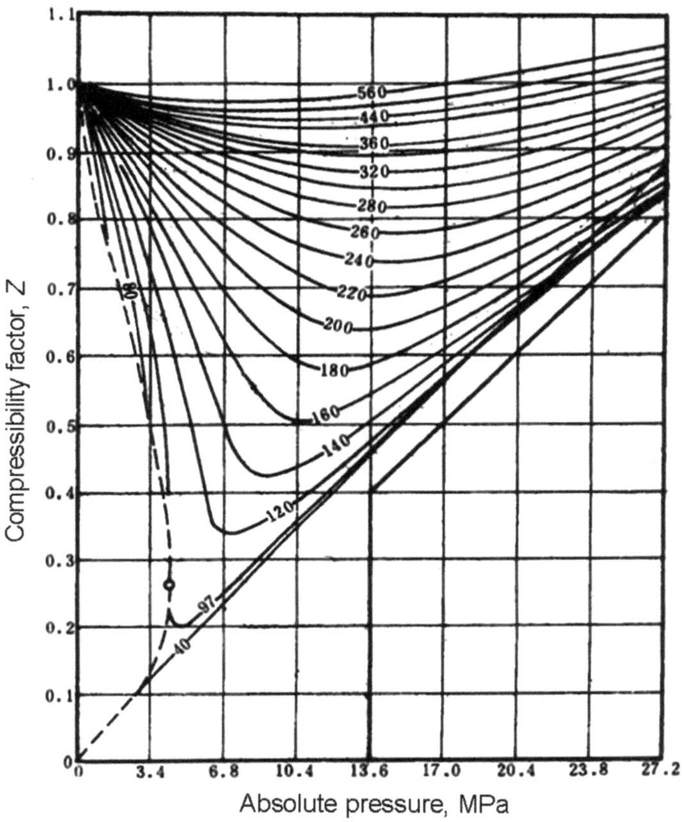

Fig. 3.23 Compressibility factor for propane [25]

$$m = \frac{pMV}{ZRT} \tag{3.13}$$

From Fig. 3.21, we obtain the Z-factor of the methane: $Z = 0.9$.
Then,

$$m = \frac{7.0 \times 16.04 \times 2.0}{0.9 \times 0.008314 \times 303} = 99.05 \text{ kg}$$

3.3.2.3 Principle of Corresponding States

In general, different gases have different physical properties (e.g., density, viscosity, compressibility, and so on) even at the same temperature and pressure. However, it has been found that almost all gases have very similar characteristics at critical point (state). The reduced properties of gases are thus defined based on the properties of gases at its critical point. In thermodynamics, the reduced properties of a fluid are a set of state variables normalized by the fluid's state properties at its critical point. The common reduced properties of gases include reduced temperature and reduced pressure, which are defined as follows:

$$T_r = \frac{T}{T_c} \quad \text{and} \quad p_r = \frac{p}{p_c}, \tag{3.14}$$

where T_r is the reduced temperature of a gas, dimensionless; p_r is the reduced pressure of the gas, dimensionless; T_c is the critical temperature of the gas, K; p_c is the critical pressure of the gas, MPa; T is the given temperature of the gas, K; p is the given pressure of the gas, MPa.

For most pure gases, it is simple to calculate the reduced temperatures and reduced pressures by Eq. (3.14) because their critical temperature and critical pressure are easily obtained from various manuals. Tables 3.13 and 3.14 list the critical properties of common hydrocarbon and nonhydrocarbon gases from Chemical Handbook.

Table 3.13 Physical properties of hydrocarbon gases

Gases	Molecular formula	Molecular weight	Boiling point (0.1 MPa) (°C)	Critical pressure, p_{ci} (MPa)	Critical temperature, T_{ci} (K)	Acentric factor
Methane	CH_4	16.043	−161.50	4.6408	190.67	0.0115
Ethane	C_2H_6	30.070	−88.61	4.8835	303.50	0.098
Propane	C_3H_8	44.097	−42.06	4.2568	370.00	0.1454
Isobutane	iC_4H_8	58.124	−11.72	3.6480	408.11	0.1756
Normalbutane	nC_4H_8	58.124	−0.50	3.7928	425.39	0.1928
Isobutane	iC_5H_{12}	72.151	27.83	3.3336	460.89	0.2273
Normalpentane	nC_5H_{12}	72.151	36.06	3.3770	470.11	0.2510

Table 3.14 Physical properties of nonhydrocarbon gases

Gases	Molecular formula	Molecular weight	Boiling point (0.1 MPa) (°C)	Critical pressure, p_{ci} (MPa)	Critical temperature, T_{ci} (K)	Acentric factor, ω
Carbon dioxide	CO_2	44.010	−78.51	7.3787	304.17	0.2250
Helium	He	4.003	−268.93	0.2289	5.278	
Hydrogen	H_2	2.016	−252.87	1.3031	33.22	−0.2234
Hydrogensulfide	H_2S	34.076	−60.31	9.0080	373.56	0.0949
Nitrogen	N_2	28.013	−195.80	3.3936	126.11	0.0355
Oxygen	O_2	31.999	−182.96	5.0807	154.78	0.0196
Aqueous vapor	H_2O	18.015	100	22.1286	647.33	0.3210

Natural gas is a mixture of gases. The critical properties of natural gases vary with their composition. In oil fields, the compositions of natural gases vary from well to well. Moreover, they may be different even from the same well only at different moments. So, each natural gas may have different critical properties from any others. Consequently, it is not easy to determine accurately the reduced properties of any natural gas by Eq. (3.14) unless its critical properties are given.

In general, the accurate critical properties of any natural gas can only be measured in the laboratory. In practice, the *pseudocritical temperature* and *pseudocritical pressure* were proposed for convenience. They are defined as

$$T_{\mathrm{pc}} = \sum_i y_i T_{ci} \quad \text{and} \quad p_{\mathrm{pc}} = \sum_i y_i p_{ci}, \tag{3.15}$$

where T_{pc} is the pseudocritical temperature of a gas mixture, K; p_{pc} is the pseudocritical pressure of the gas mixture, MPa; T_{ci} is the critical temperature of component i in the gas mixture, K; p_{ci} is the critical pressure of component i in the gas mixture, MPa; y_i is the mole fraction of component i in the gas mixture, f.

Note that pseudocritical properties are not the critical properties of a gas mixture but only an approximation of its properties. Therefore, Eq. (3.15) is often known as Kay's mixture rule.

The pseudocritical temperature and pseudocritical pressure of a natural gas may be calculated either from the gas analysis data (if available) or from the specific gravity of the gas.

Generally, the density of a natural gas is more easily determined than its composition in oil fields. In cases where the composition of a natural gas is not available, the pseudocritical properties, i.e., p_{pc} and T_{pc}, can be estimated from the specific gravity of the gas. Brown et al. [25] presented a graphical method for a convenient approximation of the pseudocritical pressure and pseudocritical temperature of gases when only the specific gravity of the gas is available [26]. The correlation is presented in Fig. 3.24. It is valid for $H_2S < 2\ \%$, $CO_2 < 2\ \%$, $N_2 < 5\ \%$, and total amount of inorganic compounds is less than 9 %.

Based on the concept of pseudocritical properties, the pseudoreduced temperature and pseudoreduced pressure are developed for mixed gases, they are written as

$$T_{\mathrm{pr}} = \frac{T}{T_{\mathrm{pc}}} \quad \text{and} \quad p_{\mathrm{pr}} = \frac{p}{p_{\mathrm{pc}}}, \tag{3.16}$$

where T_{pr} is the pseudoreduced temperature of a mixed gas, dimensionless; p_{pr} is the pseudoreduced pressure of the gas, dimensionless; T_{pc} is the pseudocritical–critical temperature of the gas, K; p_{pc} is the pseudocritical pressure of the gas, MPa; T is the absolute temperature of the gas, K; p is the absolute pressure of the gas, MPa.

According to van der Waals, the principle of corresponding states (or theorem of corresponding states) indicates that all fluids, when compared at the same reduced temperature and reduced pressure, have approximately the same compressibility

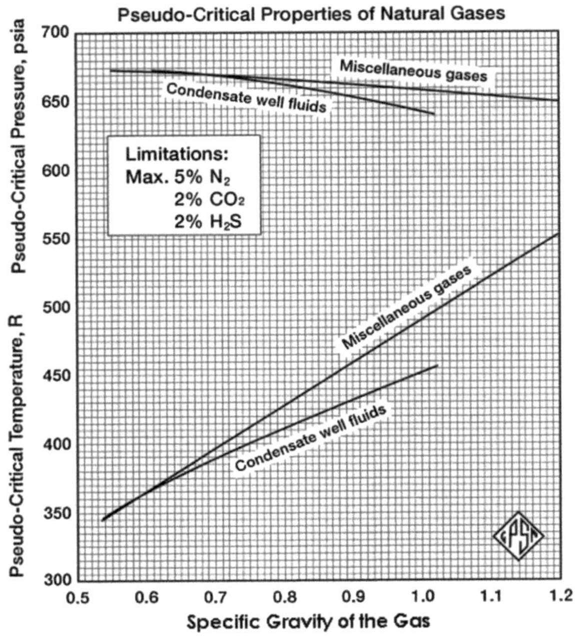

Fig. 3.24 The correlation of pseudocritical properties of natural gases [25]

factor, i.e., the same reduced volume, and all deviate from ideal gas behavior to about the same degree [27]. In the situation, all fluids are in *corresponding states*.

From real gas Eq. (3.11), we have

$$Z = \frac{pV}{nRT} \tag{3.17}$$

Substitute T, P, and V (molar volume) by $T = T_r T_c$, $p = p_r p_c$, and $V = V_r V_c$, respectively, then

$$Z = \frac{p_c V_c}{RT_c} \cdot \frac{p_r V_r}{T_r}, \tag{3.18}$$

where T_r is the reduced temperature of a gas, dimensionless; p_r is the reduced pressure of the gas, dimensionless; V_r is the reduced mole volume of the gas, dimensionless; T_c is the critical temperature of the gas, K; p_c is the critical pressure of the gas, MPa; V_c is critical mole volume of the gas, m^3; R is universal gas constant, $R = 0.008314$ MPa m^3/(Kmol K).

Let

$$\frac{p_c V_c}{RT_c} = Z_c, \tag{3.19}$$

where Z_c is termed the compressibility factor of the gas at its critical point, dimensionless.

Z_c is predicted to be a constant independent of substance by many equations of state [28]. As a result, Eq. (3.18) proves that all gases almost have the same Z-factor at the same corresponding states, namely the same reduced temperature and reduced pressure. This is highly accurate for those gases, which have similar chemical properties and critical properties (i.e., similar molecular characteristics).

Consequently, the principle can be effectively used to correlate the properties of natural gases due to the similarity in their compositions. That is to say, all natural gases almost have the same intensive properties as long as they are in the same corresponding states no matter what compositions they have.

Intensive properties are those which are independent of the amount of material; for example, density, pressure, temperature, viscosity, compressibility factor, chemical potential, and surface tension. Extensive properties depend on the amount of material; for example, area, weight, inertia, and volume.

3.3.2.4 Generalized Z-factor Chart for Natural Gas

Based on the principle of corresponding states, a generalized chart of Z-factors for natural gases can be plotted by the experimental data of any natural gas. A typical generalized chart of Z-factors is shown in Fig. 3.25. Figure 3.25 shows clearly the relations of Z-factors as a function of reduced temperature and pressure. According the principle of corresponding states, Z-factors of any natural gas can be easily determined using the generalized chart if the reduced properties of the gas are given. In practice, we read the Z-factors of natural gases from Fig. 3.25 customarily according to the pseudoreduced temperature and pseudoreduced pressure of a natural gas because of the reason described above.

Example 3.4 The composition of a natural gas is given in Table 3.15. Determine the Z-factor of the natural gas at 49 °C and 10.2 MPa. Then calculate the volume occupied by 1000 mol natural gas.

Solution

1. Calculate the pseudocritical temperature (T_{pc}) and pseudocritical pressure (p_{pc}) of the natural gas by Eq. (3.15). The results list in the last row in Table 3.15.
2. By Eq. (3.16), the pseudoreduced temperature and pseudoreduced pressure of the gas are

$$T_{pr} = \frac{T}{T_{pc}} = \frac{273 + 49}{212.7} = 1.51$$

$$p_{pr} = \frac{p}{p_{pc}} = \frac{10.2}{4.59} = 2.22$$

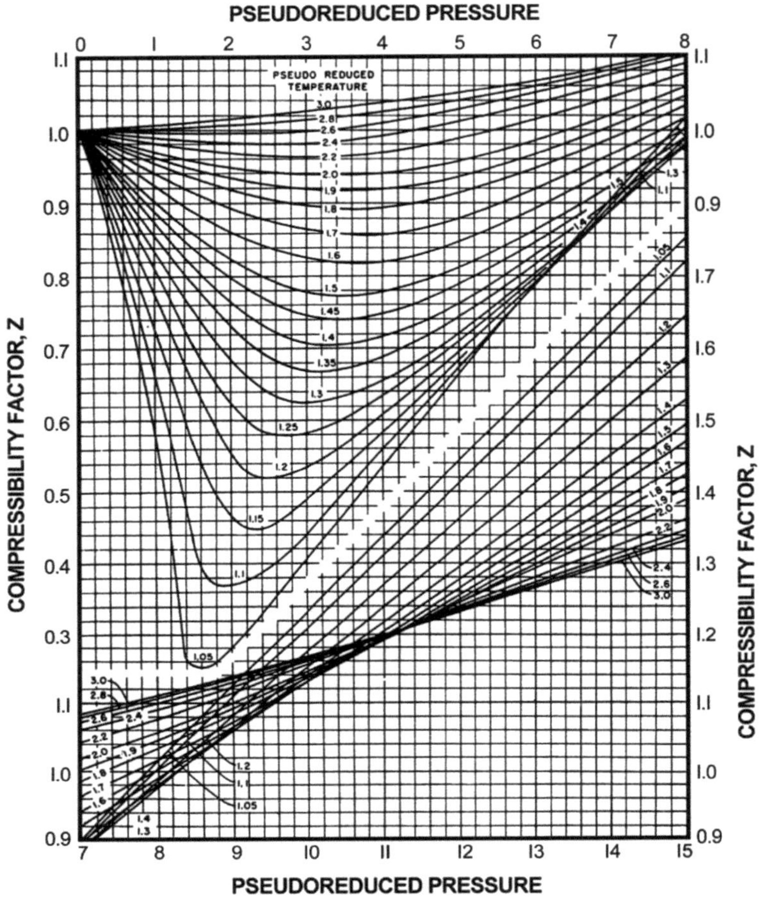

Fig. 3.25 *Z*-factors of natural gases as a function of reduced pressure and temperature [29]

Table 3.15 Mass and mole compositions of a natural gas

Components	Mole fraction, y_i	Critical temperature, T_{ci} (K)	y_iT_{ci}	Critical pressure, p_{ci} (MPa)	y_ip_{ci}
CH_4	0.85	190.5	162.0	4.59	3.90
C_2H_6	0.09	306.0	27.5	4.88	0.44
C_3H_8	0.04	369.6	14.8	4.25	0.17
nC_4H_8	0.02	425.0	8.5	3.79	0.08
Sum	1.00	$T_{pc} = \sum y_iT_{ci} = 212.7$		$p_{pc} = \sum y_ip_{ci} = 4.59$	

3. Read Z-factor in Fig. 3.25 by T_{pr} and p_{pr} calculated above: $Z = 0.82$.
4. In terms of real gas Eq. (3.11), the volume occupied by 1000 mol natural gas at 49 °C and 10.2 MPa is

$$V = \frac{ZnRT}{p} = \frac{0.82 \times 1 \times 0.008314 \times (273 + 49)}{10.2} = 0.22(\mathrm{m}^3)$$

Example 3.5 The relative density of a natural gas is 0.76. Determine the Z-factor of the natural gas at 49 °C and 10.2 MPa.

Solution

1. According to the relative density of the natural gas, the pseudocritical temperature and pseudocritical pressure are obtained from Fig. 3.42: $T_{pc} = 212$ K; $p_{pc} = 4.54$ MPa.
2. Using Eq. (3.16), the pseudoreduced temperature and pseudoreduced pressure of the natural gas are calculated as follows:

$$T_{pr} = \frac{T}{T_{pc}} = \frac{273 + 49}{212} = 1.51$$

$$p_{pr} = \frac{p}{p_{pc}} = \frac{10.2}{4.54} = 2.25$$

3. Read the Z-factor of the gas in Fig. 3.25 by T_{pr} and p_{pr} calculated above: $Z = 0.813$.

Note that the Z-factor obtained from a generalized chart (Fig. 3.25) is accurate enough only if the content of nonhydrocarbons in the gas is low enough (lower than 5 %). For example, if the content of nitrogen is less than 2 %, and the content of carbon dioxide is less than 1 %, the error of a Z-factor from the chart may be less than 3 %.

Besides the hydrogen sulfide and carbon dioxide, nitrogen is common in natural gases. Generally, the presence of nitrogen does not greatly affect the accuracy of a Z-factor from a generalized chart; Z-factor increases by about 1 % for each 5 % of nitrogen in the gas [4]. However, the presence of hydrogen sulfide and carbon dioxide in natural gases may cause large errors in Z-factors from a generalized chart. Moreover, higher content of heavier hydrocarbons (e.g., $C_7{}^+$) in natural gases may also lead to large errors in Z-factors from a chart. Higher contents of nonhydrocarbons and heavier hydrocarbons in natural gases result in large deviations in the pseudoreduced properties of natural gases by Kay's rule. Therefore, the correction to the Z-factor from a chart is necessary. The remedy to the error caused by acid gases is to adjust the pseudoreduced properties to account for the unusual behavior of these acid gases. The equations used for the adjustment are [26]:

$$T'_{pc} = T_{pc} - \varepsilon \quad (3.20)$$

$$p'_{pc} = \frac{p_{pc} T'_{pc}}{T_{pc} + y_{H_2S}(1 - y_{H_2S})\varepsilon}, \quad (3.21)$$

where T_{pc} is the pseudocritical temperature calculated by mole fraction of a natural gas, K; p_{pc} is the pseudocritical pressure calculated by mole fraction of natural gas, MPa; T'_{pc} is the pseudocritical temperature adjusted; p'_{pc} is the pseudocritical pressure adjusted, MPa; y_{H_2S} is the mole fraction of H_2S in natural gas, f; ε is the correction factor of pseudocritical temperature obtained from Fig. 3.26, Ra. Note that Eq. (3.20) is correct only when ε has been converted to Kelvin temperature.

In natural gases, *heptanes plus* (C_7^+) is commonly used. It means an aggregate of components heavier than hexane. The pseudocritical properties of heptanes plus can be determined either by Eq. (3.15) or from Fig. 3.27.

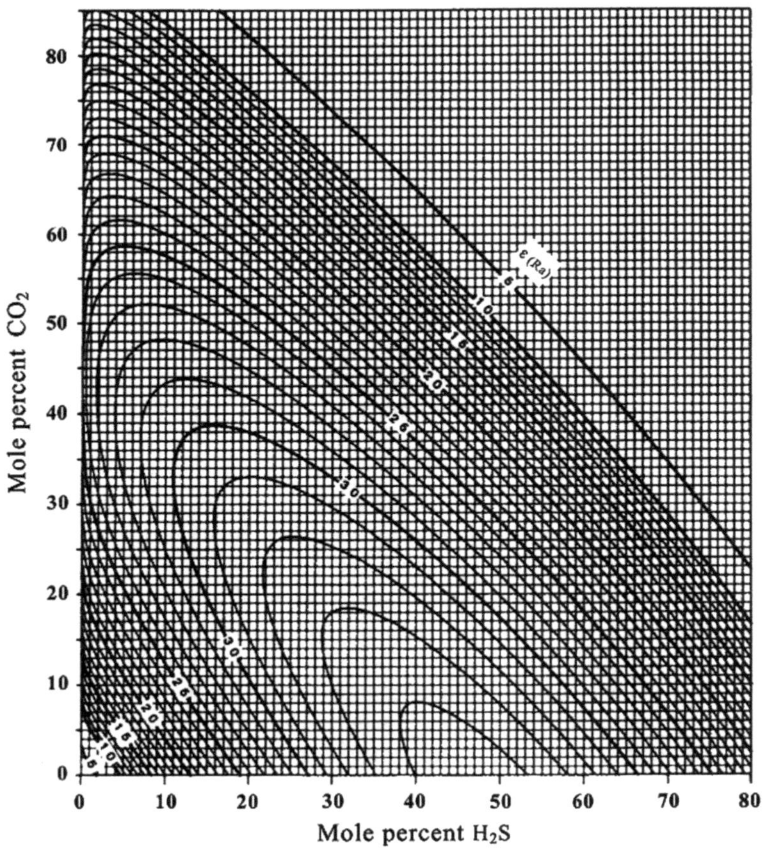

Fig. 3.26 Nonhydrocarbon component adjustment factors for the pseudocritical properties of natural gases (McCain [30])

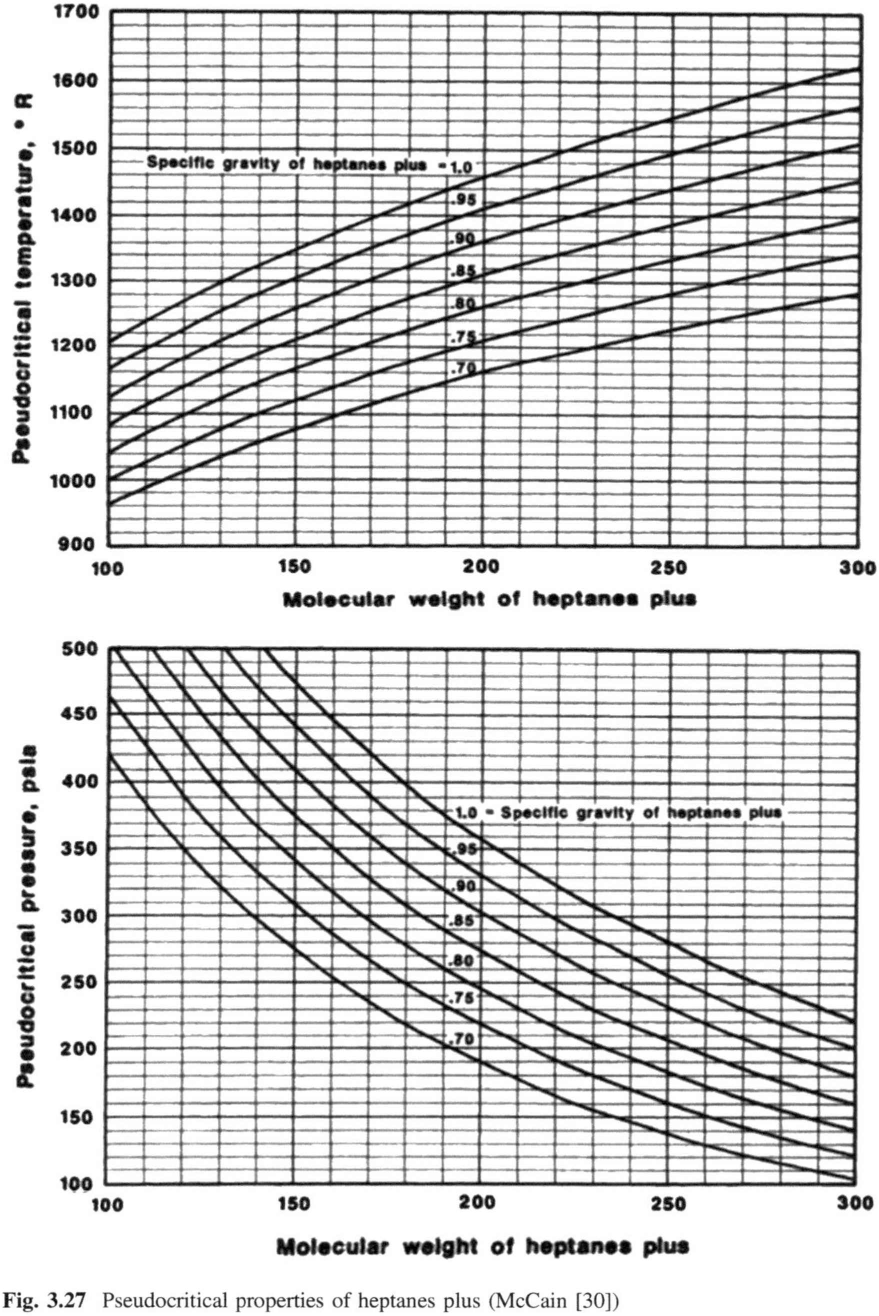

Fig. 3.27 Pseudocritical properties of heptanes plus (McCain [30])

Example 3.6 A natural gas contains some CO_2 and H_2S. Its composition lists in Table 3.16. Determine the *Z*-factor of the natural gas at 60 °C and 10 MPa.

Solution

1. According to the composition of the gas, calculate the pseudocritical temperature (T_{pc}) and pseudocritical pressure (p_{pc}) by Eq. (3.15). The results are shown in the last line in Table 3.16.
2. Determine the value of ε in Fig. 3.26 in terms of the content of CO_2 and H_2S:

$$\varepsilon = 30(\mathrm{Ra}) = 16.67(\mathrm{K})$$

3. Amend the pseudocritical temperature and pseudocritical pressure by Eqs. (3.20) and (3.21):

$$T'_{pc} = T_{pc} - \varepsilon = 250.01 - 16.67 = 233.34(\mathrm{K})$$

$$p'_{pc} = \frac{p_{pc}T'_{pc}}{T_{pc} + y_{H_2S}(1 - y_{H_2S})\varepsilon} = \frac{5.779 \times 233.34}{250.01 + 0.2(1 - 0.2) \times 16.67} = 5.336(\mathrm{Mpa})$$

4. Calculate the pseudoreduced temperature and pseudoreduced pressure of the natural gas by Eq. (3.16):

$$T_{pr} = \frac{T}{T'_{pc}} = \frac{273 + 60}{233.34} = 1.42$$

$$p_{pr} = \frac{p}{p'_{pc}} = \frac{10.0}{5.336} = 1.87$$

5. Determine the value of *Z*-factor in Fig. 3.25 by T_{pr} and p_{pr}: $Z = 0.82$.

Table 3.16 Mass and mole compositions of a natural gas

Components	Mole fraction, y_i	Critical temperature, T_{ci} (K)	y_iT_{ci}	Critical pressure, p_{ci} (MPa)	y_ip_{ci}
CO_2	0.10	304.1	30.41	7.37	0.737
H_2S	0.20	373.5	74.70	9.00	1.80
CH_4	0.60	190.5	114.30	4.59	2.754
C_2H_6	0.10	306.0	30.60	4.88	0.488
Sum	1.00	$T_{pc} = \sum y_i T_{ci} = 250.01$		$p_{pc} = \sum y_i p_{ci} = 5.779$	

3.3.3 Gas Formation Volume Factor

Gas formation volume factor is defined as the volume of free gas, measured at reservoir conditions, required to produce unit volume of the gas at surface standard conditions. Thus

$$B_g = \frac{V_R}{V_{sc}}, \tag{3.22}$$

where B_g is gas formation volume factor, m^3/m^3; V_R is the volume of a gas at reservoir conditions (temperature and pressure), m^3; V_{sc} is the volume of the gas at surface standard conditions (0.1 MPa, 20 °C), m^3.

Note that standard conditions denote 0.1 MPa and 20 °C in China, and are 14.7 psia and 60 °F in Europe and the United States. Units of gas formation volume factor are also different. Units of reservoir cubic feet per standard cubic foot, res cu ft/scf, are often used in Europe and the United States.

Formation volume factor is also known as reservoir volume factor. The reciprocal of the formation value of the factor sometimes is called *gas expansion factor* [4]. Unfortunately, the term *formation volume factor* is used occasionally when gas expansion factor is meant. One must always examine the value of the factor to be sure which is intended.

Based on the equation of stats for gases, B_g of a natural gas can be directly calculated with Eq. (3.22). In terms of the real gas equation, the volume of a natural gas at reservoir conditions can be obtained:

$$V_R = Z\frac{nRT}{P}, \tag{3.23}$$

where T and p represent reservoir temperature (K) and pressure (MPa), separately; n is the mole number of the gas.

The volume of the same gas at surface standard conditions (T_{sc}, p_{sc}) is

$$V_{sc} = \frac{nRT_{sc}}{p_{sc}}, \tag{3.24}$$

where V_{sc} is the volume of the gas at surface standard conditions, m^3; T_{sc} and p_{sc} are the standard temperature, K, and pressure at the surface, MPa, respectively; n is the mole number of the gas, Kmol; R is universal gas constant, $R = 0.008314$ MPa m^3/(Kmol K).

According to Eq. (3.22), the formation volume factor for the gas is

$$B_g = \frac{V_R}{V_{sc}} = \frac{Z\frac{nRT}{p}}{\frac{nRT_{sc}}{p_{sc}}} = \frac{ZTp_{sc}}{T_{sc}p} \tag{3.25}$$

Since T_{sc} = 273 + 20 (K), p_{sc} = 0.1 MPa and T = 273 + t (K), then

$$B_g = \frac{Z(273 + t)}{2930p}, \tag{3.26}$$

where t is reservoir temperature, °C; p is reservoir pressure, MPa.

From this, the formation volume factor for natural gases at any reservoir conditions can be easily determined with Eq. (3.26) as long as the Z-factor is given.

Generally, the Z-factors of a natural gas can be read from generalized compressibility charts as described in unit Sect. 3.3.2.4. However, the Z-factors of a gas should be measured in the laboratory if the gas has special content of nonhydrocarbons or/and heavier hydrocarbon, or is from ultra-deep formation.

Due to the great compressibility of natural gases and the high pressures in reservoirs, the formation volume factor of natural gases is generally far less than one (Fig. 3.28).

Example 3.7 Calculate the formation volume factor of a dry gas with a relative density of 0.80 at reservoir temperature of 100 °C and reservoir pressure of 30 MPa.

Solution

First, the pseudocritical properties and pseudoreduced properties should be estimated before Z-factor can be calculated.

From Fig. 3.24, we have T_{pc} = 210 K and p_{pc} = 4.57 MPa at γ_g = 0.80.

Then,

$$T_{pr} = \frac{T}{T_{pc}} = \frac{(273 + 100)}{216} = 1.73 \quad \text{and} \quad p_{pr} = \frac{p}{p_{pc}} = \frac{30}{4.57} = 6.56$$

So, from Fig. 3.25, we read Z = 0.943.

Second, calculate B_g by Eq. (3.26):

$$B_g = \frac{Z(273 + t)}{293p} = \frac{0.943 \times (273 + 100)}{293 \times 30 \times 10} = 0.004$$

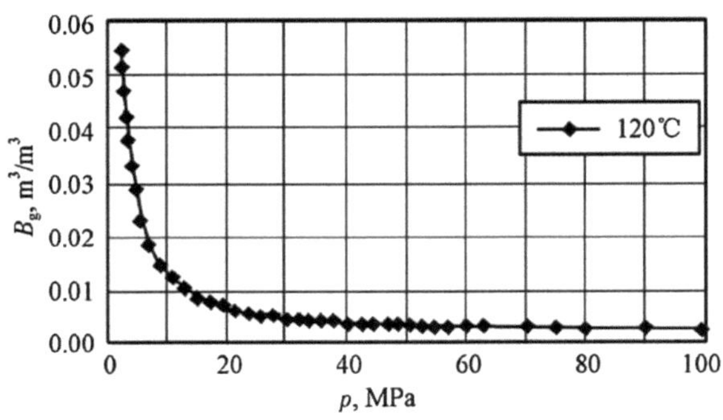

Fig. 3.28 A typical curve of gas formation volume factor as a function of pressure at reservoir temperature (杨胜来 2004)

3.3.4 Isothermal Compressibility of Gas

Gas usually is the most compressible medium in the reservoir [3]. Isothermal compressibility of gas, symbol C_g, is extensively used in transient well test analysis and other reservoir engineering problems. Gas isothermal compressibility is a measure of a change in the volume of a gas with respect to a change in reservoir pressure. Since reservoir temperature is always constant during gas production, the isothermal gas compressibility is extensively used in determining the compressible properties of gas reservoirs.

The isothermal compressibility of gas is defined as the fractional change of gas volume as pressure is changed at a constant temperature [4]:

$$C_g = -\frac{1}{V}\left(\frac{\partial V}{\partial p}\right)_T, \tag{3.27}$$

where C_g is the isothermal compressibility of gas, 1/MPa; V is the volume of a gas, m^3; p is reservoir pressure, MPa; T is reservoir temperature, K.

For convenience, C_g is simply referred to as *compressibility* or *gas compressibility*. Care should be taken so that one is not confused with the two concepts: C_g and Z because they both use the word "compressibility", and both are related to the effect of pressure on gas volume. C_g (gas compressibility) means the ability of the gas being compressed at constant temperature; whereas Z (compressibility factor or Z-factor) denotes the deviation from ideality of a gas. C_g is used for the calculation of the elastic energy of a gas reservoir, but Z-factor is used for the calculation of gas behavior in the equation of states.

In general, the compressibility of natural gas is significantly higher than that of crude oil and interstitial water in reservoirs, typically by two orders of magnitude [11]. The gas compressibility is about in the order of 720×10^{-4} MPa^{-1}, whereas typical water and oil compressibilities are about 6×10^{-4} and 15×10^{-4} MPa^{-1}, respectively [7].

Equation (3.27) is a definition of isothermal gas compressibility. It is not easy to use in practice. For convenience, it is necessary to convert the definition to a formula, with which one can directly calculate the C_g according to the attributes of a gas.

From real gas equation of state, Eq. (3.11), we have

$$V = nRT\frac{Z}{p} \tag{3.28}$$

Since Z is a function of pressure, differentiating the above equation with respect to pressure at constant temperature T gives

$$\left(\frac{\partial V}{\partial p}\right)_T = nRT\frac{p\frac{\partial Z}{\partial p} - Z}{p^2} \tag{3.29}$$

Combining Eqs. (3.27), (3.28), and (3.29) produces the following generalized expression:

$$C_g = -\frac{1}{nRT\frac{Z}{p}} \times nRT\frac{p\frac{\partial Z}{\partial p} - Z}{p^2} \tag{3.30}$$

Then

$$C_g = \frac{1}{p} - \frac{1}{Z} \cdot \frac{\partial Z}{\partial p} \tag{3.31}$$

The Z and $(\partial Z/\partial p)_T$ in Eq. (3.31) can be obtained from compressibility charts (Z-p chart) for pure gases. In detail, one should first read the Z-factor of a pure gas by the pressure (p) and temperature (T) on the compressibility chart of the gas; second, calculate the slope of tangent line through the point of the value of Z, $\Delta Z/\Delta p$, which is the value of $(\partial Z/\partial p)_T$.

Particularly, for ideal gas, $Z = 1$ and $\partial Z/\partial p = 0$; therefore, $C_g = 1/p$. It indicates that the isothermal compressibility of an ideal gas is inversely proportional to the pressure.

Example 3.8 Calculate the isothermal compressibility of methane at 20 °C and 6.8 MPa.

Solution

1. Read the Z-factor value of methane at 20 °C and 6.8 MPa from the Z-p chart of methane (Fig. 3.21): $Z = 0.89$.
2. Determine the slope of tangent line through the point of Z-factor (at 20 °C and 6.8 MPa):

$$\frac{\partial Z}{\partial p} = \frac{\Delta Z}{\Delta p} = -0.0155(\text{MPa}^{-1}).$$

3. Calculate the isothermal compressibility, C_g of methane by Eq. (3.31):

$$C_g = \frac{1}{p} - \frac{1}{Z}\frac{\partial Z}{\partial p} = \frac{1}{6.8} - \left(\frac{1}{0.89}\right)(-0.0155)$$
$$= 1645 \times 10^{-4}(\text{MPa}^{-1})$$

For natural gases (gas mixture), however, it is impossible to calculate directly the C_g by Eq. (3.31) because no proper Z-p chart of a natural gas can be used for the determination of the value of $\partial Z/\partial p$. For mixed gases, only Z-p_r charts are available. In this chart, Z-factor is a function of reduced properties of gases. Therefore, the C_g of a natural gas can be calculated only if the equation is directly related with the reduced properties of the mixed gas.

Based on the idea, substitute $p_{pr}p_{pc}$ for p in Eq. (3.31), and we have

$$C_g = \frac{1}{p_{pr}p_{pc}} - \frac{1}{Z} \cdot \frac{\partial Z}{\partial (p_{pr}p_{pc})} \tag{3.32}$$

Then,

$$C_g = \frac{1}{p_{pc}} \left(\frac{1}{p_{pr}} - \frac{1}{Z} \cdot \frac{\partial Z}{\partial p_{pr}} \right) \tag{3.33}$$

In Eq. (3.33), both C_g and Z are the functions of pseudoreduced pressure; and the value of $\partial Z / \partial p_{pr}$ can be directly determined from the Z-p_r chart.

Therefore, if the composition of a natural gas is given, the pseudocritical properties, p_{pc} and T_{pc} of the natural gas can be determined by Eq. (3.15). The pseudoreduced properties, p_{pr} and T_{pr} of the natural gas can then be calculated by Eq. (3.16). The Z-factor and $\partial Z / \partial p_{pr}$ in Eq. (3.33) can then be determined from a generalized Z-factor chart (Fig. 3.25). Finally, the isothermal compressibility of the natural gas can be calculated by Eq. (3.33).

Example 3.9 Calculate the isothermal compressibility of a natural gas at 49 °C and 10.2 MPa. The composition of the natural gas is given in Table 3.15 of Example 3.4.

Solution

1. The pseudocritical properties and pseudoreduced properties of the natural gas have been calculated in Example 3.5. They are

$$p_{pc} = 4.59(\text{MPa}); \quad T_{pc} = 212.7(\text{K});$$
$$p_{pr} = 2.22; \quad T_{pr} = 1.51.$$

2. The Z-factor of the natural gas read from Fig. 3.25 is 0.813. The slope of the tangent line through the point of Z-factor is determined as $\partial Z / \partial p_{pr} = 0.053$.
3. Then the compressibility of the natural gas can be calculated by Eq. (3.33) as follows:

$$C_g = \frac{1}{p_{pc}} \left(\frac{1}{p_{pr}} - \frac{1}{Z} \frac{\partial Z}{\partial p_{pr}} \right) = \frac{1}{4.59} \left[\frac{1}{2.22} - \frac{1}{0.813} (-0.053) \right]$$
$$= \frac{1}{4.59} \times 0.5156 = 1123 \times 10^{-4} (\text{MPa}^{-1})$$

Multiplying Eq. (3.33) by p_{pc} yields

$$C_g p_{pc} = \frac{1}{p_{pr}} - \frac{1}{Z} \cdot \frac{\partial Z}{\partial p_{pr}} \tag{3.34}$$

Let

$$C_{pr} = C_g p_{pc} \tag{3.35}$$

The term C_{pr} is called the *isothermal pseudoreduced compressibility*. It is dimensionless.

Combining Eqs. (3.34) and (3.35) gives

$$C_{pr} = \frac{1}{p_{pr}} - \frac{1}{Z} \cdot \frac{\partial Z}{\partial p_{pr}} \tag{3.36}$$

In Eq. (3.36), the pseudoreduced compressibility is a function of pseudoreduced pressure and pseudoreduced temperature. The correlations were developed by Trube [31] and are shown in Fig. 3.29. Obviously, the pseudoreduced compressibility is easily obtained from Fig. 3.29, the isothermal gas compressibility can then be readily calculated by the following expression:

$$C_g = \frac{C_{pr}}{p_{pc}} \tag{3.37}$$

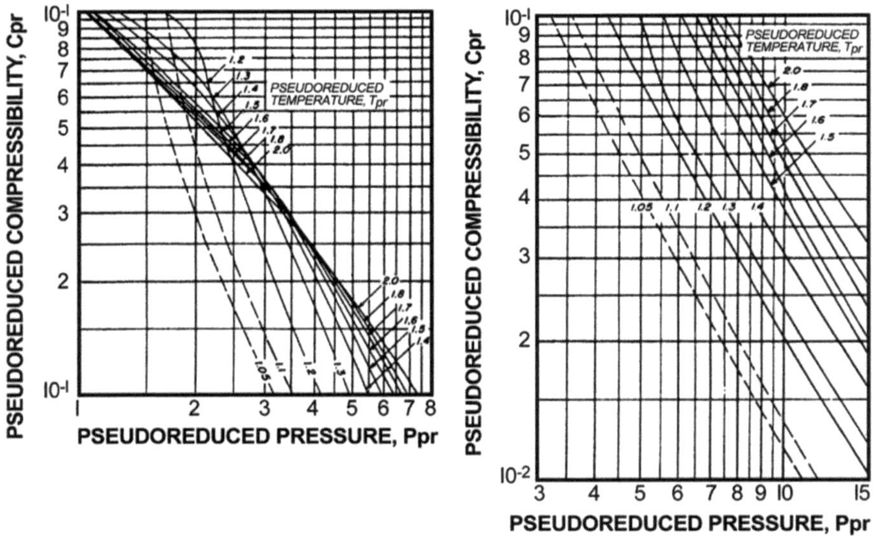

Fig. 3.29 Pseudoreduced compressibility for natural gases [31]

Example 3.10 Determine the compressibility of the natural gas given in Example 3.9.

Solution

1. The following parameters are given in Example 3.9:

$$p_{\text{pc}} = 4.59(\text{MPa}); \quad T_{\text{pc}} = 212.7(\text{K});$$
$$p_{\text{pr}} = 2.22; \quad T_{\text{pr}} = 1.51.$$

2. Determine the pseudoreduced compressibility from the chart Fig. 3.29 according to $p_{\text{pr}} = 2.22$ and $T_{\text{pr}} = 1.51$: $C_{\text{pr}} = 0.51$.
3. By Eq. (3.37), the isothermal compressibility of the natural gas is

$$C_{\text{g}} = \frac{C_{\text{pr}}}{p_{\text{pc}}} = \frac{0.51}{4.59} = 1111 \times 10^{-4}(\text{MPa}^{-1})$$

3.3.5 Viscosity of Natural Gas

The viscosity of a fluid is a measure of its internal friction resistance to flow. Gas, having significantly lower viscosity than oil and water, tends to dominate multiphase flow in the reservoir [11]. So, the knowledge of natural gas viscosity is essential for the studies on the dynamic/flow behavior of a gas through reservoirs.

If the friction between layers of the fluid is small, i.e., low viscosity, an applied shearing force will result in a large velocity gradient. As the viscosity increases, each fluid layer exerts a larger frictional drag on the adjacent layers and velocity gradient decreases (Fig. 3.30).

The viscosity of a fluid expresses its resistance to shearing flows, where adjacent layers move parallel to each other with different speeds. It is generally defined as the ratio of the shear force (internal friction) per unit area to the local velocity gradient when a relative motion between molecular layers of a fluid (i.e., a natural gas) occurs. Therefore, the viscosity of a natural gas can be defined by the following equation:

$$\mu_{\text{g}} = \frac{F/A}{\text{d}v/\text{d}Z}, \tag{3.38}$$

where μ_{g} is the viscosity of a natural gas, P; *F/A* is the shear force per unit area, dyn/cm^2; d*v*/d*Z* is the velocity gradient of the natural gas, cm/(s cm).

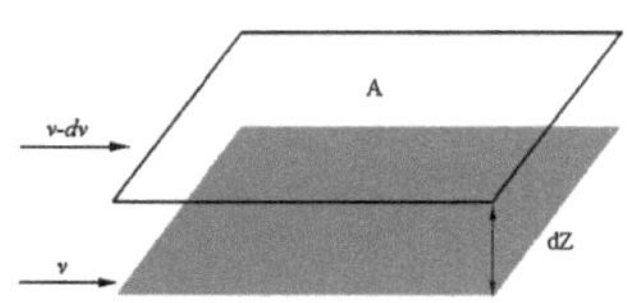

Fig. 3.30 Laminar shear of fluid between two layers of fluid

The cgs physical unit for the dynamic viscosity of a fluid may be poise (P), centipoise (cP), micropoises (μP), or mPa s. The poise (P) is rarely used in practice. The centipoise (cP) is equal to the SI multiple milliPascal seconds (mPa s). For example, water at 20 °C has a viscosity of 1.002 mPa s = 1.0020 cP.

In oil fields, centipoise is often used for liquid viscosity, while micropoise is more common for gas viscosity. However, mPa s is prevalent at present.

One poise equals a viscosity of 1 dyne s/cm^2 and can be converted into other field units by the following relationships:

1 P = 1 dyn s/cm^2 or 1 P = 1 g/cm s.
1 P = 100 cP = 1 × 10^6 μP.
1 cP = 1 mPa s.

The gas viscosity is not commonly measured in the laboratory because it can be estimated precisely from empirical correlations [26]. Like all intensive properties, the viscosity of a natural gas can be completely described by the following function:

$$\mu_g = f(p, T, y_i) \tag{3.39}$$

This relationship states that gas viscosity is a function of pressure, temperature, and composition of the gas. In different pressure ranges, the variation in gas viscosity is different. Figure 3.31 shows the pressure regions of different gas denseness.

3.3.5.1 Gas Viscosity at Low Pressures

At lower pressures, the distance between gas molecules is so far that gas viscosity is independent of pressure. According to molecular dynamics, gas viscosity is determined by the following expression:

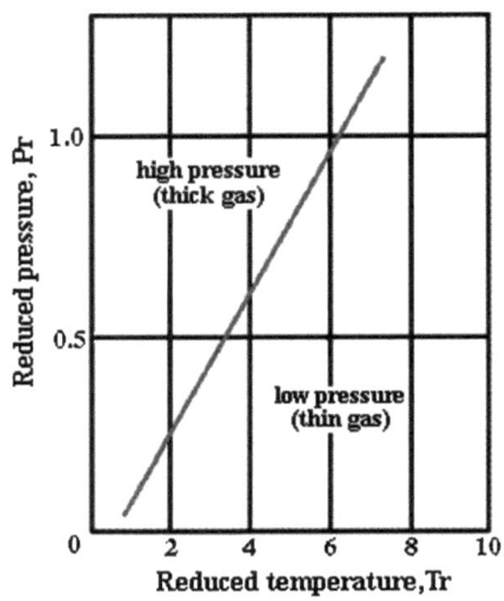

Fig. 3.31 Pressure regions for the change in gas viscosity (He [32])

$$\mu_g = \frac{1}{3}\rho v \bar{\lambda}, \tag{3.40}$$

where μ_g is gas viscosity, g/cm s; ρ is gas density, g/cm; v is the average velocity of gas molecules, cm/s; λ is the average free path of gas molecules, cm.

Equation (3.40) states that gas viscosity depends on the density, the mean velocity, and the mean free path of gas molecules. On the other hand, the mean free path is inversely proportional to the density; and the mean velocity of gas molecules is independent of pressure. So an increase in density due to an increase in pressure does not result in any change in viscosity.

However, as the temperature increases, the thermal motion and average velocity of gas molecules increase greatly. The collision frequency between molecules and the internal friction of the gas thus increase. As a result, gas viscosity shows the visible increase with the increase in temperature. The relationship between viscosity and temperature for pure gases at atmospheric pressure is given in Fig. 3.32.

Figure 3.32 shows that the viscosity of pure gases increases with the increase in temperature and the decrease in gas molecular weight at atmospheric pressure. At lower pressures, the viscosity of natural gases (gas mixtures) also has the same

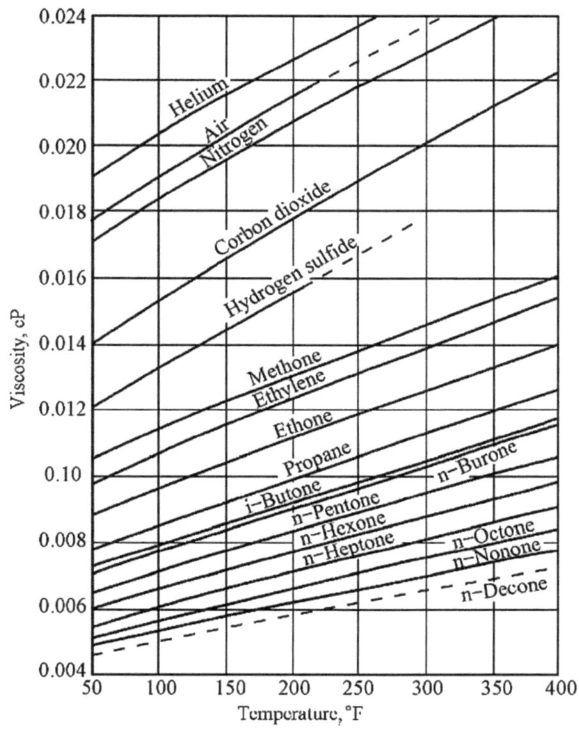

Fig. 3.32 Viscosities of pure gases at atmospheric pressure [33]

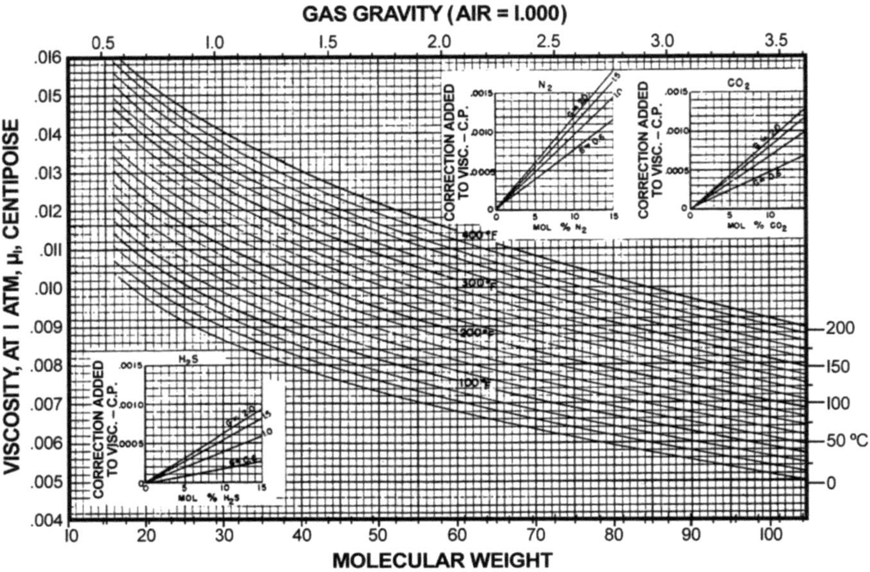

Fig. 3.33 Viscosity chart of hydrocarbon gases at atmospheric pressure, including correction for H_2S, CO_2, and N_2 [33]

behavior as pure gases. The relationship of the viscosity of hydrocarbon gases as a function of atmospheric pressure and temperature, and molecular weight is shown in Fig. 3.33. The viscosity of natural gases at surface conditions can be directly obtained from the diagram. However, the viscosity of natural gases is obviously affected by the nonhydrocarbons and their amount in the gas. Therefore, if a natural gas contains nonhydrocarbons, the viscosity read from the curves in Fig. 3.32 should be corrected using the correction factor from the insets.

3.3.5.2 Gas Viscosity at High Pressure

At high pressures, the change in gas viscosity is similar to the change in the viscosity of a liquid. Gas viscosity decreases with the decrease in temperature, pressure, and gas molecular weight. However, the effects of nonhydrocarbons on gas viscosity are the same as at low pressures. In a word, nonhydrocarbons in natural gases always result in the increase in gas viscosity.

3.3.5.3 Determination of Gas Viscosity

The natural gas viscosity is not commonly measured in the laboratory because experimental determination is difficult [4]. Usually, the petroleum engineer must rely on viscosity correlations. Gas viscosity can be precisely estimated from empirical correlations or plots available.

Determination of Gas Viscosity at Atmospheric Pressure

The following equation may be used to calculate the viscosity of a mixture of gases when the composition of the gas mixture is known and the viscosities of the components are known at the pressure and temperature of interest

$$\mu_g = \frac{\sum y_i \mu_{gi} M_i^{0.5}}{\sum y_i M_i^{0.5}}, \tag{3.41}$$

where μ_g is the viscosity of a natural gas at desired temperature and atmospheric pressure, mPa s; y_i is the mole fraction of the *i*th component in the gas, f; μ_{gi} is the viscosity of the *i*th component in the gas at desired temperature and atmospheric pressure, which can be obtained from Fig. 3.32, mPa s; M_i is the molecular weight of the *i*th component in the gas.

Example 3.11 Calculate the viscosity of a natural gas, whose composition lists in Table 3.17, at 93.3 °C and 0.101 MPa.

Solution

1. Determine the molecular weight, M_i and viscosity, μ_{gi} of each component in the natural gas form handbooks available and Fig. 3.32, and then list them in Table 3.17. The viscosity of each component is read from Fig. 3.32 by the temperature, 93.3 °C.
2. Calculate the values of $y_i M_i^{0.5}$ and $y_i \mu_{gi} M_i^{0.5}$ of each component in the natural gas, and then list them in Table 3.17.
3. According to the data determined above, calculate the viscosity of the gas at 93.3 °C and 0.101 MPa:

$$\mu_g = \frac{\sum y_i \mu_{gi} M_1^{0.5}}{\sum y_i M_1^{0.5}} = \frac{0.0537}{4.311} = 0.0125(\text{mPa s})$$

Table 3.17 The composition of a natural gas

Components	y_i	M_i	μ_{gi}	$M_i^{0.5}$	$y_i M_i^{0.5}$	$y_i \mu_{gi} M_i^{0.5}$
CH_4	**0.85**	16.0	0.0130	4.00	3.40	0.0442
C_2H_6	**0.09**	30.1	0.0112	5.48	0.493	0.0055
C_3H_8	**0.04**	44.1	0.0098	6.64	0.266	0.0026
n-C_4H_{10}	**0.02**	58.1	0.0091	7.62	0.152	0.0014
Sum	–	–	–	–	$\sum = 4.311$	$\sum = 0.0537$

If the composition of a gas is not given, Fig. 3.33 can be used to obtain the viscosities of gas mixtures of hydrocarbons at atmospheric pressure. Inserting plots in Fig. 3.33 is used to correct the effect of nonhydrocarbons (hydrogen sulfide, nitrogen, or carbon dioxide) in the gas on the gas viscosity. In short, each of nonhydrocarbon gases in the gas increases the viscosity of the natural gas, which is superposed by each of the nonhydrocarbon.

Example 3.12 A natural gas contains 10 % mole of H_2S, and its relative density is 0.6. Determine the viscosity of the natural gas at 0.1 MPa and 43.3 °C.

Solution

1. Read the viscosity of the gas from Fig. 3.33 by its relative density, 0.6 and temperature, 43.3 °C: $\mu_{g1} = 0.0113$ (mPa s).
2. Read the modified value of H_2S from the insert plot at the bottom left corner of Fig. 3.33 according to the content of H_2S, 10 % and the relative density of the natural gas, 0.6: $m = 0.0002$(mPa s)
3. Calculate the viscosity of the gas at 0.1 MPa:

$$\mu_g = \mu_{g1} + m = 0.0113 + 0.0002 = 0.0115(\text{mPa s})$$

In numerical simulations, only correlations can be expediently used for the estimation of gas viscosity. The following equations are commonly used to determine the viscosity of gas mixtures at lower pressures:

$$\mu_g = \begin{cases} (3.40)T_{pr}^{8/9}\Big/\xi, & T_{pr} < 1.5 \\ 16.68 \times \left(0.1338T_{pr} - 0.0932\right)^{5/9}, & T_{pr} > 1.5 \end{cases}, \tag{3.42}$$

where μ_g is the viscosity of a natural gas (gas mixture), 10^2 mPa s; T_{pr} is the pseudoreduced temperature of the gas. The coefficient, ξ is determined by the following equation:

$$\xi = \frac{T_{pc}^{1/6}}{p_{pc}^{2/3}\left(\sum y_i M_i\right)^{0.5}}, \tag{3.43}$$

where T_{pc} is the pseudocritical temperature of the gas, K; p_{pc} is the pseudocritical pressure of the gas, MPa.

This correlation is called "corresponding state correlation" owing to it based on the principle of corresponding states.

Calculation of Gas Viscosity at High Pressure

In most instances, the petroleum engineer is concerned with the viscosity of natural gases at high pressures. The viscosity of natural gases at high pressures is often determined based first on the viscosity of the gas at atmospheric pressure.

Figure 3.34 is a diagram of the viscosity ratio as a function of pseudoreduced properties of hydrocarbon, which was developed in terms of the principle of corresponding states. Using this chart, one can easily estimate gas viscosity under high pressures. First of all, a viscosity ratio, μ_g/μ_{g1} of a natural gas can be determined from the chart (Fig. 3.34) according to the pseudoreduced properties of the gas. Then, the viscosity ratio multiplied by the gas viscosity at one atmosphere gives the viscosity of the gas at given high pressure.

That is to say, the following steps should be followed in order to obtain the gas viscosity at high pressures:

1. Using natural gas gravity or molecular weight ($M = \gamma_g \times 28.97$), determine the viscosity of the gas at atmospheric pressure and reservoir temperature, μ_{g1} in Fig. 3.33;

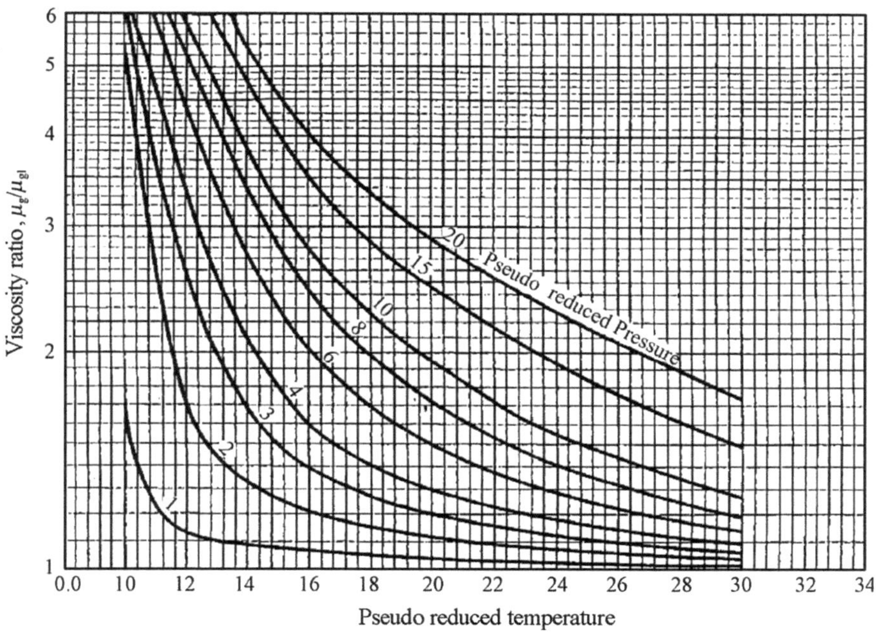

Fig. 3.34 Viscosity ratios for hydrocarbon gases at elevated pressure [33]

2. Correct the viscosity (μ_{g1}) for impurity gases, if any, using insets in Fig. 3.33. The correction value $\mu_{g1} = \mu_{g1}$ (uncorrected) + N_2 correction + H_2S correction + CO_2 correction.
3. Read the viscosity ratio (μ_g/μ_{g1}) of the gas in Fig. 3.34 in terms of its pseudocritical properties (T_{pr}, p_{pr}).
4. Calculate the viscosity of the gas at reservoir pressure by the following equation:

$$\mu_g = \frac{\mu_g}{\mu_{g1}} \times \mu_{g1}, \tag{3.44}$$

where μ_g is the gas viscosity at reservoir conditions, mPa s; μ_{g1} is the gas viscosity at atmosphere pressure and reservoir temperature, mPa s; μ_g/μ_{g1} is the viscosity ratio obtained from Fig. 3.34, dimensionless.

During the production of condensate gas reservoirs, the composition of condensate gas varies with reservoir pressure when reservoir pressure is lower than the upper dew point pressure because condensate oil condenses continually from the gas phase in the reservoir. The change in condensate gas composition finally causes the change in gas viscosity. Customarily, the determination of condensate gas viscosities at different pressures often relies on the correlations. There are many correlations available in petroleum engineering. One of the most common correlations is called "*residual viscosity*", and expressed as follows:

$$(\mu - \mu_0)\eta = 1.08\left[\exp(1.439\rho_{pr}) - \exp\left(-1.111\rho_{pr}^{1.858}\right)\right], \tag{3.45}$$

where μ is the viscosity of a condensate gas at high pressure, 10^2 mPa s; μ_0 is the viscosity of condensate gas at atmosphere pressure, 10^2 mPa s; η is a coefficient, calculated by the following formula: $\eta = T_{pc}^{1/6}/\left(M^{1/2}p_{pc}^{2/3}\right)$; $\rho_{pr} = \rho/\rho_{pc}$ is the pseudoreduced density of the gas, dimensionless; $\rho_{pc} = p_{pc}/\left(ZRT_{pc}\right)$ is the pseudo-critical density of the gas, mol/m^3; M is the molecular weight of the gas; Z is the Z-factor of the gas, dimensionless; R is gas constant, R = 0.008314 MPa m^3/(Kmol K).

3.3.6 Dissolution of Natural Gas in Crude Oil

Dissolution may occur in any of the three states (gas, liquid, or solid) of matter. In general, fluids that mix with or dissolve in each other in all proportions are said to be miscible fluids (Fig. 3.35). If two fluids do not mix but, rather, form two layers, they are said to be immiscible. Oil with dissolved gas, which is a mixture of hydrocarbon gas and liquid, is thus an example of miscible fluids.

The dissolution and separation between natural gas and crude oil are very common in the process of reservoir formation and reservoir production. They are different phase transitions occurring in hydrocarbon systems. They may occur when

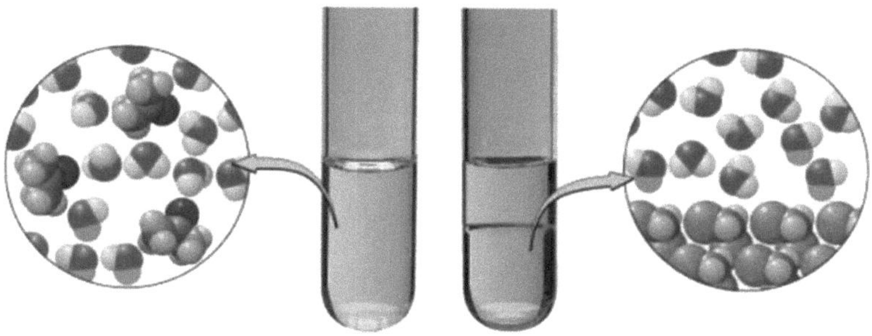

Left: a miscible fluid of acetone and water;
Right: water and methylene chloride are immiscible and for two layers.

Fig. 3.35 Immiscible and miscible liquids [34]

reservoir pressure changes. Actually, they are a vital foundation in oil production because they can help us optimize the method of oil–gas separation in oil production, and optimize the proportion of gas injection in miscible or immiscible flooding.

3.3.6.1 Solubility and Solubility Factor of Gas

The amount of a substance dissolving in a solvent depends on both the substance and the solvent. We describe the amount that dissolves in terms of solubility.

Gas solubility means the amount of a gas that dissolves in a unit liquid at certain temperature and pressure. Gas solubility varies with temperature and pressure. The variation of gas solubility with pressure can be a very useful property.

From general chemistry, we know that the solubility of a pure gas in a liquid obeys Henry's law. Namely, the solubility of a gas (solute) in a liquid (solvent) is proportional to the partial pressure of the gas above liquid at given temperature. Expressed mathematically, Henry's law is

$$R_s = \alpha p, \tag{3.46}$$

where R_s is the solubility of the gas, (SC) m^3/m^3 (volume of solute per unit volume of solvent); p is the partial pressure of the gas, MPa; α is Henry's law constant for the gas at given temperature, (SC) $m^3/(m^3$ MPa). It is also called solubility factor of the gas.

Solubility factor (α) means the volume increment of a gas dissolved in unit liquid with a unit increase in pressure at given temperature. It characterizes the ability of a gas dissolving in a liquid.

Solubility factor relates with both gas and liquid properties. At given temperature, the solubility factor of a gas in a given liquid depends on the properties of the gas. Figure 3.36 shows several solubility curves of pure gases. It can be seen that

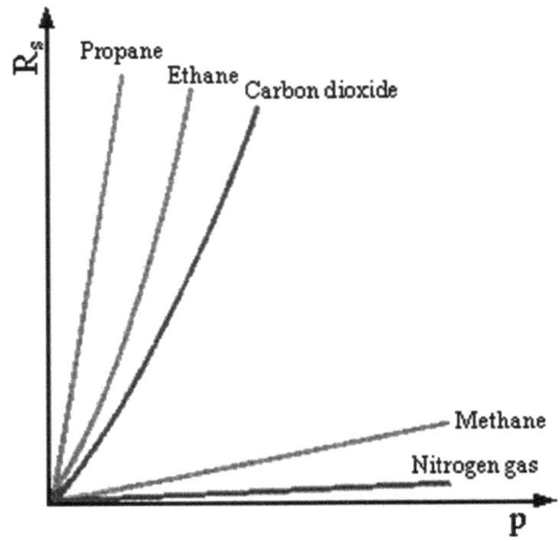

Fig. 3.36 Solubility curves of pure gases (Kotyakhov [35])

the gases with large molecular weights have large solubility factor, and thus dissolve more easily in a liquid.

3.3.6.2 Dissolution of Natural Gas in Crude Oil

In Henry's law, the solubility factors of pure gases are constant, thus the relationship between gas solubility and pressure is linear. However, natural gas is a gas mixture, which chiefly consists of hydrocarbons. Its solubility factor is thus never constant. Figure 3.37 shows a typical solubility curve of a natural gas in a crude oil. Obviously, it is not a straight line. Apparently, the behavior of a nature gas dissolving in a crude oil does not obey Henry's law.

Figure 3.37 shows that the solubility factor of the gas decreases with the increase in pressures when pressure is lower than p_A. This indicates that the ability of the gas dissolving in the oil decreases with the increase in pressure. As is well known, the majority of a natural gas is hydrocarbon gases; and the proportion of each hydrocarbon component in the natural gas decreases with the increase in the molecular

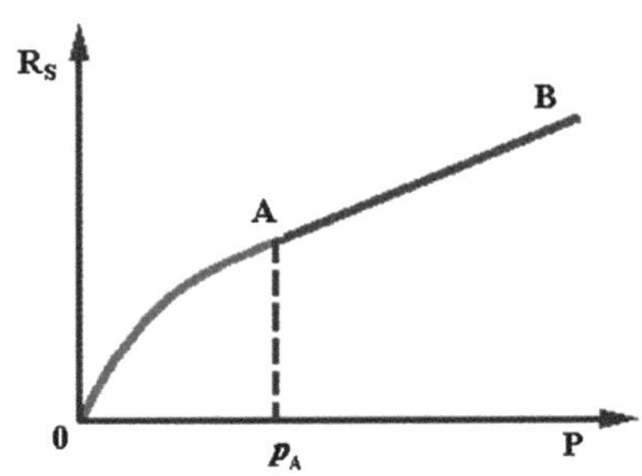

Fig. 3.37 Typical solubility curve of a natural gas in a crude oil

Table 3.18 The average solubility factor of a natural gas in an oil with relative density of 0.873

Pressure range (MPa)	Average solubility factor [$m^3/(m^3$ MPa)]
0–0.7	34.00
0.7–2.0	4.95
2.0–3.8	4.46
3.8–5.4	3.33
5.4–7.2	2.13
7.2–9.0	1.73

weight of the component. The larger the molecular weight is, the smaller the solubility factor of the component is in the oil.

In terms of the general rule, "*like dissolves like*", the heaviest component in the natural gas, such as C_{7+}, dissolves first in crude oil; then the heavier components, as hexane (C_6), pentane (C_5), and so on; finally, the lightest, i.e., methane. Apparently, the lighter the component is, the higher pressure is needed for the component dissolving the oil. Therefore, at lower pressures, the solubility factors decrease with the increase in pressure; and the solubility behavior of the natural gas is an R_s-p curve. After that, when pressure reaches p_A in Fig. 3.37, there is only methane (a few amount of nitrogen gas, sometimes) left in gas phase. At this time, the solubility behavior turns into a long straight line. It is the solubility curve of methane in the oil, so, the solubility factor is a constant.

For evaluating the behavior of a natural gas dissolving in crude oil, the term *average solubility factor* is used. It is calculated by the following expression:

$$\bar{\alpha} = \frac{\Delta R_s}{\Delta p}, \tag{3.47}$$

where $\bar{\alpha}$ is the average solubility factor of a natural gas in unit pressure difference Δp, (SC) $m^3/(m^3$ MPa); Δp is pressure difference, MPa; ΔR_s is the increment of solubility of the natural gas in per unit change in pressure Δp, (SC) m^3/m^3.

Table 3.18 lists the average solubility factors of a natural gas in a crude oil with relative density of 0.873. These data prove that the average solubility factors of the gas decrease with the increase in pressure; thus, the ability of the natural gas dissolving in the oil decreases with the increase in pressure.

It can be seen that the lighter the natural gas is, the lower its average solubility factor is, and thus the lower the gas solubility is. So, the solubility of a dry gas in an oil is less than that of a wet gas in the same oil.

3.3.6.3 Factors Affecting the Solubility of Natural Gas

Many factors may affect the dissolution of natural gases in crude oils, such as the pressure, temperature, the properties of natural gases and oils. The effect of pressure and the properties of natural gases have been described above. Next, we will analyze and simplify the influence of the following factors: temperature, the properties of oil and the dissolving mode, and the solubility of natural gases in oils.

Properties of Crude Oil

Due to *like dissolves like*, the lighter the crude oil is, the higher the solubility of a natural gas in the oil is. The reason is that the lighter the crude oil is, the smaller the difference between the properties (e.g., density) of oil and gas. Therefore, the solubility of a natural gas in heavy oil is less than in light oil (Fig. 3.38).

Temperature

Experiments have proved that temperature will decrease the dissolution of a natural gas in an oil (Fig. 3.39). The higher the temperature is, the molecules of the gas move faster and gas molecules escape from oil more easily. Thereby, high temperature results in low solubility of a gas in an oil. Moreover, Fig. 3.39 shows that the higher the pressure is, the more quickly the gas solubility decreases with the increase in temperature.

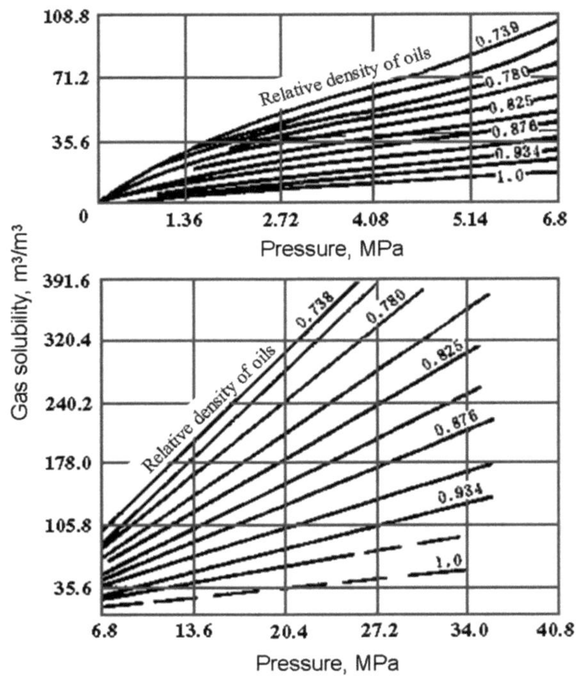

Fig. 3.38 Solubilities of natural gas in crude oils with different densities (Beal 1964)

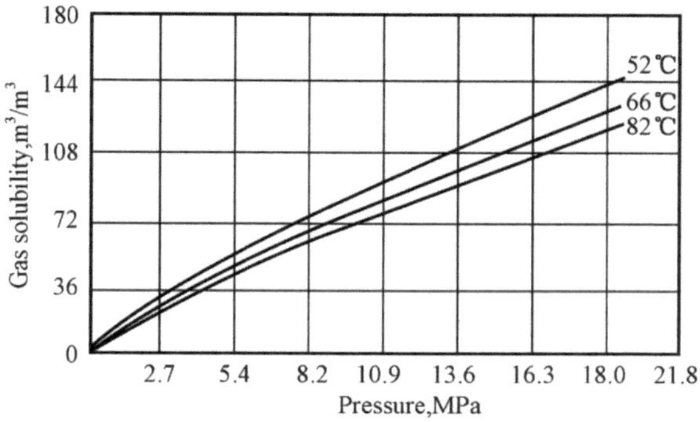

Fig. 3.39 Solubility curves of a natural gas at different temperatures (Kotyakhov [35])

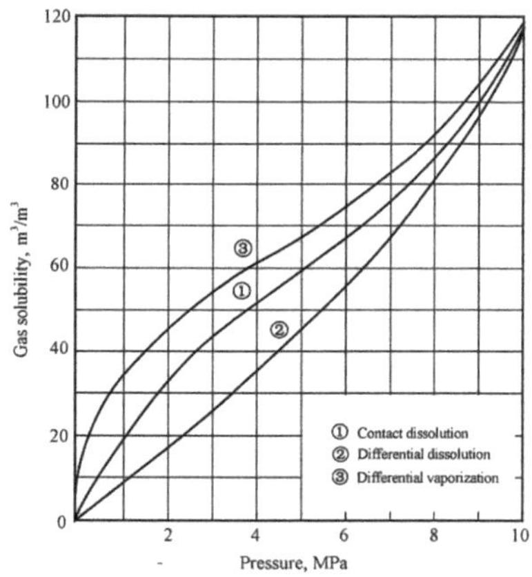

Fig. 3.40 Solubility curves of a natural gas in an oil (He [32])

Dissolving Mode

The gas evolved from an oil can dissolve into the oil again at the same temperature–pressure conditions. However, different dissolving modes lead to different gas solubilities in the oil even for the same oil and gas at the same conditions. Figure 3.40 shows the visible differences of gas solubility curves between contact dissolution and differential dissolution.

Contact dissolution means that during the process of gas dissolving in oil, the gas and the oil contact all the time with each other. The total composition of the oil–gas system keeps constant. Differential dissolution refers to the process in which each few amount of gas can enter and dissolve in the oil at each infinitesimal increment in pressure, and during the process, let pressure increase successively till no gas can dissolve in the oil. The most important is that the total composition of the oil–gas system varies with the pressure.

In Fig. 3.40, curve ① is the solubility curve of contact dissolution, and curve ② is the curve of differential dissolution. Obviously, at any given pressure, the solubility of contact dissolution is higher than that of differential dissolution. Because each increment in pressure in differential dissolution is so infinitesimal that extremely few amount of heavier hydrocarbons in a gas can dissolve in the oil. Finally, the total amount of gas dissolved in the oil in differential process is thus less than that in contact process.

In a word, the lighter the oil is and the heavier the gas is, the larger the solubility of a natural gas in oil is. The lower the temperature of a system is and the higher the pressure is, the larger the gas solubility is.

3.4 Properties of Crude Oil

In petroleum engineering, the reservoir engineer is primarily interested in the properties of crude oil. A detail description of physical properties of crude oil is necessary to the theoretical study and application of petroleum reservoir engineering in oil production.

The nature of crude oil changes greatly. It is dependent on the composition of the oil, namely the proportion of hydrocarbons and nonhydrocarbons in the oil. Moreover, the properties of in-place oil are also different from those of oil at surface conditions owing to the great difference in temperature and pressure. In reservoirs, in-place oil often contains a great deal of dissolved gas. When produced to the surface, the in-place oil will shrink and become more viscous because of the release of dissolved gas from the oil. Therefore, there are many differences between in-place oil and the oil at the surface conditions.

3.4.1 General Properties of Crude Oil

The value of a crude oil is dependent on its properties. The general properties of a crude oil means the physical properties of the oil at surface conditions, mainly including the density, viscosity, pour point, color, and so on. For the better understanding of the in-place oil' nature, several general properties of oil are introduced here.

3.4.1.1 Density

The density of a crude oil is defined as the mass per unit volume of the oil:

$$\rho_o = \frac{m}{V}, \tag{3.48}$$

where ρ_o is the density of a crude oil, g/cm^3; m is the mass of the oil, g; V is the volume of the oil, cm^3.

Generally, the density of a crude oil is associated with the reference pressure and temperature values. They are the standard conditions at the surface (20 °C) and 0.1 MPa, in China; 60 °F (15.6 °C) and 14.7 psia (0.1 MPa) aboard [11]. The density of in-place oil is often lower than that of degassed crude at the surface, i.e., stock tank oil or dead oil, due to the effect of dissolved gas.

3.4.1.2 Relative Density

In China and the former Soviet Union, relative density means the ratio of the density of a stock tank oil (a crude oil at 20 °C, 0.1 MPa) to the density of pure water at 4 °C and 0.1 MPa.

$$d_4^{20} = \frac{\rho_o}{\rho_w}, \tag{3.49}$$

where d_4^{20} is the relative density of a stock tank oil, dimensionless; ρ_o is the density of the oil, g/cm^3; ρ_w is the density of water at 4 °C and 0.1 MPa, g/cm^3.

In general, the relative density of stock tank oils ranges from 0.7927 (47°API) to 1.0000 (10°API)

However, *API* (American Petroleum Institute) *gravity*, expressed in degrees, is more prevalent in Europe and the United States. API gravity is also a measure of oil density, and is related to specific gravity by the following equation [11]:

$$\text{API} = \frac{141.5}{\gamma_o} - 131.5, \tag{3.50}$$

where γ_o is the specific gravity of crude oil, dimensionless.

The specific gravity of a crude oil (γ_o) is defined as the ratio of the oil density over the water density, both measured at the same reference temperature and pressure. A standard temperature of 60 °F (15.6 °C) and atmospheric pressure are customarily specified for petroleum engineering.

The API gravity of a crude oil is inversely proportional to its specific gravity, γ_o. Generally, the specific gravity of crude oil varies between 0.8 and 0.97 in most instances [11]. The corresponding API gravity varies from 14.3°API to 45.4°API. Obviously, the higher the API gravity is, the lighter the crude oil; and vice versa. Normally, the price, which a producer receives for his/her oil, depends on its gravity, the less dense (higher API) being the most valuable.

One should note the small difference between the standard conditions specified for the specific gravity and the relative density of a crude oil. Therefore, the value of the specific gravity γ_o of an oil does not equal the value of its relative density d_4^{20}. In general, $\gamma_o \approx 1.002–1.004\ d_4^{20}$ [1].

3.4.1.3 Pour Point

The pour point of a crude oil is the lowest temperature at which it is observed to pour or flow under prescribed conditions [4]. It is a rough indication of the lowest temperature at which oil is readily pumpable. In addition, the pour point can be defined as the minimum temperature of a crude oil, after which, on decreasing the temperature, the oil ceases to flow.

The pour point of a crude oil depends on the contents of wax, resin-asphaltene, and light hydrocarbons in the oil. The pour point of a crude oil increases with the increase in the content of heavier components, especially the wax classes, in the oil. Usually, the pour point of crude oil ranges from −56 to 50 °C. If the pour point of a crude oil is higher than 40 °C, the oil is known as *high pour point crude*.

The crude oil produced in China is predominantly paraffin-base crude, thus has higher wax content and higher pour point. For instance, the wax content of the crude oils from Daqing oil field is about 25–30 %, and their pour point is about 25–30 °C. The pour point of a crude oil from Liaohe oil field is over 40 °C.

The properties of stock tank oils from the oil fields in China and other countries are listed in Tables 3.19 and 3.20, respectively.

Table 3.19 The properties of stock tank oils from oil fields in China [18]

Properties / Production place	Relative density	Viscosity (mPa s) 50 °C	70 °C	Pour point (°C)	Wax content (%)	Resin content (%)	Asphalt content (%)	Sulfur content (%)
S zone of Daqing field	0.8753	17.4	–	24	28.6	13.3	–	0.15
T zone of Shengli field	0.8845	37.69	17.69	33	17.9	18.3	3.1	0.47
Horizon G of Gudao field	0.9547	427.5	157.5	−12	0	27.5	6.6	2.25
Horizon M of Daguang field	0.9174	51.97	25.55	−12	6.2	14	6.27	0.13
Kelamayi field	0.8699	19.23	–	−50	2	12.6	0.01	0.13
Horizon L of Yumen field	0.858	12.9	–	−16	8.3	22.6	–	–
W zone of Jianghan Field	0.9744	–	62.2	21	3.8	51	9.6	11.8
C zone of Liaohe field	0.9037	37.4	–	−7	4.7	17.6	0.15	0.26
Chuanzhong field	0.8394	12.3	–	30	18.1	3.4	–	–
Horizon P_s of Renqiu field	0.8893	63.5	–	33	22.6	20.7	–	2.35

Table 3.20 The properties of stock tank oils from the oil fields around the world [18]

Properties / Production place	Specific gravity	Viscosity (mPa s)	Pour point (° C)	Wax content (%)	Resin content (%)	Asphalt content (%)	Sulfur content (%)
Hassi Messaoud oil field, Algeria	0.804	2.76 (20 ° C)	−45.6	2.4	0	0.06	0.13
Kirkuk field, Iraq	0.844	4.16 (20 ° C)	−36	3.9	–	1.5	1.95
Cinta oil field, Indonesia	0.855	23.7 (50 ° C)	43.3	29.3	–	19.5	0.08
Baku oil field, Nigeria	0.78–0.950	4.4 (38 ° C)	17.8	5.1	–	–	0.16
Aghajari oil field, Iran	0.852	6.56 (21 ° C)	–	Medium	–	0.6	1.42
Parma oilfeld, Canada	0.838	–	10	–	–	–	0.29
East Texas oil field, U.S.	0.827–0.835	3.4	–	–	–	–	0.21
Romashkinskoys oil field, Russia	0.868	7.5 (50 ° C)	29	4.3	–	–	1.61

3.4.2 Separation of Natural Gas from Crude Oil

In reservoirs, crude oil usually contains a great deal of dissolved gas due to high pressure. When produced out from the hot, high-pressure reservoir to the surface, crude oil experiences the reduction in pressure and temperature. During the process of pressure deplete, the dissolved gas in oil tends to escape from the oil. At surface conditions, a great deal of gas evolves from the oil. For the convenience of transportation, the dissolved gas must be completely separated from the oil at surface conditions. The physical separation or oil–gas separation of the crude oil from wells is one of the basic operations in crude oil production.

Essentially, oil–gas separation is a phenomenon of gas evolving from crude oil because of pressure deplete. It may occur in reservoirs, within wellbores, and in surface production. The separation of oil and gas in surface production are primarily concerned here. In surface production, based on the way of pressure deplete, two methods for gas–oil separation are developed: *flash separation* and *stage separation*.

3.4.2.1 Flash Separation

The problem of separating the gas from a crude oil mixture (well-stream fluid) falls into the well-known problem of flashing a well-stream fluid into two streams: vapor and liquid. A "*flash*" is a distillation in which a well-stream feed is partially vaporized to give a vapor that is rich in volatile components, e.g., methane and ethane.

Flash separation means the method used for distilling the liquefied petroleum gas from crude oil. The principle of this separation is *flash vaporization*. Flash vaporization means the vaporization in which an oil mixture is partially vaporized to produce the oil and gas phases in a pressure drop. Namely, in flash separation, the pressure of a feed of crude mixture is first declined to the pressure which is lower than the bubble point pressure of the system at given temperature in a closed vessel. Then, the mixture is partially vaporized at lower pressure to produce a liquid phase and a gas phase in equilibrium. Finally, the gas over the liquid is completely removed from the vessel. Therefore, flash separation is also known as *flash liberation*.

Figure 3.41a illustrates the flash vaporization test for an oil mixture in a laboratory PVT cell. When a given amount of mercury is removed from the cell (as

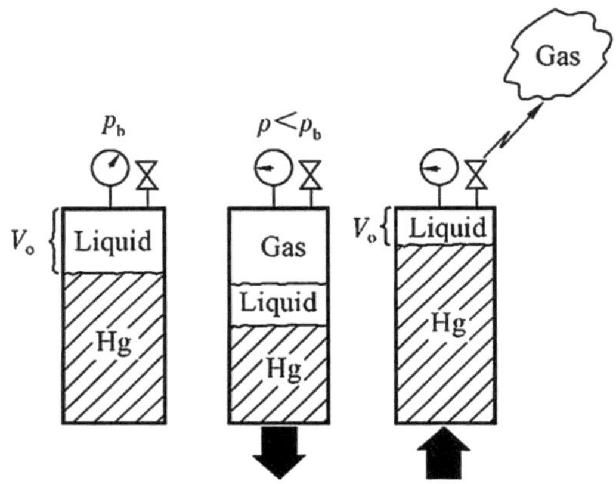

(a) Illustration of flash vaporization (McCain [30])

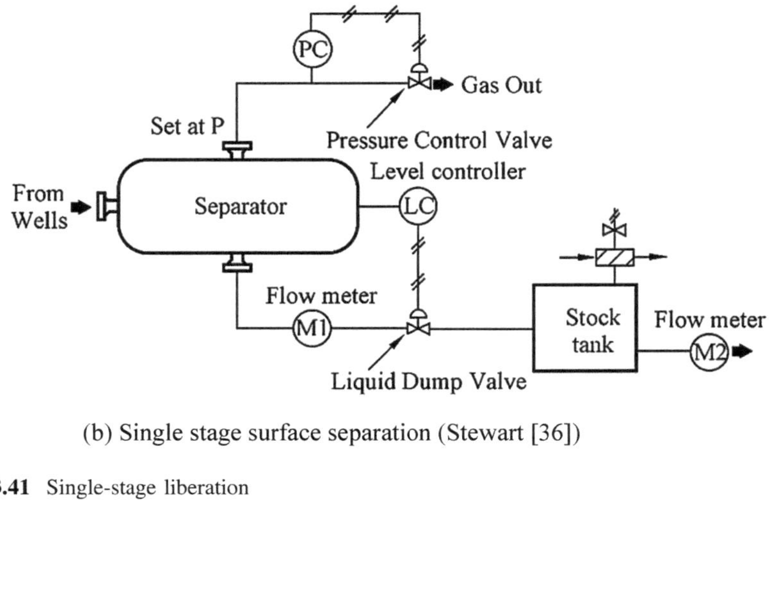

(b) Single stage surface separation (Stewart [36])

Fig. 3.41 Single-stage liberation

shown in Fig. 3.41), the pressure in the cell is declined. Then, the light components in the oil vaporize to gas. When the gas and the liquid are in equilibrium, open the valve on the top of the cell and reject the mercury into the cell again to maintain the pressure in the cell at a constant, and to let the gas escape through a valve.

Figure 3.41a shows that the gas liberated from the oil keeps in contact with the oil before removed, and the total composition of the system is constant. Therefore, flash vaporization is also called *single-stage liberation*. A flash vaporization is a phase transition process of constant in composition and variation in volume.

Figure 3.41b shows a single-stage process. A fluid mixture is flashed in the separator, and then the liquids from the separator enter a stock tank. In the stock tank, the liquids may be flashed again if necessary. Stock tank is not normally considered a separate stage of separation, though it most assuredly is [33].

In a flash vaporization test, the pressure of the crude oil mixture is declined by expanding the mixture volume. Figure 3.42 shows the relationship between the pressure and the volume of the mixture, measured in the test; namely, the *p*–*V* curves of flash experiment. Clearly, the volume of the system at pressures less or greater than the pressure, p_3, is a linear function of pressure. However, the two straight lines have different slopes.

In the region of pressure higher than p_3, there is only liquid in the cell, and the decline in pressure only results in oil expansion. So, the slope of the line ($\Delta V/\Delta p$) is small. When the pressure is lower than p_3, however, the decline in pressure causes both the dissolved gas evolving from the oil and the oil and gas expansion. Moreover, the gas expands more easily than the oil. So, the slope of the line ($\Delta V/\Delta p$) is larger than that at pressures higher than p_3.

Therefore, the two lines intersect at the point, p_3. The pressure, p_3, is thus the bubble point pressure of the oil mixture. That is to say, the bubble point pressure of a hydrocarbon mixture can be determined by the "*break*" of the *p*–*V* curves of flash test. However, it must be emphasized that the linear of *p*–*V* relationship only exists near the bubble point. If the pressure regions are far from the bubble point, the linear may not exist.

According to the volumes of the gas and the oil from the flash separation of a crude mixture, the gas–oil ratio of separation can be calculated by the following expression:

$$R = \frac{V_g}{V_o}, \tag{3.51}$$

where R is the gas–oil ratio of oil–gas separation, m^3/m^3; V_g is the volume of the gas separated from the crude mixture, m^3, V_o is the volume of the oil left in separator after the gas is removed, i.e., the volume of dead oil, m^3. Both are measured at standard conditions (20 °C and 0.1 MPa).

In surface production, flash separation can separate more gas from a well-stream crude, thus produces a higher gas–oil ratio. Moreover, because of more light-hydrocarbon components evolving from the oil in flash separation, the density of the gas separated from the crude oil is heavier. Therefore, more light-hydrocarbon components of the crude oil are lost from the oil. The stock tank oil thus becomes heavier.

In order to decrease the loss of light oil in oil–gas separation, stage separation is thus developed.

3.4.2.2 Stage Separation

Stage separation is also used to separate well-stream hydrocarbon mixture into gas and liquid phases by two or more equilibrium flashes at consecutively lower pressures. Its principle is the differential vaporization, but not flash vaporization.

Differential vaporization is the process in which the gas is removed as fast as it is liberated from the liquid mixture. A procedure of constant temperature differential vaporization is illustrated in Fig. 3.43.

Suppose a mixture in a cell is originally at p and T. Reduce the pressure of the mixture to p_1 that is less than its bubble point pressure (Fig. 3.43), and then gas liberates from the mixture till the gas and the liquid are in equilibrium. If it were possible to remove the free gas from the cell without disturbing the liquid, a new system having a bubble point of p'_b would remain. Continued stepwise removal of small increments of equilibrium gas would eventually produce a series of new systems having different bubble point pressures and compositions, which finally brings the cell to atmospheric pressure. So, in differential vaporization, the composition of the system remained in the cell varies with the pressure in the cell.

Theoretically, differential vaporization requires that each gas bubble should be removed from the cell or separator as it is formed. However, it is inoperable in practice for the test in the laboratory or surface production. A substitute for this separation is a series of batch or simple distillations in practice. Thereby, a series of stepwise flash vaporizations are used to approximate the differential vaporization in surface production. This stepwise process of flash vaporization can be called *stage separation*. Generally, two, three, or more stage separators can be used in series at successively lower pressures to obtain a maximum amount of stable stock tank

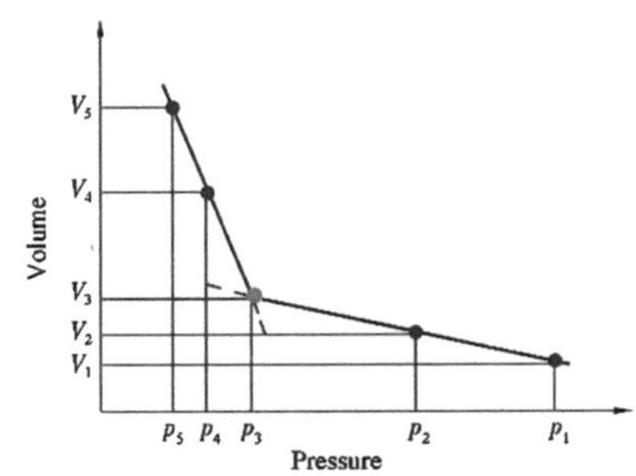

Fig. 3.42 Pressure–volume relation of flash separation

liquid. Therefore, stage separation means this kind of oil–gas separation by passing the oil mixture through several separators at successively lower pressures.

Figure 3.44 shows a two-stage separation process. The well-stream mixture is flashed in an initial separator, and then the liquids from the initial separator are flashed again in the next low-pressure separator (second-stage separator). Then fluids from second-stage separator finally enter the stock tank. In a two-stage separation, the equilibrium pressure in second-stage separator is just the atmospheric pressure.

Figure 3.45 illustrates a three-stage separation process. The well-stream mixture is first flashed at initial pressure; and then the liquids from the first stage are flashed at successively lower pressure two times before entering the stock tank. Flash calculation has proved that more light hydrocarbons can be stabilized in the liquid in stage separation than in flash separation.

The gas–oil ratio of stage separation is calculated by

$$R = \frac{\sum V_{gi}}{V_o}, \tag{3.52}$$

where V_o is the volume of the stock tank oil, m^3; ΣV_{gi} is the total volume of gas evolved from the liquid at each equilibrium pressure, m^3; V_{gi} is the volume of gas from ith stage separator. Likewise, the volumes of oil and gas are both measured at standard conditions, normally 20 °C and 0.1 MPa.

Stage separation can decrease the loss of light hydrocarbons from crude mixtures and obtain more amount of stock tank oil. However, in this separation, the investment in surface equipment will be increased, and the process is more complex. Therefore, it is usually less than three stages for oil–gas separation in surface

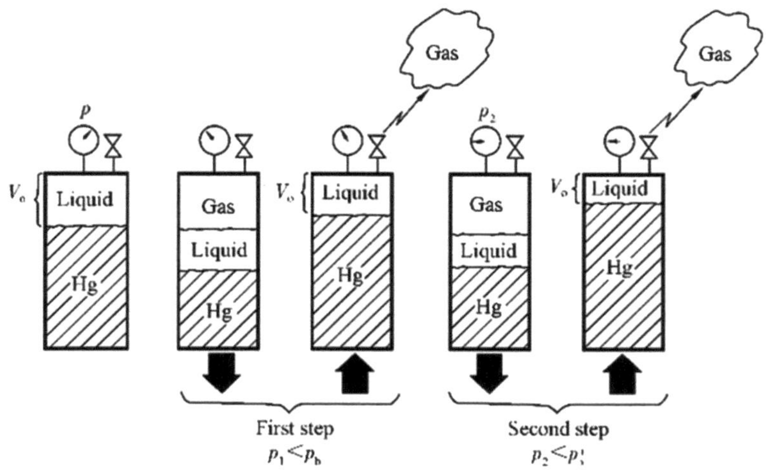

Fig. 3.43 Laboratory differential vaporization procedure (McCain [30])

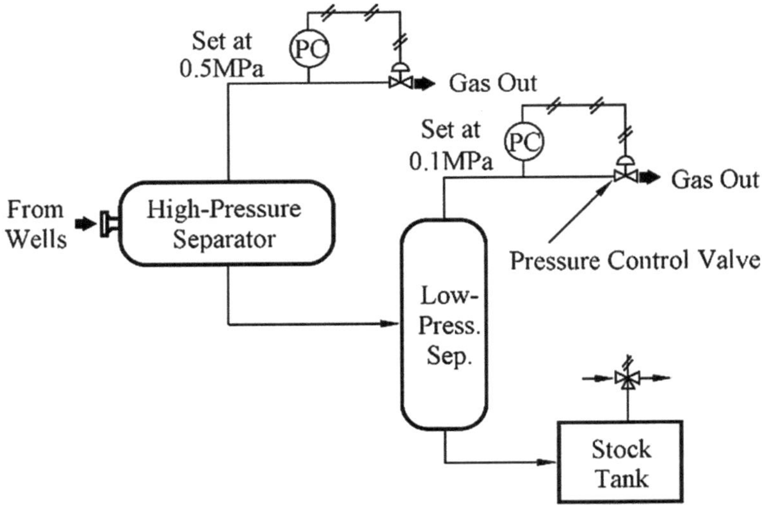

Fig. 3.44 Two-stage surface separation (after Stewart and Arnold [36])

production. In addition, the natural gas from flash separation may be sent to an NGL recovery plant to extract further the heavier hydrocarbons in the gas because the gas from flash separation is usually heavier than that from stage separation.

In a word, the optimum separation conditions (pressure, temperature, and stages) must be carefully demonstrated according to the attributes of the well-stream mixture from wells. The detailed calculations of gas–liquid equilibrium in oil–gas separation are described in Sect. 3.6.

3.4.2.3 Differences Between the Results of Flash and Stage Separations

Theoretical and experimental investigates have proved that the separation results depend on not only the separation conditions but also the separation method. Tables 3.21 and 3.22 list the results from flash separation and stage separation for different samples.

From Table 3.21, we can see that the amount of gas obtained from flash process is larger than that from stage separation; and contrarily, the amount of oil from flash vaporization is less than that from stage separation. The gas–oil ratio of flash separation is thus higher than that of stage separation. Table 3.22 shows the similar differences between the two separation methods.

In Table 3.22, even at the same final pressure, the gas–oil ratios (273 m^3/m^3) of an oil sample in flash separation is higher than that (239.3 m^3/m^3) in stage separation. The gas and oil obtained from flash separation are also heavier than those from stage process (Table 3.22). In a word, the gas obtained from stage separation is less and dryer than that from flash separation, and the stock tank oil from stage separation is more and lighter than that from flash separation.

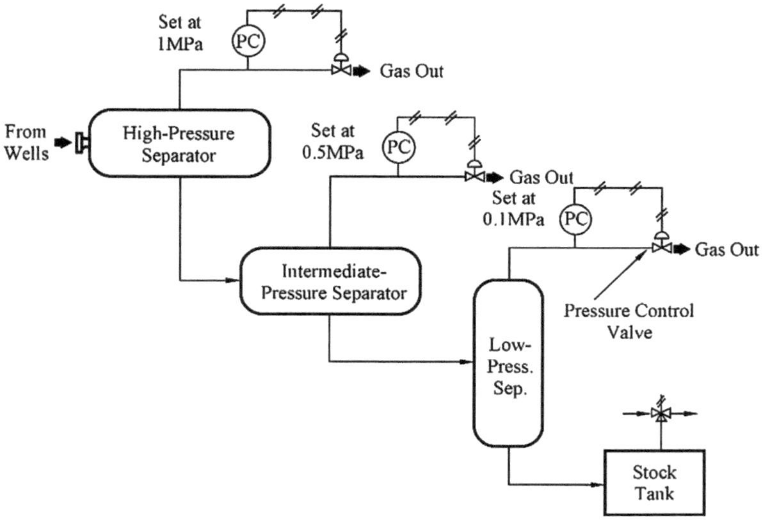

Fig. 3.45 Three-stage surface separation (after Stewart and Arnold [36])

Why the differences between the two methods occur? The fundamental reason is the difference of pressure drops between the two methods. In stage separation, each decrease in pressure is very small. It is much less than that in flash separation. Under small pressure drop, only the lightest components in the mixture can flash into the gas phase at each stage. In flash separation, on the contrary, a large pressure drop may cause the heavier components to flash into the gas phase, and then a large amount of gas may escape from the crude mixture. Therefore, all

Table 3.21 Effect of separation method on the amount of gas evolved (He [32])

Oil samples from bottom hole	Gas–oil ratio (m^3/m^3)		Relative error of the gas volumes obtained by the two methods (%)
	Flash separation	Stage separation	
Romashkin, Russian	59.6	41.4	30.6
Cibuersic, Russian	83.0	71.5	13.9
New Dimitr, Russian	175.4	144.5	17.6

Table 3.22 Comparison between results of separation for an oil sample (He [32])

p_1 (MPa)	T (°C)	Gas–oil ratio (m^3/m^3)		γ_o (15.5 °C)	γ_g (15.5 °C)
		Flash at p_1	Flash at 0		
0	25	273	–	0.7807	0.9733
0.4	25	229	10.3	0.7711	0.8559

the gases from each stage in stage separation are still lighter and less than that from flash separation; and more light components thus remained in the oil in stage separation.

3.4.2.4 Vaporization Processes in Reservoirs and Tubings

When reservoir pressure is below bubble point pressure, the dissolved gas will escape from the crude oil in reservoirs. In this case, the liberation of dissolved gas from the oil is also a vaporization process. This vaporization lies roughly between differential vaporization and flash vaporization. In general, the liberation of gas from the oil in reservoirs with any appreciable vertical permeability is similar to a differential process; whereas the liberation in reservoirs having continuous shales (or other impermeable strata) approaches a flash process. In a word, the oil–gas separation occurring in reservoirs are more complicated than that in the laboratory or in surface production.

In addition, the process of crude oil flowing through the tubing also approaches a flash process. When crude oil flows from the bottom hole to the wellhead, the gas liberates continuously from the oil along the tubing due to the decline in pressure. In the process of the oil flowing through the tube, the liberated gas and the oil are in contact with each other. So, it is similar to a flash vaporization.

3.4.3 Solution Gas–Oil Ratio of Formation Crude

In reservoirs, crude oil usually contains a quantity of dissolved gas. When crude oil is brought to the surface, the dissolved gas (light hydrocarbons) will evolve from the oil and remains in the reservoir. Solution gas–oil ratio is a measure of the gas evolved from the crude oil in the reservoir. It can indicate indirectly the volatility of a crude oil.

Solution gas–oil ratio of formation crude, symbol R_s, means the amount of gas (measured at standard conditions) that evolves from the oil as the oil is transported from the reservoir to surface conditions [4]. It is mathematically expressed as

$$R_s = \frac{V_g}{V_s}, \tag{3.53}$$

where R_s is the solution gas–oil ratio of in-place oil, (SC) m^3/m^3; V_g is the volume of gas separated from the oil at surface standard conditions, (SC) m^3; V_s is the volume of the oil at standard conditions, m^3.

In Europe and the United States, solution gas–oil ratio indicates the total standard cubic feet of dissolved gas evolved per barrel of stock tank oil (scf/stb).

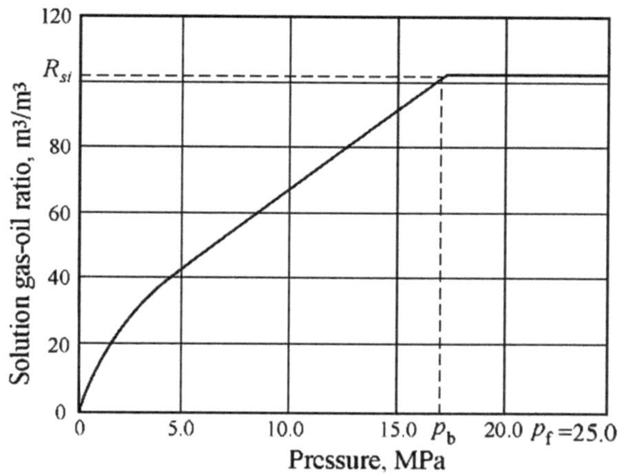

Fig. 3.46 Solution gas–oil ratio curve of a typical in-place oil (flash separation)

Obviously, solution gas–oil ratio of in-place oil is a measure of the amount of the gas dissolved in the oil in reservoirs. The larger the solution gas–oil ratio is, the more the gas dissolved in the oil is.

Figure 3.46 shows a typical solution gas–oil ratio curve for an undersaturated crude oil. The curve is the solution gas–oil ratio as a function of pressure at reservoir temperature. From Fig. 3.46, we can see that the solution gas–oil ratio is constant at pressures above the bubble point pressure. When reservoir pressure is higher than the bubble point pressure, no gas evolves from in-place oil to the reservoir. Above bubble point pressure, therefore, the solution gas–oil ratio remains constant and is referred to as *initial solution gas–oil ratio*, expressed in R_{si}. Below bubble point pressure, the dissolved gas liberates partly from the in-place oil, leaving less gas dissolved in the liquid. The solution gas–oil ratio is thus less than the initial solution gas–oil ratio; and it decreases with the decline in pressure.

From Sect. 3.4.2, we know that the gas–oil ratios of an oil may be different if separated with different methods even at the same separation conditions. So, the values of the solution gas–oil ratio of an in-place oil depends on the method of separating it at surface. For the comparison between different in-place oils, the solution gas–oil ratio curve is customarily measured by flash separation.

For a given system, separation and dissolution between oil and gas at given temperature are two opposite processes. For example, below the bubble point pressure, the value of the solution gas–oil ratio of an in-place oil is equal to the value of the solubility of the gas dissolved in the oil in contact dissolution. Therefore, the solution gas–oil ratio is sometimes referred to as *gas solubility* [4, 8].

However, differential separation and differential dissolution are not reversible because the compositions of the system are different in the two processes at each pressure. In Fig. 3.40, the curve ③ is the degassing curve of an oil–gas mixture in

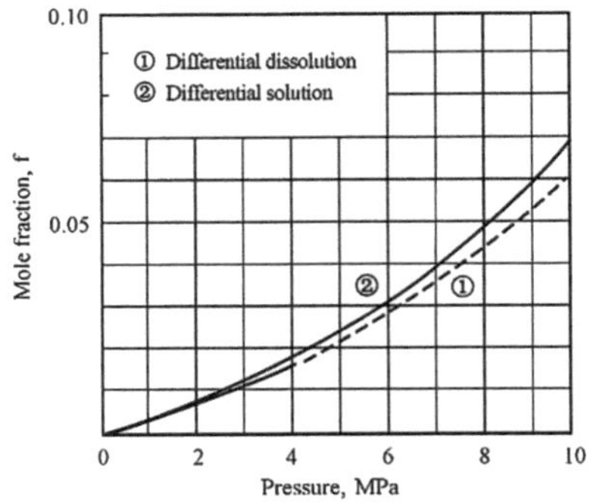

Fig. 3.47 Solubility curve and degassing curve of methane in an oil (He [32])

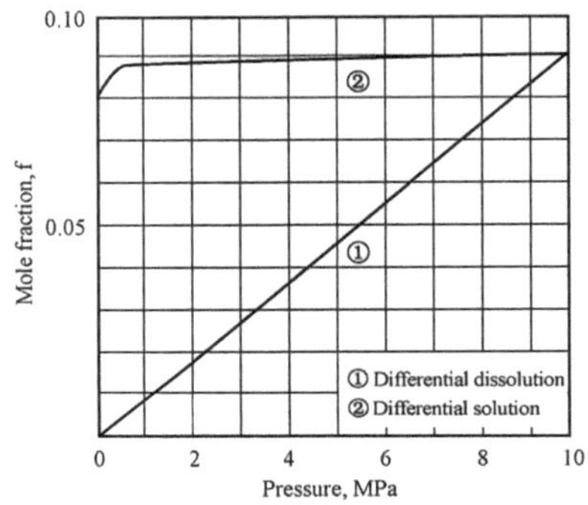

Fig. 3.48 Solubility curve and degassing curve of pentane in an oil (He [32])

differential vaporization while the curve ② denotes the solubility curve of the gas in the oil in differential solution. It is clear that the two curves are different despite they are from the same oil–gas mixture. The degassing curve and solubility curve of methane in an oil are shown in Fig. 3.47; and the similar curves of pentane are shown in Fig. 3.48. Figure 3.47 shows that there is a little difference between the two curves of methane, but in Fig. 3.48, the two curves of pentane are completely different.

Generally, the solution gas–oil ratio of in-place oils is customarily obtained in the laboratory, but it may be estimated from correlations [26].

In addition, the solution gas–oil ratio of in-place oil increases with the decrease in oil's density and reservoir temperature. Therefore, the solution gas–oil ratio of in-place oil varies from oil to oil. Table 3.23 lists the initial solution gas–oil ratios and other properties of in situ oils from different oil fields in China and abroad.

The solution gas–oil ratio of in-place oil varies in wide range. Some crude oils contain only several cubic meters of gas, such as the oils from Renqiu oil field in

Table 3.23 Physical properties of formation crude from oil fields in China and other countries [18]

Oil fields	R_{si} (SC) m^3/m^3	B_o	C_o (10^{-4} MPa^{-1})	Oil fields	R_{si}, (SC) m^3/m^3	B_o
Horizon P of Daqing, China	48.2	1.13	7.7	Gebil-Mary oil field, Romania	1.1	1.05
West Dagang, China	37.3	1.09	7.3	Oil field in Midland, U. S.	11.0	1.07
Shengli, China	70.1	1.22	–	Bolivar coastal oil field, Venezuela	85.1	1.26
Gudao, China	27.5	1.10	7.3	Pembina oil field, Canada	89.0	1.25
Horizon Ps of Renqiu, China	7.0	1.10	10.35	Aghajari oil field, Iran	190.0	1.42
Horizon L of Yumen, China	65.8	1.16	9.6	Ekofisk oil field of the North Sea, Norwegian	580.0	1.78

China and from Gebil-Mary oil field in Romania; whereas some oils may contain up to several hundred cubic meters of gas, such as the oil from North Sea oil field in Norway.

Note that a distinction must be made between the solution gas–oil ratio and the producing gas–oil ratio [11]. Oil reservoirs are encountered with or without a gas cap at discovery. A gas cap may exist on top of the oil zone, as shown in Fig. 1.35 in Chap. 1. In this case, the produced gas may come from two parts: free gas from the gas cap and the evolution of light hydrocarbons from the crude oil. Hence, the producing gas–oil ratio is expected to be greater than the solution gas–oil ratio of the in-place oil. Furthermore, the cumulative gas–oil ratio is defined as the cumulative gas production over the cumulative oil production from a well or a reservoir to a certain period during production [11].

3.4.4 Formation Volume Factor for in-Place Oil

3.4.4.1 Oil Formation Volume Factor

As is well known, an oil shrinks after dissolved gas liberates from the oil at surface conditions. Therefore, there must be a correlation between the volume of the oil at reservoirs and the volume of the oil at stock tank conditions. In fact, this correlation is represented by a term, the *oil formation volume factor* of in-place oil.

Oil formation volume factor is a measure of the shrinkage or reduction in the volume of in-place oil as it is produced. Accurate evaluation of oil formation volume factor is of prime importance as it relates directly to the calculation of petroleum reserve and oil production at the surface [11].

In general, the volume of a stock tank oil is less than the volume of the oil in reservoirs. The change in oil volume, which accompanies the change from reservoir conditions to surface conditions, is due to the following factors [4]. The most important factor is the liberation of gas from the oil as pressure decreases from reservoir pressure to surface pressure. This will cause a rather large decrease in the volume of the oil especially when there is a significant amount of dissolved gas in the oil. On the other hand, the reduction in pressure also causes the remaining oil to expand slightly, but it is somewhat offset by the shrinkage of the oil due to the decrease in temperature. On all accounts, the change in oil volume from reservoir to surface can then be expressed in terms of oil formation volume factor.

Oil formation volume factor, symbol B_o, and abbreviated as oil FVF, is defined as the ratio of the volume of oil at the prevailing reservoir pressure and temperature to the volume of the same oil at standard conditions at the surface:

$$B_o = \frac{V_o}{V_{os}}, \tag{3.54}$$

where B_o is the formation volume factor for in-place oil, m^3/m^3; V_o is the volume of in-place oil at reservoir temperature and pressure, m^3; V_{os} is the volume of the oil at standard conditions (0.1 MPa, 20 °C), m^3.

In Europe and the United States, oil FVF means the volume of reservoir oil required to produce one barrel of oil in stock tank (bbl/stb).

Oil FVF relates a volume of reservoir oil to the volume of the oil at the surface. Reservoir oil usually includes oil and gas dissolved in the oil; but the stock tank oil does not contain any light component that can liberate from the oil at standard conditions. So, the dissolved gas is considered as a part of the reservoir oil. If no gas is evolved from a crude oil at surface conditions, it is said to be *a dead oil.*

Although crude oil could be compressed at high pressures, oil FVF is usually larger than unity. The reason is that oil expansion caused by dissolved gas and high temperature is more significant than oil compression caused by high pressure. The volume of reservoir oil is thus often larger than the volume of the oil at the surface.

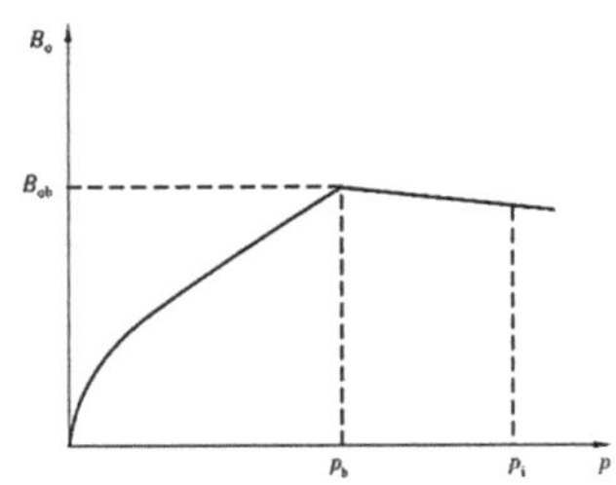

Fig. 3.49 Relation between volume factor of in-place oil and pressure

Obviously, the more the dissolved gas in the oil is, the larger the formation volume factor of the oil is, as shown in Table 3.21.

Figure 3.49 shows a typical curve of oil FVF as a function of pressure for an undersaturated crude oil. In the diagram, the initial pressure of the reservoir oil is higher than its bubble point pressure. As pressure decreases from initial pressure to bubble point pressure, the formation volume factor of the oil increases slightly owing to the expression of the oil in the reservoir.

At bubble point pressure, p_b, the oil reaches its maximum expansion and consequently attains a maximum value of B_{ob}.

Below the bubble point pressure, a further reduction in pressure results in the evolution of gas from the oil in the reservoir. Thereby, the amount of gas dissolved in the reservoir oil decreases with decline in pressure; and then the formation volume factor of the oil decreases.

In theory, if the reservoir pressure could be reduced to atmospheric pressure and the temperature, 20 °C, the value of the formation volume factor would nearly equal 1 m^3/m^3 [4].

In general, oil FVF is usually between 1.02 and 1.2 for black oils (see Table 3.23); but volatile crude tends to exhibit a relatively high oil FVF, as significant amounts of gas come out of the oil when pressure decreases. Typical values of oil FVF could range between 1.5 and 3.7 for volatile oil reservoirs. Besides, it may be close to unity for low-shrinkage crude.

Oil FVF is also dependent on reservoir temperature. Generally, higher temperatures lead to less dissolved gas in an oil. The oil FVF thus decreases with the increase in temperature. Furthermore, oil FVF is also affected by the oil–gas separation method since there is a difference between the gas volumes from flash separation and stage separation.

Oil FVF may be obtained either from experiments in the laboratory (assuming a representative sample has been obtained) or from the correlations (see Sect. 3.7).

3.4.4.2 Shrinkage Factor

As describe above, the volume of an in-place oil will decrease when it is produced to the surface. This phenomenon is known as the *shrinkage* of crude oil, which can be measured by the shrinkage factor of crude oil.

Oil shrinkage factor (abbreviated as SF) is defined as the reciprocal of oil formation volume factor, donated by the symbol b_o:

$$b_o = \frac{1}{B_o} = \frac{V_{os}}{V_o}, \tag{3.55}$$

where b_o is the shrinkage factor of an in-place oil, m^3/m^3; B_o is the formation volume factor of the oil, m^3/m^3; V_o is the volume of the oil in reservoirs, m^3; V_{os} is the volume of the oil at standard conditions (0.1 MPa, 20 °C), m^3.

Table 3.24 Petroleum fluid spectrum [37]

Fluid type	Typical initial		Stock tank liquid	
	Shrinkage factor (stb/res bbl)	GOR (scf/stb)	°API	Color
Low-shrinkage crude oil (low GOR, black, or ordinary oil)	>0.5	<2000	<45	Very dark, often black
High-shrinkage crude oil (high GOR or volatile oil)	<0.5	2000–3300	>40	Colored, usually brown

Oil shrinkage factor indicates the remaining fraction/percent of unit volume of reservoir oil produced at the surface at standard conditions. Oil shrinkage factor mainly depends on how much gas dissolved in the reservoir oil. It generally varies between 0.9 and 0.6.

Note that there is another parameter, *oil shrinkage*, which is different from the oil shrinkage factor in physical meaning. Oil shrinkage means the percentage of reduction in the volume of reservoir oil, mathematically expressed as

$$\beta = \frac{B_o - 1}{B_o} = \frac{V_o - V_{os}}{V_o}, \tag{3.56}$$

where β is the shrinkage of an in-place oil, m^3/m^3; B_o is the formation volume factor of the oil, m^3/m^3; V_o is the volume of the oil in reservoirs, m^3; V_{os} is the volume of the oil at standard conditions (0.1 MPa, 20 °C), m^3.

Essentially, oil shrinkage truly indicates the reduction in the volume of a crude oil when produced from reservoir to the surface. If the quantity of dissolved gas is not too large (shrinkage factor >0.5), the oil is said to be a low-shrinkage, ordinary, or black oil [4]. On the contrary, the oil is said to be a high-shrinkage or volatile oil (Table 3.24).

Note that the values of oil shrinkage factor and oil shrinkage for an oil are different in virtue of they have different physical significance. Quote the two parameters with caution.

The formation volume factor of an oil divided by its volume in the reservoir can produce the volume of the oil in stock tank. The shrinkage factor of an oil multiplied by its volume in the reservoir can also find the corresponding stock tank volume. Both terms are in use, but petroleum engineers have adopted universally the formation volume factor of oil [4].

3.4.4.3 Total Formation Volume Factor

When reservoir pressure is below the bubble point pressure, oil and gas can coexist in the reservoir due to the evolution of dissolved gas from the oil in the reservoir. To describe the variation in total volume of the oil and gas in the reservoir with the pressure, the term, *total formation volume factor*, is used.

The total formation volume factor of reservoir oil and gas, denoted B_t, is defined as the ratio of the total volume of the oil and gas (if free gas presents) at the prevailing pressure and temperature to the volume of the oil at standard conditions, mathematically expressed as

$$B_t = \frac{V_o + V_g}{V_{os}}, \tag{3.57}$$

where B_t is the total formation volume factor, m^3/m^3; V_o is the volume of the oil at prevailing reservoir pressure, m^3; V_g is the volume of the gas evolved at prevailing reservoir pressure, m^3; V_{os} is the volume of stock tank oil at standard conditions (0.1 MPa, 20 °C); m^3.

Because naturally occurring hydrocarbon mixtures usually exist in either one or two phases in reservoirs, the term *two-phase formation volume factor* has become synonymous with the total formation volume [10]. This parameter is required in transient well test analysis when the flow rate consists of both oil and gas, and in material balance calculations.

From Eq. (3.57), we have

$$\begin{aligned} B_t &= \frac{V_o + V_g}{V_{os}} \\ &= \frac{V_o}{V_{os}} + \frac{V_g}{V_{os}} = B_o + \frac{V_g}{V_{os}} \end{aligned} \tag{3.58}$$

In Eq. (3.58), the volume of oil at prevailing reservoir pressure that is lower than bubble point pressure is V_o; V_g is the volume of the gas evolved from the oil at prevailing reservoir pressure. Namely, it is the volume of the gas that cannot be dissolved in the oil at prevailing reservoir pressure. According to the definition of the solution gas–oil ratio of in-place oil and the FVF of oil and gas, the quantity (V_g) of the gas evolved from the oil at prevailing reservoir pressure can be determined by the following expression:

$$V_g = (R_{si} - R_s) \times V_{os} \times B_g, \tag{3.59}$$

where R_{si} is the initial solution gas–oil ratio of in-place oil, m^3/m^3; R_s is the solution gas–oil ratio of in-place oil at prevailing reservoir pressure, m^3/m^3; B_g is gas formation volume factor, m^3/m^3.

Substitute Eq. (3.59) for V_g in Eq. (3.58), then

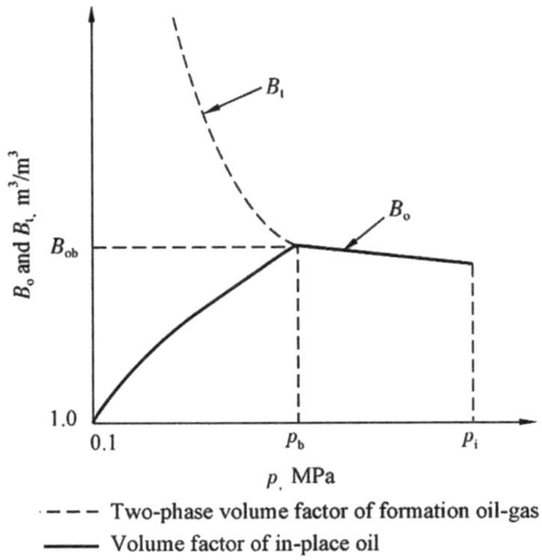

Fig. 3.50 B_t and B_o versus pressure

$$B_t = B_o + \frac{(R_{si} - R_s)V_{os}B_g}{V_{os}} \tag{3.60}$$

So,

$$B_t = B_o + (R_{si} - R_s)B_g \tag{3.61}$$

As all of these properties, R_s, B_o, and B_g, are a function of pressure, B_t is thus a function of pressure. A typical curve of B_t as a function of pressure is shown in Fig. 3.50, and the oil FVF curve is also illustrated in the diagram.

Above the bubble point pressure, the total formation volume factor is equal to the oil formation volume factor since no gas is liberated from the oil at these pressures. Below the bubble point pressure, the total formation volume factor increases rapidly with the decline in pressure due to the evolution of gas and its expansion. At the bubble point pressure, p_b, the total formation volume factor is considered to equal the oil FVF because the evolved gas can be ignored.

3.4.5 Isothermal Compressibility of Oil

Above the bubble point pressure, the quantity of dissolved gas in oil reservoirs is constant, thus in-place oils can be compressed if the reservoir pressure increases. This property of in-place oil can be described by a parameter, known as *isothermal compressibility of oil*. In thermodynamics, the isothermal compressibility of oil is a measure of the relative change in the volume of a crude oil as a response to a pressure change.

In ordinary oil production, pressure depletion can normally be assumed to be isothermal. The isothermal compressibility of oil is defined as the fractional change of oil volume as pressure is changed at a constant temperature; mathematically expressed as

$$C_o = -\frac{1}{V_o}\left(\frac{\partial V_o}{\partial p}\right)_T, \tag{3.62}$$

where C_o is the isothermal compressibility of oil, 1/MPa; V_o is volume of the oil, m^3; p is reservoir pressure, MPa; T is reservoir temperature, K. The subscript T denotes a constant temperature, i.e., reservoir temperature.

The isothermal compressibility of oil is also called the coefficient of isothermal compressibility of oil; or *oil compressibility* for short [4].

The curve of oil compressibility as a function of pressure for typical black oil is shown in Fig. 3.51. Below the bubble point, reservoir oil cannot be compressed because of the expansion of dissolved gas in the oil. Equation (3.62) can only apply at pressures equal to or higher than the bubble point pressure. Above the bubble point pressure, p_b, oil compressibility decreases with the increase in pressure. At the point of p_b, oil compressibility reaches its maximum value (Fig. 3.51).

Oil compressibility is a measure of the elastic energy of reservoir oil. If a reservoir contains low-compressibility oil, namely less dissolved gas in this oil, the reservoir will suffer from large pressure drops only after limited production.

Oil compressibility depends on the content of the dissolved gas in an oil and the temperature and pressure of the reservoir. In general, the isothermal compressibility of in-place oils varies between 10×10^{-4} and 150×10^{-4} MPa^{-1}; whereas the isothermal compressibility of stock tank oils commonly ranges from 4×10^{-4} to 7×10^{-4} MPa. The higher the reservoir temperature is, and the lower the reservoir pressure and the density of the oil are, oil compressibility is thus the larger.

Because oil compressibility is a function of pressure, the average compressibilities of a reservoir oil in different pressure regions are different, as shown in Table 3.25. The average oil compressibility near the bubble point pressure is clearly higher than that far from the bubble point pressure. At lower pressures, reservoir oil is less density, and then easily compressed.

In terms of Eq. (3.62), we have

$$C_o = -\frac{1}{V_f}\left(\frac{\partial V_f}{\partial p}\right)_T \approx -\frac{1}{V_f}\frac{\Delta V_f}{\Delta p} \tag{3.63}$$

The curve of oil formation volume factor in Fig. 3.49 shows that reservoir oil reaches its maximum volume, V_f at bubble point pressure p_b, namely, $V_{fmax} = V_b$. Then, Eq. (3.63) can be written as

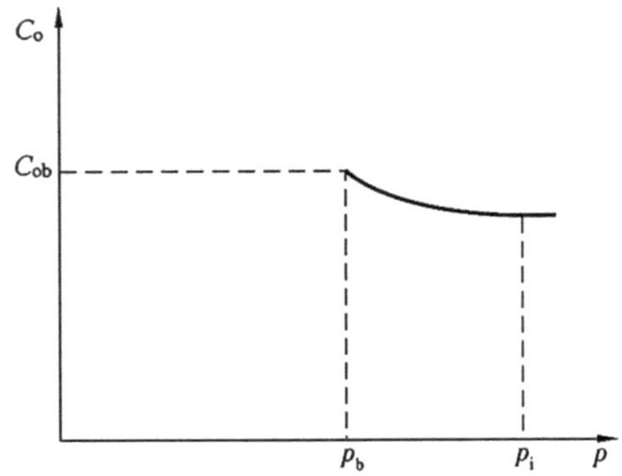

Fig. 3.51 Typical curve of oil compressibility as a function of pressure at constant reservoir temperature

$$C_o = -\frac{1}{V_f}\frac{V_b - V_f}{p_b - p} \tag{3.64}$$

Based on Eq. (3.64), obviously, one can determine the isothermal compressibility of an oil if p_b and V_b are given. So, it is a formula of measuring oil compressibility in the laboratory.

In addition, according to the definition of oil FVF, Eq. (3.64) can be equivalently written as

$$C_o = -\frac{1}{B_o}\frac{B_{ob} - B_o}{p_b - p} \tag{3.65}$$

From Eq. (3.65), the oil compressibility can be calculated with oil FVF.

Similar to oil FVF, the isothermal compressibility of oil may be obtained either from the laboratory or from the correlations (see Sect. 3.7).

3.4.6 *Oil Viscosity*

Oil viscosity is a vital property of crude oil. Suppose a crude oil in flow is a series of liquid layers superimposed on each other, the oil viscosity is a measure of the

Table 3.25 The average compressibilities of an in-place oil in different pressure regions

Pressure region (MPa)	Average oil compressibility, C_o (10^{-4} MPa^{-1})
p_b = 19.0	–
19.0–19.4	38.9
19.4–24.2	36.0
24.2–29.2	30.2
29.2–34.4	24.7

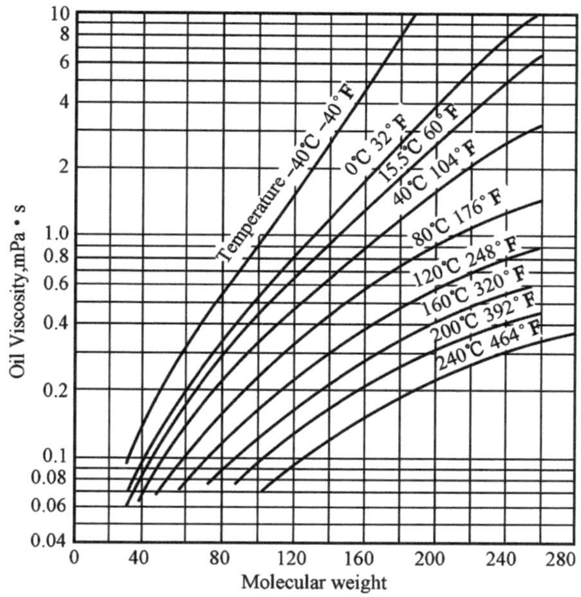

Fig. 3.52 Relation between oil viscosities and molecular weights [25]

resistance between the liquid layers. So, the same definition of viscosity can be applied to oil as gas. A high viscosity implies a high resistance to flow while a low viscosity indicates a low resistance to flow. Therefore, oil viscosity can characterize the flow ability of crude oils. In oil production, oil viscosity affects the output of oil wells and the development effect of oil reservoirs. Viscous crude oils would require more energy to flow toward the wellbore than low-viscosity oils. Superhigh-viscosity oils could make oil wells off production. Bitumen, having extremely high viscosity, would not flow at all in reservoirs unless thermal recovery methods are used to reduce the viscosity of the oil. So, oil recovery tends to be lower in reservoirs containing high-viscosity oil.

In petroleum engineering, oil viscosity is a vital parameter required in predicting the fluid flow, both in reservoir and in surface facilities, since it affects the velocity with which the oil will flow under a given pressure drop.

The viscosity of crude oils from fields around the world varies in a widely region. At reservoir conditions, oil viscosity may range from less than 1 mPa s to thousands mPa s. The viscosity of stocktank oils may also change from less than 1 mPa s to several ten thousands mPa s. At surface conditions, it can be seen that some of crude oils are as thin as water while some crudes are as thick as the semisolid plastic.

Oil viscosity, like other physical properties of crude oil, is affected by many factors. The factors of interest of the petroleum engineer mainly include oil composition, temperature, dissolved gas, and pressure.

Table 3.26 The variation of oil viscosity with the content of resin-asphaltene in the oil [18]

Well	Xin-1	Tuo-2	Guan-12	Xin-2	Xin-6
Content of resin-asphaltene (%)	27.6	36.7	41.9	49.5	64.5
Viscosity of stock tank oil (mPa s)	58.5	490	885	5120	6400

Table 3.27 Relationship between oil viscosity and the initial solution gas–oil ratio of in-place oil [18]

Producing horizon	Initial solution gas–oil ratio [(SC) m^3/m^3]	Oil viscosity (mPa s)
Horizon G of Gudao, China	27.5	14.2
Horizon M of Daguang, China	37.3	13.3
Horizon P of Daqing, China	48.2	9.30
Horizon L of Yumen, China	68.5	3.20
Ying 4 well in Shengli, China	70.1	1.88

Essentially, the viscosity of a crude oil depends on its composition. The higher the average molecular weight of a crude oil is, the higher the oil viscosity is, as shown in Fig. 3.52. In crude oils, the most important components affecting oil viscosity are the large-molecule compounds, such as resin and asphaltene, in the oil. These compounds have involuted molecular structure. When crude oil flows, the involuted structures of large-molecule compounds will greatly increase the internal friction of the oil, and then result in higher oil viscosity.

Table 3.26 shows the variation of oil viscosity with the content of resin-asphaltene in the oil from Shengli oilfield, China. Clearly, oil viscosity increases greatly with the increase in the content of resin-asphaltene in the oil. Study shows that a higher content of the resin-asphaltene and a lower temperature will result in the formation of colloform texture of large-molecule compounds in the oil; and then the colloform texture will cause the behavior of the oil like a non-Newtonian fluid, such as the thixotropy, static shear stress τ_0, and oil viscosity decreasing with the increase in shear rate and so on. The high-viscous oils often show these features.

The dissolved gas in reservoir oil is another important factor affecting oil viscosity. Generally, the less the gas dissolved in reservoir oil, the higher the viscosity of the oil, as shown in Table 3.27. The dissolved gas, including lighter hydrocarbons, tends to increase the distance between the molecules of liquid hydrocarbons. As the gas is released from the oil, the molecules of liquid hydrocarbons clump together, which thus increases the viscosity of the oil.

Figure 3.53 shows the oil viscosity as a function of temperature and pressure. As a whole, oil viscosity is very sensitive to temperature, and decreases fast with the increase in temperature. However, the viscosities of different oils have different sensitivity of temperature, as shown in Fig. 3.54.

A typical curve of oil viscosity as a function of pressure at constant temperature is shown in Fig. 3.55. Above bubble point, oil viscosity decreases as the pressure

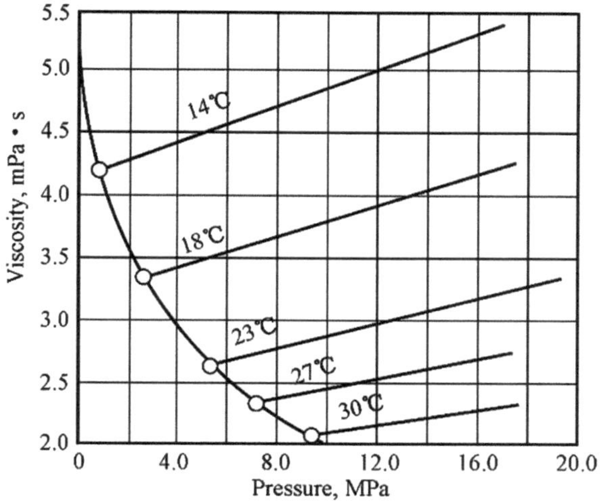

Fig. 3.53 Relationship of oil viscosity to pressure or temperature (Kotyakhov [35])

decreases. In reservoir production, the viscosity of reservoir oil decreases with the decline in pressure due to oil expansion as long as reservoir pressure remains above bubble point; however, below the bubble point, the viscosity of reservoir oil increases fast with the decline in pressure due to the successional release of dissolved gas from the oil. At bubble point pressure, oil viscosity reaches its minimum value as the oil is in its maximum volume.

At surface conditions, stock tank oil is found to have higher viscosity owing to the release of all volatile components from the oil.

Consequently, in the process of a reservoir production, reservoir pressure should be maintained over the bubble point as possible as we can. If reservoir pressure is lower than bubble point pressure, the increase in oil viscosity will cause the decrease in the flowability of oil. As a result, the difficulty of producing the oil increases greatly and the oil recovery is obviously decreased.

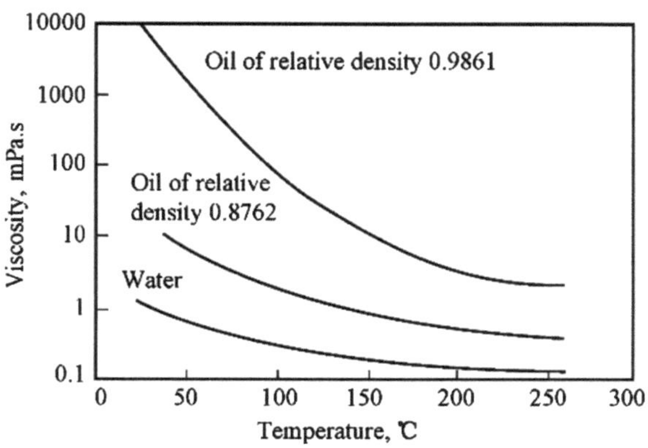

Fig. 3.54 Relationship between oil viscosity and temperature for different oils [1]

Oil viscosity may also be obtained from laboratory tests, or from the correlations (see Sect. 3.7).

3.4.7 Oil Density at Reservoir Conditions

Oil density is necessary for most reservoir engineering calculations. The density of crude oil at reservoir temperature and pressure can be calculated when the following information is available from laboratory studies:

(a) Density of crude oil at standard conditions;
(b) Quantity of dissolved gas in the oil at reservoir conditions;
(c) Density of gas at standard conditions;
(d) Change in crude oil volume when produced.

The downhole density of oil at reservoir conditions can be calculated from the surface density using the following equation, if the solution gas–oil ratio and oil formation volume factor are given [7]:

$$\rho_{\mathrm{orc}} = \frac{\rho_{\mathrm{o}} + R_{\mathrm{s}}\rho_{\mathrm{g}}}{B_{\mathrm{o}}}, \tag{3.66}$$

where ρ_o is the oil density at reservoir conditions, kg/m^3; ρ_o is the oil density at standard conditions, kg/m^3; ρ_g is the density of gas at standard conditions, kg/m^3; R_s is the solution gas–oil ratio, sm^3/stm^3; B_o is the oil formation volume factor, rm^3/stm^3.

Standing (1981) [38] proposed an empirical correlation for estimating the oil FVF as a function of the solution gas–oil ratio (R_s), the specific gravity of stock tank oil (γ_o), the specific gravity of solution gas (γ_g), and the system temperature T. By coupling the mathematical definition of the oil formation volume factor with Standing's correlation, the density of a crude oil at a specified pressure and temperature can be calculated from the following expression [26]:

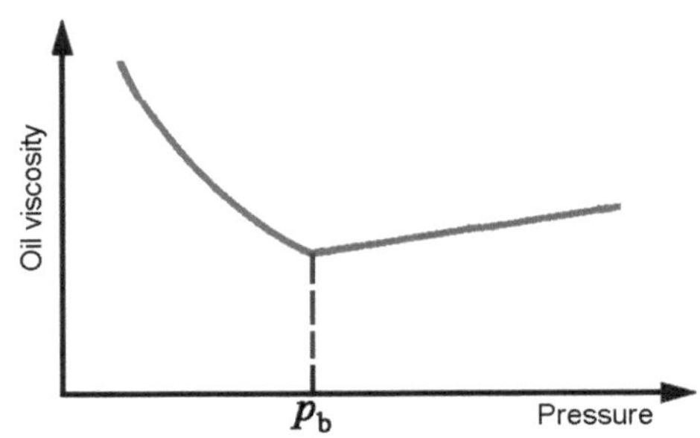

Fig. 3.55 Oil viscosity as a function of pressure at constant temperature

$$\rho_o = \frac{62.4\gamma_o + 0.0136R_s\gamma_g}{0.972 - 0.000147\left[R_s\left(\frac{\gamma_g}{\gamma_o}\right)^{0.5} + 1.25(T - 460)\right]^{1.175}}, \tag{3.67}$$

where γ_o is specific gravity of the stock tank oil, dimensionless; R_s is the solution gas–oil ratio, scf/stb; ρ_o is oil density, lb/ft^3; T is system temperature, °R.

Above the bubble point, the density of oil in reservoir can be given by Tarek [26]:

$$\rho_{orc} = \rho_{ob}\exp(C_o(p - p_b)), \tag{3.68}$$

where ρ_{orc} is the density of the oil at pressure p, kg/m^3; ρ_{ob} is the density of oil at bubble point pressure, kg/m^3; p is the reservoir pressure above the bubble point p_b, MPa; C_o is the isothermal compressibility coefficient at average pressure, MPa^{-1}.

The density of reservoir oil depends on the solution gas–oil ratio, reservoir temperature, and pressure. Generally, the higher the solution gas–oil ratio and reservoir temperature are, the lager the volume of the reservoir oil is, the density of the reservoir oil is then the lower. The typical curve of reservoir oil's density as a function of pressure is shown in Fig. 3.56. Visibly, the curve is similar to the curve of oil viscosity. Above the bubble point, the density of reservoir oil decreases with the decrease in pressure due the expansion of oil. Below the bubble point pressure, the density of reservoir oil increases with the increase in pressure due to the evolution of dissolved gas from the oil. At the bubble point pressure, the density of reservoir oil reaches its minimum value.

3.5 Properties of Formation Water

Formation water, also called oil field water here, means the water that occurs naturally within the pores of rocks. Water from fluids introduced to formations through drilling or other interference, such as mud and seawater, does not constitute formation water. Formation water, or interstitial water or connate water, might have been the water present when the rock originally formed.

Formation water is tightly associated with hydrocarbon fluids (crude oil or natural gas) in reservoirs. The energy from edge water or bottom water in oil reservoirs is probably stronger than the energy from gas cap or solution gas. Water compressibility is an important factor of describing the elastic energy of formation water. Some physical and chemical properties of formation water, such as viscosity and the ability of washing oil, make itself better than gas as oil displacement agents. In addition, in water injection operation, the chemical composition of formation water is often used to ascertain the potential of formation plugging due to the incompatibility between the external injecting water and the formation.

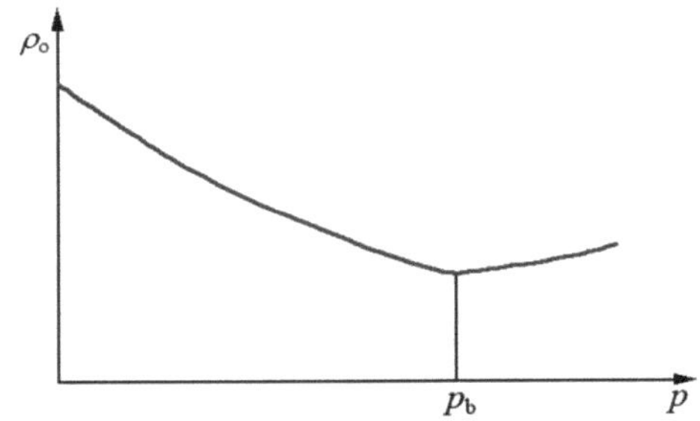

Fig. 3.56 The density of reservoir oil as a function of pressure at constant temperature

Knowledge of formation water properties is needed in various reservoir studies along with oil and gas properties. Of interest to petroleum reservoir engineers are the salinity, viscosity, density, and compressibility of formation water. Since hydrocarbon gas is soluble in water at reservoir temperature and pressure. The solubility of gas in water (or the solution gas-water ratio) and water formation volume factor are necessary for conducting reservoir studies, such as well test analysis, multiphase fluid flow simulation, and so on. Water found in subsurface formations usually has a high salt content due to the dissolution of minerals. In addition, subsurface water is found to be in compatible with surface water (fresh water) that is injected into the reservoir during water flooding. Surface water has a lower salinity, and contains no hydrocarbon gases. The chemical compositions of formation water can also be served to clarify the geological histories of oil accumulations.

3.5.1 Composition of Formation Water

All formation waters contain dissolved solids, primarily sodium chloride. The formation water sometimes is thus called *brine* or *salt water*. In some cases, the oil field water appear to be little similar to seawater, since both contain dissolved solids. However, oil field water is independent of seawater, either in concentration of solids or in the distribution of the ions present. Generally, oil field waters contain much higher concentrations of solids than seawater does. Formation waters have been reported with total solid concentrations ranging from as little as 200 ppm to saturation, which is approximately 300,000 ppm. Seawater contains about 35,000 ppm total solids.

The dissolved cations commonly found in oil field waters are sodion Na^+, calcium ion Ca^{2+}, and magnesium ion Mg^{2+}, shown in Table 3.28. Occasionally, K^+, Ba^{2+}, Li^+, Fe^{2+}, and Sr^{2+} are present. The most common anions are chloride ion, Cl^-, sulfate ion, $SO_4{}^{2-}$, and bicarbonate ion, $HCO_3{}^-$. Also, $CO_3{}^{2-}$, $NO_3{}^-$, Br^-, I^-, $BO_3{}^{2-}$, and S^{2-} are often present.

The common metallic bases sodium (Na^+) and potassium (K^+), and the alkaline earths calcium (Ca^{2+}) and magnesium (Mg^{2+}), are grouped as alkalies, and all are positive (+) radicals. The acids or negative (−) radicals are grouped into two—the strong acids, sulfates, and chlorides (Cl^-), and the weak acids, carbonates ($CO_3{}^{2-}$), bicarbonates ($HCO_3{}^-$), and so on.

Besides the dissolved salts, microorganisms of different species and stains are usually present in formation water. These organisms cause the corrosion in the wellbore and the formation plugging during water injection operations. The very hardy anaerobic sulfate-reducing bacteria are considered to be the most troublesome organisms commonly found in formation water.

The point is that the water contained in a producing formation has different ions composition than any other brine, even those in the immediate vicinity of that formation.

The content of solids dissolved in formation water is usually expressed by the weight concentrations of the anions and cations in parts per million, ppm, or milligram per liter, mg/L. Parts per million refers to grams of solids per one million grams of brine. Milligram per liter refers to milligram solid per liter of brine.

The total concentration of solids dissolved in formation water is referred to as "*salinity*", which represents the total content of the anions and cations in brine.

The salinities of formation waters vary greatly from oil field to oil field. It may range from several thousand mg/L to several hundred thousand mg/L, as described above. Formation water is characterized by its high salinity. In most cases, the salinity of oil field water is even higher than that of sedimentary water which occur in the strata of the same age, type, and depth where not associated with oil or gas. For example, the salinity of oil field water sometimes is up to 3×10^5 mg/L. At initial reservoir conditions, the high-salinity brine may be a saturated solution. When reservoir fluids (oil, gas, and water) flow to surface in production, solids will educe from brine due to the decrease in temperature and pressure. Badly, it could be salting out in wellhole, which makes production difficult.

When the interaction between ions is discussed, the concentration expressed in ionic milligram equivalent is customarily used because the reaction between ions of anions and cations is in equivalent weight, namely the anions of a certain equivalent weight reacts with the cations of the same equivalent weight. The concentration of ionic milligram equivalent equals the concentration of an ion divided by the equivalent weight of the ion, expressed in milligram equivalent per liter, meq/L. For example, if the concentration of chloride ion, Cl^{-1}, is 7896 mg/L, and its equivalent weight is 35, the concentration of milligram equivalent for chloride ion is 225.6

Table 3.28 Analysis results of formation water in Dingbian field, China

Wells	Cations				Anions		Salinity (mg/L)	Water types
	K^+ Na^+	Ca^{2+}	Mg^{2+}	Ba^{2+}	Cl^-	$HCO_3{}^-$		
3039	38,174	7653	1126	752	75,772	480	123,957	$CaCl_2$
5310	31,932	1958	534	269	59,511	298	97,502	$CaCl_2$

(i.e., 7896/35) meq/L. The equivalent weights of common ions in reservoirs are shown in Table 3.29.

In addition, the hardness of formation water is often used for evaluating the content of divalent cation in formation water, especially in tertiary oil recovery. In most cases, divalent cation in formation water exerts great effects on chemical flooding agent.

Hardness means the content of divalent cation, such as Ca^{2+}, Mg^{2+}, dissolved in formation water per liter.

3.5.2 *Types of Formation Water*

Most of the classifications proposed for formation water are based on the content of common anions and cations and the compounding types of these ions. The most commonly used is Sulin's classification that classified the formation water into four types: Na_2SO_4, Na HCO_3, $MgCl_2$, and $CaCl_2$ as shown in Table 3.30.

The type of formation water can be determined in terms of the reaction coefficient of chemical compounds. The reaction coefficient is determined by the combination order between main ions in formation water. The combination order is determined by the affinity of ions. For example, the chloride ions (Cl^-) tend to combine first with univalent sodions, next with divalent magnesium ions, finally the calcium ions (Fig. 3.57). Therefore, according to the equivalence ratio of Na^+ and Cl^- and the reaction coefficient of chemical compounds in formation water, Sulin proposed the classification of formation water, as shown in Table 3.30.

Table 3.29 The equivalent weight of common ions in reservoir waters

Cation	Equivalent weight	Anion	Equivalent weight
Ba^{2+}	68.7	CO^{3-}	30
Ca^{2+}	20	HCO_3^-	61
H+	1	Cl^-	35.3
Fe^{3+}	18.6	OH^-	17
Fe^{2+}	27.9	O^{2-}	8
Mg^{2+}	12.2	SO_4^{2-}	48
Na^+	23	S^{2-}	16
Sr^{2+}	43.8	SO_3^{2-}	40

3.5.3 *Properties of Formation Water*

3.5.3.1 Solubility of Gas in Formation Water

The formation water, associated with hydrocarbon reservoirs, commonly contains certain amount of dissolved gas at reservoir temperature and pressure. The amount of gas in oil field water can be described by the term "*solubility of gas in water*". Solubility of gas in formation water is defined as the volume of reservoir gas dissolved in the formation water which has unit volume at standard conditions, and expressed in m^3/m^3.

The solubility of gas in formation water depends on temperature, pressure, and the salinity of formation water (Fig. 3.58). As the temperature increases, the solubility of gas in formation water first decreases, then increases slightly, and reaches its minimum value at about 66 °C (Fig. 3.58a) [39]. According to Fig. 3.58a, pressure has a large influence on the solubility of gas in water. For example, at 66 °C, the gas solubility is about 0.7 m^3/m^3 at 3.4 MPa, and increases to about 2.0 m^3/m^3 at 13.6 MPa. The solubility of gas in water also is influenced by the amount of dissolved salts (Fig. 3.58b). Increasing salinity decreases the solubility of gas in formation water. The effect of salinity on gas solubility can be corrected using Fig. 3.58b.

Example 3.13 Estimate the solubility of natural gas in reservoir water with 30,000 ppm sodium chloride at a reservoir temperature and pressure of 100 °C and 23.8 MPa, respectively.

Table 3.30 Types of formation water [32]

Equivalence ratio	Reaction coefficient	Water type	Environment
$\frac{Na^+}{Cl^-} > 1$	$\frac{Na^+ - Cl^-}{SO_4^{2-}} < 1$	Na_2SO_4	Continental erosion environment (surface water)
	$\frac{Na^+ - Cl^-}{SO_4^{2-}} > 1$	$NaHCO_3$	Continental environment (oil field, gas field water)
$\frac{Na^+}{Cl^-} < 1$	$\frac{Cl^- - Na^+}{Mg^{2+}} < 1$	$MgCl_2$	Marine environment (sea water)
	$\frac{Cl^- - Na^+}{Mg^{2+}} > 1$	$CaCl_2$	Deep enclosed environment (gas field water)

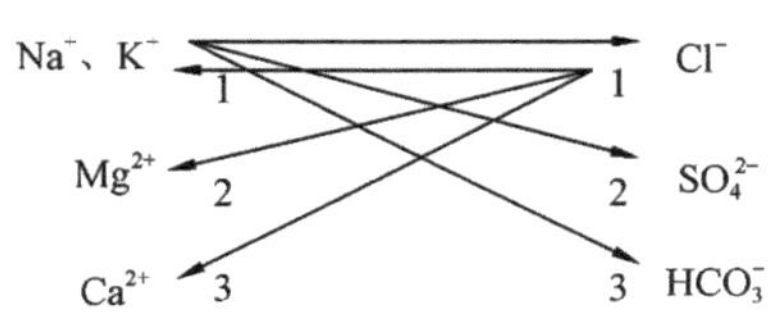

Fig. 3.57 Illustrative diagram of combination order of main ions

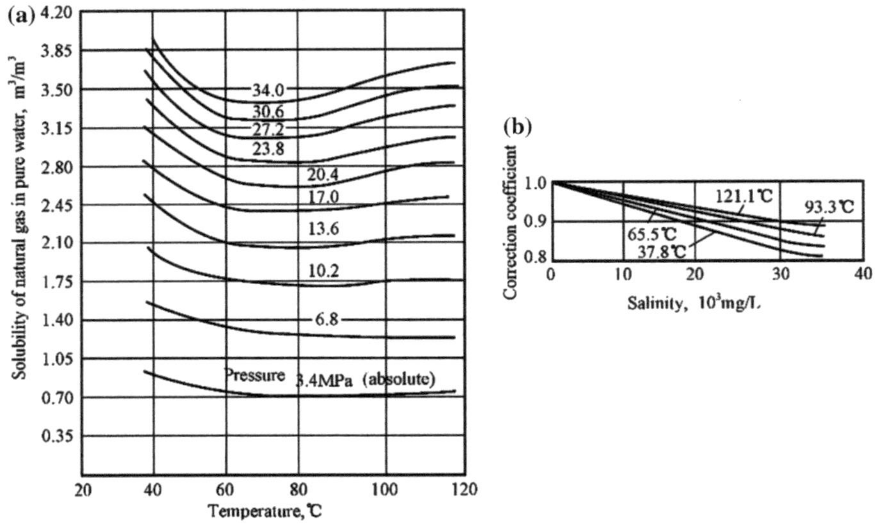

(a) Solubility of natural gas in pure water as a function of temperature and pressure;
(b) Correction of natural gas solubility for salt content

Fig. 3.58 Solubility of natural gas in water as a function of temperature and pressure [40]

Solution

According to the reservoir temperature and pressure, read R_{swp} from Fig. 3.58a: R_{swp} = 2.98 m³/m³.

Using Fig. 3.58b, read the correction factor of R_{swp} for salinity: C_{Rsw} = 0.89.

Then the solubility of gas in reservoir water is

$$R_{sw} = R_{swp} \times C_{Rsw} = 2.98 \times 0.89 = 2.56\ \mathrm{m}^3/\mathrm{m}^3$$

3.5.3.2 Volume Factor of Formation Water

Volume factor of formation water, symbol B_w, is defined as the ratio of the volume of formation water at reservoir temperature and pressure to the volume of the same water at standard conditions:

$$B_w = \frac{V_w}{V_{ws}}, \tag{3.69}$$

where B_w is the volume factor of reservoir water, m³/m³; V_w is the volume of formation water at reservoir temperature and pressure, m³; V_{ws} is the volume of formation water at standard conditions (0.1 MPa, 20 °C); m³.

Figure 3.59 shows the variation of volume factors of pure water and gas-saturated water with temperature and pressure. Obviously, both the volume factors of pure water and gas-saturated water are not sensitive to temperature and pressure. Therefore, the volume factor of formation water is almost a constant, i.e., unity, in reservoir production.

3.5.3.3 Isothermal Compressibility of Formation Water

Isothermal compressibility of formation water is defined as the relative volume change of formation water as a response to a pressure change at isothermal condition, i.e.,

$$C_{\mathrm{w}} = -\frac{1}{V_{\mathrm{w}}}\left(\frac{\partial V_{\mathrm{w}}}{\partial p}\right)_T, \tag{3.70}$$

where C_{w} is the isothermal compressibility of formation water, 1/MPa; V_{w} is the volume of formation water, m^3; p is reservoir pressure, MPa; T is reservoir temperature, K.

The isothermal compressibility of formation water is a function of pressure and temperature, as shown in Fig. 3.60a. At any given pressure, the water isothermal compressibility decreases as the temperature is increased from ambient, reaching a minimum compressibility at about 55 °C [39]. Then, the isothermal compressibility increases continuously with temperature increase. At any given temperature, the compressibility decreases as the pressure increases. The compressibility can be corrected for dissolved gas content (if any) using Fig. 3.60b.

Example 3.14 Estimate the isothermal compressibility of a formation water. Its salinity is 30,000 mg/L at a reservoir temperature (93 °C) and pressure (20.8 MPa).

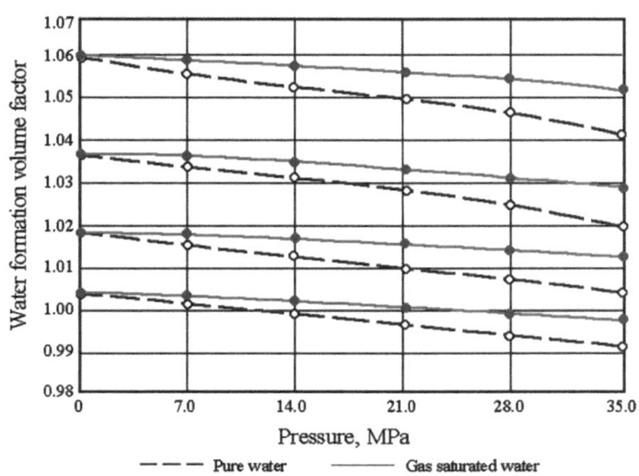

Fig. 3.59 Water formation volume factor as a function of pressure and temperature [40]

Solution

1. According to the reservoir temperature 93 °C and reservoir pressure 20.8 MPa, estimate gas solubility R_{swp} from Fig. 3.58a as: R_{swp} = 2.7 m^3/m^3.
   Using Fig. 3.58b, read the correction factor of R_{swp} for salinity: C_{Rsw} = 0.88.
   The solubility of gas in formation water is: $R_{sw} = R_{swp} \times C_{Rsw}$ = 2.7 × 0.88 = 2.7 m^3/m^3.
2. According the temperature 93 °C and pressure 20.8 MPa, estimate C_{wp} for formation water from Fig. 3.60a as: $C_{wp} = 4.53 \times 10^{-4}$ MPa^{-1}.
3. Correct C_{wp} for solution gas using Fig. 3.60b:

By the gas solubility R_{sw}, read the correction factor C'_w/C_w from Fig. 3.60b: $C'_w/C_w = 1.13$.
Correct C_{wp}:

$$C_w = C_{wp} \times C'_w/C_w = 4.53 \times 10^{-4} \times 1.13 = 5.12 \times 10^{-4}\ \text{MPa}^{-1}$$

3.5.3.4 Viscosity of Formation Water

Likewise, the viscosity of formation water controls the flow of formation water through permeable formations. Water viscosity is important in the prediction of aquifer response to pressure drops in reservoirs.

The viscosity of formation water as a function of temperature, pressure, and salinity is showed in Fig. 3.60. Water viscosity is chiefly controlled by temperature. As for liquids in general, water viscosity reduces with the increasing temperature. The influence of pressure and salinity on water viscosity is light (Fig. 3.61). Water viscosity is in the order of 0.5–1.0 mPa s, and is usually lower than that of oil [7].

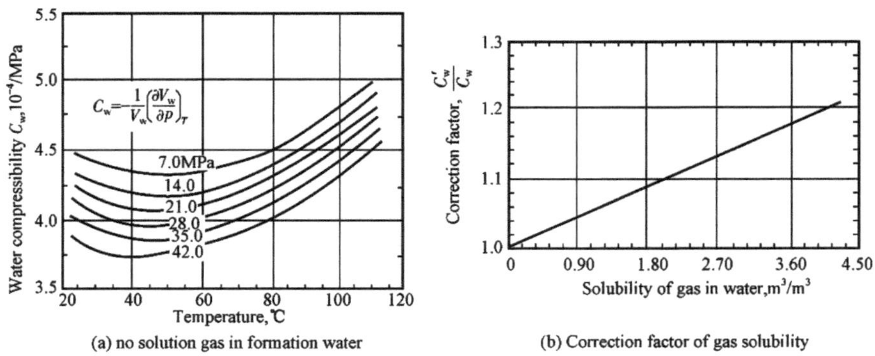

Fig. 3.60 Effect factors of the compressibility factor of formation water [40]

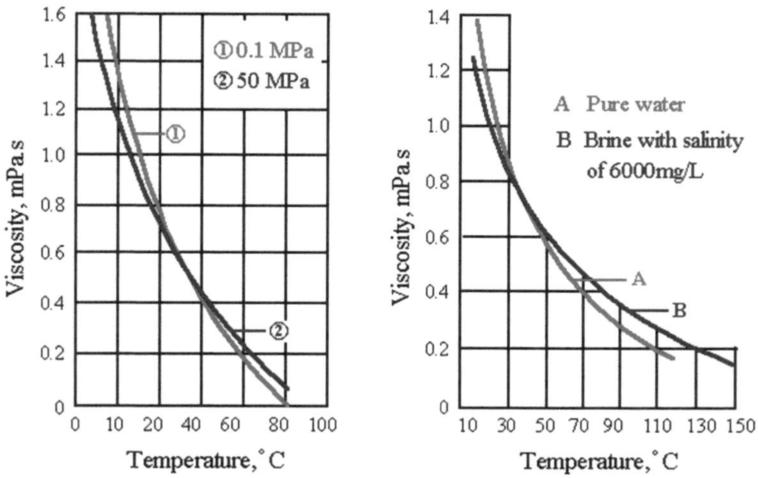

Fig. 3.61 The viscosity of formation water as a function of temperature, pressure, or salinity (Kotyakhov [35])

3.6 Gas–Liquid Equilibrium Calculation for Reservoir Fluids

All petroleum reservoir fluids undergo pressure and temperature changes during the process of production. The changes in pressure and temperature frequently result in the equilibrium gas and liquid phases occurring in petroleum production. The conditions at which gas and liquid phases coexist, the quantities, compositions, and properties of these phases are matters of considerable practical importance in petroleum reservoir engineering calculations, compositional simulation, and in the design of surface separation facilities. Despite water is also commonly present as an additional liquid phase, the effect of water on hydrocarbon phase behavior can be ignored in most cases.

In the *p*–*T* diagram of hydrocarbon mixtures, equilibrium gas and liquid phases coexist on and inside the envelop. In two-phase region of *p*–*T* diagram, different temperature–pressure conditions represent different amounts, compositions of gas and liquid phases in equilibrium.

The most accurate and reliable method of obtaining the data of phases and phase equilibrium conditions is laboratory investigation, including compositional analysis and reservoir fluid study (i.e., PVT test). Experimental measurements may be visual, as in a windowed cell, or based on break points in the pressure–volume data [37], but these experiments are extremely time consuming and expensive. The calculation methods are thus developed for predicting phase behavior. Besides, calculations for phase behavior can be relied upon in the absence of laboratory data.

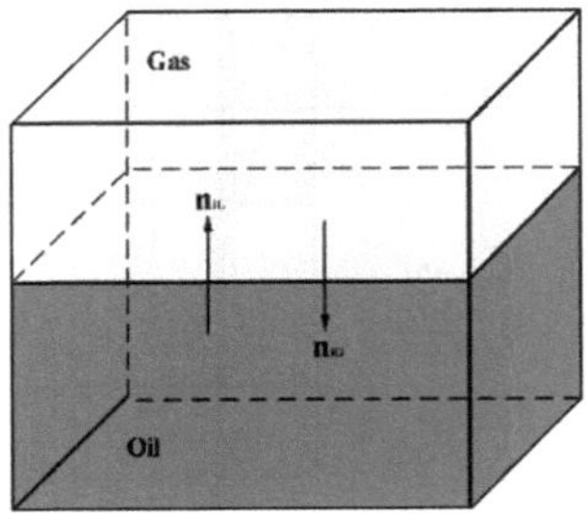

Fig. 3.62 An illustration for gas–liquid equilibrium

These calculations for equilibrium gas and liquid phases are called *phase equilibrium* or *vapor–liquid equilibrium* (VLE) calculations.

The basic data for phase equilibrium calculations include the total composition of reservoir fluids, pressure, temperature, and the properties of all individual components, including pseudocomponents. Based on these data, phase equilibrium calculations typically involve the following calculations: bubble points, dew points, equilibrium-phase compositions, and so on. In petroleum production, three basic phase equilibrium calculations are involved: (1) dew point calculation; (2) bubble point calculation; (3) flash vaporization calculation—phase split at specified conditions. These calculations enable us to predict reservoir fluid behavior and to design the conditions for light-hydrocarbon processing (NGL recovery) and oil–gas separation at surface production.

In general, bubble point calculations are useful to estimate reservoir conditions, but they are not particularly useful in surface-operation calculations except for reboiled stabilizers and separators [37]. Dew point calculations are not useful for crude oil process studies.

Equilibrium flash vaporization calculations can estimate the percentage vaporized and the equilibrium-phase compositions for mixtures that are partly vaporized or condensed, as in phase separators. These calculations are important in the simulation of separator trains used in the stabilization of crude oil.

In this section, the method and procedure of calculation for the vapor–liquid equilibrium of hydrocarbon mixtures will be presented.

3.6.1 Equilibrium Constant

The reasonably accurate way to predict gas–liquid equilibria is through correlations based on experimental observations of gas–liquid equilibrium behavior factors [4]. These correlations usually involve the *equilibrium constant*, K_i.

Phase equilibrium is thermodynamics equilibrium. In a vapor–liquid system shown in Fig. 3.62, for any one component at any moment, if the number of molecules from vapor phase into liquid phase is just equal to the number of molecules from liquid phase into vapor phase, the multiphase system then reaches its phase equilibrium. In the vapor–liquid phase equilibrium, the proportion of each component in each phase is a dynamic equilibrium.

In a vapor–liquid equilibrium, the distributions of any component i in the vapor and liquid phases can be described by the term, *equilibrium constant* of the component, symbol K_i. The equilibrium constant of ith component in a gas–liquid mixture is defined as the ratio of the mole fraction of the component in the vapor phase (y_i) to the mole fraction of the component in the liquid phase (x_i); mathematically expressed as

$$K_i = \frac{y_i}{x_i}, \tag{3.71}$$

where K_i is the equilibrium constant of ith component in a gas–liquid mixture, dimensionless; y_i is the mole fraction of the component in the vapor phase, f; x_i is the mole fraction of the component in the liquid phase, f.

Note that the word "*constant*" used here is a misnomer, but is widely used in the petroleum industry. In fact, "*equilibrium ratio*" proposed by Muskat [41] is a more exact term to describe the equilibrium distribution of a component between gas and liquid phases because of the ratio is not solely a function of pressure and temperature. As the habit of using, however, the term, equilibrium constant, will be used here despite equilibrium ratio is more meaningful than "equilibrium constant". In addition, equilibrium ratios are sometimes called *equilibrium vapor–liquid distribution ratios*, *distribution coefficients*, or simply, *K-factors* or *K-values* [4].

The equilibrium constant of each component in a mixture is a function of pressure, temperature, and the total composition of the mixture. At lower pressure, the effect of the composition on K-factors is very light; whereas it is considerably notable above the pressure of 7 MPa. In a word, equilibrium constants depend on the equilibrium state of a vapor–liquid system.

3.6.2 Equations for Phase Equilibrium Calculation

The use of equilibrium constant makes it possible to calculate the equilibrium conditions and compositions of equilibrium gas and liquid phases. The mathematical expressions used in phase equilibrium calculations are called *phase equilibrium equations*. Actually, phase equilibrium equations are a set of expressions which describe the relationship between the phases and the composition of multiphase mixtures in equilibrium.

The equations for gas–liquid equilibrium calculations can be obtained from the principle of material balance and phase equilibrium.

Consider F mole of a hydrocarbon mixture or a feed consisting of N components. The total composition of the mixture is n_i, $i = 1, 2, \ldots, N$. In gas–liquid equilibrium, there are N_L moles of liquid of composition x_i and N_g moles of vapor of composition y_i. Suppose K_i is the equilibrium constant of ith component. In the equilibrium system, according to the principle of material balance, we have

$$N_L + N_g = 1 \tag{3.72}$$

$$x_i N_L + y_i N_g = n_i, \tag{3.73}$$

where the total composition n_i, liquid-phase composition x_i, and gas-phase composition y_i are wholly expressed in mole fraction.

Equation (3.72) means a total material balance in mole fraction of gas and liquid phases. Equation (3.73) is a total material balance for each component in gas and liquid phases.

As the composition of each component is given in mole fractions, the following constraints should be satisfied:

$$\sum_{i=1}^{N} n_i = 1 \tag{3.74}$$

$$\sum_{i=1}^{N} x_i = 1 \tag{3.75}$$

$$\sum_{i=1}^{N} y_i = 1 \tag{3.76}$$

In the equilibrium system, the equilibrium distribution of component i in gas and liquid phases depends on its equilibrium constant:

$$K_i = \frac{y_i}{x_i} \tag{3.77}$$

Combining Eqs. (3.72), (3.73), (3.77) and eliminating y_i (or x_i) in these equations give

$$x_i = \frac{n_i}{K_i - (K_i - 1)N_L} \tag{3.78}$$

$$y_i = \frac{n_i K_i}{1 + (K_i - 1)N_g} \tag{3.79}$$

Applying the constraints of Eqs. (3.75)–(3.78), and Eqs. (3.76)–(3.79), we have

$$\sum_{i=1}^{N} x_i = \sum_{i=1}^{N} \frac{n_i}{K_i - (K_i - 1)N_L} = 1 \tag{3.80}$$

$$\sum_{i=1}^{N} y_i = \sum_{i=1}^{N} \frac{n_i K_i}{1 + (K_i - 1)N_g} = 1 \tag{3.81}$$

Equations (3.80) and (3.81) are the generalized phase equilibrium equations used for the gas–liquid equilibrium calculations of hydrocarbon mixtures. They are also called *flash equations*.

Both Eqs. (3.80) and (3.81) can be utilized to calculate the equilibrium compositions of gas and liquid phases in a hydrocarbon mixture at different conditions whatever the mixture is in reservoirs or in surface separator. In either case, a trial and error solution is required for solving above equilibrium equations due to the equilibrium constant K_i unknown. The equilibrium constant K_i needs to be first determined by empirical formula or other methods before any equilibrium calculation.

3.6.3 Calculation for Vapor–Liquid Equilibrium

Based on Eq. (3.80) or (3.81), these calculations for dew point, bubble point, and equilibrium flash can be done.

3.6.3.1 Bubble Point Calculation

This calculation is based on the idea that the first bubble formed in the liquid mixture does not visibly change the composition of the initial mixture and the mixture is in gas and liquid equilibrium. In terms of Eq. (3.72), as the quantity of the gas phase in the mixture is infinitesimal, we have: $N_g \approx 0$; $N_L \approx 1$. Then the composition of the gas phase (the bubble) is determined by Eq. (3.81),

$$y_i = n_i K_i \tag{3.82}$$

For all components in gas phase, the sum of their mole fractions must be identically unity:

$$\sum_{i=1}^{N} y_i = \sum_{i=1}^{N} n_i K_i = 1 \tag{3.83}$$

In Eq. (3.83), the equilibrium constant K_i for each component should be first determined by the bubble point pressure p_b and the given temperature T.

A trial and error method can be used here to determine the bubble point pressure at a given temperature. The initial bubble point pressure at a given temperature is supposed by selecting a trial value of pressure, based on which initial equilibrium constants can be determined by experiential formulas. Then calculate the sum of $n_i K_i$ in Eq. (3.82), and judge whether or not it is equal to unity within an acceptable

Table 3.31 The composition of a well stream and its trail value of p_b [18]

Composition of well stream, n_i (mole fraction)		Trail value of pressure: P = 21.0 MPa		Trail value of pressure: P = 22.45 MPa		Trail value of pressure: P = 23.2 MPa	
		K_i	n_iK_i	K_i	n_iK_i	K_i	n_iK_i
C_1	0.4404	2.150	0.9469	2.06	0.9072	2.020	0.8896
C_2	0.0432	1.030	0.0445	1.025	0.0433	1.020	0.0441
C_3	0.0405	0.672	0.0272	0.678	0.0274	0.680	0.0275
C_4	0.0284	0.440	0.0125	0.448	0.0127	0.450	0.0128
C_5	0.0174	0.300	0.0052	0.316	0.0055	0.323	0.0056
C_6	0.0290	0.215	0.0062	0.320	0.0067	0.239	0.0069
$C_7{}^+$	0.4011	0.024	0.0096	0.025	0.0100	0.026	0.0104
Sum	1.000		1.0521		1.0138		0.9969

tolerance of error. If the sum of n_iK_i is less than 1, then select a lower trial pressure and repeat above calculations at new trial pressure; if the sum of n_iK_i is higher than unity, a higher trial pressure is selected and repeat above calculations at the trial pressure till the sum is unity in an acceptable tolerance of error.

On the other hand, if pressure is given instead of temperature, the trial and error method is used to determine the bubble point temperature in the same way. Namely, a temperature should be first assumed at the given pressure p, then the K_i's are estimated and Eq. (3.83) is checked.

In detail, the trial and error procedure is as follows:

(a) A trial value of bubble point pressure is assumed for the given T, or is estimated by the following formula:

$$p_b = \sum_{i=1}^{N} n_i p_{bi}, \tag{3.84}$$

where p_b is an estimation of the bubble point pressure of the mixture, MPa; p_{bi} is the bubble point pressure of the ith component in the mixture, MPa. Although Eq. (3.84) is common expression of estimating the bubble point pressure of an ideal solution mixture at a given temperature, it is also a good initial trial value of bubble point pressure of a real liquid.

(b) Estimate the initial values of K_i by K-value charts or empirical Eq. (3.94) according to the trial bubble point pressure and given temperature.

(c) Calculate the mole fraction y_i of each component in equilibrium liquid phase by Eq. (3.82); and then make the summation of y_i.

(d) Check up if the sum of y_i is equal to one in an acceptable tolerance of error. If it is true, the trial pressure is the bubble point pressure of the mixture at given temperature.

(e) If not, a new trial value of pressure should be selected and repeat above calculations till the sum of y_i equals unity in an acceptable tolerance of error.

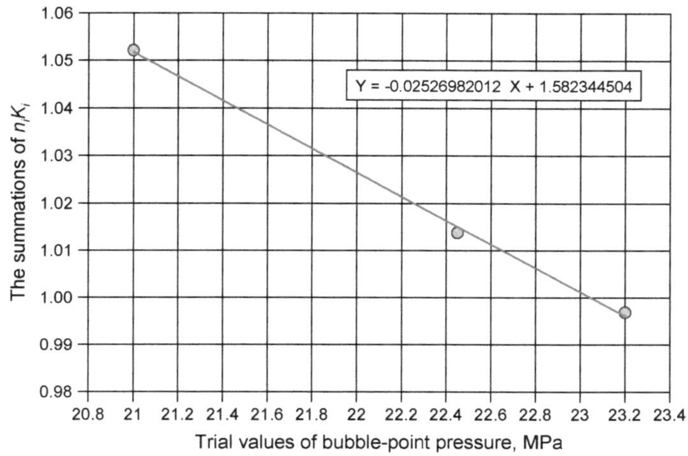

Fig. 3.63 $\sum n_iK_i$ versus trial values of pressure

Example 3.15 The composition of a well stream is listed in Table 3.31. Calculate the bubble point pressure of the well stream at reservoir temperature of 103 °C.

Solution

Using the trial and error method, determine the bubble point pressure by Eq. (3.81), within an acceptable tolerance.

1. Determine the first trial value of pressure by Eq. (3.84): 21.0 MPa
2. Estimate the initial values of K_i by empirical Eq. (3.94) in terms of the trial pressure and the given reservoir temperature of 103 °C. The results are shown in column 3 of Table 3.31.
3. Calculate the values of n_iK_i, and show them in column 4 of Table 3.31.
4. Calculate the sum of n_iK_i. It is 1.0521. The difference between the sum of n_iK_i and unity is 0.0521, which is higher than the acceptable tolerance of error. So, a new trail pressure should be selected.
5. Try a second trial pressure 22.45 MPa, which results in a sum of 1.0138. Further trial results in a sum of n_iK_i 0.9969. All results of calculations are given in Table 3.31.

The accurate value of pressure can be determined by linear interpolation or extrapolation because the relationship between Σn_iK_i and pressure is essentially linear.

The linear relationship between the trial pressures and the corresponding summations (Σn_iK_i) is plotted in Fig. 3.63. Therefore, the bubble point pressure of the well stream is then determined as 23.05 MPa by interpolating on the line according to unity Σn_iK_i.

3.6.3.2 Dew Point Calculation

Dew point calculation is used for the equilibrium calculation of a gas mixture. In dew point calculations, the vapor composition is given. We wish to know at what pressure the gas (or vapor) will begin to condense at a given temperature. When the first droplet of liquid is condensed, the initial composition of the gas is supposed to be constant. The liquid droplet and gas is in equilibrium. Therefore, we have, by Eq. (3.72), $N_L \approx 0$ and $N_g \approx 1$ at the dew point of the gas mixture.

The mole fraction of each component in the first droplet (liquid phase) is obtained by Eq. (3.80):

$$x_i = \frac{n_i}{K_i} \tag{3.85}$$

Then, the summation of the x_i's must be satisfied by the following expression:

$$\sum_{i=1}^{N} x_i = \sum_{i=1}^{N} \frac{n_i}{K_i} = 1 \tag{3.86}$$

In the same way, a trial and error is required for the determination of an accurate dew point pressure. A dew point pressure at given temperature is estimated and the corresponding K_i's are then found. The summation of x_i's is then calculated and compared to one. The pressure is adjusted as necessary to satisfy the dew point requirement, Eq. (3.86).

In the calculations of dew point pressure with a trial and error procedure, the initial trial value of pressure may be obtained by the following expression:

$$p_d = \frac{1}{\sum_{i=1}^{N} n_i / p_{\mathrm{d}i}}, \tag{3.87}$$

where p_{d} is an estimation of the dew point pressure of the mixture, MPa; $p_{\mathrm{d}i}$ is the dew point pressure of the ith component in the mixture, MPa.

Similarly, this equation is mainly used for the calculation of the dew point pressure of an ideal gas mixture at a given temperature. However, it can also provide a good initial trial value of dew point pressure for a real gas.

A dew point temperature can be calculated similarly, if the pressure is at first specified instead of the pressure.

3.6.3.3 Flash Vaporization Calculations

The usual flash calculation is the so-called *isothermal flash* [37]. This usage does not mean the mixture (or a feed) is flashed isothermally from its initial condition, but that the temperature and pressure of the flash are specified. That is to say, equilibrium flash calculation means the calculation of the properties and compositions of the equilibrium vapor and liquid phase of a mixture at specified temperature and pressure. Therefore, the conditions of a valid two-phase equilibrium flash are Eq. (3.80) for liquid phase, or Eq. (3.81) for gas phase.

In this case, the composition of the mixture, n_i, is given; and the corresponding equilibrium constants, K_i's can be determined according to the specified temperature and pressure. Only the mole fraction of equilibrium liquid phase (N_L) or gas phase (N_g), and the corresponding compositions (y_i or x_i) of equilibrium liquid phase or gas phase are needed to determine. Therefore, first assume an N_g (N_L), then the N_L (N_g) can be determined by Eq. (3.72), and then the y_i's (x_i's) for each component can be calculated by Eq. (3.79) (Eq. (3.78)). Finally, determine whether or not the sum of the x_i's (y_i's) is unity in an acceptable tolerance of error. If not, adjust N_g (N_L) slightly and repeat above calculations.

In flash calculations, be sure the mixture is indeed flashed at the specified p and T before making the trial and error computation. Check to be sure whether the mixture is above its bubble point or below its dew point temperature. Use the following relation [37]:

$$\sum_{i=1}^{N} \frac{n_i}{K_i} > 1 \text{ (guarantees liquid is present)} \tag{3.88}$$

$$\sum_{i=1}^{N} n_i K_i > 1 \text{ (guarantees vapor is present)} \tag{3.89}$$

If both sums are larger than one, proceed with the flash; otherwise, only a single phase is present. If the sum of n_i/K_i is not larger one, only a gas is present in the mixture. If the sum of n_iK_i is not larger one, similarly, only a liquid is present in the mixture.

3.6.4 Determination of Equilibrium Constant

In all calculations above, the first problem is how to determine the initial equilibrium constant for each component. The validity and speed of a phase equilibrium calculation also depends on the accuracy of the initial equilibrium constants.

The accurate values of equilibrium constants (K-factors) can only be expected from the test in the laboratory. However, it is not convenient for phase equilibrium calculations. In phase equilibrium calculations, the initial values of K-factor are often estimated by other methods, such as Raoult's and Dalton's laws, empirical correlations, K-value charts, or the fugacities of pure components in vapor and liquid phases.

3.6.4.1 Raoult's and Dalton's Laws

For ***ideal solutions**** at equilibrium, Raoult's law and Dalton's law may be combined to calculate the concentration of each component in the vapor and liquid phases.

****Idea solution***

An ideal liquid solution is a solution in which mutual solubility results when the components are mixed, no chemical interaction occurs upon mixing and the intermolecular forces of attraction and repulsion are the same between unlike as between like molecules.

The properties of ideal solutions lead to two practical results. First, there is no heating effect when the components of an ideal solution are mixed. Second, the volume of the ideal solution equals to the summation of the volumes that the components would occupy as pure liquids at the same temperature and pressure.

Raoult's law states that the partial vapor pressure of each component in an ideal mixture of liquids is equal to the vapor pressure of the pure component multiplied by its mole fraction in the mixture. It is mathematically expressed as

$$p_i = xp_i^0, \tag{3.90}$$

where p_i is the partial vapor pressure of the component i in vapor phase, MPa; p_i^0 is the vapor pressure of the pure component i at the temperature of the mixture, MPa; x is the mole fraction of the component i in the mixture, f.

Dalton's law states that in a mixture of nonreacting gases, the partial pressure of a component in the vapor mixture is equal to the total pressure of the mixture (exerted by the vapor) multiplied by its mole fraction in the mixture, mathematically expressed as

$$p_i = yp_{\mathrm{T}}, \tag{3.91}$$

where p_i is the partial pressure of the component i in a vapor mixture, MPa; p_{T} is the total pressure of the mixture, MPa; y is mole fraction of the component i in the gas mixture, f.

Combining Eqs. (3.90) and (3.91), we have

$$xp_i^0 = yp_{\mathrm{T}} \tag{3.92}$$

According to the definition of K-factor (Eq. (3.71)), arranging Eq. (3.92) gives the following expression:

$$K_i = \frac{y_i}{x_i} = \frac{p_i^0}{p_{\mathrm{T}}} \tag{3.93}$$

K_i is the equilibrium constant of the component i at system temperature and total pressure. Apparently, for an ideal solution, regardless of the total composition of the mixture, the equilibrium constant of each component simply depends on the temperature and pressure of the system. In addition, the vapor pressure of a component is only a function of system temperature.

However, the following two factors restrict the application of determining the equilibrium constants using Raoult's and Dalton's laws.

(a) A pure substance has no vapor pressures at temperatures above its critical temperature. Thus, Raoult's law is limited to temperatures less than the critical temperatures of all components in a system. For example, the equilibrium constant of methane cannot be evaluated using Raoult's law at temperatures above 46.7 °C.

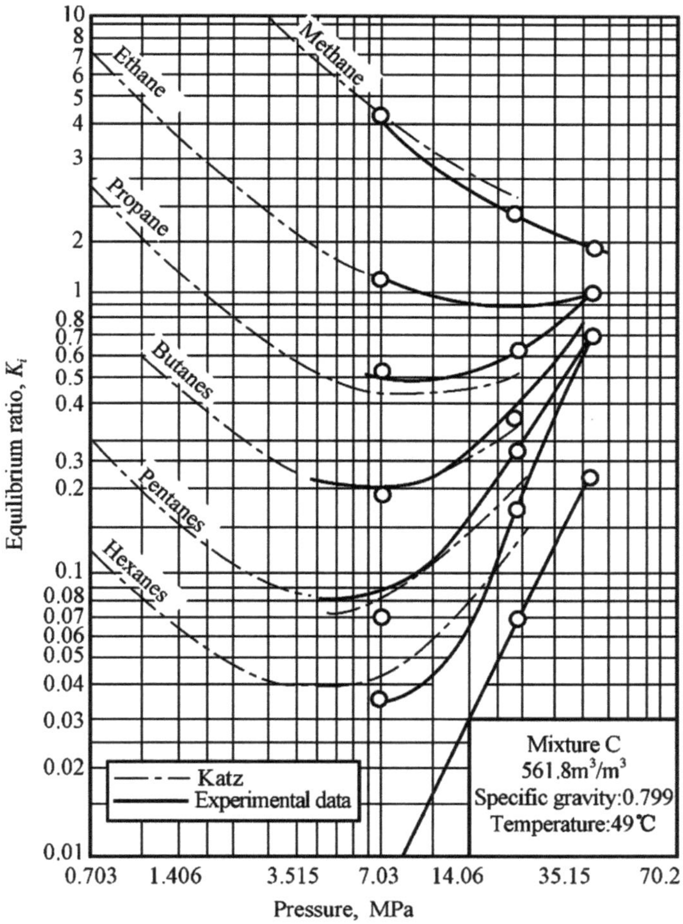

Fig. 3.64 *K*-values versus pressure at a given temperature and composition

(b) Dalton's law of partial pressures assumes that each component in the vapor behaves as an ideal gas. This assumption thus places an upper limit of pressure of about 0.4 MPa in practical use.

Therefore, this method is limited to certain temperature and pressure. It is primarily used to determine the equilibrium constants of components at normal temperature and pressure.

3.6.4.2 Correlations

Due to the effect of composition, Wilson presented the following correlation based on a great deal of experimental data:

$$K_i = \frac{\exp[5.37(1+\omega_i)(1-1/T_{ri})]}{p_{ri}}, \tag{3.94}$$

where K_i is the equilibrium constant of the *i*th component, dimensionless; T_{ri} is the reduced temperature of the component, dimensionless; p_{ri} is the reduced pressure of the component, dimensionless; ω_i is the acentric factor of the component, dimensionless.

Equation (3.94) is customarily used to determine the initial value of equilibrium constant for each component in a mixture in the gas–liquid equilibrium calculations.

3.6.4.3 *K*-value Charts

The *K*-value method is at first designed for hand calculation [42].

Determining in the laboratory the *K*-values of the components of a given hydrocarbon system at various pressures and at a given temperature, one obtains families of curves similar to that shown in Fig. 3.64. It is a typical *K*-value chart for a hydrocarbon system at a given composition and temperature. Here the *K*-values for the various components are plotted versus the pressure on double logarithmic paper. At low pressures, this plot yields a straight line with a slope of minus one. As pressure increases, all *K*-values decrease further and, with the exception of methane, reach a minimum value. At higher pressure, the *K*-values of all curves tend to end in one point, lying on the point $K_i = 1$ and characterized by a certain pressure. The pressure thus is known as the so-called *convergence pressure*, symboled by p_{cv}, which depends on the temperature and the composition of the mixture.

This appears to suggest that the composition of both phases should be the same at the convergence pressure due to $y_i = x_i$. As the similarity of both phases only occurs at the critical point, it implies that for any petroleum mixture, the critical temperature may be selected arbitrarily, which cannot be correct [8]. The fact is that the convergence pressure does not physically exist, unless the temperature at which

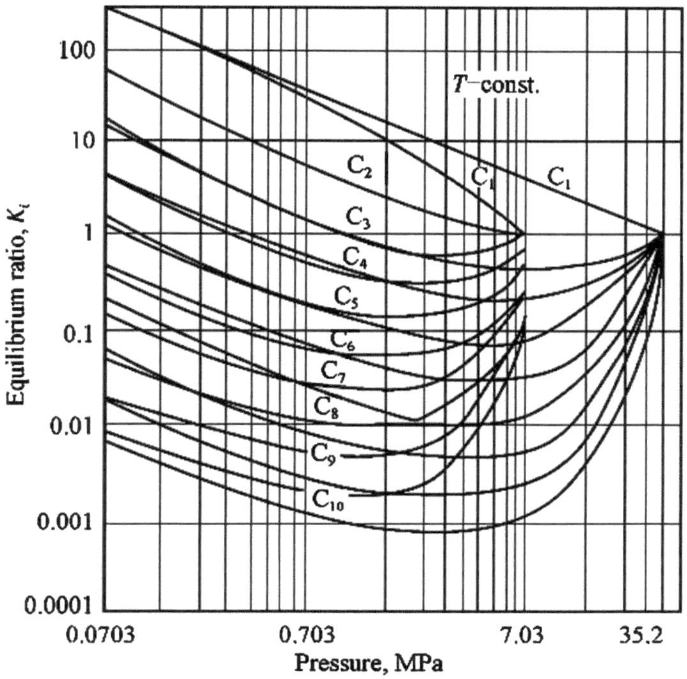

Fig. 3.65 Equilibrium ratios of components of hydrocarbon systems of 7.0 and 35 MPa, apparent convergence pressure [43]

the K-values are determined is the critical temperature of the mixture [8]. That to say, if the laboratory experiment is performed at the critical temperature of the system, then the convergence pressure agrees with the critical pressure. At any other temperature, the convergence is rarely apparent, because the system has its bubble point at a pressure lower than the convergence pressure as the system pressure increases; and there is just one phase instead of two at the point of convergence. Therefore, convergence pressure cannot be reached or determined in experiment at any other temperature. Actually, in the region between the bubble point pressure and the apparent convergence pressure, the equilibrium ratio lacks its physical meaning: the corresponding curve sections and the point of convergence itself are merely the results of extrapolation according to the tendency of these experimental curves.

The pressure, at which K-factor reaches one by the extrapolation of K-value curves, is thus referred to as "*apparent convergence pressure*". Obviously, the apparent convergence pressure is indeed different from the convergence pressure of a system. Both of them, however, are simply called "*convergence pressure*" in engineering.

The processes, using charts to determine *K*-values, relate the composition of the system to convergence pressure. It is assumed that if two systems of hydrocarbons agree as to convergence pressure, then the equilibrium ratios of their components will likewise be equal at the given pressure and temperature. Hence, in order to find K_i's of a system, it is necessary to know the convergence pressure of the system and the equilibrium ratios belonging to various convergence pressures.

Figure 3.65 shows two families of *K*-value curves for two systems. The convergence pressure is 7.0 MPa for the first family shown in the figure, and 35.0 MPa for the second (Amyx et al. 1960). At pressures below 0.7 MPa, perceptibly, the system's composition and knowledge of its exact convergence pressure lose much of their importance because the equilibrium ratios for each component of two systems are in good agreement. However, visible differences between the equilibrium ratios of each component in the two systems occur at pressures higher than 0.7 MPa. Knowledge of the composition and convergence pressure of a system is necessary for the determination of *K*-values. In conclusion, at lower pressure (<0.7 MPa), *K*-values are almost independent of the composition and convergence pressure of a system while they do depend on the composition and convergence pressure at higher pressure (>0.7 MPa).

K-values for common hydrocarbon components at different convergence pressures (4.2, 5.6, 7.0, 21.0, 28.5, 35.0, 70.0, and 140.0 MP) may be found in the NGAA (Natural Gas Association of American) Equilibrium Ratio Data Book (1957) [44]. To save page space, just two charts of propane at 35.0 MPa convergence pressure and heptane plus ($C_7{}^+$) are shown here in Figs. 3.66 and 3.67.

Now, the problem is which convergence pressure chart should be used to determine the *K*-values of components for a given system. Therefore, the convergence pressure of a system should be first determined. Its importance is substantial especially at high pressures. The convergence pressure of a given multicomponent system of hydrocarbons can be determined by various means [44, 45]. They are omitted here for page space.

In general, the operating pressure of oil–gas production at the surface is considerably less than the convergence pressure of a system, the *K*-values estimated from charts with various convergence pressures thus vary slightly, and an error in the estimate of convergence pressure has little effect on the resulting calculation. Therefore, in establishing the *K*-values required for equilibrium calculations in low-pressure separators, a satisfactory accuracy will be achieved if convergence pressure is simply assumed to equal 35 MPa, and this has been proved by experiments [44]. However, as operating pressure approaches convergence pressure, such as the calculation for reservoir mixture, *K*-values become very sensitive to the convergence pressure used [4]. Then care must be taken in the selection of correct value of convergence pressure. Sometimes other methods, such as trial and error method, are needed for the determination of correct convergence pressure of a system at high pressure. In addition, it is suggested that the pressure at least 10 %

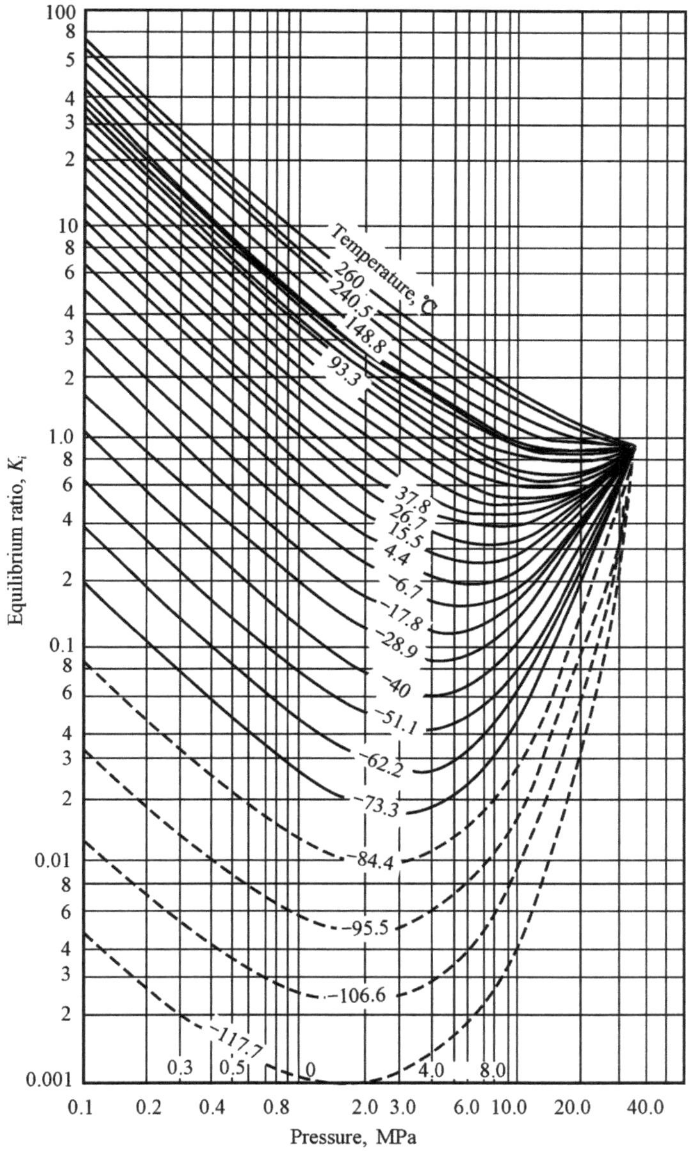

Fig. 3.66 Equilibrium ratios of propane at 35.0 MPa convergence pressure [43]

higher than the bubble point or dew point pressure of a system can be used as the convergence pressure if the bubble point or dew point pressure is given.

At higher pressure, a trial and error method is necessary to estimate convergence pressure of a mixture. Someone believed that the convergence pressure for a

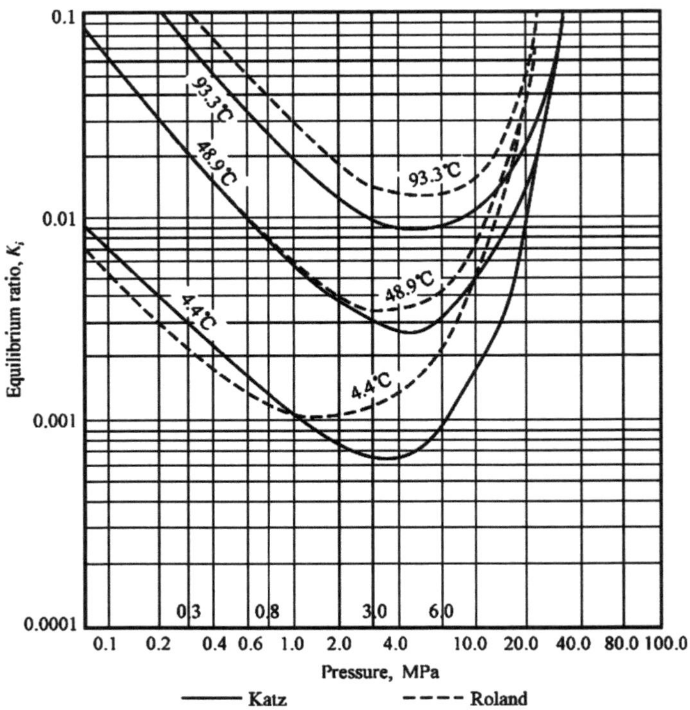

Fig. 3.67 Equilibrium ratios of $C_7{}^{+}$ according to Katz and Roland (Amyx et al. 1960)

mixture is at least higher 10 % than the bubble point pressure or dew point pressure of the mixture if they are given. In general, black oils have convergence pressures of about 69 MPa, retrograde gases about 35 MPa, and volatile oils about 48 MPa4.

Based on the *K*-value charts, the equilibrium ratios of C_3–C_6 components generally have an error lower than 10 %, whereas larger errors occur in the *K*-values of CH_4 and C_{7+} components. The error for *K*-values of CH_4 is even higher than 30 %. The most useful method for the estimation of the *K*-values of CH_4 and C_{7+} is calculation.

3.6.4.4 Fugacity Method

A better approximation of actual equilibrium constants can be obtained in terms of the fugacities of pure components. This method is based on thermodynamic model, and is developed for the calculation on computer. At present, various methods and algorithms based on the fugacity of pure components have been developed for *K*-value calculation at home and abroad.

The fugacity of a component can be regarded as a vapor pressure modified to represent properly the escaping tendency of the molecules from one phase into the other.

In an ideal solution, the fugacity of any component in the system is equal to fugacity of the pure component multiplied by its mole fraction:

$$f_{i\mathrm{l}} = f_{i\mathrm{l}}^{0} x_i \tag{3.95}$$

$$f_{i\mathrm{g}} = f_{i\mathrm{g}}^{0} y_i, \tag{3.96}$$

where $f_{i\mathrm{l}}$ and $f_{i\mathrm{g}}$ represent the fugacities of component i in the liquid and gas phases, separately; $f_{i\mathrm{l}}^{0}$ is the fugacity of the pure component i in liquid state; $f_{i\mathrm{g}}^{0}$ is the fugacity of the pure component i in gas state.

At the gas–liquid equilibrium, the fugacities of component i in liquid and vapor are the same:

$$f_{i\mathrm{l}} = f_{i\mathrm{g}} \tag{3.97}$$

Combining Eqs. (3.96) and (3.96), Eq. (3.97) is then written as

$$f_{i\mathrm{l}}^{0} x_i = f_{i\mathrm{g}}^{0} y_i \tag{3.98}$$

In terms of the definition of equilibrium constant, we have

$$K_i = \frac{y_i}{x_i} = \frac{f_{i\mathrm{l}}^{0}}{f_{i\mathrm{g}}^{0}} \tag{3.99}$$

Based on the definition of fugacity coefficient φ_i, Eqs. (3.96) and (3.96) can be written as

$$f_{i\mathrm{l}} = x_i \varphi_{i\mathrm{l}} p \tag{3.100}$$

$$f_{i\mathrm{g}} = y_i \varphi_{i\mathrm{g}} p, \tag{3.101}$$

where p is the equilibrium pressure of the system; $\varphi_{i\mathrm{l}}$ is the fugacity coefficient of the ith component in liquid phase; $\varphi_{i\mathrm{g}}$ is the fugacity coefficient of the ith component in gas phase.

Combining Eqs. (3.99)–(3.101) gives

$$K_i = \frac{y_i}{x_i} = \frac{f_{i\mathrm{l}}^{0}}{f_{i\mathrm{g}}^{0}} = \frac{f_{i\mathrm{l}}/x_i}{f_{i\mathrm{g}}/y_i} = \frac{\varphi_{i\mathrm{l}}}{\varphi_{i\mathrm{g}}} \tag{3.102}$$

Equation (3.102) can be used to calculate equilibrium constants. The fugacity coefficients, $\varphi_{i\mathrm{l}}$ and $\varphi_{i\mathrm{g}}$, in Eq. (3.102) depend on the thermodynamic properties of component i in the mixture, such as temperature, pressure, and the composition of the system. The rigorous integral equations which are used for the calculation of fugacity coefficients ($\varphi_{i\mathrm{l}}$, $\varphi_{i\mathrm{g}}$) can be obtained from the equation of state (EOS).

The following equations relate the fugacity to the measurable properties of P, V, and T of a system:

$$RT\ln\,\varphi_{i\mathrm{l}} = \int\limits_{V_\mathrm{L}}^{\infty}\left[\left(\frac{\partial p}{\partial n_{i\mathrm{l}}}\right)_{V_\mathrm{L}Tn_{i\mathrm{l}}} - \frac{RT}{V_\mathrm{L}}\right]\mathrm{d}V_\mathrm{L} - RT\ln Z_\mathrm{L} \tag{3.103}$$

$$RT\ln\,\varphi_{i\mathrm{g}} = \int\limits_{V_\mathrm{g}}^{\infty}\left[\left(\frac{\partial p}{\partial n_{i\mathrm{g}}}\right)_{V_\mathrm{g}Tn_{i\mathrm{g}}} - \frac{RT}{V_\mathrm{g}}\right]\mathrm{d}V_\mathrm{g} - RT\ln Z_\mathrm{g} \tag{3.104}$$

As long as Eqs. (3.103) and (3.104) can describe accurately the PVT behavior of equilibrium gas and liquid phases, fugacity coefficients ($\varphi_{i\mathrm{l}}$, $\varphi_{i\mathrm{g}}$) of each component can be determined. Note that the values of Z-factor in Eqs. (3.103) and (3.104) should be calculated using the equation of state (EOS). For example, the PR (Peng–Robinson) equation of state in the form of Z-factor can be used for Z-factor calculation:

$$\begin{aligned} &Z_m^3 - (1 - B_m)Z_m^2 + \left(A_m - 2B_m - 3B_m^2\right)Z_m \\ &\quad - \left(A_mB_m - B_m^2 - B_m^3\right) = 0 \end{aligned}, \tag{3.105}$$

where A_m and B_m are the coefficients of the equation, and determined by the following expressions:

$$A_m = \frac{a_m p}{(RT)^2},\quad B_m = \frac{b_m p}{RT}$$

Substituting Z-factor in Eqs. (3.103) and (3.104) for Z-factor in Eq. (3.105) gives the expressions of fugacity coefficients:

$$\begin{aligned} \ln\,\varphi_i = {} & \frac{b_i}{b_m}(Z_m - 1) - \ln(Z_m - B_m) \\ & - \frac{A_m}{2\sqrt{2}B_m}\left(\frac{2\psi_i}{a_m} - \frac{b_i}{b_m}\right)\ln\left(\frac{Z_m + 2.414\,B_m}{Z_m - 2.414\,B_m}\right) \end{aligned} \tag{3.106}$$

In Eq. (3.106), the factor P_i is calculated by the following expression:

$$\psi_i = \sum_i^n x_i\left(a_i a_j \alpha_i \alpha_j\right)^{0.5}\left(1 - k_{ij}\right), \tag{3.107}$$

where x_i is the liquid-phase composition if it is for the calculation of fugacity coefficient $\varphi_{i\mathrm{l}}$ of a liquid component; otherwise, x_i is the gas-phase composition when the fugacity coefficient φ_{ig} of a gas component is calculated; a_i, b_i, a_m, b_m, α_i, and k_{ij} in Eqs. (3.106) and (3.107) are coefficients of PR equation of state. The details for these parameters can be available from the relevant books which introduce EOS.

Likewise, a trial and error method is necessary for K-values estimation based on fugacity coefficients.

3.6.5 *Procedures of Phase Equilibrium Calculation Based on EOS*

In phase equilibrium calculations, iteration or trial and error method is a required algorithm, generally including stepwise iteration and Newton–Raphson iteration. For simplicity, a brief introduction to iterative method in gas–liquid equilibrium calculation is presented here.

3.6.5.1 Procedure for the Calculation of Dew Point or Bubble Point [18]

Using Eq. (3.85), the procedure for dew point calculation is as follows:

(a) Input the basic properties of each component: n_i, p_{ci}, T_{ci}, ω_i, and k_{ij};
(b) Assume an initial dew point of the system: T and p_d;
(c) Estimate the initial value of equilibrium constants of the components by Eq. (3.94);
(d) Calculate the initial composition of equilibrium liquid phase, x_i, by Eq. (3.85). In this case, $y_i = n_i$;
(e) Determine the coefficients of PR-EOS: a_i, b_i, a_m, b_m, A_m, and B_m;
(f) Find the compressibility factors of liquid and gas phases using Eq. (3.105): Z_L and Z_g;
(g) Find the fugacity coefficients (φ_{il}, φ_{ig}) and fugacities (f_{il}, f_{ig}) of the components in liquid and gas phases;
(h) Check if $|f_{il} - f_{ig}| \leq \varepsilon$; here ε is the tolerance of error; $\varepsilon = 10^{-3}$;
(i) If the difference between the fugacities (f_{il}, f_{ig}) of each component in liquid phase and gas phase is higher than the tolerance of error, the iterations of the K-values then proceed using the following iterative equation:

$$K_{i(j+1)} = K_{ij}\frac{f_{il,j}}{f_{ig,j}}$$

Then return to Step (d);
(j) If the error of fugacity is acceptable, calculate F_{j+1} by the following equation:

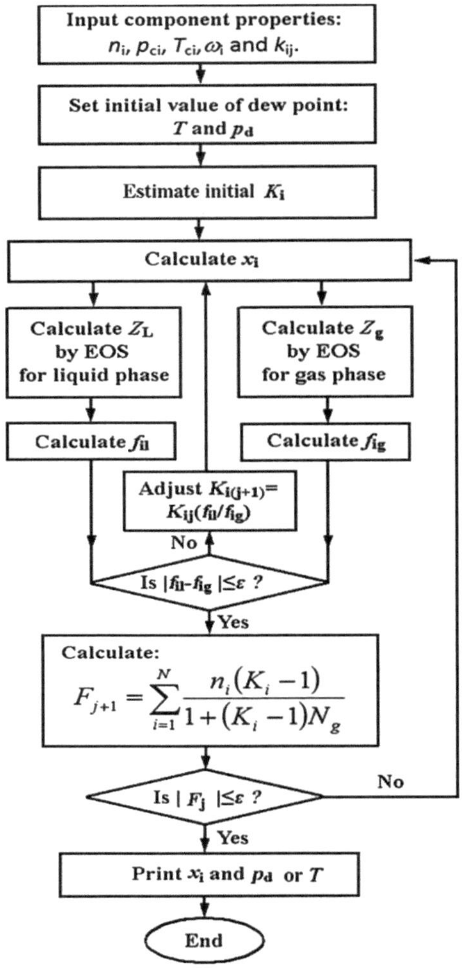

Fig. 3.68 Flow chart for dew point calculation using Newton–Raphson iteration

$$F_{j+1}=\sum_{i=1}^{N}y_i=\sum_{i=1}^{N}\frac{n_i(K_i-1)}{1+(K_i-1)N_{\mathrm{g}}}$$

(k) Check if $|F_j|\le\varepsilon$; here $\varepsilon=10^{-3}$;

(l) If the error above is not acceptable, adjust slightly p_d or T with the following equation

$$p_{j+2}=p_{j+1}\frac{F_{j+1}(p_{j+1}-p_j)}{F_{j+1}-F_j},$$

then return to Step (d);

(m) If the error above is acceptable, print the results: x_i and p_d if changed or T if changed;
(n) Proceed to next calculation by repeating Steps (b) through (m).

The procedure for dew point calculation is also suitable for bubble point calculation if substituting Eq. (3.82) for Eq. (3.85), and x_i for y_i.
The flow chart of dew point calculation is shown in Fig. 3.68.

3.6.5.2 Procedure for Flash Vaporization Calculation [18]

(a) Input the basic properties of each component: n_i, p_{ci}, T_{ci}, ω_i, and k_{ij};
(b) Set an initial value of temperature and pressure for flash vaporization: T and p;
(c) Estimate the initial value of K-values of the components by Eq. (3.94);
(d) Calculate N_g by Newton–Raphson iteration using the following iterative equation:

$$N_{g(j+1)} = N_{gj}\frac{F(N_{gj})}{F'(N_{gj})}, \quad \text{where } F(N_{gj}) = \sum_{i=1}^{N}\frac{n_i(K_i-1)}{1+(K_i-1)N_{gj}};$$

(e) Check if $|F(N_{gj})| \leq \varepsilon$; here $\varepsilon = 10^{-3}$;
(f) If the error above is not acceptable, go back to Step (d);
(g) If the error above is acceptable, go back to Step (h);
(h) Calculate the composition of liquid phase, x_i, by Eq. (3.80); and the composition of gas phase, y_i, by Eq. (3.81);
(i) Find the coefficients of PR-EOS: a_i, b_i, a_m, b_m, A_m, and B_m;
(j) Determine the compressibility factors of liquid and gas phases with Eq. (3.105): Z_L and Z_g;
(k) Find the fugacity coefficients (φ_{il}, φ_{ig}) and fugacities (f_{il}, f_{ig}) of components in equilibrium liquid and gas phases;
(l) Check if $|f_{il} - f_{ig}| \leq \varepsilon$; here $\varepsilon = 10^{-3}$;
(m) If the error above is not acceptable, proceed the iterations of the K-values using the following equation:

$$K_{i(j+1)} = K_{ij}\frac{f_{il,j}}{f_{ig,j}},$$

and then return to Step (d);
(n) If the error above is acceptable, print the results: x_i, y_i, N_g, and N_L.
(o) Proceed to next calculation by repeating Steps (b) through (n).

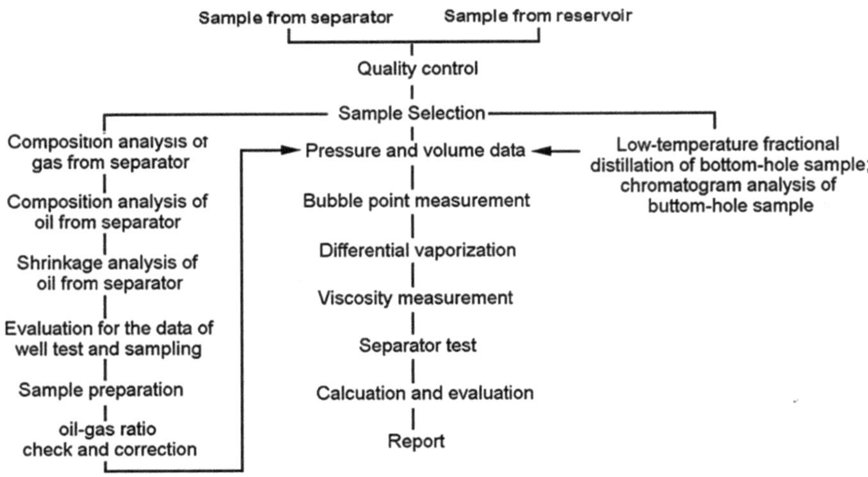

Fig. 3.69 Typical procedure of crude oil analysis

3.7 Experimental Measurement of Reservoir Fluid's Properties

In petroleum production, the properties of reservoir fluids are indispensable data to the geological and engineering studies and calculations, such as petroleum reserves, reservoir numerical simulation, surface productions, and so on. How can we obtain them?

The phase behavior and physical properties of hydrocarbon mixtures can be accessed through the researches of the relationship between pressure, volume, and temperature (P–V–T). As described above, the PVT behavior of gas mixtures can be analyzed with various equations of state, such as the compressibility equation of state ($PV = ZnRT$), van der Waals' cubic equation of state, and so on. However, the PVT behavior of reservoir crudes can hardly research with the equation of state because the appropriate equation of state, which can describe accurately the PVT behavior of crude oils, had not been found. Then, how can we obtain the property parameters of crude oils at reservoir conditions?

At present, the prevailing method of measuring the properties of reservoir oils is PVT tests of reservoir fluids. It is the most accurate method. The appropriate procedures for crude oil and natural gas analysis, from sampling to final report, have been formed at home and abroad [18]. The typical flow chart of crude analysis is shown in Fig. 3.69.

Fig. 3.70 RUSKA-601 PVT apparatus (U.S.)

Fig. 3.71 DBR-JEFRI PVT apparatus (Canada)

3.7.1 PVT Apparatus and Fluid Sample Preparation

3.7.1.1 PVT Apparatus

The PVT apparatus is designed to measure the volume of a given amount of gas or liquid as a function of its temperature and pressure. The apparatus is completely automated and is computer controlled in order to yield precise data. Despite of the differences between pressure-transmitting medium, the ranges of pressure and temperature, and measuring instruments of the PVT apparatus designed and used by China and other countries, the basic principle and structure of PVT apparatus are similar.

In the 1990s, most of PVT apparatus used mercury as the pressure medium. Because of high density, mercury does not dissolve in oil, and can fill almost the entire chamber excepting that is occupied by fluids in PVT apparatus, which results in small "*dead volume*" in test process. Consequently, mercury is customarily used at that time. However, mercury is strongly volatile at high temperatures, and the mercury vapor is toxic. Once leaked, mercury can lead human to chronic poisoning. Therefore, since the 1990s, the American RUSKA and Canada DBR companies have successfully developed the piston mercury-free PVT apparatus for the first time, which are shown in Figs. 3.70 and 3.71. These instruments mainly consist of a pressure generation, control, and measurement system, a temperature control and measurement system, a stainless steel gas chamber, and a Windows-based software for ease of operation and accuracy of results.

Our Haian Petroleum Instrument Co. Ltd. and Huabao Petroleum Instrument Co. Ltd in Yangzhou also developed one after the other the piston PVT apparatus. Figure 3.72 is the schematic diagram of the PVT system designed and manufactured in China. The major components are as follows:

(a) PVT cell: it a vessel for holding oil and gas, and is used for phase equilibrium test of hydrocarbon systems at high pressure and temperature.
(b) High-pressure metering pump: it is used for pressurizing the PVT system;, and for measuring the volume of oil and gas at high pressures.
(c) Accessory equipment: it includes high-pressure viscometer, downhole sampler, the tank containing pressure medium, and so on.
(d) Temperature control system: it is used for heating the oil and gas, and maintaining a constant temperature for the system. Temperature control system includes temperature-controlled water bath (oil bath), PVT cell, high-pressure viscometer, electric control box, and so on.

The new-type PVT apparatus developed at home and abroad have had a number of improvements in instruments, such as using stepless speed pump, high-pressure view cell; adding laser or ultrasonic measuring device, image acquisition system which makes observation easy and leads to more visible and accurate test results. Therefore, the volume of PVT cell (chamber) of the apparatus can be determined by calibration. The program automatically can change the temperature and pressure to the desired set values. The change in volume can then be recorded automatically.

3.7.1.2 Fluid Sample Preparation

Sampling

Reservoir engineers are primarily concerned with what is happening in the reservoirs. It is highly desirable to have a fluid sample closely resembling the original reservoir fluids as much as possible.

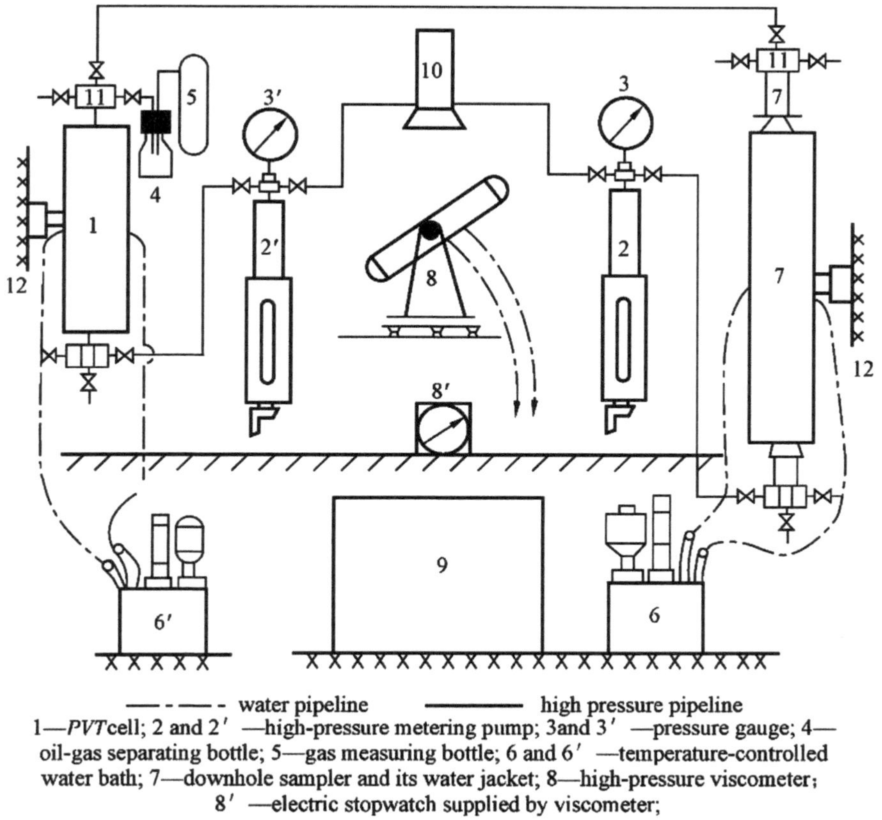

Fig. 3.72 Schematic diagram of PVT system [18]

Sampling means obtaining a representative sample of reservoir fluids. The representativeness of the fluid sample determines whether the test results can be used or not. A representative fluid sample indicates that the sample should remain the original attributes of the fluids in reservoirs as much as possible, for example, the temperature, pressure, phase behavior, and the composition of the fluids. Generally, sampling a representative fluid should be strictly finished according to the standard of petroleum industry.

Fluid samples can be obtained from downhole and/or from surface separator. There are two options to obtain a representative sample of reservoir fluids. The first option is obtaining a fluid sample from the bottom hole of well. This method can remain the original features of a reservoir fluid. The next option is using surface sampling, namely, sampling fluids (gas and liquid) from surface separators and then recombine the fluids according to the producing GOR.

Transferring Sample

Transferring sample means the process that the downhole fluid sample is transferred into PVT cell at reservoir pressure and temperature from the sampler after sampling. PVT experiments can be carried out only when the sample is transferred into PVT cell. The key of transferring samples is that the attributes of the fluid sample cannot be altered during the process of sample transferred from the sampler to the cell. To achieve this, maintaining the reservoir temperature and pressure is necessary.

Before the sample being transferred, the sampler, 7 in Fig. 3.72 is inserted in the temperature-controlled water bath, 6 in Fig. 3.72. Then, the temperature and pressure of the sample in the sampler can be maintained to the reservoir conditions using the water bath 6 and the high-pressure metering pump 2. And then control the pressure medium (e.g., brine) in the PVT cell to the same temperature and pressure as the sample using another water bath 6′ and another high-pressure metering pump 2′. After that, using the metering pump 2 pump the brine to displace the sample out from the sampler. At the same time, withdraw the metering pump 2′ to maintain the pressure of the sample constant, which is indicated on pressure gauges 3 and 3′. As a result, the fluid sample is successfully transferred into the PVT cell from the sampler. The volume of the sample is thus measured by the metering pump, which can be read from the scale of the metering pump. Then the sample preparation is over.

3.7.2 *Items of PVT Measurements*

After the sample preparation, various PVT tests can be performed to determine the phase behavior and physical properties of reservoir fluids. Generally, the items of PVT tests include the following content.

3.7.2.1 Tests of Phase Behavior

Despite the equation of state can describe the PVT behavior of gas mixtures in reservoirs, experimental test is still necessary owing to the accuracy of experimental data, especially for the determination of the critical point, dew point, or bubble point of a hydrocarbon system. In the PVT test, the phase behavior of an oil–gas system can be accurately measured, such as the locus of dew point, the locus of bubble point, the critical point pressure and temperature, and so on.

3.7.2.2 *Z*-factor Measurement for Special Natural Gases

The *Z*-factor of a natural gas is usually obtained from *Z*-factor charts, but it is less accurate for special natural gases, such as the ultra-deep natural gases, the gases containing high-content heavier hydrocarbons or/and nonhydrocarbons. In these cases, the Z-factor of the natural gases should be determined with PVT experiments.

3.7.2.3 Measurement of the Properties of In-place Oils

The properties of reservoir oils, needed to measure, include oil compressibility, oil formation volume factor, saturation pressure, oil viscosity, relative density, and so on. Pressure–volume relations can be measured at reservoir temperature by the expansion experiment. Generally, the following property parameters of in-place oils can be measured by PVT experiments:

(a) Bubble point pressures
(b) Isothermal compressibility of in-place oil
(c) Flash GOR, namely solution gas–oil ratio of in-place oil
(d) Formation volume factor of in-place oil
(e) Oil viscosity.

3.7.2.4 Measurement of the Properties of Condensate Gases

For condensate gas reservoirs *constant volume depletion test* (CVD test) is one of the most important experiments. It is designed for the simulation of production performances of depletion-drive condensate gas reservoirs. The aim of this experiment is to predict the change in the content of condensate oil, the composition of well stream, and the condensate recovery in the reservoir production.

In addition, the solution, separation, and miscibility experiments between oil and gas can also be carried out using PVT apparatus. For example, the optimal stage of oil–gas separation, gas solubility, and minimum miscibility pressure can be determined in these experiments. These parameters are of important practical significance in reservoir engineering studies for EOR.

3.8 Correlations for Physical Properties of Crude Oil

The data of the physical properties of crude oils are usually determined by laboratory experiments that are performed on samples of actual reservoir fluids. In the absence of experimentally measured properties of crude oils and for the convenient of data input in reservoir simulation, it is necessary for the petroleum engineer to determine the properties from empirically derived correlations [26].

3.8.1 Correlations for Oil Formation Volume Factor

Most of the published empirical B_o correlations utilize the following generalized relationship: $B_o = f(R_s, \gamma_g, \gamma_o, T)$.

3.8.1.1 Standing's Correlation

Standing (1947) [46] presented a graphical correlation for estimating the oil formation volume factor with the solution gas–oil ratio, gas gravity, oil gravity, and reservoir temperature as the correlating parameters. This graphical correlation originated from examining a total of 105 experimental data points on 22 different California hydrocarbon systems. An average error of 1.2 % was reported for the correlation [26].

Standing (1981) showed that the oil formation volume factor can be expressed more conveniently in a mathematical form by the following equation [38]:

$$B_o = 0.9759 + 0.000120\left[R_s\left(\frac{\gamma_g}{\gamma_o}\right)^{0.5} + 1.25(T - 460)\right]^{1.2}, \tag{3.108}$$

where T is temperature, °R; γ_o is the specific gravity of the stock tank oil, dimensionless; γ_g is the specific gravity of the solution gas, dimensionless.

3.8.1.2 Vasquez–Beggs Correlation

Vasquez and Beggs (1980) [47] developed a relationship for determining B_o as a function of the solution gas–oil ratio, gas gravity, oil gravity, and reservoir temperature. The proposed correlation was based on 6000 measurements of B_o at various pressures [26]. Using the regression analysis technique, Vasquez and Beggs found the following equation to be the best form to reproduce the measured data:

$$B_o = 1.0 + C_1 R_s + (T - 520)\left(\frac{\mathrm{API}}{\gamma_{gs}}\right)[C_2 + C_3 R_s], \tag{3.109}$$

where R_s is the solution gas–oil ratio of oil, scf/stb; T is temperature, °R; γ_{gs} is the gas corrected specific gravity related to the separator gas gravity by Danesh [8]:

$$\gamma_{gs} = \gamma_g\left[1.0 + 5.912\left(10^{-5}\right)(\mathrm{API})\left(T_{sep} - 460\right)\log\left(\frac{p_{sep}}{114.7}\right)\right], \tag{3.110}$$

where γ_g is the gas gravity at the actual separator conditions of p_{sep} and T_{sep}; p_{sep} is actual separator pressure, psia; T_{sep} is actual separator temperature, °R.

Values for the coefficients C_1, C_2, and C_3 in Eq. (3.109) are given in Table 3.32.

Vasquez and Beggs reported an average error of 4.7 % for the proposed correlation.

Table 3.32 Values for the coefficients C_1, C_2, and C_3 in Eq. (3.109)

Coefficient	API $\leq$ 30	API > 30
C_1	4.677×10^{-4}	4.670×10^{-4}
C_2	1.751×10^{-5}	1.100×10^{-5}
C_3	1.811×10^{-8}	1.337×10^{-9}

3.8.1.3 Glaso's Correlation

Glaso (1980) proposed the following expressions for calculating the oil formation volume factor:

$$B_o = 1.0 + 10^A, \tag{3.111}$$

where

$$A = -6.58511 + 2.91329 \log B_{ob}^* - 0.27683\left(\log B_{ob}^*\right)^2 \tag{3.112}$$

and B_{ob}^* is a *correlating number*. B_{ob}^* is defined by the following equation:

$$B_{ob}^* = R_s\left(\frac{\gamma_g}{\gamma_o}\right)^{0.526} + 0.968(T - 460), \tag{3.113}$$

where T is temperature, °R; γ_o is specific gravity of the stock tank oil, dimensionless.

The above correlations were originated from studying PVT data on 45 oil samples. The average error of the correlation was reported at −0.43 % with a standard deviation of 2.18 % [26].

Sutton and Farshad (1984) [48] concluded that Glaso's correlation offers the best accuracy when compared with the Standing and Vasquez–Beggs correlations. In general, Glaso's correlation underpredicts formation volume factor. Standing's expression tends to overpredict oil formation volume factors greater than 1.2 bbl/stb. The Vasquez–Beggs correlation typically overpredicts the oil formation volume factor [26].

It should be noted that all the correlations could be used for any pressure equal to or below the bubble point pressure [26].

3.8.2 *Correlations for Oil Viscosity*

According to the pressure, the viscosity of crude oils can be classified into three categories [26]:

Dead oil viscosity: the dead oil viscosity is defined as the viscosity of crude oil at atmospheric pressure (no gas in solution) and system temperature.

Saturated oil viscosity: the saturated (bubble point) oil viscosity is defined as the viscosity of the crude oil at the bubble point pressure and reservoir temperature.
Undersaturated oil viscosity: the undersaturated oil viscosity is defined as the viscosity of the crude oil at a pressure above the bubble point and reservoir temperature.

Oil viscosity is often estimated using empirical correlations developed by a number of investigators including Beal [49], Beggs and Robinson (1975) [50], Standing (1981) [38], Glaso (1985) [51], Khan (1987) [52], and Ahmed (1989) [53]. A summary of these correlations is given by Ahmed (1989) [53]. Engineers should select and validate a correlation with measurements before it is used. A brief description of several correlations that are widely used in estimating the viscosity of above three oils is given below.

3.8.2.1 Estimation of Dead Oil Viscosity

Several empirical methods are proposed to estimate the viscosity of the dead oil, including: Beal's correlation, Beggs–Robinson correlation; Glaso's correlation [26].

Beal's Correlation

From a total of 753 values for dead oil viscosity at and above 100 °F, Beal (1946) developed a graphical correlation for determining the viscosity of the dead oil as a function of temperature and the API gravity of the crude [49]. Beal proposed the correlation in order to estimate the dead oil when the API gravity of crude oil and reservoir temperature are known. Standing (1981) expressed the proposed graphical correlation in a mathematical relationship as follows [38]:

$$\mu_{\text{od}} = \left(0.32 + \frac{1.8 \times 10^7}{\text{API}^{4.5}}\right)\left(\frac{360}{t + 200}\right)^A, \tag{3.114}$$

where

$$A = 10^{\left(0.43 + \frac{8.33}{API}\right)} \tag{3.115}$$

and μ_{od} is the viscosity of dead oil as measured at 14.7 psia and reservoir temperature, mPa s; T is reservoir temperature, °R.

Beggs–Robinson Correlation

Beggs and Robinson (1975) developed an empirical correlation for determining the viscosity of the dead oil. The correlation originated from analyzing 460 dead oil viscosity measurements. The proposed relationship is mathematically expressed as follows:

$$\mu_{\text{od}} = 10^{X} - 1, \tag{3.116}$$

where $X = Y\,(T - 460)^{-1.163}$; $Y = 10^{Z}$; $Z = 3.0324 - 0.02023°\text{API}$.

An average error of −0.64 % with a standard deviation of 13.53 % was reported for the correlation when tested against the data used for its development. Sutton and Farshad (1984) reported an error of 114.3 % when the correlation was tested against 93 cases from the literature [26].

Glaso's Correlation

Glaso (1980) also proposed a generalized mathematical relationship for estimating the dead oil viscosity. The relationship was developed from experimental measurements on 26 crude oil samples. The correlation has the following form:

$$\mu_{\text{od}} = 3.141 \times 10^{10}(T - 460)^{-3.444}[\log(\text{API})]^{a}, \tag{3.117}$$

where the coefficient a is given by: $a = 10.313\,[\log(T - 460)] - 36.447$.

The above expression can be used within the range of 50–300 °F for the system temperature and 20–48° for the API gravity of the crude [26]. Sutton and Farshad (1984) concluded that Glaso's correlation showed the best accuracy of the three previous correlations [48].

3.8.2.2 Estimation of Saturated Oil Viscosity

Since crude oil would contain dissolved gas under reservoir conditions, a correction for saturated oil viscosity is necessary. Chew and Connally (1959) [54] proposed a graphical correlation to adjust the dead oil viscosity according to the solution gas–oil ratio at saturation pressure. The correlation was developed from 457 crude oil samples, and for saturated oil at the bubble point pressure [26]. Standing (1977) [55] expressed the graphical correlation in a mathematical form as follows:

$$\mu_{\text{ob}} = 10^{a}(\mu_{\text{od}})^{b} - 1, \tag{3.118}$$

where μ_{ob} is viscosity of oil at the bubble point, cp; μ_{od} is the viscosity of dead oil; and

$a = R_s\,[2.2(10^{-7})\,R_s - 7.4(10^{-4})),$
$b = 0.68/10^e + 0.25/10^d + 0.062/10^e,$
$c = 8.62(10^{-5})R_s,$
$d = 1.1(10^{-3})R_s,$
$e = 3.74(10^{-3})R_s$, and R_s is the solution gas–oil ratio at bubble point, scf/stb.

This correlation is valid over the usual range of reservoir pressure and temperature, with dead oil viscosity of 0.377–50 mPa s and solution gas–oil ratio of 51–3544 scf/stb [11].

3.8.2.3 Estimation of Undersaturated Oil Viscosity

Vasquez and Beggs Correlation

If the reservoir pressure was above the bubble point, another correction would be necessary for the undersaturated oil as it is compressed further. Vasquez and Beggs proposed the following correlation when the viscosity of oil at the bubble point pressure is available [56]:

$$\mu_o = \mu_{ob}(p/p_b)^m, \tag{3.119}$$

where p_b is bubble point pressure, psia; $m = 2.6p^{1.187}\exp(-11.513 - 8.98 \times 10^{-5}p)$; p is pressure of interest, psia.

The correlation is valid under a wide range of reservoir conditions, including API gravity of 15.3° to 59.5°, oil viscosity of 0.117–148 cp, gas gravity of 0.511–1.351, and solution gas–oil ratio of 9.3–2199 scf/stb [11].

Beggs–Robinson Correlation

Beggs and Robinson proposed a correlation to estimate the live oil viscosity according to the solution gas–oil ratio at saturation pressure and the density of dead oil at the surface:

$$\mu_o = A\mu_{od}^B \tag{3.120}$$

with

$$\left.\begin{aligned} A &= 4.4044(\rho_o R_s + 17.7935)^{-0.515} \\ B &= 3.0352(\rho_o R_s + 26.6904)^{-0.338} \end{aligned}\right\}, \tag{3.121}$$

where μ_o is the viscosity of oil at pressures lower than bubble point pressure, mPa s; μ_{od} is the viscosity of dead oil, mPa s; ρ_o is the density of dead oil, g/cm^3; R_s is the solution gas–oil ratio, m^3/t.

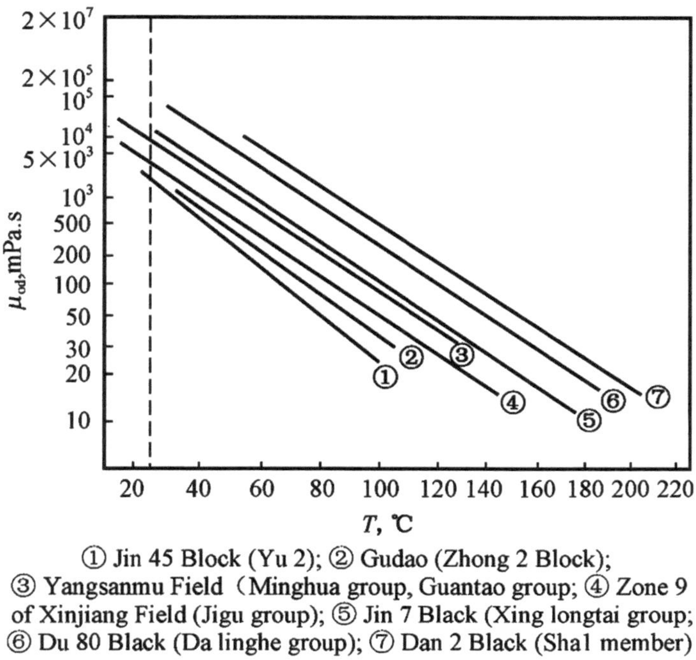

Fig. 3.73 The curves of viscosities of viscous oils from typical viscous oil zones in China versus temperature [18]

It is reported that the correlation shows good accuracy to both black oil and heavy oil when tested in China's oil fields.

To test viscous oils, a lot of PVT data, from Liaohe, Shengli, and Xingjiang oil fields in China, are analyzed. Based on the solution gas–oil ratios and densities of viscous oils from oil fields above, the values of the coefficients A and B in Eq. (3.120) are determined by Eq. (3.121). The viscosities of dead oils at reservoir temperatures are then found in terms of the corresponding viscosity–temperature curves. As a result, the viscosities of viscous oils at reservoir conditions are estimated by Eq. (3.120). The results are shown in Table 3.32. Plot these viscosities to reservoir temperature in Fig. 3.73, which shows a good linear relationship between viscosity and temperature. The relationship can be expressed mathematically as [18]

$$\lg \mu_o = -0.9704 + 1.101911 \lg \mu_{og} \tag{3.122}$$

The correlation coefficient of expression above is 0.9650.

It can be seen from Table 3.33 that the correlation shown a good accuracy compared the viscosities estimated from correlation to the viscosities measured from experiments. The average error is 1.17 %, which can satisfy engineering requirements.

3.8.3 Correlations for Oil Compressibility

There are several correlations that are developed to estimate the oil compressibility at pressures above the bubble point pressure. Three of these correlations are presented below [26]: Vasquez–Beggs correlation, Petrosky–Farshad correlation, and McCain's correlation.

3.8.3.1 Vasquez–Beggs Correlation

From a total of 4036 experimental data points used in a linear regression model, Vasquez and Beggs (1980) [47] correlated the isothermal oil compressibility coefficients with R_{sb}, T, °API, γ_g, and p. They proposed the following expression [26]:

$$C_o = \frac{-1433 + 5R_{sb} + 17.2(T - 460) - 1180\gamma_{gs} + 12.61\text{API}}{10^5 p}, \tag{3.123}$$

where T is temperature, °R; p is pressure above the bubble point pressure, psia; R_{Sb} is solution gas–oil ratio at the bubble point pressure, scf/stb; γ_{gs} is corrected gas gravity as defined by Eq. (3.110).

Table 3.33 The estimated and measured viscosities of viscous oils from part viscous oil zones in China [18]

Oil field	Zone	GOR (m^3/t)	Coefficient		At reservoir temperature		μ_o (mPa s)
			A	*B*	μ_{od} (mPa s)	μ_o (mPa s)	
Area Shul of Liaohe	Black 1-6-12	31	0.6075	0.7799	268.2	46.9	37.9
West Dagang	Area 2	31	0.6069	0.7794	320	54.4	69.5
Yangsanmu		31	0.6052	0.7782	260	45.8	37
West Dagang	Area 4	32	0.6016	0.7757	90	20	27
Area Shul of Liaohe	Black 1-6-12	33	0.5914	0.7683	432	62.6	100
West Dagang	Area 1	35	0.5836	0.7627	160	28	38.1
Area Shul of Liaohe	Du 84 Black	36	0.5791	0.7594	383	53	80
Gaosheng	Area Gao 3	39	0.5610	0.7460	300	39.5	74
Area Shul of Liaohe	Black 1-8-16	39	0.5673	0.7506	150	24.5	25
Gaosheng	Area Gao 1	50	0.5200	0.7143	310	31.3	38.5

3.8.3.2 Petrosky–Farshad Correlation

Petrosky and Farshad (1993) proposed a relationship for determining the oil compressibility for undersaturated hydrocarbon systems [57]. The equation has the following form [26]:

$$C_o = 1.705 \times 10^{-7} R_{sb}^{0.69357} \gamma_g^{0.1885} \mathrm{API}^{0.3272} (T - 460)^{0.6729} p^{-0.5906}, \quad (3.124)$$

where T is temperature, °R; R_{sb} is solution gas–oil ratio at the bubble point pressure, scf/stb; p is pressure above the bubble point pressure, psia.

3.8.3.3 McCain's Correlation

Below the bubble point pressure, McCain et al. (1988) [58] correlated the oil compressibility with pressure p, the oil API gravity, solution gas–oil ratio at the bubble point R_{sb}, and the temperature T in °R. Their proposed relationship has the following form [15]:

$$C_o = \exp(A), \quad (3.125)$$

where the correlating parameter A is given by the following expression:

$$\begin{aligned} A = 7.573 - 1.450 \ln(p) - 0.383 \ln(p_b) + 1.402 \ln(T) \\ + 0.256 \ln(\mathrm{API}) + 0.499 \ln(R_{sb}) \end{aligned} \quad (3.126)$$

if the bubble point pressure is known.

Using the correlation, reservoir pressure and temperature, bubble point, oil API gravity, and solution gas–oil ratio at bubble point are needed to estimate the compressibility of saturated oil.

3.9 Application of Reservoir Fluid Properties

In petroleum production, the properties of reservoir fluids find wide application in many areas, such as in reservoir simulation and reservoir engineering calculations. Various property parameters of reservoir fluids are the most basic and the essential for evaluating oil-bearing formation and designing the development plan of hydrocarbon reservoirs.

In the development of oil reservoirs, the commonly important applications of the property parameters of reservoir fluids are hydrocarbon reserve calculation and reservoir performance analysis.

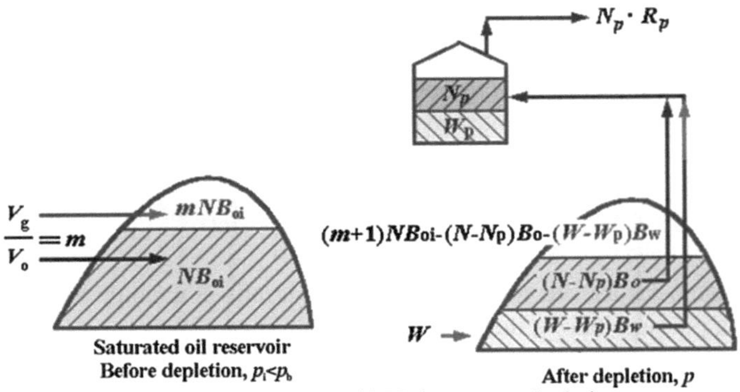

(a) Volumes at initial pressure, (b) Volumes at a depletion pressure, p.

Fig. 3.74 Volume changes in the reservoir associated with a finite pressure drop

Let us take the example of an oil reservoir with gas cap and water influx, realize the application of property parameters of reservoir fluids through the derivation of material balance equation of the reservoir.

3.9.1 Material Balance in Oil Reservoirs

The material balance method is a fundamental reservoir engineering tool for the evaluation of past and future global reservoir performance [59]. It is based on the Law of Mass Conservation applied to a reservoir at large by considering it as a large tank at a uniform pressure. Applied to past performance, a material balance analysis provides insight in the prevailing production mechanism and allows the estimation of the hydrocarbons initially in-place. In its predictive mode, the material balance can be used to generate future reservoir performance and to estimate the potential recovery of the hydrocarbons in-place.

In its simplest form, the material balance states that at any time the amount of hydrocarbons in-place must be equal to the amount of the hydrocarbons initially in-place minus the amount of the hydrocarbons produced.

Considering an oil reservoir, part of oil and gas has been produced at finite pressure drop. In this case, material balance equation can be obtained from gas or oil separately. For example, considering the material balance of the gas, the basic material balance is that the amount of gas produced and gas remained in reservoirs should be equal to the initial amount of gas in reservoirs before pressure depletion. The equation then expressed as

$$\begin{aligned}&\text{The amount of gas initial-in-place in reservoirs}\\&\quad= \text{the cumulative gas production}\\&\qquad+ \text{the amount of gas remained in reservoirs}\end{aligned}$$

This is the material balance of gas phase in petroleum reservoirs. In petroleum production, material balance must always be met no matter in what reservoirs, such as the saturated reservoirs, undersaturated reservoirs with a gas cap, and the reservoirs with edge water or/and bottom water, even the reservoirs accompanied by water injection or gas injection.

3.9.2 Derivation of the Material Balance Equation

The general form of the material balance equation was first presented by Schilthuis' in 1941. The equation is derived as a volume balance which equates the cumulative observed production, expressed as an underground withdrawal, to the expansion of the fluids in the reservoir resulting from a finite pressure drop [60].

To obtain a general form of the material balance equation for a hydrocarbon reservoir, consider a depletion oil reservoir with active aquifer and a gas cap, as shown in Fig. 3.74. The total hydrocarbon volume in this reservoir is the pore volume occupied by hydrocarbons in the reservoir (HCPV). In the reservoir, given that

HCPV is the pore volume occupied by hydrocarbons in the reservoir, m^3;
N is the OOIP (original oil in-place) in the reservoir, m^3;
G is the initial hydrocarbon volume of the gas, m^3;
m is the ratio of $HCPV_{gas}$ to $HCPV_{oil}$ under initial conditions as shown in Fig. 3.74, and is a constant;
N_p is the cumulative production of stock tank oil at reservoir pressure p, (SC) m^3;
G_p is the cumulative gas production at reservoir pressure p, (SC) m^3;
R_{si} is the initial solution gas–oil ratio, m^3/m^3;
R_p is the cumulative/average gas–oil ratio (GOR), defined as the ratio of cumulative gas production (m^3) to cumulative oil production (m^3);
R_s is the solution gas–oil ratio at reservoir pressure p, (SC) m^3/m^3;
W is the cumulative water influx when reservoir pressure declines to p, m^3;
W_p is the cumulative water production at reservoir pressure p, (SC) m^3;
B_{oi} is the formation volume factor of oil in-place at initial reservoir pressure p_i;
B_o is the formation volume factor of oil in-place at reservoir pressure p;
B_{gi} is the formation volume factor of gas at initial reservoir pressure p_i;
B_g is the formation volume factor of gas at reservoir pressure p;
B_{ti} is the total formation volume factor at initial reservoir pressure p_i;
B_t is the total formation volume factor at reservoir pressure p;
B_t is the formation volume factor of formation water at reservoir pressure p.

Due to a gas cap in the oil column (Fig. 3.74), the oil is saturated by gas in the reservoir. The initial pressure in the oil column is equal to the bubble point pressure of the reservoir, i.e., $p_i = p_b$. Therefore, even a slight drop in the pressure will cause the unceasing liberation of the dissolved gas in the reservoir.

Suppose that the reservoir pressure depleted from its initial value p_i to current value p because of reservoir production; and a certain amount of oil, gas, and water

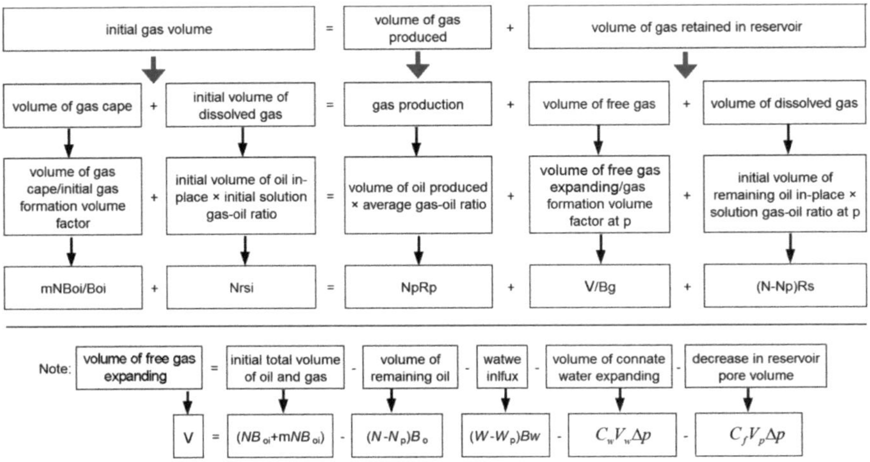

Fig. 3.75 A block diagram of the material balance equation for gas in oil reservoir

have been produced at pressure p. The cumulative volumes of produced fluids measured at surface conditions include N_p (cumulative oil), W_p (cumulative water), and G_p (cumulative gas). Then, the volumetric material balance of gas is expressed, at standard conditions, as

$$\begin{aligned}&\text{Initial hydrocarbon volume of gas}\\&\quad=\text{the cumulative volume of gas produced}\\&\qquad+\text{the volume of gas retained in the reservoir}\end{aligned}$$

In order to better understand the equation, it is broken down to a block diagram, as shown in Fig. 3.75. Each part of this balance equation is evaluated as follows.

3.9.2.1 Left-Hand Side (Initial Hydrocarbon Volume of Gas)

Initial hydrocarbon volume of gas includes two parts: the volume of free gas in the gas cap and the volume of dissolved gas in the oil.

The initial volume of free gas is equal to the volume occupied by gas cap:

$$V_{\text{free-gas}} = mNB_{\text{oi}}/B_{\text{gi}} \tag{3.127}$$

The initial volume of dissolved gas is determined by the initial volume of oil:

$$V_{\text{dissolved-cap}} = NR_{\text{si}} \tag{3.128}$$

Then, the total initial volume of gas is given by

$$G = V_{\text{free-gas}} + V_{\text{dissolved-gas}} = mN\frac{B_{\text{oi}}}{B_{\text{gi}}} + NR_{\text{si}} \tag{3.129}$$

3.9.2.2 Right-Hand Side (Gas Produced Plus Gas Retained)

Gas Produced

The volume of gas produced can be given by the oil production and the cumulative gas–oil ratio:

$$G_{\text{p}} = N_{\text{p}}R_{\text{p}} \tag{3.130}$$

Then the volume of gas retained in the reservoir also includes two parts. One is free gas retained in the reservoir, another is dissolved gas still retained in oil at reservoir pressure p.

Dissolved Gas

The volume of dissolved gas retained in oil at reservoir pressure p is given by the volume of retained oil and the solution gas–oil ratio at p:

$$\left(N - N_{\text{p}}\right)R_{\text{s}} \tag{3.131}$$

The calculation of the volume of free gas remaining in reservoir at pressure p is slightly complicated. As the pressure drops from p_i to p, the fluids in the reservoir expand and the water is allowed to encroach in the reservoir. Then, consideration must be given to everything that can change the volume of the fluids in the reservoir during the depletion: the expansion of the oil, gas cape, connate water, the cumulative water influx. Each factor above may be accounted for as follows:

Remaining Oil

The volume of oil remaining in the reservoir is

$$V_{\text{oil-atreservoirs}} = \left(N - N_{\text{p}}\right)B_{\text{o}} \tag{3.132}$$

Encroached Water

If the reservoir is subjected to a water drive (edge water, bottom water, or injecting water), water enters the pore space originally occupied by gas as the reservoir pressure decreases. The volume occupied by encroached water is

$$V_{\text{encroached-water}} = \left(W - W_{\text{p}}\right)B_{\text{w}} \tag{3.133}$$

Connate Water

Although water compressibility is smaller, huge amounts of connate water in rock pores make water expansion cannot be ignored during depletion. If S_{wi} is the average connate water saturation, the volume of water expansion can be calculated according to the definition of isothermal compressibility to the water:

$$\mathrm{d}V_{\mathrm{w}} = C_{\mathrm{w}} V_{\mathrm{w}} \Delta p, \tag{3.134}$$

where V_{w} is the total volume of water:

$$V_{\mathrm{w}} = V_{\mathrm{p}} S_{\mathrm{wi}} = \frac{(NB_{\mathrm{oi}} - mNB_{\mathrm{oi}})}{1 - S_{\mathrm{wi}}} S_{\mathrm{wi}}, \tag{3.135}$$

where V_{p} is the total reservoir pore volume: $V_{\mathrm{p}} = (NB_{\mathrm{oi}} + mNB_{\mathrm{oi}})/(1 - S_{\mathrm{wi}})$.

Reservoir Pore Volume

As pore volume decreases with the decline in reservoir pressure, a reduction in total reservoir pore volume can be determined based on the definition of rock compressibility as

$$-\mathrm{d}V_{\mathrm{p}} = C_{\mathrm{f}} V_{\mathrm{p}} \Delta p \tag{3.136}$$

Free Gas Then, the volume of free gas at pressure p can be determined by the following expression:

$$V = \frac{(NB_{\mathrm{oi}} + mNB_{\mathrm{oi}}) - (N - N_{\mathrm{p}})B_{\mathrm{o}} - (W - W_{\mathrm{p}})B_{\mathrm{w}} - \frac{(NB_{\mathrm{oi}} + mNB_{\mathrm{oi}})}{1 - S_{\mathrm{wi}}} \Delta p(C_{\mathrm{f}} + C_{\mathrm{w}} S_{\mathrm{wi}})}{B_{\mathrm{g}}} \tag{3.137}$$

Finally, equating this left-hand side to the right-hand side: Eqs. (3.130), (3.131), and (3.137), give the general expression for the material balance as

$$\begin{aligned} mN\frac{B_{\mathrm{oi}}}{B_{\mathrm{gi}}} + NR_{\mathrm{si}} &= N_{\mathrm{p}} R_{\mathrm{p}} + (N - N_{\mathrm{p}})R_{\mathrm{s}} \\ &+ \frac{(NB_{\mathrm{oi}} + mNB_{\mathrm{oi}}) - (N - N_{\mathrm{p}})B_{\mathrm{o}} - (W - W_{\mathrm{p}})B_{\mathrm{w}} - \frac{(NB_{\mathrm{oi}} + mNB_{\mathrm{oi}})}{1 - S_{\mathrm{wi}}} \Delta p(C_{\mathrm{f}} + C_{\mathrm{w}} S_{\mathrm{wi}})}{B_{\mathrm{g}}} \end{aligned} \tag{3.138}$$

or

$$\begin{aligned} (GB_{\mathrm{gi}} + NB_{\mathrm{oi}})\left[1 - \frac{\Delta p(C_{\mathrm{f}} + C_{\mathrm{w}} S_{\mathrm{wi}})}{1 - S_{\mathrm{wi}}}\right] &= (G - G_{\mathrm{p}})B_{\mathrm{g}} \\ + \left[NR_{\mathrm{si}}(N - N_{\mathrm{p}})R_{\mathrm{s}}\right]B_{\mathrm{g}} + (N - N_{\mathrm{p}})B_{\mathrm{o}} &+ (W - W_{\mathrm{p}})B_{\mathrm{w}} \end{aligned} \tag{3.139}$$

Yet another slight modification can be made by writing: $B_{\mathrm{t}} = B_{\mathrm{o}} + (R_{\mathrm{si}} - R_{\mathrm{s}})B_{\mathrm{g}}$, $B_{\mathrm{ti}} = B_{\mathrm{oi}}$ and $C_{\mathrm{t}} = \frac{C_{\mathrm{f}} + C_{\mathrm{w}} S_{\mathrm{wi}}}{1 - S_{\mathrm{wi}}}$ in which case.

$$N = \frac{N_p\left[B_o + \left(R_p - R_s\right)B_g\right] - (W - W_p)B_W}{(B_t - B_{ti}) + mB_{ti}\left(\frac{B_g - B_{gi}}{B_{gi}}\right) + (1 + m)C_t\Delta pB_{ti}} \tag{3.140}$$

Equation (3.140) is the general form of material balance for oil reservoirs. Equation (3.140) can be simplified according to different hydrocarbon reservoir types as follows:

1. If $m = 0$, i.e., undersaturated oil reservoirs, the denominator of Eq. (3.140) can be simplified.
2. If no edge water, bottom water, or injecting water, W and W_p are zero, and the numerator of Eq. (3.140) can be simplified.
3. If $N = 0$, it is a gas reservoir, and the left of Eq. (3.140) is zero.

When W is zero, values of N and m are relatively easy to determine, if $N = 0$ (gas field) or $m = 0$ (undersaturated oil), an analysis of the production data will usually give the volume of gas or oil initially in place together with some indication of the strength of the water drive: these cases are discussed below.

It can be seen that Eq. (3.140) includes three group data, which can be obtained by the following different methods.

The first group is the parameters of PVT properties of reservoir fluids, including R_{si}, R_s, B_{gi}, B_g, B_{oi}, B_o, B_{ti}, B_t, B_w, and so on, which are determined generally by experiment or chart. The key of determining such kind of parameters is how to find a more accurate prevailing reservoir pressure which should be first known before the experiment for the determination of the parameters above.

The second group is production data, such as N_p, W_p, R_p. Generally, the data of oil production is relatively accurate in field data because it is vital sign of petroleum production. However, the gas, separated from oil, is often burned out or used as fuel; and the water is often bled off or injected into the reservoir. The data of gas and water production are less accurate due to less attention; error of these data may be up to 10 % or more. Therefore, special attention should be paid on data processing.

The final group is geologic parameters, such as m and W. The value of m is generally determined by geology and well logging data. The water influx W may be determined according to core experiments, analytic method of fluid flowing through porous medium or statistical analysis.

It should be noted that material balance equation can only be used for the hydrocarbon reservoirs which have a very long history of production, enough production data, and an evident decline in reservoir pressure.

Exercise

Question

1. What is the chemical composition of crude oil?
2. What is the saturated vapor pressure?
3. What is the critical point of a hydrocarbon system?

4. What is dew point pressure?
5. What is retrograde condensation?
6. What is the solubility of natural gas in crude oil?
7. What is equilibrium constant?
8. What is the solution gas–oil ratio of in-place oil?
9. What is the apparent molecular weight of natural gas?
10. What is the relative density of natural gas?
11. Based on the properties of hydrocarbon mixtures, how many types are hydrocarbon reservoirs usually divided into? Describe the major changes in fluid characteristics that may accompany production in each type of reservoir.
12. Describe the typical fluid characteristics that would be expected in a condensate gas reservoir. What phenomenon may occur during the production of a condensate gas reservoir? Discuss a development plan to attain optimum recovery from a condensate gas reservoir.
13. What are the major hydrocarbons of crude oil? What are the major hydrocarbons of natural gas? Briefly describe the composition of chemical compounds typically found in petroleum fluids, including common impurities.
14. What is the ideal gas law, and how it is modified to account for the behavior of real gases? How are the pseudoreduced properties of natural gas calculated?
15. Why is knowledge of the dew point necessary to effectively develop and produce a gas condensate reservoir? In a condensate gas reservoir, where would condensation be expected to occur first—near the reservoir boundary or at the producing wells?
16. A discovered reservoir does not appear to have a gas cap. What fluid properties would be important to know, and why?
17. Name the most significant phenomenon that can take place in an undersaturated oil reservoir as the pressure declines with production. Define the following fluid properties and discuss their significance in effectively producing the reservoir:

    (a) Bubble point pressure
    (b) Viscosity of the oil
    (c) Solution gas–oil ratio
    (d) Oil gravity
    (e) Oil isothermal compressibility and gas isothermal compressibility.

    Does the rate of change in pressure in an undersaturated reservoir usually accelerate once the pressure declines below the bubble point? Why or why not?
18. Describe the changes in the following fluid properties both above and below the bubble point pressure:

    (a) Liquid composition
    (b) Viscosities of the oil and gas
    (c) Oil formation volume factor
    (d) Total formation volume factor
    (e) Liquid density.

Are these properties interrelated?

19. Once the reservoir pressure decreases below the bubble point, would oil be produced in as much quantity as above the bubble point? Explain by identifying the fluid properties mentioned in question (18), which are responsible for any changes in fluid flow dynamics within the reservoir.
20. Why is the study of vapor–liquid equilibrium important in reservoir engineering? Define convergence pressure and explain its role in these calculations.
21. What are the gas deviation factor and the gas formation volume factor? Discuss their significance in calculating fluid properties and estimating reserves.

Calculation

1. The composition of a natural gas is given in Table 3.34.
   (a) Evaluate the mole fraction of each component;
   (b) Determine the molecular weight and relative density of the gas.
2. A gas composition is given in Table 3.35. Determine the apparent molecular weight, relative density, pseudocritical pressure, and pseudocritical temperature of the gas.
3. Compute the mass of methane contained in a vessel of 2 m^3 at 5 MPa and 27 °C.
4. Calculate the compressibility factor of a natural gas with relative density of 0.743 at 13.5 MPa and 93 °C.

Table 3.34 The composition of a given natural gas

Composition	Mass fraction, G_i	Molecular weight, M_i
Methane	0.70	16.043
Ethane	0.14	30.070
Propane	0.09	44.097
Butane	0.07	58.124

Table 3.35 The composition of a given natural gas

Component	Mole fraction
C_1	0.765
C_2	0.073
C_3	0.021

Table 3.36 Pressure and oil volume measured in a test

Pressure (MPa)	34.5	31.0	27.5	24.0	22.8	21.4	20.8(p_b)
Oil volume (cm^3)	192.10	193.38	194.79	196.30	196.90	197.62	198.00

5. Calculate the isothermal compressibility of a natural gas with relative density of 0.862 at 20 MPa and 80 °C.
6. Calculate the formation volume factor of a natural gas with relative density of 0.743 at 15.5 MPa and 99 °C.
7. Estimate the density of a dead oil with relative density of 93 at 38 °C.
8. What is the relative density of a stock tank oil with gravity of 46.5 API?
9. A liquid sample from an oil reservoir had a volume of 10 cm^3 in a laboratory vessel at reservoir temperature and bubble point pressure. The liquid was expelled through laboratory equipment which is the equivalent of the field two-step separator system. The oil volume collected in the stock tank was 8.5 cm^3. The first separator produced 45.5 cm^3 of gas, and the second produced 7.5 cm^3 of gas. Calculate the formation volume factor of the oil and the solution gas–oil ratio.
10. The reservoir pressure of a black oil reservoir appears to be above the bubble point pressure of the oil. At initial conditions, the oil of 12 m^3/days enters the wellbore. The oil is processed through a separator into a stock tank. The stock tank accumulates 8.7 m^3 of oil with relative density of 0.85 each day. The separator produces 900 m^3/days of gas with relative density of 1.2. What is the formation volume factor of the oil? What is the solution gas–oil ratio?
13. A sample of reservoir oil was placed in a laboratory cell at reservoir temperature of 120 °C. Pressure was varied and oil volume measured in the test lists in Table 3.36.
    Determine the values of the isothermal compressibility of the oil at bubble point pressure and between 31.0 and 27.5 MPa.
14. The solution gas–oil ratio of a crude oil is 800 m^3/m^3 at 20 MPa and 50 °C. Given the following PVT data:

    (a) Bubble point pressure: 17.5 MPa;
    (b) Oil relative density: 0.85;
    (c) Gas relative density: 0.77 (air = 1).

    Estimate densities and viscosities of the crude oil at 50 °C and 17.5 MPa, and 17.5 MPa.
15. Given that the reducible water saturation of a sandstone oil reservoir is 25 %. The oil-bearing volume of the reservoir is 14.4×10^{-7} m^3. The average porosity of the reservoir is 20 %. The isothermal compressibilities of the oil and water are 10×10^{-4} and 4×10^{-4} l/MPa, respectively. The rock compressibility is 1×10^{-4}l/MPa. If the average oil formation volume factor is 1.2 between the initial reservoir pressure (27 MPa) and the saturation pressure of the oil (20 MPa), estimate the recoverable reserves of the oil reservoir.

References

1. Yang S et al (2004) Reservoir physics. Petroleum Industry Press, Beijing
2. Sylvain JP (1958) Oil reservoir engineering. McGraw-Hill, New York
3. Saeid M, William AP, James GS (2006) Handbook of natural gas transmission and processing. Gulf Professional Publishing, Houston
4. William DM (1990) The properties of petroleum fluids. Penn Well Books, Oklahoma
5. Balsaraf VM (2009) Applied chemistry, vol 2. I.K. International Pvt Ltd., Delhi
6. Norman JH (1991) Dictionary of petroleum exploration, drilling and production
7. Frank J, Mark C, Mark G (2008) Hydrocarbon exploration and production. Elsevier, Amsterdam
8. Danesh A (1998) PVT and phase behavior of petroleum reservoir fluids. Elsevier, Amsterdam
9. James GS (2011) An introduction to petroleum technology, economics, and politics. Wiley, New York
10. Tarek A (2009) Working guide to reservoir rock properties and fluid flow. Gulf Professional Publishing, Houston
11. Abdus S, Ghulam MI, James LB (2008) Practical enhanced reservoir engineering: assisted with simulation software. Penn Well Books, Oklahoma
12. GB/T 26979-2011 (2011) The classification of natural gas pool. General administration of quality supervision, inspection and quarantine of People's Republic of China and national standardization administration of China
13. Whitson CH, Brule MR (2000) Phase behavior. SPE monograph. J20. Society of Petroleum Engineers, Richardson
14. Speight JG (2007) The chemistry and technology of petroleum, 4th edn. CRC Taylor & Francis Group, Boca Raton
15. Satinder C (2010) Heavy oils: reservoir characterization and production monitoring. SEG Books
16. Enrico F (2012) Thermodynamics. Courier Corporation. ISBN 978-0-486-13485-7
17. William CL, Gary JP (2005) Standard handbook of petroleum and natural gas engineering. Gulf Professional Publishing, Houston
18. He GS, Tang H (2011) Reservoir physics. Petroleum Industry Press, Beijing
19. Ye Z (2007) Principle of enhanced oil recovery. Petroleum Industry Press, Beijing
20. Marshall BS (1951) Volumetric and phase behavior of oil field hydrocarbon systems: PVT for engineers. California Research Corp
21. Bloomer OT et al (1953) Institute of Gas Technology. Research Bulletin J22
22. Brown OT et al (1947) Natural gas and the volatile hydrocarbon, NGAA
23. Paul G (2001) Reservoir fluid. http://docs.ma3hd.net/dtail.php?search=Paul+Glover+RESERVOIR+FLUID&type=pdf&file=http://www2.g
24. http://iptibm1.ipt.ntnu.no/~jsg/undervisning/naturgass/parlaktuna/Chap3.pdf
25. Brown GG, Katz DL et al (1948) Natural gasoline and volatile hydrocarbons, Tulsa, OK
26. Tarek A (2010) Reservoir engineering handbook. Gulf Professional Publishing, Houston
27. Tester JW, Modell M (1997) Thermodynamics and its applications. Prentice Hall, Englewood Cliffs. ISBN 0-13-915356-X
28. Goodstein D (1985) Critical phenomena and phase transitions. In: States of matter, 1st edn. General Publishing Company, Ltd.
29. Standing MB, Katz DL (1942) Density of natural gases. AIME Trans 146(01):140–149
30. McCain W (1990) The properties of petroleum fluids. PennWell Publishing Company, Tulsa, OK.
31. Trube AS (1957) Compressibility of natural gases. AIME Trans 210
32. He GS (2011) Reservoir physics. Petroleum Industry Press, Beijing
33. Carr NL, Kobayashi R, Burrows DB (1954) Viscosity of hydrocarbon gases under pressure. AIME Trans 264
34. Darrell E, Steven DG (2010) General chemistry. Cengage Learning

35. Kotyakhov FI (1958) Основы физики нефтяного пласта (Foundation of Reservoir Physics). Petroleum Industry Press, China
36. Stewart M, Arnold K (2008) Gas-liquid and liquid-liquid separators. Gulf Professional Publishing, Houston
37. Francis SM, Richard ET (1995) Oilfield processing of petroleum: crude oil. Penn Well Books, Oklahoma
38. Standing MB (1981) Volumetric and phase behavior of oil hydrocarbon systems, 9th edn. SPE, Dallas
39. Djebbar T, Erle CD (2011) Petrophysics: theory and practice of measuring reservoir rock and fluid transport properties. Gulf Press, Houston
40. Dodson CR, Standing M (1944) Pressure-volume-temperature and solubility relations for natural gas-water mixtures. In: Drilling and production practice. American Petroleum Institute
41. Muskat M (1949) Physical principles of oil production, 1st edn. Mc-Graw-Hill Book Company, Inc., New York
42. John JC (2010) Acid gas injection and carbon dioxide sequestration. Wiley, New York
43. Natural Gasoline Association of America (1957) Natural Gasoline Association of America's (NGAA) equilibrium ratio data book
44. Szilas AP (1986) Production and transport of oil and gas: gathering and transportation. Elsevier, Amsterdam
45. William DM (1990) The properties of petroleum fluids. Penn Well Books, Oklahoma
46. Standing MB (1947) A pressure-volume-temperature correlation for mixtures of California oils and gases. Drill and Prod Prac, API (275)
47. Vasquez M, Beggs D (1980) Correlations for fluid physical properties prediction. JPT, June, pp 968–970
48. Sutton RP, Farshad FF (1984) Evaluation of empirically derived PVT properties for Gulf of Mexico crude oils. SPE paper 13172, presented at the 59th annual technical conference, Houston, Texas.
49. Beal C (1946) The viscosity of air, water, natural gas, crude oil and its associated gases at oil field temperatures and pressure
50. Beggs HD, Robinson JR (1975) Estimating the viscosity of crude oil systems. JPT, September, pp 1140–1141
51. Glaso O (1980) Generalized pressure-volume-temperature correlations. JPT, May, pp 785–795
52. Khan SA et al (1987) Viscosity correlations for Saudi Arabian crude oils. SPE 15720
53. Ahmed TH (1989) Hydrocarbon phase behavior. Gulf Publishing, Houston
54. Chew J, Connally Jr (1959) A viscosity correlation for gas-saturated crude oils. Trans AIME 216:23–25
55. Standing MB (1977) Volumetric and phase behavior of oil field hydrocarbon systems. Dallas: Society of Petroleum Engineers, pp 125–126
56. William CL (2009) Working guide to reservoir engineering. Gulf Professional Publishing, Houston
57. Petrosky GE, Farshad F (1993) Pressure-volume-temperature correlations for Gulf of Mexico crude oils. SPE Paper 26644, presented at the 68th annual technical conference of the SPE in Houston, Texas
58. McCain J, William D, Rollins et al (1988) Coefficient of isothermal compressibility of black oils at pressures below the bubblepoint. Source: SPE Formation Evaluation 3(3):659–662. ISSN: 0885923X
59. Jacques H (1988) Fundamentals of gas reservoir engineering. Elsevier, Amsterdam
60. Dake LP (1978) Fundamentals of reservoir engineering. Elsevier, Amsterdam
61. Amyx J, Bass D, Whitney R (1960) Petroleum reservoir engineering. McGraw-Hill Book Company, New York
62. Clark NJ (1969) Elements of petroleum reservoir. SPE

Chapter 4
Microscopic Mechanism of Multiphase Fluids Flowing Through Rocks

Xuetao Hu and Shuyong Hu

In general, it is rare that there is only one phase fluid in petroleum reservoir. Petroleum reservoirs contain at least two fluid phases in the porous rock: either oil and water or gas and water. In most cases, two-phase or multiphase immiscible fluids, such as oil and water, or gas and water, or oil, gas and water, coexist in reservoir. When more than one immiscible fluid phase is present in reservoir rock, a thin film develops at the boundary between the two fluid surfaces. The "thin film" or the boundary region between two immiscible fluids, or adjacent bulk phases, is called *interface*. Substance at an interface usually has different physical properties and energy characteristics from that in the bulk phases. The physical or chemical phenomena occurring at the interface between phases are known as *interfacial phenomenon*.

In reservoir rock, there are also a large number of small pores that displays huge specific surface area. Because the reservoir fluids in pores contact directly with the surfaces of rock particles, many interfaces form between fluids or between fluid and solid in reservoir pores. Various physicochemical reactions occur at these interfaces. Diverse interfacial phenomena, such as interfacial tension, interfacial adsorption, interfacial viscosity, wetting and capillary pressure, etc., are thus extremely prominent in petroleum reservoir. As the effect of interfacial phenomenon, the distribution and flowing of fluids in rock pores are obviously different from that of fluids in pipeline or in channel. The flowing of fluid through porous medium, such as soil, loose sediments and consolidated rock, etc., is generally called "*seepage*".

In the process of fluid flowing through rock pores, the interfacial phenomena between different bulk phases will induce various problems, such as wetting

X. Hu (✉) · S. Hu
School of Oil & Natural Gas Engineering, Southwest Petroleum University, Chengdu, Sichuan, China
e-mail: huxt@swpu.edu.cn

S. Hu
e-mail: hushuyong@swpu.edu.cn

X. Hu et al. (eds.), *Physics of Petroleum Reservoirs*, Springer Geophysics,
DOI 10.1007/978-3-662-55026-7_4

hysteresis, viscous fingering and the additional resistance, and so on. The displacement efficiency of water flooding is then decreased and the irreducible oil has remained in rock pores. Only after the mechanism of fluids flowing through porous medium is well recognized, the macroscopic distribution of oil, gas, or water in reservoir be effectively determined based on a good understanding about the distributions and flow of oil, gas, or water in rock pores.

The distribution and flow of fluids in the reservoir have a significant impact on oil production. Laboratory measurements of multiphase fluid flowing in porous media show that fluid behavior depends on the properties of the solid matrix and the fluids that are present in medium. Then, this chapter will mainly focus on the representation of the mechanism of oil and water flowing in rock pores and rock–fluid interactions using physical parameters such as relative permeability and capillary pressure. This chapter is the foundation of enhanced oil recovery.

4.1 Interfacial Phenomena in Reservoir Rocks

The discussion of interfacial phenomenon begins by addressing the question: What is an interface or surface?

The physical boundary between two adjacent bulk phases is defined as an *interface*, where the properties differ significantly from those of the bulk phase it separates. The surface of a phase is a particularly simple type of interface at which the phase is in contact with the surrounding world, i.e., the atmosphere or, in the ideal case, the vacuum. Sometimes, people use the word *surface* in order to define the physical boundary of only one of these phases, such as solid surface, liquid surface, etc. In many standard physical chemistry, surface chemistry and surface physics textbooks, the words *surface* and *interface* are used interchangeably because the authors neglect the small differences between the air phase and absolute vacuum conditions. On the other hand, some authors define and use the word surface for only the solid–gas and liquid–gas phase contacts, and the word *interface* for all of other phase contacts, but this has no scientific basis and should be abandoned [1].

In order to obtain a good understanding about the interfacial phenomena in reservoir rock, a general review of some basic concepts, such as interfacial tension, interfacial adsorption and interfacial viscosity, in physical chemistry is given as follows.

4.1.1 Interfacial Tension

In a highly dispersed multiphase system, there are many interfaces, and the total area of interface of the system is huge. The behavior of a multiphase system is often determined by the interface characteristics of the system. Some interface characteristics of a multiphase system are absolutely vital to the whole system. For example, interfacial free energy and interfacial tension are two important interface characteristics of a multiphase system.

4.1.1.1 Interfacial Free Energy and Interfacial Tension

Interfacial Free Energy

In a multiphase system, the interfacial phenomenon depends on the interaction between the molecules of fluids or the molecules of fluid and rock. In dealing with the problems of multiphase system, one needs to consider the effect of the forces at the interface when two immiscible fluids are in contact.

Figure 4.1 shows a typical gas–liquid system. At the interface between the liquid and gas, the behavior of liquid molecules is different from the similar molecules in the bulk phase.

In the bulk phase, the cohesion forces keep the molecules close to each other, and translational and rotational motion of molecules takes place within the liquid with considerable freedom. Each molecule in the bulk is surrounded by other molecules on all sides and is attracted in all directions, which results in a zero net force of all short- and long-range forces. The acting at any point in the bulk is the same (Fig. 4.1b). Intermolecular forces are symmetrical and in equilibrium in the bulk liquid.

However, the molecules close to the interface have different neighbors and these forces acting on the molecules are not compensated for. The molecules at the interface range are attracted inward and also to each side by their neighbors, but there is no enough outward attraction to balance the inward pull, because there are very few gas molecules outside. Hence every liquid molecule at the interface is subject to a strong inward attraction perpendicular to the interface, which leads to a

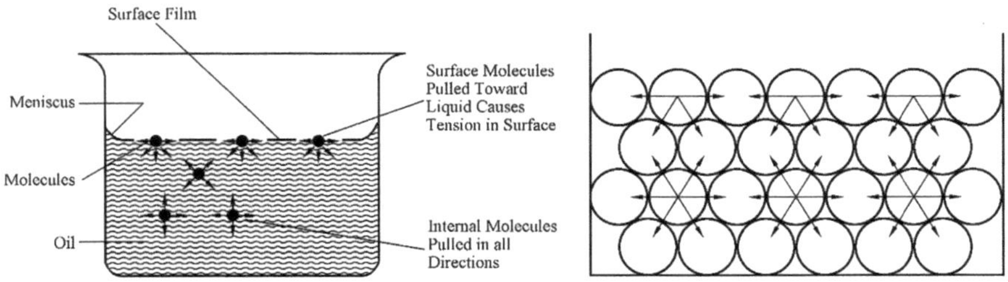

(a) (b)

a. After Clark, N. J., Elements of Petroleum Reservoirs, SPE, 1969;
b. from Dynarowicz 1993

Fig. 4.1 Simple illustration of the asymmetric forces acting at a molecule close to the interface

tendency of interface molecules moving toward the inside of liquid. In the process of interface forming, the interface molecules are moving inward more rapidly than interior molecules, which move upward to take their places. This process decreases the number of molecules in the interface, and the liquid surface area is then reduced. This surface contraction continues until the interior accommodates the maximum possible number of molecules. This means that the interface molecules are in a state of higher free energy than those in the bulk liquid. That to say, the molecules at the interface have higher energy. The excess energy is called the *interfacial free energy*, or, more correctly, the *excess interfacial free energy*. It is also called *excess surface free energy* if one phase is vapor. It must be emphasized that this interfacial free energy is not the total free energy of interfacial molecules but the excess free energy that the molecules possess by virtue of their being in the interface.

Due to the excess free energy of interfacial molecules, it takes energy to put molecules on the interface, whereas the molecules in liquid can move freely without work consumed. Since it takes energy to create the interfacial surfaces of a system, the system will try to minimize the total interfacial area for a given volume because a system naturally tries to find a state of minimum potential energy. Hence we can see spherical droplets, meniscus, etc. (Fig. 4.2).

Interface free energy is the internal energy of a system. It is the volume-based property and increases with the increase in interface area of a system. It is also a state function of a system and varies with the phase or phase states of a system. That is to say, if the temperature or/and pressure of a system changes, the interfacial free energy of the system changes. In a word, any factor that causes any change in the interface of a system will induce the change in the interfacial free energy of the system.

Despite the molecules located at two-phase boundaries (i.e., solid–gas, solid–liquid, liquid-gas and $liquid_1$–$liquid_2$) behave differently to those in the bulk phase, this rule generally does not apply for solid–solid and gas–gas interfaces where atomic and molecular bonding in the solid structure restrict the reorientation of

Fig. 4.2 Water beading on a leaf

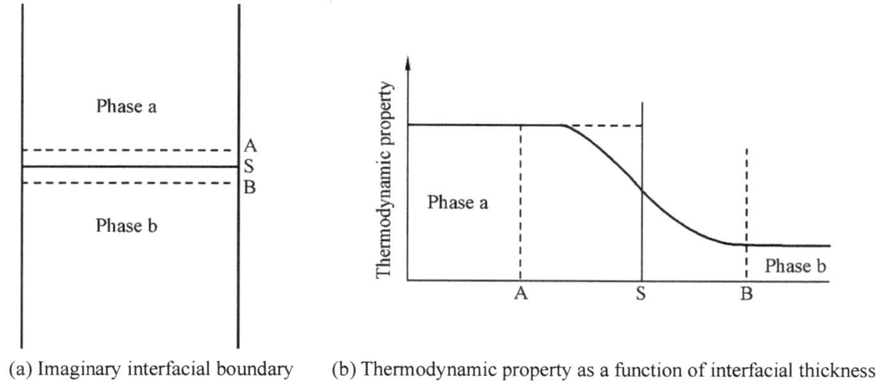

Fig. 4.3 Illustration of the property variation of molecules at the interface between phases

interfacial molecules for the former, and the ease of miscibility of different gas molecules in free space does not allow any interface formation for the latter [1].

There is an orientation effect for molecules at fluid interfacial: the molecules at or near the fluid interfacial will be orientated differently with respect to each other than the molecules in the bulk phase. Any molecule at the fluid interfacial would be under an asymmetrical force field, resulting in interfacial tension. The nearer the molecule is to the interfacial boundary, the greater the magnitude of the force due to asymmetry. As the distance from the interfacial boundary increases, the effect of the force field decreases and vanishes after a certain length. Thus, in reality there is no very sharp interfacial boundary between phases, rather there is a molecular gradient giving a change in the magnitudes of both density and orientation of interfacial molecules (Fig. 4.3).

We should remember that this transition layer is very thin, usually between one and five monomolecular layers for liquid–vapor interfaces. The location of the so-called *mathematically dividing surface* is only a theoretical concept to enable scientists to apply thermodynamics and statistical mechanics [1]. As a result, an interface may be considered, from a thermodynamic standpoint, either as a mathematical plane or as a distinct phase having a finite thickness [2].

The interfacial free energy term is preferred in the literature and it is equivalent to the work done to create reversibly fresh interfacial area of a system at a given temperature, pressure and composition of the system. Namely,

$$\text{Interfacial free energy} = -W_{\text{reversible}} \tag{4.1}$$

Here, the unit of the interfacial free energy is J/m in the SI system.

Specific Interfacial Free Energy

Specific excess interfacial free energy means the interfacial free energy per unit interface area.

$$\sigma = \left(\frac{\partial U}{\partial A}\right)_{T,p,n} \tag{4.2}$$

where σ is the specific interfacial free energy, J/m^2; A is the interfacial area of a system, m^2; U is the interfacial free energy of the system, J.

The specific interfacial free energy of a system may be defined as the work required to create per unit interfacial area of a system, such as the immiscible system of oil and water. It is equivalent to the surface tension of an equilibrium isotropic surface.

$$\sigma = \frac{-W}{\Delta A} \tag{4.3}$$

where ΔA is the increment of the interfacial area of a system, m^2; W is the work required to create per unit interfacial area, J.

Specific interfacial free energy is measured in SI system as joules per square meter (J/m^2) and in the CGS system of units as ergs per cm^2.

Interfacial Tension

Interfacial tension (IFT) refers to the tension between liquids at a liquid–liquid interface. Surface tension refers to the tension between fluids at a gas–liquid interface.

Interfacial/surface tension may be defined in terms of the energy change that is required to produce the fresh interfacial/surface area of a system. To help us define the surface tension, we use the apparatus shown in Fig. 4.4. It consists of a C-shaped wire frame, on which is mounted a wire *ab* that can slide with negligible friction and has a length L. The frame and sliding wire contain a thin film of liquid. Because surface tension causes the liquid surface to contract, a force F is needed to move the slider to the right and extend the surface.

Suppose the distance of the slider moving in the direction perpendicular to itself is Δx; the increment in surface area of the liquid film is ΔA; and the work required to create the surface area of the liquid film is W.

So, the change in the surface free energy of the system ΔU:

$$\Delta U = -W \tag{4.4}$$

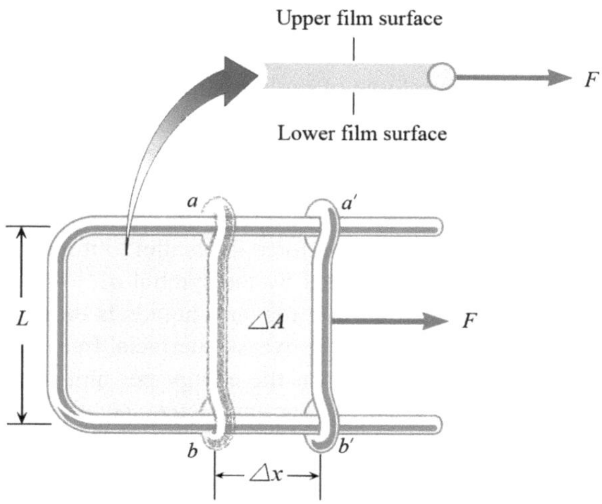

Fig. 4.4 Schematic diagram of a C-shaped wire frame and a wire slider

The work required to increate per unit surface area of the system is:

$$\frac{-W}{\Delta A} = \frac{\partial U}{\Delta A} = \sigma \tag{4.5}$$

As the slider *ab* has been moved a distance Δx by the force F, then:

$$-W = F \times \Delta x \tag{4.6}$$

For the specific case illustrated in Fig. 4.4, there is an upper surface and a lower surface, as the blow-up drawing indicates. Thus, the change in film area is:

$$\Delta A = 2L \times \Delta x \tag{4.7}$$

From Eq. (4.5), we have:

$$-W = \Delta A \times \sigma \tag{4.8}$$

Combine Eqs. (4.6), (4.7) and (4.8):

$$F \times \Delta x = 2L \times \Delta x \times \sigma$$

Then:

$$\sigma = \frac{F}{2L} \tag{4.9}$$

where U is the interfacial free energy of the system, J; W is the work required to create per unit interfacial area, J; ΔA is the increment of the interfacial area of a system, m^2; σ is the specific interfacial free energy, J/m^2; L is the height of wire frame, m; Δx is the distance of the slider moving in the direction perpendicular to itself, m; F is a force of pulling slider, N.

As a result, the surface free energy, which is denoted by the Greek letter sigma σ and indicated by Eq. (4.9), is the magnitude of the force F per unit length over which it acts.

Interfacial tension (IFT)/surface tension is thus defined as the force along a line of unit length, where the force is parallel to the interfacial but perpendicular to the line. It is also represented by the symbol σ.

The interfacial tension for pure liquids is dimensionally equivalent and numerically equal to the specific excess interfacial free energy. It has the dimension of the force per unit length or of the energy per unit area. The two are equivalent—but when referring to energy per unit of area, people use the term interfacial free energy—which is a more general term in the sense that it applies also to solids and not just liquids. Both are intensive thermodynamic properties, whereas the interfacial free energy is an extensive property and is dependent on the interfacial area of the system.

Interfacial tension is the force per unit length and is caused by the unbalanced attraction force between the interfacial molecules. So, interfacial tension causes a reduction in interfacial surface and creates a membrane-like surface. Actually, if carefully placed, a needle will float on the surface of the liquid, supported by the thin membrane even though it is considerably denser than the liquid.

The units of IFT are typically expressed in milli-Newtons per meter (mN/m) in the SI system or the equivalent dynes per centimeter (dyn/cm) in the cgs system. One dyn/cm corresponds to 0.001 N/m. That is, 1 dyn/m = 1 mN/m.

The equivalence of measurement of energy per unit area to force per unit length can be proven by dimensional analysis. For a liquid, the surface tension (force per unit length) and the specific surface free energy (surface energy density) are identical. Water has a surface tension of 72.8 mN/m and a specific interfacial free energy of 72.8 J/m^2. It must be stressed that surface tension and specific interfacial free energy are numerically equal only in systems where no adsorption occurs at interface [2].

In general, if we mention the interfacial tension of one substance, it should be specified exactly what are the two phases. Without instructions, one of the two phases is necessarily considered as air. For example, generally speaking, the surface tension of water is equal to 72.8 mN/m, it means the value between water and air.

The values of the interfacial/surface tension for some typical liquids are summarized in Table 4.1.

Besides, it needs to say something about the surface behavior of solids, such as metals, polymers, semiconductors, glasses, ceramics, and rocks. All these solid substrates have a surface tension like a liquid. Unlike liquids, the intermolecular forces of solids are so strong (e.g., by lattice forces) that the shape of their surface is not determined by the corresponding surface tension. Above the melting point, on the condition that the solid can melt, such systems have the properties of a typical liquid.

Table 4.1 Interfacial tensions of common liquids

Liquid	Surface tension (liquid–air, 20 °C) (mN/m)	Interfacial tension (liquid–water, 20 °C) (mN/m)
Mercury	484.0	375
Water	72.8	
Transformer oil	39.1	45.1
Crude oil from Dewy Yilmaz field	27.2	30.3
Normal hexane	18.4	51.1
Normal octane	21.8	50.8
Benzene	28.9	35.0
Toluene	28.4	–
Ether	17.0	10.0
Carbon tetrachloride	26.9	
Carbon bisulfide	33.5	
Carbon dichloride	28.5	
Dichloroethane	32.5	
Ethanol	22.3	

4.1.1.2 Factors of Affecting Interfacial Tension

Interfacial tension is a characteristic property of interface. The equilibrium state of an interface/surface is characterized by certain conditions that include the validity of equilibrium thermodynamics. Many factors affecting the interfacial tension of an immiscible system are thus involved.

The interfacial tension between liquids varies greatly for different liquids, or for the same liquid system but at different composition or at different temperatures and pressures. That is to say, the value of IFT depends on the compositions and properties of the two fluids and the temperature and pressure of the system. For example, the interfacial tension of an oil–water system varies according to the chemical composition of the oil; the presence or absence of surfactants and/or gas in solution, the pH of the water, as well as pressure and temperature.

From Table 4.1, it can be seen that the surface tension of mercury is larger than the interfacial tension between mercury and water. The reason is that the surface free energy of a system increases with the difference between polarities of two-phase molecules. Among liquids, the polarity of water molecule is the greatest, whereas air's polarity is the lowest. So, the interfacial tension between water and air is the highest. The interfacial tension between ether and water is lower than that between crude oil and water (Table 4.1). The polarity difference between molecules of oil and water is larger than that between molecules of ether and water despite oil and ether both are organic substance.

With an increase in temperature, the kinetic energy of the molecules increases and cohesive forces between them decrease, resulting in a decrease in interfacial tension. For example, the surface tension of water decreases from 72.6 to 58.9 mN/m, with an increase in temperature from 20 to 100 °C.

4.1.1.3 Interfacial Tension Between Reservoir Fluids

Despite there has been but little direct quantitative application of the numerical values of surface- and interfacial tension data of reservoir fluids to production problems, except with relation to capillary phenomena (see Sect. 4.3), an appreciation of their magnitudes is helpful in an understanding of many of the details of production mechanisms.

By the reason of the complex of hydrocarbon composition and the variation of reservoir temperature–pressure conditions of different oil/gas reservoirs, the interfacial tensions between reservoir fluids vary in wide range. The interfacial tensions of oil/water or oil/gas in different reservoirs are very different.

Interfacial Tension of Oil–Gas System

In reservoir, crude oil generally contains a certain amount of dissolved gas. When reservoir pressure is lower than the saturation pressure of oil reservoir, a certain amount of dissolved gas will escape from oil and oil–gas two phases coexist in reservoir. It has been proved that the interfacial tension of oil–gas system in reservoir is lower than that on the ground.

The variations of interfacial tensions of several oil–gas systems with pressure are shown in Fig. 4.5. It shows the effects of dissolved gas on the interfacial tensions of oil–gas systems. It is visible that the surface tension of crude oil is higher even under high pressure. The reason is that 80 % of air is nitrogen and its solubility in oil is very low. However, the presence of dissolved gas in oil greatly reduces the surface tension of crude oil as shown in Fig. 4.5. When a large amount of natural gas dissolves in crude oil, oil expands and the distance between oil molecules increases. As a result, the difference between the densities of oil and gas decreases and it lead to a decrease in specific interfacial free energy of oil–gas system. By reason of "like dissolves like", the heavy the natural is, the more the amount of natural gas dissolves in oil. Therefore, the amount and nature of gas determines the magnitude of the reduction.

The variation of oil–gas interfacial tension with pressure is shown in Fig. 4.5, where the interfacial tension decreases with increasing pressure. The reason is that the higher the pressure is, the larger the amount of dissolved gas in oil is.

It has been proved that the interfacial tension is very small for near critical mixtures and approaches zero as the critical point is approached [4]. Hence the effect of temperature on IFT depends on the relative position to the critical point. For a gas condensate, IFT is expected to decrease by decreasing temperature, where the opposite is expected for an oil sample [2] (Fig. 4.6).

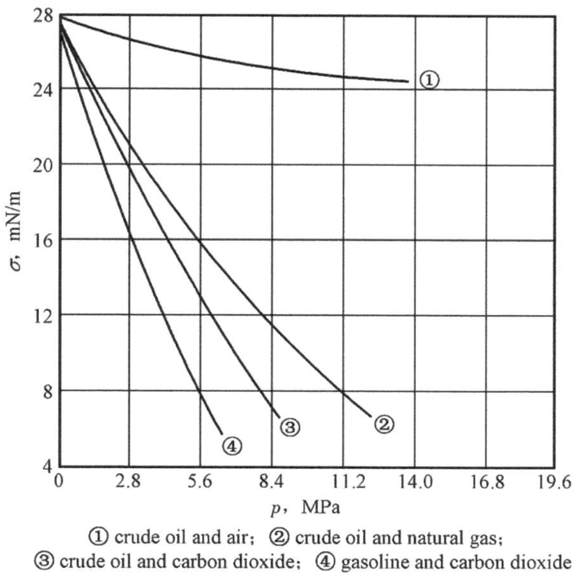

Fig. 4.5 Interfacial tensions of several oil–gas systems [3]

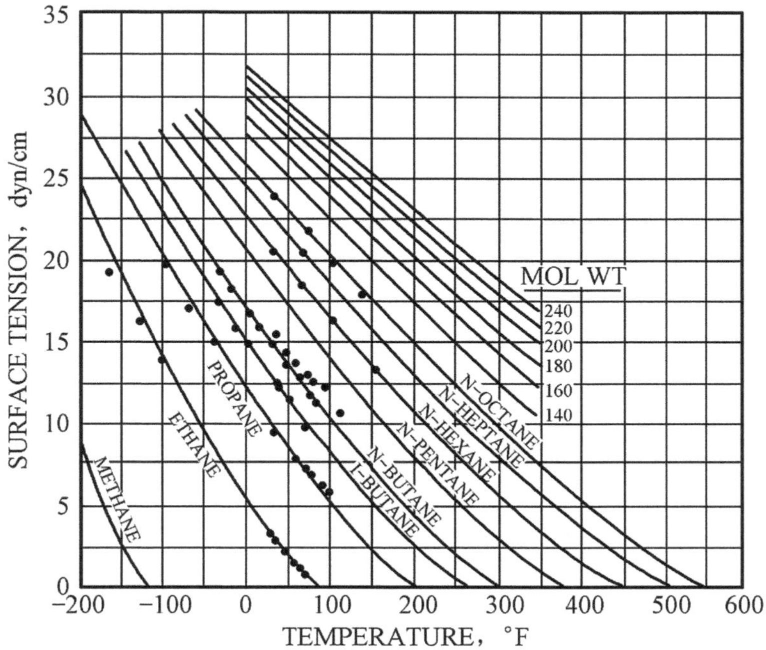

Fig. 4.6 Surface tensions of several hydrocarbons (Katz et al. [5])

At a given temperature, surface tension of hydrocarbons in equilibrium with the atmosphere or their own vapor increases with increasing molecular weight 4 (Fig. 4.6). For a given hydrocarbon, surface tension decreases with increasing temperature (Fig. 4.6).

Interfacial Tension of Oil–Water System

Figure 4.7 shows the effect of dissolved gas and pressure on the interfacial tension of three oil–water systems [6]. For each system, interfacial tension increases as pressure increases, but drops slightly as pressure is increased above the bubble point. Below the bubble point, the amount of dissolved gas in oil increases with increasing pressure, which leads to an increase in the density difference between oil and water. For this reason, the difference of forces on oil–water interfacial molecules thus increases. Above the bubble point, on the contrary, oil is compressed as pressure increases. The density difference between oil and water thus decrease with increasing pressure, and the difference of forces on interfacial molecules decreases.

The effect of temperature on interfacial tensions for some oil–water systems is shown in Fig. 4.8 [7]. The interfacial tension reduces with increasing temperature. Increasing temperature will cause the increase in molecular motion, which thus leads to a decrease in the difference of forces on oil–water interfacial molecules. The reduction in interfacial tension with increasing temperature is usually somewhat more pronounced than is observed for surface tension [8].

Although no quantitative relation is observed, the general trend suggests lower interfacial tensions of oil–water for the higher API gravity crudes [2]. However, in studies with a crude oil containing large amounts of resins and asphaltenes, different effects of temperature on interfacial tension were observed when measurements

Fig. 4.7 Effect of dissolved gas and pressure on interfacial tension between crude oil and water (Hocott [6])

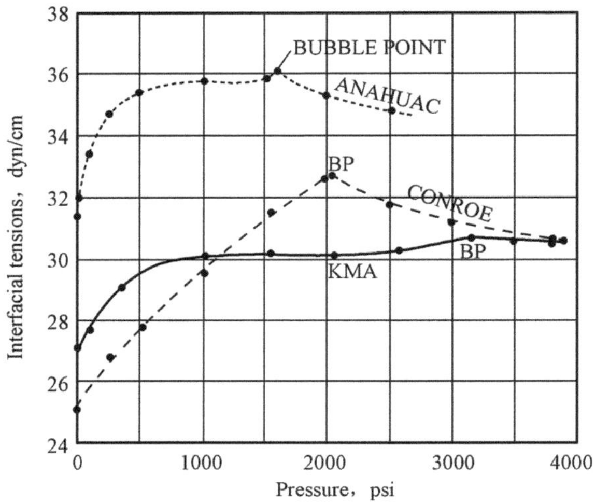

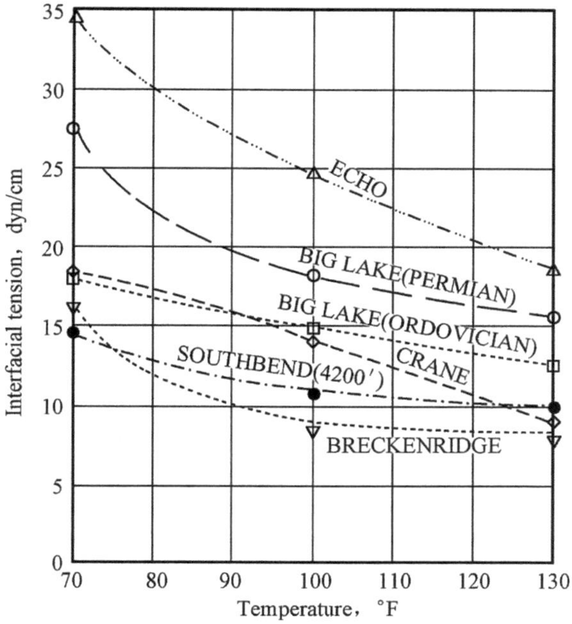

Fig. 4.8 Influence of temperature on interfacial tensions of oil-water systems (Livingston [7])

made at aerobic conditions were compared to anaerobic tests [9]. Interfacial tension between the crude and reservoir brine showed a decrease with an increase in temperature under aerobic conditions, whereas at anaerobic conditions, interfacial tension increased with increasing temperatures [2]. This difference in behavior was attributed to oxidation of the stock tank oil in the aerobic tests. At conditions of reservoir temperature and pressure, interfacial tension of the live reservoir oil was higher than the stock tank oil. The study concluded that live reservoir crude should be used in measurements of interfacial properties and that if stock tank oil is used, at least the temperature should correspond to reservoir conditions [2].

On the other hand, the increase in the salinity of water will cause an increase in the interfacial tension of oil–water system because of the increase in the difference of molecular polarity between oil and water.

In a word, in consideration of the complex of reservoir fluids, the interfacial tension between reservoir fluids varies differently. Table 4.2 lists the interfacial tensions of oil–water systems from oil fields at home and aboard.

4.1.1.4 Determination of Interfacial Tension

Interfacial tension proves itself in various effects, which offers a number of paths to its measurement. The commonly used methods of measuring interfacial tension include Wilhelmy plate method, pendant drop method, and spinning drop method. Which method is optimum depends upon the nature of the system being measured,

Table 4.2 Interfacial tensions of oil–water systems from oil fields at home and aboard

Oil field	Interfacial tension of oil-water (mN/m)	Condition
Dewy Yilmaz, Russian	30.2	Surface
Romashkin, Russian	25.6	Surface
34 oilfields in Texas, U.S.	13.6–34.3	Surface
Shengli, China	23–31	70 °C
Liaohe, China	9–24	45–85 °C
Daqing, China	30–36	Reservoir
Changqing, China	28.6	51 °C
Renqiu, China	40	Surface

the conditions under which its tension is to be measured, and the stability of its interface when it is deformed.

Generally, Wilhelmy plate method is often used for the measurement of higher interfacial tension (1–10^2 mN/m); pendant drop method for moderate interfacial tension (10^{-1}–10^2 mN/m); whereas spinning drop method especially for the measuring of low or ultralow interfacial tension (10^{-3}–10^{-1} mN/m) of microemulsion or between oil and water.

At present, the widely used methods in China for the measurement of the interfacial tension of oil–water or oil–gas system are pendant drop method and Wilhelmy plate method. They are also the measuring methods specified by China's petroleum industry standard.

Wilhelmy Plate Method

A Wilhelmy plate is a thin plate that is used to measure equilibrium surface or interfacial tension at an air–liquid or liquid–liquid interface. In this method, the plate is oriented perpendicular to the interface and the force exerted on it is measured.

The Wilhelmy plate consists of a thin plate usually on the order of a few centimeters square (Fig. 4.9). The plate is often made from glass or platinum which may be roughened or pre-wetted with a hydrophilic coating to ensure complete wetting. The plate is cleaned thoroughly and attached to a scale or balance via a thin metal wire.

This technique is based on the geometry of a fully wetted plate in contact with, but not submerged in, the heavy phase.

In this method, a vertical plate of known perimeter is suspended at the interface. The test liquid is brought into contact with the bottom of the plate, causing the plate to be pulled down into the liquid by the surface tension force. The plate is thus wetted. The contact angle of the liquid with the 'high surface energy' plate is assumed to be zero hence no corrections for additional forces are required.

In test, the plate is maintained in this position by a force F which balances the meniscus force (interfacial tension) of the liquid acting on the lower edge of the

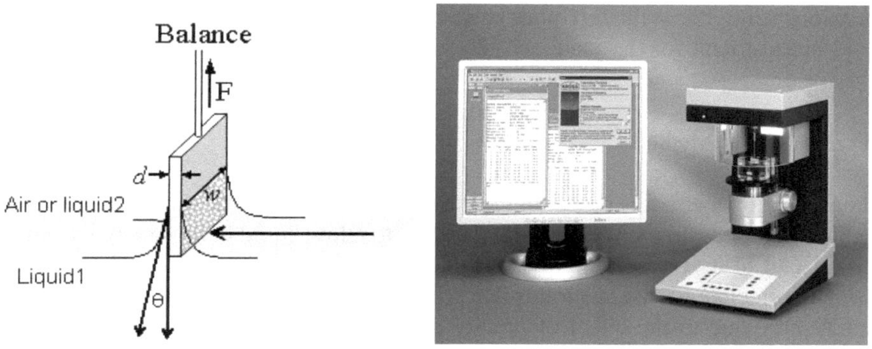

(a) Schematic diagram (b) Tensiometer

Fig. 4.9 Wilhelmy plate method (from Wikimedia)

plate (Fig. 4.9). This force F, measured by the microbalance or a tensiometer, is used to calculate the interfacial tension using the Wilhelmy equation:

$$\sigma = \frac{F}{2\cos\theta} \tag{4.10}$$

where σ is interfacial tension, mN/m; F is balance force on the plate, 10^{-2} mN; L is the wetted perimeter (2w + 2d), i.e., the three-phase contact line, of the Wilhelmy plate, cm; θ is the contact angle between the liquid phase and the plate. In practice the contact angle is rarely measured, instead either literature values are used, or complete wetting ($\theta = 0$) is assumed.

The method is suitable for the measurement of the interfacial tension of oil–water system whose density is less than 0.4 g/cm^3 and interfacial tension is higher, ranging 5–100 mN/m.

Because the plate is not moved during measurements, the Wilhelmy plate allows accurate determination of surface kinetics on a wide range of timescales and it displays low operator variance. In a typical plate experiment, the plate is lowered to the surface being analyzed until a meniscus is formed, and then raised so that the bottom edge of the plate lies on the plane of the undisturbed surface. If measuring a buried interface, the second (less dense) phase is then added on top of the undisturbed primary (denser) phase in such a way as to not disturb the meniscus. The force at equilibrium can then be used to determine the absolute surface or interfacial tension.

Modern instruments use plates of standard dimensions so that measurements of the plate size and its weight are not required. At present, CBVP A-3 tensiometer is often used at home and the dimension of matching plate is 2.4 × 2.4 × 0.03 cm.

Note that adsorption of organic compounds from the laboratory environment or test solutions can be a major source of experimental error when measuring surface tensions using the Wilhelmy plate method. In experiment, therefore, strict cleaning and pretreatment for the plate are necessary.

Pendant Drop Method

Pendant drop method is based on the measurement of the static shape (geometry) of a liquid drop. A simple way to form a drop is to allow liquid to flow slowly from the lower end of a vertical tube of small diameter, such as a syringe tip (Fig. 4.10).

The principle of this method is that a liquid drop may be formed from the tip of a small tube due to the effect of gravity and interfacial tension. Gravity makes the liquid drip from the tip, whereas the interfacial tension of the liquid causes the liquid to hang from the tip, forming a pendant. The shape of a drop of liquid hanging from a syringe tip is determined by the balance of gravity and interfacial tension. A video system (Fig. 4.11) is used to capture the shape of a drop, and determine the geometric parameter of the pendant drop (Fig. 4.10).

In this method, two parameters of the drop should be experimentally determined by a goniometer (Fig. 4.11). They are the equatorial diameter d and the diameter d_i at the distance d from the top of the drop (Fig. 4.10). The interfacial tension of the liquid is then calculated from the following equation:

$$\sigma = \frac{\Delta \rho g d^2}{H} \quad (4.11)$$

where σ is the interfacial tension, mN/m; g is the acceleration of gravity, 980 cm/s^2; $\Delta\rho$ is the density difference between the two fluid phases, g/cm^3; The shape-dependent parameter (H) depends on a value of the "shape factor" S. S is a function of diameters d and d_i. It may be obtained from related table.

The pendant drop technique, as other interfacial tension measurement techniques, requires extreme cleanliness to obtain good quality and reproducible results.

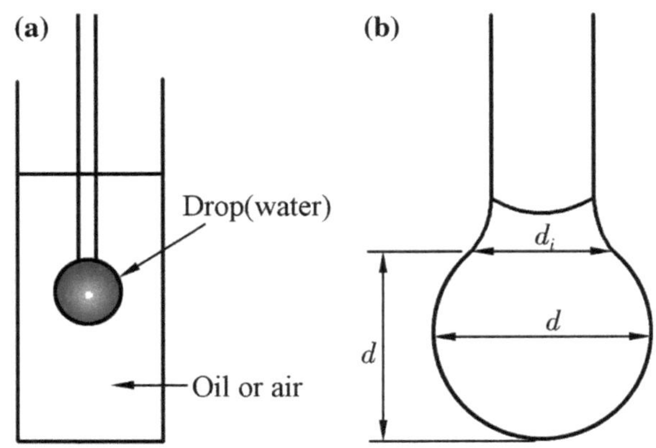

Fig. 4.10 Pendant drop profile

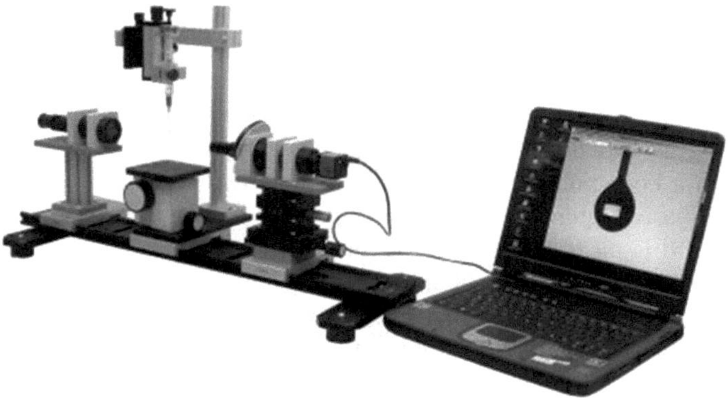

Fig. 4.11 A video system used for drop shape measurement

Here, the needle used for hanging the drop should be well cleaned and the climbing of the interface over the outer surface of the needle should be avoided.

It is recommended that needles with a diameter that is less than 0.7 mm may be used when common liquid is measured. If measured liquid is viscous, needle with a diameter between 1.5 and 2.0 mm may be chosen depending on the viscous of the liquid. In a word, the diameter of the needle should not be too small because it reduces the value of *d* and, consequently, the precision of interfacial tension determination. In addition, if the density of measured liquid is less than that of medium, L-shaped needles may be used for measurement (Fig. 4.12)

Fig. 4.12 L-shaped needle

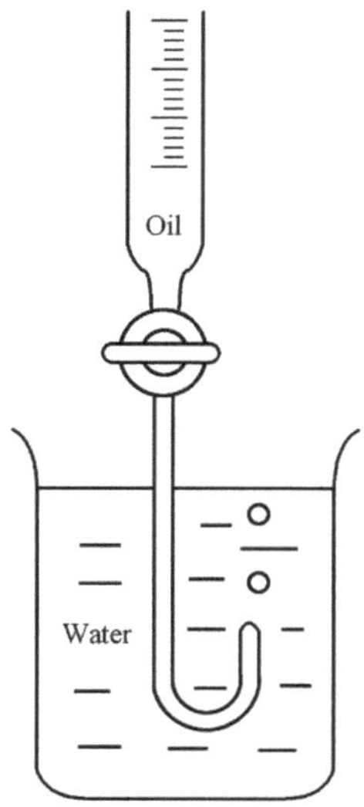

Spinning Drop Method (Rotating Drop Method)

This technique is ideal for measuring very low interfacial tensions.

This technique relies on the fact that gravitational acceleration has little effect on the shape of a fluid drop suspended in a liquid, when drop and the liquid are contained in a horizontal tube spun about its longitudinal axis (Fig. 4.13). In this method, the basic required parameter is the radius or diameter of a drop within a heavy phase. It is measured while drop and liquid are rotated.

Measurements are carried out in a rotating horizontal tube which contains a dense fluid. One drop of a less dense liquid is placed inside the transparent fluid. Since the rotation of the horizontal tube creates a centrifugal force toward the tube walls, the liquid drop starts to be elongated and this elongation stops when the interfacial tension and centrifugal forces are balanced. Values obtained at this equilibrium point are used to estimate surface tension of the particular liquid by using appropriate correlations. A device called "*spinning drop tensiometer*" (Fig. 4.14) is generally utilized for this purpose.

At low rotational velocities (ω), the fluid drop will take on an ellipsoidal shape (Fig. 4.13), but when ω is sufficiently large, it will become cylindrical. Under this latter condition, the radius (R) of the cylindrical drop is determined by the interfacial tension, the density difference ($\Delta\rho$) between the drop and the surrounding fluid, and the rotational velocity of the drop. As a result, the interfacial tension is calculated by following equation:

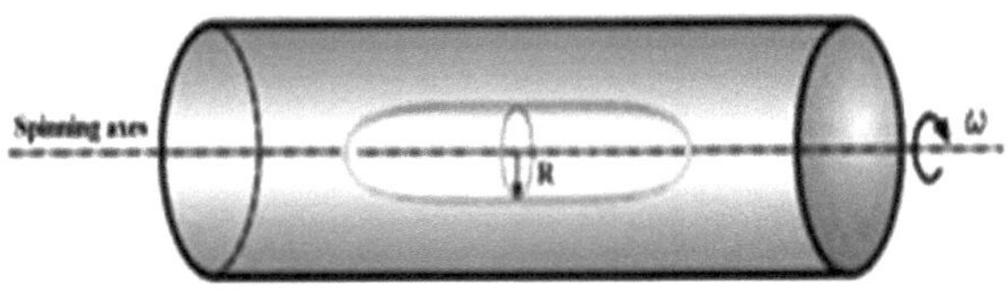

Fig. 4.13 Sketch map of Spinning drop method to measure interfacial tension

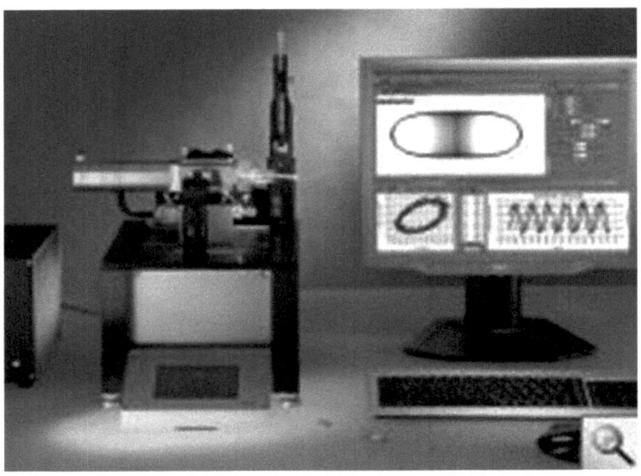

Fig. 4.14 Spinning drop tensiometer

$$\sigma = \frac{\Delta\rho\omega^2}{4} R^3 \tag{4.12}$$

where $\Delta\rho$ is the density difference between the drop and the surrounding fluid, g/cm^3; and σ is the interfacial tension, mN/m; R is radius of the drop in circular cylinder, m; ω is an angular velocity, r/min.

Spinning drop method is not expected to give accurate results for the high surface tension measurements since the centrifugal force that is required to maintain a cylindrical shape of the drop is much higher in the case of liquids that has high interfacial tensions. Spinning drop method is preferred for the accurate measurements of interfacial tensions below 10^{-2} mN/m (or $10^{-3}.10^{-1}$ mN/m).

Nomograph

The nomograph for the interfacial tension of oil–water system in reservoir conditions is shown in Fig. 4.15. The interfacial tension of oil–water system in reservoir can be directly determined from Fig. 4.15 if reservoir temperature and the interfacial tension of oil–water at surface are given.

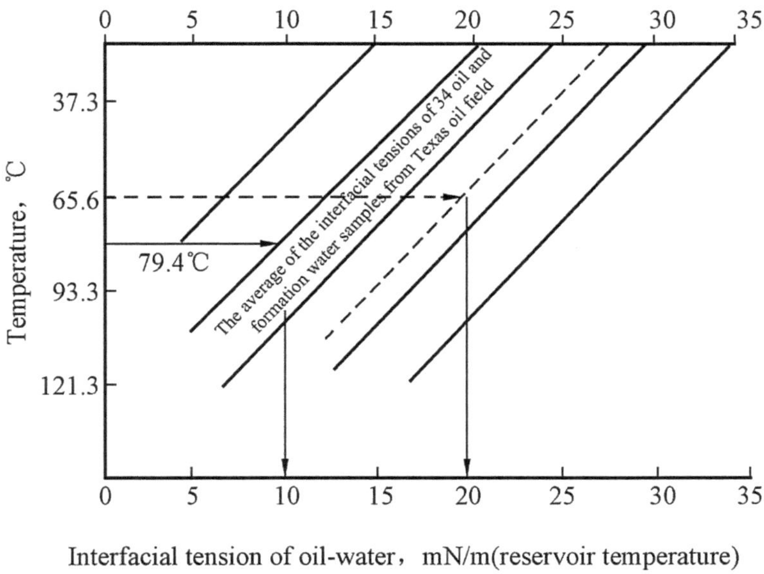

Fig. 4.15 Nomograph for IFT of oil–water system in reservoir conditions [3]

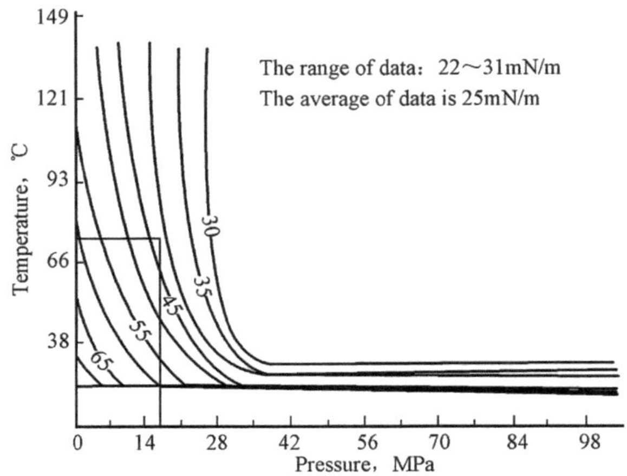

Fig. 4.16 Nomograph for IFT of methane–water system in reservoir conditions [3]

The nomograph for the interfacial tension of methane–water system in reservoir conditions is shown in Fig. 4.16. In a dry gas reservoir, the interfacial tension of methane–water system in reservoir can be directly determined from Fig. 4.16 if reservoir temperature and pressure are given.

4.1.2 *Interfacial Adsorption*

Adsorption is a very important interfacial phenomenon. It is a technical term which expresses the taking up of gas, vapor, or liquid by a surface or interface. Hence adsorption is a surface phenomenon. As mentioned earlier, the molecules on the surface of liquid or solid experience a net attraction downward and have higher energy than those inside. The surface of liquids or solids, therefore, tends to attract and retain other molecules when brought in contact with a gas or solution.

4.1.2.1 Adsorption

Matter that accumulates at the interface during the adsorption process is termed *adsorbate*, whereas the surface where it accumulates is called the *adsorbent*.

The phenomenon of attracting and retaining the molecules of an adsorbate substance on the surface of adsorbent (a liquid or solid) resulting into higher concentration of adsorbate molecules on the surface is termed as *adsorption* [10]. In short, adsorption is the adhesion of atoms, ions, or molecules from a gas, liquid, or dissolved solid to a surface [11]. The inverse process, *desorption*, refers to the loss of matter from the interface, typically by diffusion into the bulk of one of the adjoining phases. Adsorption is spontaneous process because it results in a decrease of interfacial energy. As the molecules remain only at the surface and do not go deeper into the bulk of the solid, the concentration of adsorbed gas or liquid is more at the surface than in the bulk.

The phenomenon of adsorption can occur at all surfaces and five types of interfaces can exist: gas–solid, liquid–solid, liquid–liquid, solid–solid, and gas–solid. The gas–solid interface has probably received the most attention in the literature and is the best understood. The liquid–solid interface is now receiving much attention because of its importance in many electrochemical, biological systems [12] and reservoir fluids–rock system.

For example, when a finely divided active charcoal is stirred in a dilute solution of a dye, the intensity of color in the solution is decreased. Dye is adsorbed on the solid surface. When finely divided solid is exposed to gas at low pressure, the pressure of gas is decreased considerably because of the adsorption of molecules of gas on solid. Charcoal, silica gel, alumina gel, clay, reservoir rock, etc., are very good adsorbents because they have highly porous structures and hence large surface areas. Colloids, on account of their extremely small dimensions, possess enormous surface area per unit mass and, are, therefore, good adsorbents.

Similar to surface tension, adsorption is a consequence of surface energy. In a bulk material, all the bonding requirements (be they ionic, covalent, or metallic) of the constituent atoms of the material are filled by other atoms in the material. However, atoms on the surface of the adsorbent are not wholly surrounded by other adsorbent atoms and therefore can attract adsorbates. The exact nature of the bonding depends on the details of the species involved. According to the nature of interaction, adsorption can be divided into two main classes: (a) *physisorption* or *physical adsorption*, where the forces and processes are reversible, nonspecific, and of relatively low energy, and (b) *chemisorption* or *chemical adsorption*, where the forces and processes of adsorption are specific, irreversible, and of higher energy [12].

Physisorption (by van der Waals and Electrostatic Forces)

Physisorption is the most common form of adsorption. It is a process in which the electronic structure of the atom or molecule is barely perturbed upon adsorption [13]. This is thus generally a reversible process. In physisorption, the physical forces are responsible for holding adsorbate molecules on the surface of adsorbent. All the physical forces are weak forces, such as van der Waals and electrostatic forces. Therefore, it is characterized by a low heat of adsorption, usually of the order of 40 kJ per mole. This value is of the same order of magnitude as the heat of vaporization of the adsorbate and lends credence to the concept of a weak 'physical' bonding [12]. Physisorption generally takes place at low temperature and decreases with increase in temperature. In physisorption multimolecular layers may be formed on the surface of adsorbent [10].

The fundamental interacting force of physisorption is mainly caused by van der Waals force. Even though the interaction energy is very weak (thermal energy at room temperature $\sim$26 meV), physisorption plays an important role in nature.

Chemisorption (by Chemical Bonding)

If the chemical forces hold adsorbate to the surface of the adsorbent, the adsorption is known as chemisorption [12]. Chemisorption is a subclass of adsorption, driven by a chemical reaction occurring at the exposed surface. In this case the adsorbate undergoes a strong chemical interaction with the unsaturated surface and gives rise to a high heat of adsorption, usually of the order of 400 kJ per mole. The chemical bonds between the adsorbate and the surface molecules of adsorbent, such as a solid, are formed; and this attraction may be stronger than the forces holding the adsorbent (solid) together. A new chemical species is thus generated at the adsorbent surface (e.g., corrosion, metallic oxidation). The strong interaction between the adsorbate and the substrate surface creates new types of electronic bonds—ionic or covalent, depending on the reactive chemical species involved [14]. Due to specificity, the nature of chemisorption can greatly differ from system to system, depending on the chemical identity and the surface structure. Chemisorption is often characterised by taking place at elevated temperatures. It differs from physical adsorption both in the strength of the bond between the adsorbed species and the surface and in terms of the specificity of the binding forces.

Differences Between Physisorption and Chemisorption

In comparison with chemisorption, physisorption, generally speaking, can only be observed in the environment of low temperature. In practice, the categorization of a particular adsorption as physisorption or chemisorption depends principally on the binding energy of the adsorbate to the substrate. The most important distinguishing features between physisorption and chemisorption may be summarized as follows [12, 15]:

(a) Physisorption is a general phenomenon and occurs in any solid/fluid or solid/gas system. Chemisorption is characterized by chemical specificity.
(b) In physiorption, perturbation of the electronic states of adsorbent and adsorbate is minimal. For chemisorption, changes in the electronic states may be detectable by suitable physical means.
(c) The elementary step in physisorption from a gas phase does not involve activation energy. Chemisorption often involves activation energy. Typical binding energy of physisorption is about 10–100 meV. Chemisorption usually forms bonding with energy of 1–10 meV.
(d) The energy of chemisorption is the same order of magnitude as the energy change in a comparable chemical reaction. Physisorption is always exothermic, but the energy involved is generally not much larger than the energy of condensation of the adsorptive. However, it is appreciably enhanced when physisorption takes place in very narrow pores.

(e) Under appropriate conditions, e.g., at high relative pressures, physisorption generally occurs as a multilayer. Chemisorbed molecules are linked to reactive parts of the surface by valence bonds and only form monolayer adsorption.
(f) An activation energy is often involved in chemisorption and at low temperature, the system may not have sufficient thermal energy to attain thermodynamic equilibrium. It may sometimes be quite slow depending upon the nature of chemical reaction involved. Physisorption systems generally attain equilibrium fairly rapidly, namely instantaneous, but equilibration may be slow if the transport process is rate-determining.
(g) Physisorption equilibrium is reversible and is rapidly established. Chemisorption is irreversible. Physically adsorbed layer can be removed very easily by changing pressure or concentration. On the other hand, the removal of a chemisorbed layer, however, requires much more rugged conditions such as high temperatures, etc.
(h) A physisorbed molecule keeps its identity and on desorption returns to the fluid phase in its original form. If a chemisorbed molecule undergoes reaction or dissociation, it loses its identity and cannot be recovered by desorption.
(i) Physical adsorption as condensation can occur with any gas–solid system provided only that the conditions of temperature and pressure are suitable. The chemisorption will take place only if the gas is capable of forming a chemical bond with the surface atoms.
(j) To the adsorption of gases on solids, the rate of physical adsorption increases with the increase of pressure of the adsorbate, whereas the rate of chemisorption decreases with the increase of pressure of the adsorbate.
(k) Physical adsorption occurs to an appreciable extent at temperatures close to those required for liquefaction of adsorbed gases. Generally, chemisorption occurs at high temperatures. In many cases, the physical adsorption at low temperatures changes to chemisorption when the temperature is raised.

It should be noted that in some cases both physisorption and chemisorption occur simultaneously in a system, such as the system of fluid rock.

4.1.2.2 Adsorption of Gas by Solid

The study of the gas–solid adsorption process has excited the interest of both academic and industrial scientists for many years, and the reactions are not hard to find. Industrially, it is known that this phenomenon plays an essential role in the catalytic process. It is generally believed that all gases or vapors are adsorbed on the surface of all solids with which they are in contact. The phenomenon was first described in 1773 by Scheele who discovered the uptake of gases by charcoal [12].

A common theory about the adsorption of gas on solid was proposed by Irving Langmuir in 1916. Langmuir's theory postulated that the surface of the solid consists of a fixed number of adsorption sites per unit area of the surface and each site could adsorb only one gas molecule. Hence the solid surface could be covered by only a single layer or monolayer of gas molecules [16].

The relation between the amount of gas adsorbed by a solid and the equilibrium pressure at a given temperature is called an adsorption isotherm. Langmuir's adsorption isotherm of gas on solid is represented by the following equation:

$$V = V_{\infty}\frac{bp}{1+bp} \tag{4.13}$$

where V is the mole number of gas adsorbed on adsorbent, mol; V_{∞} is the max mole number of gas adsorbed on adsorbent, mol; p is the pressure of gas, Mpa; b is Langmuir parameter, 1/MPa.

It is worthwhile noticing that at low pressures, the value of bp is much lower than 1 and $V \approx V_{\infty}bp$. V is directly proportional to the pressure of gas. On the contrary, at high pressures, $bp \gg 1$ and then $V = V_{\infty}$. This indicates that the amount of gas adsorbed on solid does not increase with the increase of pressure because the surface of adsorbent has been completely covered by a single layer of gas molecules.

Besides Langmuir's theory of monolayer adsorption, the theory of multilayer adsorption proposed by other investigator can also explain some adsorption phenomena.

In a word, adsorption of gases by solids is quite common. But it is more pronounced on solid adsorbents with a large surface area [16], such as charcoal, silica gel, and reservoir rock, etc.

Historically, the research and industrial applications of adsorption began with the adsorption of solid. However, due to the complexity of the surface of a solid, the adsorption mechanism at the surface of a solid has not yet been fully understood.

In addition, the various factors that influence the adsorption of gases on solid surfaces include temperature, pressure, nature, and the surface area of the adsorbent and the nature and critical temperature of the adsorbate. The amount of gas adsorbed on a given solid surface decreases with an increase in temperature. This is explained on the basis that a true equilibrium is established between the gas in contact with the solid and the adsorbed gas at any given temperature and pressure. Adsorption of the gas on the solid surface and desorption of adsorbed gas molecules from the surface take place simultaneously setting up a dynamic equilibrium. The adsorption equilibrium is influenced by temperature. As the temperature increases the amount of gas adsorbed decreases as is expected for an exothermic process. At constant temperature adsorption of a gas onto the solid surface increases with increasing pressure. In general, the extent of adsorption increases with an increasing surface area of the solid and also with increasing pressure at a given temperature. Thus solids with highly porous structures have large surface areas and hence are good adsorbents. Adsorption of different gases on a given solid adsorbent takes place to different extents. The amount adsorbed has been found to be more for a gas with a higher critical temperature.

4.1.2.3 Adsorption at Solid–Liquid Interfaces

Adsorption at the liquid-solid interface is of great importance in industry and everyday life (e.g., in detergency, adhesion, lubrication, flotation of minerals, water treatment, oil recovery, and in pigment and particle technology) [17].

Generally, the factors that influence the adsorption of liquid onto solid adsorbents include temperature, nature of the adsorbent and its surface area, and nature of liquid. Adsorption in a given adsorbent–adsorbate system decreases with increase in temperature.

The surfaces of solid adsorbents are often physically and/or chemically heterogeneous and bonding energies may vary widely from one site to another, which leads to different adsorption behavior at different site. In general, nonpolar substances are adsorbed more effectively on nonpolar adsorbents such as activated carbon while inorganic solids such as silica, alumina, or rock adsorb polar substances preferentially [16].

In reservoir, oil and water often coexist in rock pores. Oil or water directly contacts with the surface of rock. Water is a polar liquid. There are some active and polar components in crude oil. Oil and water are thus easily absorbed on the surface of rock particles. It can be considered that the adsorption of oil by rock mainly depends on the amount of active/polar substances in oil.

Based on the selectivity of adsorption of liquid on rock, the liquid contained *sacrificial agent* should be injected into reservoir before surfactant solution in surfactant flooding. The adsorption of sacrificial agent on the surface of rock particles can decrease the loss of surfactant due to the adsorption by rock. The displacement efficiency of surfactant flooding is then greatly improved. At present, adsorption theory has found wide use in EOR.

The adsorption of liquid on solid is more complicated than that of gas by solid. Up to now, there are several adsorption isotherms applicable for adsorption of solutes from solution. One of commonly encountered adsorption equations is Langmuir type. Langmuir isotherm is also applicable in many adsorptions at solid–liquid interface. The isotherm is expressed as [16]:

$$C_a = \frac{C_0 S}{k + S} \tag{4.14}$$

where C_a is the amount of solute adsorbed per unit amount of adsorbent (g/g); C_0 and k are constants determined from experimental data; S is the concentration of the solute in the liquid phase, g/L.

4.1.2.4 Adsorption at Gas–Liquid Interfaces

When the concentration of any component in a solution or liquid is higher at the surface than in the bulk, the phenomenon is call adsorption of the component at liquid surface or gas–liquid interface. An example is the adsorption of surfactant molecules at the interface between gas and oil.

The adsorption at liquid surface depends on the nature of solute. When a few of surfactants are added in pure water, it can be found that the surface tension of the water is remarkably reduced. This is responsible for the adsorption of surfactant at gas–water interface.

Surfactants, also known as wetting agents, lower the surface tension of a liquid, allowing easier spreading, and the interfacial tension between two liquids. The term surfactant is a contraction of "*Surface-active agent.*" Surfactants are usually organic compounds that are amphipathic, meaning they contain both hydrophobic groups (their "tails") and hydrophilic groups (their "heads") (Fig. 4.17). They are typically sparingly soluble in both organic solvents and water. Surfactant molecules will spontaneously gathered in the interface of a gas–liquid system (Fig. 4.18). Surfactants can thus reduce the surface tension of water by adsorbing at the air–water interface. They can also reduce the interfacial tension between oil and water by adsorbing at the liquid–liquid interface. Several common surfactants are shown in Fig. 4.19.

The excess amount of solute adsorbed at per unit surface area is referred to as specific adsorption and expressed as *G*. At given temperature, the relation between

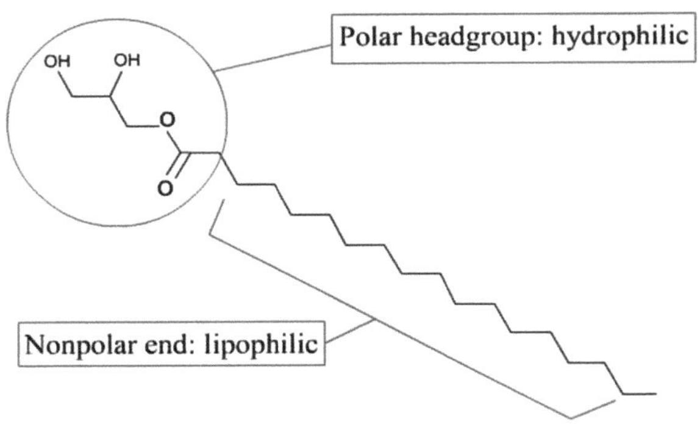

Fig. 4.17 Structure of surfactant; note the two distinct sections of the structure

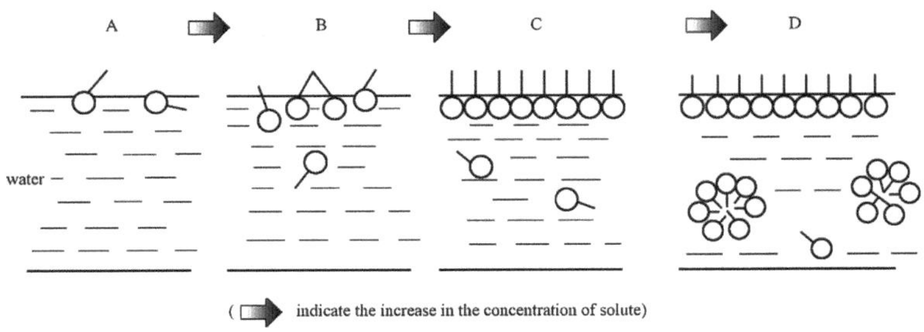

Fig. 4.18 Concentration variation of solute adsorbed at a two-phase interface

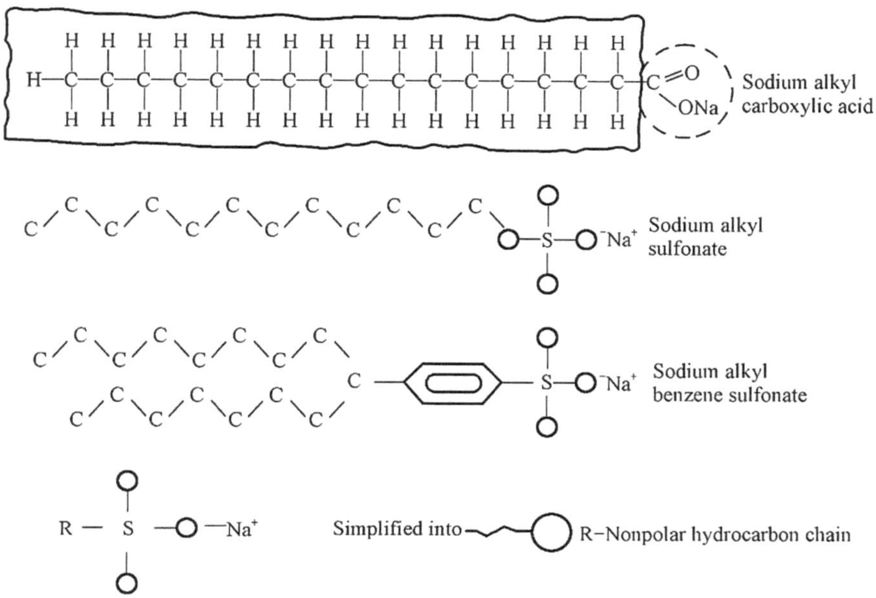

Fig. 4.19 Several common surfactants

specific adsorption and the concentration of solute can be described by following *Gibbs*'s adsorption isotherm:

$$G = -\frac{1}{RT}C\left(\frac{\partial \sigma}{\partial C}\right)_T \tag{4.15}$$

where G is the excess moles of the solute adsorbed per unit area of gas–liquid interface, mol/m^2; C is the equilibrium concentration of the solute in bulk solution, dimensionless; $\left(\frac{\partial \sigma}{\partial C}\right)_T$ is the variation of surface tension with surfactant concentration at constant temperature. It is also called surface activity; T is absolute temperature, K; R is universal gas constant, 8.3144 J/(mol · K).

4.1.2.5 Differences Between Adsorption, Absorption, and Sorption

Adsorption process creates a film of the adsorbate on the surface of the adsorbent. Adsorption differs from absorption, in which a fluid (the absorbate) permeates or is dissolved by a liquid or solid (the absorbent). In absorption, the substance is not only retained on the surface but passes through the surface to become uniformly distributed throughout the body of a solid or liquid to form a solution or a compound [12]. For example, when a hot crucible is allowed to cool in air, a film of moisture is formed at the surface. This is the case of adsorption of water vapor on the surface of a crucible. When sponge is put into water, it takes up water. It is an example of absorption.

Note that adsorption is a surface-based process while absorption involves the whole volume of the material. Therefore, adsorption is a surface phenomenon, whereas absorption is a bulk phenomenon.

In absorption, the concentration of the adsorbed molecules is always found to be greater in the immediate vicinity of the surface (adsorbent) than in the free phase (adsorbate). On the contrary, absorption involves bulk penetration of the molecules into the structure of the solid or liquid by some process of diffusion. In case of adsorption, the equilibrium is easily attained in a very short time, whereas in absorption the equilibrium takes place slowly.

In many examples, the initial rapid adsorption is followed by a slow process of absorption of the substance into the interior of the solid. In these cases, the effects of absorption cannot be distinguished from those of adsorption. Therefore, a new term called *sorption* was introduced. The term sorption encompasses both processes. That is, sorption generally means the process in which both adsorption and absorption take place simultaneously [12].

4.1.3 Interfacial Viscosity

The adsorption of substance at an interface not only changes the interfacial tension or interfacial pressure but can also affect the interfacial rheology, namely interfacial viscosity and elasticity.

Substance adsorbed at interfaces form two-dimensional surface phases that may be gaseous, expanded liquid, condensed liquid, or solid. Interfacial viscosity is the ratio between shear stress and shear rate in the plane of the interface, i.e., it is a two-dimensional viscosity (the unit for interfacial viscosity is surface Pa s or surface poise) [18]. A liquid/liquid interface has viscosity if the interface itself contributes to the resistance to shear in the plane of the interface [19].

Usually, the interfacial viscosity is higher than the bulk viscosity. The high viscosity of the adsorbed film can be accounted for in terms of the orientation of the molecules at the interface. For example, surfactants orient at the oil/water interface with the hydrophobic portion pointing to (or dissolved in) the oil and the polar group pointing to the aqueous phase as shown in Fig. 4.20. Most surfactants (and mixtures) and macromolecules adsorbed at the interface are viscous, showing a high induced interfacial viscosity [18]. It has been proved that the high interfacial viscosity in crude oil–brine systems containing a mixture of natural surfactants in the oil phase can be attributed to the presence of polar asphaltenes and resins [20]. The addition of commercial surfactants to a brine–crude oil system can remove some of the natural surfactants from the crude oil–brine interface and reducing the interfacial viscosity [20].

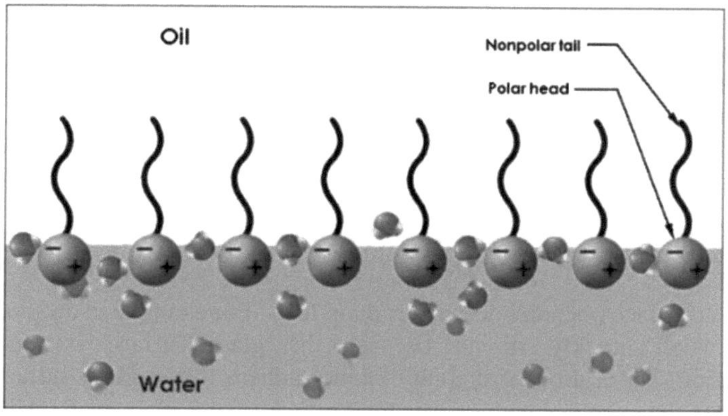

Fig. 4.20 Surfactant molecules arranged on an oil-water interface

The surface viscosity can change by more than an order of magnitude at a transition from one surface phase to another. This is analogous to the change in viscosity of bulk fluids at phase transitions.

The interface is connected to the substrate on either side of it and so is the interfacial viscosities coupled to the bulk viscosities. Therefore, it becomes laborious to determine purely interfacial viscosities without the influence of the surroundings [21].

4.2 Wettability of Reservoir Rock

In reservoir, oil, and water are usually present in rock pores before water flooding. Does water adhere to the surface of rock particles to pick up the oil or water only imbibe into the pores and displaces oil from the pore central when water is injected into reservoir? All these depend on rock wettability.

The wettability of a rock–fluid system is an overall average characteristic of a heterogeneous system with microscopic relative wetting throughout the porous medium. The rock pore surfaces have preferential wetting tendencies toward water or oil leading to establishment of the various states of overall wettability. This overall wettability has a dominant influence on the fluid flow and electrical properties of the water–hydrocarbon–rock system. It controls the capillary pressure and relative permeability behavior and thus the rate of hydrocarbon displacement and ultimate recovery [22]. It is generally believed that the rock wettability and capillary pressure are static characteristics of rock–fluid, while the relative permeability is dynamic characteristics. But whether they are static or dynamic characteristics, they are related to the microscopic distribution and the initial distribution of fluids (oil and water) in the rock pores.

When water is injected into reservoir, the rock wettability is of great significance to judge whether the injected water can wet the rock surface well or not, an analysis of water flooding efficiency in water-flooding, a selection of enhanced oil recovery methods and reservoir performance simulation.

4.2.1 Basic Concepts About Wetting

Wetting is a surface phenomenon in our daily lives. For example, a drop of water on the glass will quickly spread because the glass surface is preferentially water-wetting. But a drop of mercury cannot adhere to the glass surface, so it is thought that the mercury cannot wet the surface of the glass well. All those are wetting phenomenon.

Wetting is a phenomenon of fluid wetting the surface of solid due to the action of interfacial tension. It is also an ability of a liquid to maintain contact with a solid surface, resulting from intermolecular interactions when the two are brought together. The degree of wetting is determined by a force balance between adhesive and cohesive forces.

The property of fluid wetting the surface of a solid is called *wettability*. It is defined as a preferential tendency of one fluid to spread on or adhere to a solid surface when two or more immiscible fluids coexist on the surface of the solid [23]. Wettability is used to describe the relative ability of two fluids adhering to a solid surface.

4.2.1.1 The Reason for Reservoir Rock Exhibiting Wettability

If a core is immersed into mercury, the mercury will not spread over the core surface. While if a core is put into the water, its surface will be wetted by water. Assume that the interfacial tension between water and rock surface is σ_{ws} when they are in contact, while the interfacial tension between rock surface and air is σ_{gs} before the core is immersed into the water. The surface area of the core, A, does not change during the process, but the surface energy does change. Let ΔU denote the surface energy change during the process and then we can get:

$$\Delta U = \sigma_{ws} \cdot A - \sigma_{gs} \cdot A = \left(\sigma_{ws} - \sigma_{gs}\right)A \tag{4.16}$$

where ΔU is the variation of surface energy, J; σ_{ws} is the interfacial tension between water and rock block, J/m^2; σ_{gs} is the interfacial tension between air and rock block, J/m^2; A is the superficial area of rock, m^2.

In general, $\sigma_{gs} > \sigma_{ws}$, so the surface energy change $\Delta U < 0$, which indicates that surface energy decreases. The decrease of surface energy is the fundamental cause of the wetting phenomena. As mentioned above, all the surface phenomena have the tendency to minimize its surface energy. Therefore, wettability is also defined as a process of the surface energy decrease when solid and liquid are in contact. This can also explain why mercury is a non-wetting fluid, for $\sigma_{Hgs} > \sigma_{gs}$ which leads to a positive value of ΔU.

4.2.1.2 Evaluation of Wettability

Contact Angle (or Wetting Angle)

Contact angle, symbol θ, is a quantitative measure of the wetting of a solid by a liquid. It is defined geometrically as the angle formed by a liquid at the three phase boundary where a liquid, gas, and solid intersect as shown below. Namely, when gas and liquid phase are in intimate contact with a solid surface, the interface between gas and liquid phase intersects the solid surface at an angle θ, which is called contact angle or wetting angle. Note that the contact angle θ is always measured through the denser liquid phase. Figure 4.21 shows the contact angles at air–water–glass surface and air–mercury–glass surface.

The contact angle is determined by the resultant between adhesive forces, which result from the molecular interactions between the liquid and the solid, and cohesive forces of the liquid molecules among themselves (Fig. 4.22). The tendency of a drop to spread out over a flat solid surface increases as the contact angle decreases. Thus, the contact angle provides an inverse measure of wettability. Therefore, different values of the contact angle of a fluidsolid system indicate different degree of wetting.

Consider an oil–water–rock system as shown in Fig. 4.23. When a drop of water is placed on a surface immersed in oil, a contact angle may be formed that range from 0° to 180° [24].

When $\theta = 0°$, it means perfect or full wetting or strong wetting. The rock has extreme water wettability, called extreme water wetting.

When $\theta < 90°$, it indicates partial wetting. The rock has better water wettability, generally called water wetting.

When $\theta = 90°$, it denotes moderate wetting. The rock has neutral wettability, also called intermediate wet.

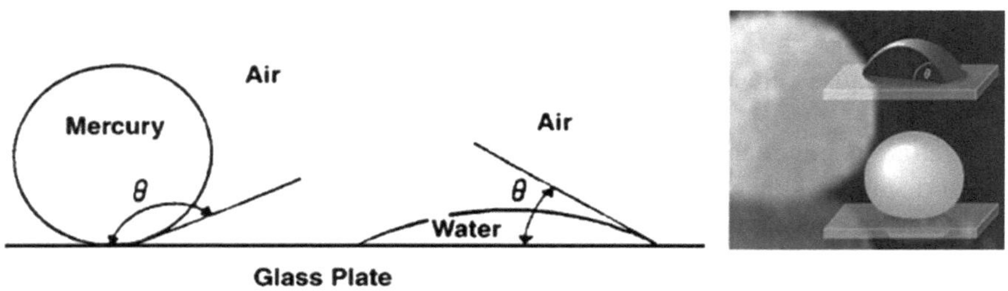

Fig. 4.21 Contact angle at air–water–glass surface and air–mercury–glass surface

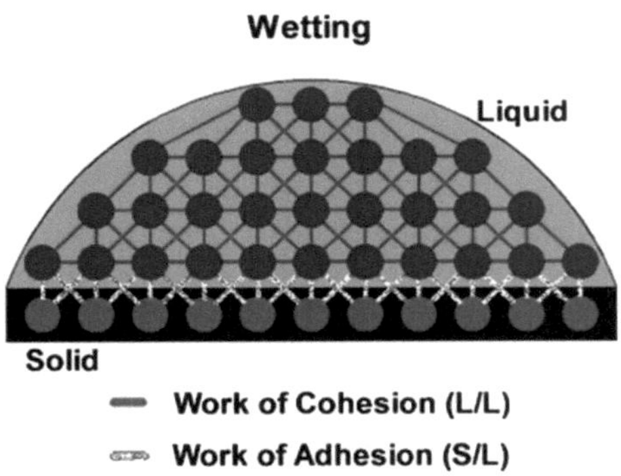

Fig. 4.22 Schematic diagram of cohesion force and adhesive force (from ramé-hart instrument co.)

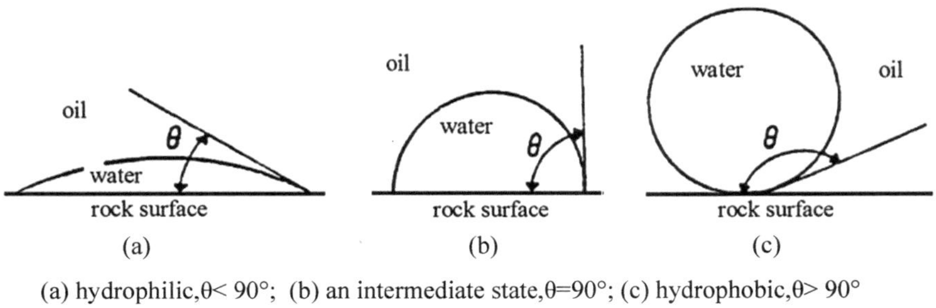

Fig. 4.23 Contact angle at oil–water–rock surface

When $\theta > 90°$, it is poor wetting or partial wetting. The rock has poor water wettability or better oil wettability, generally called oil wetting.

When $\theta = 180°$, it shows completely or perfectly non-wetting. The rock has extreme oil wettability, called extreme oil wetting.

Therefore, lower contact angle (less than 90°) usually indicates that the wetting of a solid surface is very favorable, and the fluid can spread over a large area of the surface. Higher contact angle (greater than 90°) generally means that the wetting of the surface is unfavorable, so the fluid will minimize contact with the surface and form a compact liquid droplet.

Note that a solid may show different wettability for different fluid and a fluid may also cause different wetting degree to different solid surface as shown in Fig. 4.24.

Wetting is important in the bonding or adherence of two materials. Wetting and the surface forces that control wetting are also responsible for other related effects, including so-called capillary effects (see Sect. 4.3).

As shown in Fig. 4.25 where gas and liquid are associated with a solid surface, the relationship between three different interfacial tensions can be described by Young's equation below:

Fig. 4.24 Different wetting of same fluid on different solids

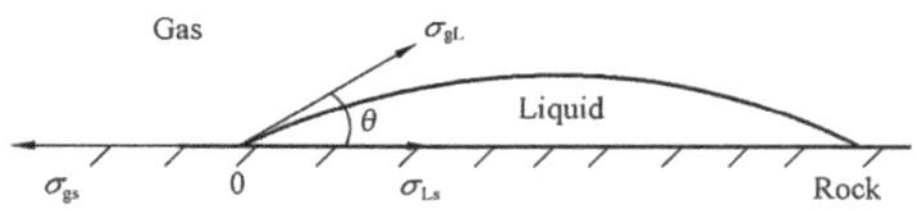

Fig. 4.25 Relationship between contact angle and surface tension

$$\sigma_{gs} = \sigma_{Ls} + \sigma_{Ls} \cos\theta \tag{4.17}$$

Therefore,

$$\cos\theta = \frac{\sigma_{gs} - \sigma_{Ls}}{\sigma_{Ls}} \tag{4.18}$$

Then,

$$\theta = \arccos\left(\frac{\sigma_{gs} - \sigma_{Ls}}{\sigma_{Ls}}\right) \tag{4.19}$$

where θ is contact angel, (°); σ_{gL} is the interfacial tension between gas and liquid, J/m^2; σ_{gs} is the interfacial tension between gas and solid, J/m^2; σ_{ls} is the interfacial tension between liquid and solid J/m^2.

According to Eq. (4.19), contact angle can be calculated if σ_{gs}, σ_{Ls}, σ_{gL} are known. However, direct measurement of solid–fluid surface tensions, σ_{gs} and σ_{Ls}, is not possible. So it is preferable to directly measure the contact angle other than to calculate by the equation.

Fig. 4.26 Adhesion work

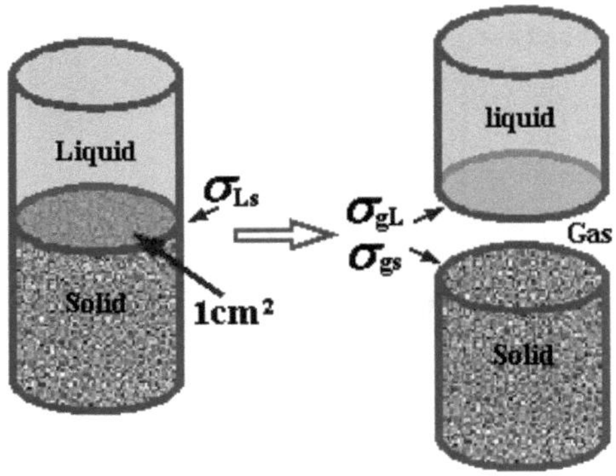

Work of Adhesion

Adhesion work is conventionally defined as the work that is required to build a unit area (such as 1 cm^2) of particular surface (Fig. 4.26). The surface energy change in this process is:

$$\Delta U_s = \left(U_{\text{new}} - U_{\text{original}}\right) \times 1 = \left[\left(\sigma_{\text{gL}} + \sigma_{\text{gs}}\right) - \sigma_{\text{Ls}}\right] \times 1 \qquad (4.20)$$

where ΔU_s is the change of surface energy, J; σ_{gs} is the interfacial tension between gas and solid, J/m^2; σ_{ls} is the interfacial tension between liquid and solid J/m^2; σ_{gl} is the interfacial tension between gas and liquid J/m^2.

Based on the concept of interfacial tension, the sum of σ_{gL} and σ_{gs} is greater than σ_{Ls}, so that ΔUs must be greater than zero, which means the increase of the surface energy. The ΔUs increment of surface energy equals adhesion work which is represented by W:

$$W = \Delta U = \left(\sigma_{\text{gL}} + \sigma_{\text{gs}}\right) - \sigma_{\text{Ls}} = \left(\sigma_{\text{gL}} - \sigma_{\text{Ls}}\right) - \sigma_{\text{gL}} \qquad (4.21)$$

where W is the work required to create per unit interfacial area, J; ΔU is the change of surface energy, J; σ_{gs} is the interfacial tension between gas and solid, J/m^2; σ_{ls} is the interfacial tension between liquid and solid J/m^2; σ_{gl} is the interfacial tension between gas and liquid J/m^2.

According to Young's Equation Eq. (4.17), can get:

$$\sigma_{\text{gL}} - \sigma_{\text{Ls}} = \sigma_{\text{gL}} \cos\theta \qquad (4.22)$$

Therefore, the relationship between contact angle and adhesion work can be obtained:

$$W = \sigma_{gL}(1 + \cos\theta) \tag{4.23}$$

Equation (4.23) indicates that the smaller the contact angel θ, the greater work of adhesion, W, which means better wettability.

There are the other standards to measure wetting degree; some (such as wetting heat) have little relationship with petroleum reservoir; and some (such as oil parallelism and water parallelism, etc.) will be discussed later in this section related to wettability measurement.

Wettability Reversal

Wettability is perhaps the most important factor that affects the oil recovery and the residual oil saturation, which is the target of enhanced oil recovery technology. Wettability controls the rate and amount of spontaneous imbibition of water and the efficiency of oil displacement by injecting water, with or without additives.

The study of the effects of wettability on oil recovery is facilitated by using additives to treat the rock surface so as to produce direct changes through all degrees of wettability from water-wet, neutral, mixed, or fractional to strongly oil-wet. In addition, water- and oil-soluble additives are used to change, or establish, a particular state of wettability.

For instance, a hydrophilic solid surface can be changed into a lipophilic surface due to the absorption of surfactant, and vice versa (Fig. 4.27a, b). The phenomenon of reciprocal transformation between water wettability and oil wettability on the surface of a solid is called wettability reversal.

Sandstones (mainly silicate) in the reservoirs are originally believed to be water-wet, so the oil on the rock surface can be easily displaced. However, the wetting characteristics often change due to the absorption of surfactant, which means the wettability reversal occurs. Now a substantial part of the reservoir sandstones is oil-wet, and oil on the surfaces is not easy to be washed off, which is

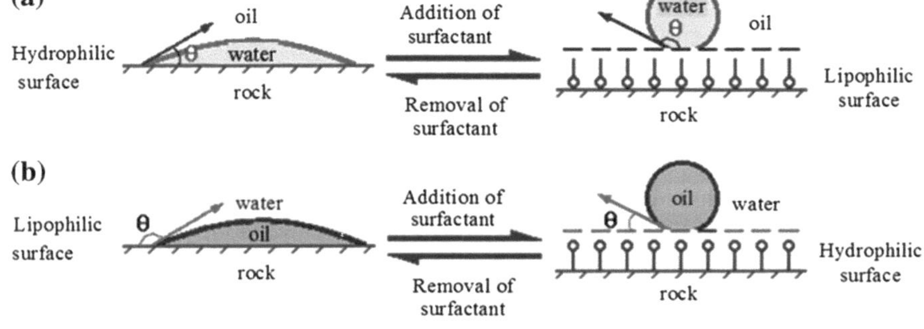

Fig. 4.27 Wettability reversal caused by the absorption of surfactant

one reason of low oil recovery. Some enhanced oil recovery methods are based on the wettability reversal. For example, the injection of active water into reservoirs, in which the surfactant adheres to the second layer by similar polarity rule, which offset the effect of original surfactant (Fig. 4.27c), and makes the lipophilic surface of sandstone convert into hydrophilic surface again. Thus, oil will be easy to be displaced and oil recovery is improved.

4.2.2 *Wettability of Reservoir Rock and Its Affecting Factors*

4.2.2.1 Types of Wettability

Reservoirs can have different types of wettability depending on the adherence of the liquid with the pore surfaces. The following are the different types of wettability [25].

Fully Water Wet

In this case water occupies the small pores and contacts the majority of the rock surface. Perfect wetting state is the extreme wettability state also known as extremely water wet rock.

Fully Oil-Wet

In oil-wet system the position of fluid is reversed from as in case of water-wet rock. In this case, oil occupies the smaller pores and spreads over the majority of the rock surface, while the water occupies the larger pores.

Intermediate or Neutrally Wet

This state exists when the rock has no strong preference for either oil or water.

Fractional Wettability

Fractional wettability means that scattered areas throughout the rock are oil-wet, whereas the rest of the area is water-wet. Fractional wettability also known as heterogeneous, spotted, or Dalmatian wettability. It occurs when the surfaces of the rocks are composed of many minerals that have very different surface chemical properties, leading to variations in wettability throughout the internal surfaces of the pores. In fractional wettability, crude oil components are strongly adsorbed in certain areas of the rock, so a portion of the rock is strongly oil-wet, while the rest is strongly water-wet.

The crude oil components which are adsorbed on rock surface are heavy oil components, known as surface-active components [26], like asphaltenes, which can precipitate due to several reasons like, depletion and adding a large volume of low molecular weight hydrocarbon to the crude oil, are some of them.

Partial wetting also is known as porphyritic (spot, patch) wetting, which means oil-wet or water-wet surface without a specific location. In terms of a single pore, part of the surface is strongly water-wet, the rest is strongly oil-wet, and oil-wet surface may be not continuous. Partial wetting "core" of mixing water-wet particles (such as glass balls) and oil-wet particles (such as Teflon balls) in a certain proportion is used in experiment (Fig. 4.28a).

The term "mixed wettability" commonly refers to the condition where the smaller pores are occupied by water and are water-wet, whereas the larger pores of the rock are oil-wet and a continuous filament of oil exists throughout the core in the larger pores (Fig. 4.28b).

4.2.2.2 The Factor Affecting Wettability of Reservoir Rock

Is the surface of reservoir rock water-wet or oil-wet? All petroleum reservoirs were originally believed to be water-wet because of the sedimentary environment and most minerals are water-wet. After analyzing the composition of crude oil, it is found that crude oils contain surface-active compounds that can alter the wettability of the reservoir. The surface-active compounds present in the crude oil will move to the fluid and rock interfaces and governs the wettability of the reservoir. More and more researches show that many rock surfaces are heterogeneous wetting, i.e., fractional wettability.

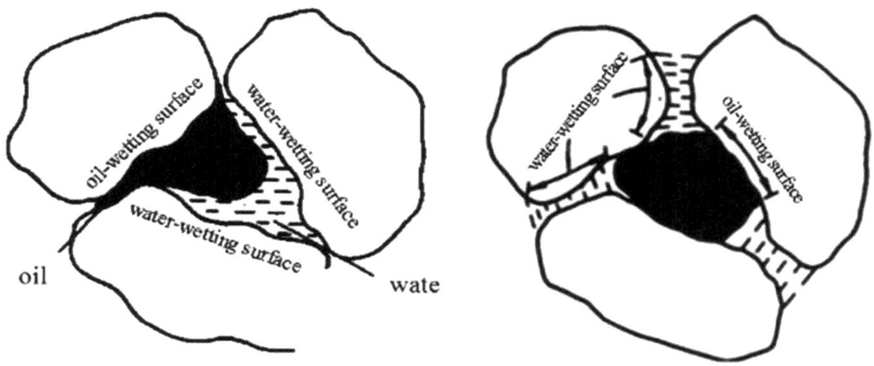

(a) Speckled wetting; (b) Mixed wetting

Fig. 4.28 Schematic diagram of speckled wetting and mixing wetting [27]

Why the wettabilities of different beds in a reservoir or a section in lateral will change? Do the surfaces of different particles in a rock show such heterogeneity in macro as well? The reason is that various factors will affect wettability. The factors are described as follows:

Mineralogical Composition of the Rock

The reservoir rocks are mainly sandstone and carbonate rocks. The composition of the latter is relatively simple, mainly calcite and dolomite. Sandstone is composed of silicate minerals with different crystals and different nature, such as feldspar, quartz, mica, clay minerals, and sulfate. Because of the diversity of the mineral composition, the sandstone wettability is more complex than the carbonate.

According to the values of contact angle of water droplets on solid surfaces, rocks can be classified into two groups. One group often has the contact angle less than 90° and are called water-wet rocks, including quartz, silicates, glass, carbonate, and aluminosilicate. Another group, such as hydrocarbon organic solids and the metal sulfides in minerals, etc., has the contact angle larger than 90° and are called oil-wet rocks. Most mineral composition of reservoir rock is water-wet, but the degrees of water wettability are different. The most hydrophilic mineral is clay composed of hydromica, the other minerals arranged by the strength order of hydrophilic are as follows: quartz, limestone, dolomite, feldspar.

Clay minerals have significant influence on the wettability of rocks. Some clay minerals, especially montmorillonite is hydrophilic, and argillaceous cement will increase the hydrophilicity of rock. Some clay minerals contain iron such as oolitic chlorite clay ($Fe_3AL_2Si_2O_{10} \cdot 3H_2O$), and iron is able to adsorb surfactants from crude oil. Iron can make partially rock surface become oil-wet when it covers the surface of rock grains. All these show that different minerals have different wettability, while the reservoir rocks have a wide range of sedimentary resource, and mineral components are very complicated, so there exists significant difference between rocks both in macroscopic and microscopic.

The Effects of Reservoir Fluid Composition

The impact of reservoir fluid composition on wettability includes: (1) the main components, different hydrocarbons (nonpolar) in crude oil; (2) the polar substances (a variety of compounds of O_2, S, N_2) contained in crude oil; (3) active substances in crude oil.

As to the non-polar hydrocarbon system in crude oil, different hydrocarbons have different carbon atom numbers, and show different nonpolarity. The advanced contact angles between polytetrafluoroethylene (PTFE) and pentane, hexane or octane are illustrated in Table 4.3. It is shown that the contact angle increases as the number of carbon atoms increases.

Table 4.3 Advancing angle of different hydrocarbons on PTFE surface

Hydrocarbon	Pentane (C_5H_{12})	Hexane (C_6H_{14})	Octane (C_8H_{18})	Twelve alkyl ($C_{12}H_{26}$)
Advanced angle(°)	0	8	26	42

Crude oil also contains polar substances in different extent besides hydrocarbon non-polarity. Polar compounds have certain impact on the mineral wettability; some can completely change the wettability of rock and make wettability reversal. Sometimes the affection is little, which depends on the nature of polar substances. The contact angle is changed when oil compositions are different in the same surface of quartz, as illustrated in Fig. 4.29. The contact angle may be greater than 90° or less than 90°.

This indicates that oil of different properties on the same surface demonstrate different wettability. Asphalt, a kind of polar substances, is easy to attach on the rock surface and make the surface change into oil-wetting. Its adsorption ability is so strong that the conventional cleaning cannot remove it. Johansen and Dunning [28] found that asphaltenes were responsible for changing some crude oil/water/glass systems from water-wet to oil-wet. The system was oil-wet when the crude was used but water-wet when the deasphalted crude was used. The addition of a very small amount (0.25 %) of the whole crude to the deasphalted crude restored the system oil-wetting.

The Effects of Surface-Active Compounds

The impact of surfactants on rock wettability is more significant than the above-mentioned polar substances. A certain amount of adsorption of active agent on huge rock surface makes wettability reversal. At present, the addition of surfactant in water flooding to reduce oil–water interfacial tension and change wettability is an effective method to improve oil displacement efficiency. The surfactants in formation water are adsorbed on rock surface and the number of adsorption will reduce with the increase of dissolved electrolytes. In addition, some metal ions in water can also change the wettability. Adding 10 mg/L copper ion in water will make some crude oil wettability change into lipophilicity from hydrophilicity.

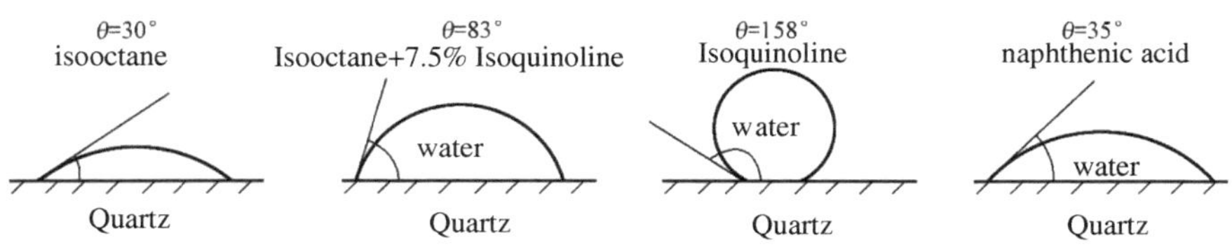

Fig. 4.29 Contact angles of different oil and water systems on quartz and calcite surface

In a word, the wettability of rock is the result of interaction of the matrix composition and formation fluid composition, and wettability is synthetic performance of rock and fluid instead of matrix's nature.

The Effects of Surface Roughness of Mineral

According to the assumptions in deriving Young's equation, smooth solid surface and uniform surface energy (interfacial tension), the measurement of contact angle requires smooth rock surface. But rock surface is rough and surface energy is nonuniform in actual reservoirs so that rock wettability is different in different positions, and so the rock displays partial wetting. In particular, the protruding parts and sharp edges and corners of mineral particles have a special impact on wettability. Many experiments show that sharp edges are difficult to overcome for wetting perimeter because wetting is blocked as wetting perimeter reaches the edges and corners. At the moment, the determination of the contact angle should take the shape angle τ into account (Fig. 4.30). The bigger shaped angle τ is, the greater resistance of corners for three-phase perimeter moving along surface is. As a result, the wetting angle measured at this time is not true.

The research on the effects of pore structure, temperature, and pressure on wettability are relatively little, but some studies have shown that temperature and pressure have little impact on wettability.

4.2.3 Wetting Hysteresis

Wetting hysteresis, a phenomenon in the process of water flooding, directly affects the shape and position of capillary pressure curves and relative permeability curves in different displacement processes (details will be discussed in the last two sections of this chapter). Wetting hysteresis is discussed alone in this section.

As shown in Fig. 4.31, the oil–water–solid interface (A and B) cannot move forward immediately when the original horizontal solid surface is set at an inclined angle α, but only the oil–water interface deforms, so that the original contact angle θ changes. At point A, advanced contact angle θ_1 forms due to part of original space of oil occupied by water, and $\theta_1 > \theta$; while at point B, oil displace water to form receding angle θ_2 forms, and $\theta_2 < \theta$.

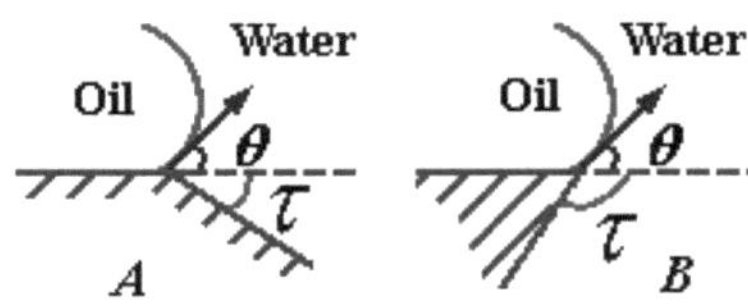

Fig. 4.30 Relationship between wetting angle θ and shaped angle τ

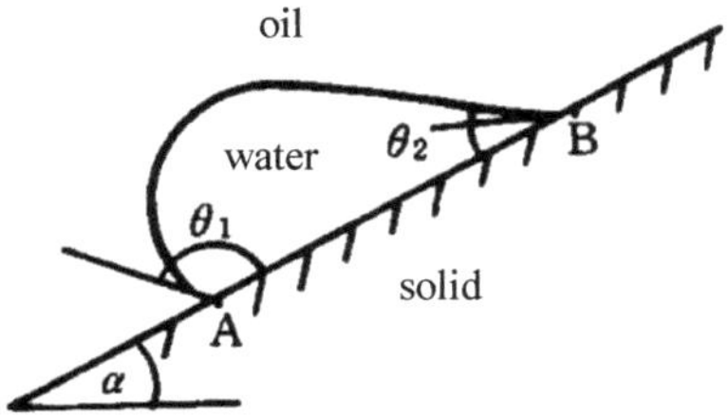

Fig. 4.31 Advanced angle θ_1 and receding angle θ_2 caused by wetting hysteresis

Wetting hysteresis is defined as the difference between advancing and receding contact angles. According to different causes, wetting hysteresis can be divided into static hysteresis and dynamic hysteresis.

4.2.3.1 Static Wetting Hysteresis

Static wetting hysteresis is the phenomenon caused by different order of oil and gas contacting with the solid, i.e., oil first or water first contacting with the solid. For example, when a water droplet is dropped on the surface of a mineral immersed in an oil, the contact angle, $\theta_{w,}$ of this system is found to be always greater than that angle, θ, in the reverse order. This phenomenon is also called *contact angle hysteresis*.

In general, θ_1 is advanced angle and θ_2 is receding angle, the difference $\Delta\theta = {}_1 - \theta_2$ expresses the degree of contact angle hysteresis ($\Delta\theta$ can sometimes be greater than 60°). It should be taken into consideration when measuring contact angle that θ obtained after equilibrium is the right one.

There are three causes leading to contact angle hysteresis: surface roughness; surface heterogeneity; immobility of surface macromolecular scale (such as the adsorption layer of surfactant on solid surface).

It, in general, is more difficult for three-phase boundary to move, and wetting hysteresis is more obvious, as the surface roughness and heterogeneity of rock grains increase.

The adsorption layer on rock surface formed by surfactant in oil greatly enhances wetting hysteresis, and wetting hysteresis reaches strongest when the adsorption layer is fully saturated. That is, because the closer the oil and water to the rock surface is, the greater the flowing resistance is. For example, if the rock surface is water-wet, the adsorption of surfactant in crude oil will affect the moving speed of oil film, and the more firmly oil film attached on the pore surface, the more difficult oil film moves, and the stronger hysteresis is. Studies have shown that contact angle is one of the main causes that lead to capillary force hysteresis.

4.2.3.2 Dynamic Wetting Hysteresis

Dynamic wetting hysteresis is the change of wetting angle caused by the lag of three-phase boundary moving along solid surface in displacement of oil by water or water by oil. Take the water-wet capillary, for example, which is similar in the reservoir (Fig. 4.32).

In static equilibrium, the contact angle is θ; three-phase boundary will not move immediately while water displaces oil, then the meniscus deforms and contact angle increases to θ_1; the contact angle will reduce to θ_2 while oil displaces water. Similarly, the contact angle formed in displacement of oil by water is called advanced angle or increasing angle θ_1; and the contact angle formed in displacement of water by oil is called receding angle or reducing angle θ_2. When three-phase boundary stops moving, the contact angle measured is called equilibrium angle θ. The relationship between them is $\theta_1 > \theta > \theta_2$.

As shown in Fig. 4.33, the value of advanced angle (or receding angle) is not a constant but related to moving speed of the three-phase boundary. The advanced angle increases (or receding angle decreases) with increase of moving speed of meniscus in porous channel. If the surface of porous channel is water-wet and meniscus moves at a low speed, the advancing angle θ_1 may be less than 90° at first, but when the moving speed of meniscus increases to a certain degree, the advancing angle θ_1 may be greater than 90° which indicates wettability inverse. This means

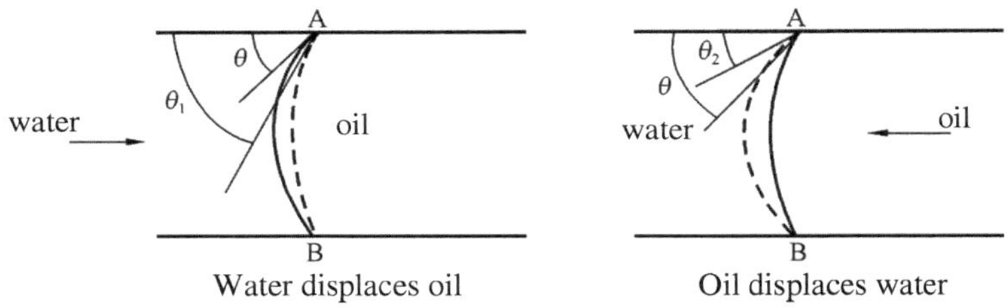

Fig. 4.32 Dynamic wetting hysteresis in pores

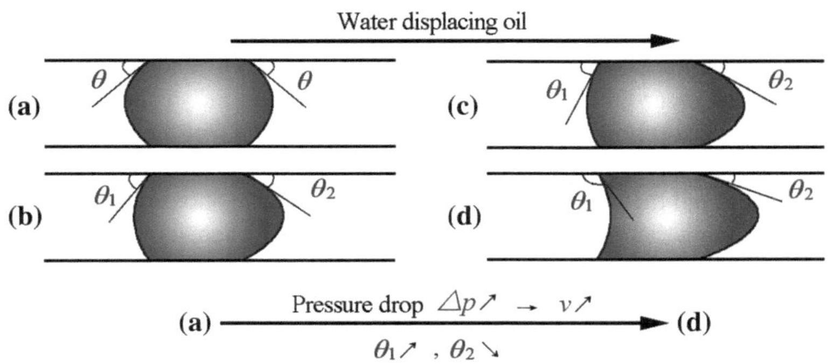

Fig. 4.33 Dynamic wetting hysteresis and wettability reversal

that the reservoir rock is water-wet under static or low speed condition, but if the displacement velocity is too large, the moving speed of the meniscus will exceed the critical velocity at which water is the wetting phase. Then the contact angle becomes greater and wettability reversal occurs. There will be an oil film left on rock surface after water flooding, which is harmful to oil recovery.

4.2.4 Effect of Rock Wettability on Water Flooding Process

Studies have shown that rock wettability has an impact on water flooding process; and rock wettability can be changed in turn after longtime water flooding.

The effect of wettability on water flooding is complicated. Three main aspects will be discussed here: (1) wettability determines the microscopic distribution of oil and water in rock porous channels and the distribution of the residual oil; (2) wettability determines the value and direction of capillary pressure in porous channel; (3) wettability of rock grains has an impact on the fines migration pattern.

Wettability determines the microscopic distribution of oil and water in rock porous channels.

The distribution of oil and water in porous media depends on the rock wettability. The water-wet part of rock surface is covered by water film, while the oil-wet part of rock surface is covered by oil film. Oil–water distribution in porous channels is illustrated in Fig. 4.34.

Under the function of various interfacial tensions in the porous media, wetting phase always tends to adhere to grain surface, and tries to occupy the narrow pores and corners by displacing the non-wetting phase to the middle of pore space. This fluid distribution occurs because it is the most energetically favorable. For a strongly water-wet rock initially at irreducible water saturation, any oil placed in the small pores would be displaced into the center of the large pores by spontaneous water imbibition because this would lower the energy of the system.

It is illustrated that water will adhere to grain surface if the surface is water-wet. Figure 4.35a represents that water will surround the contact point between grains to form pendular rings, called annular distribution, if water saturation is low.

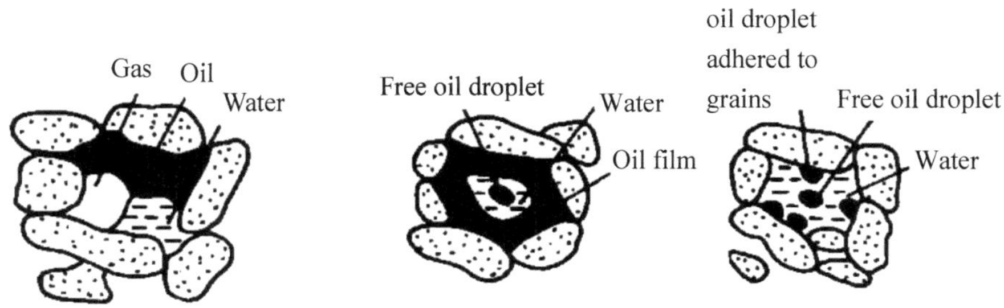

Fig. 4.34 Distribution of oil–water in porous channels

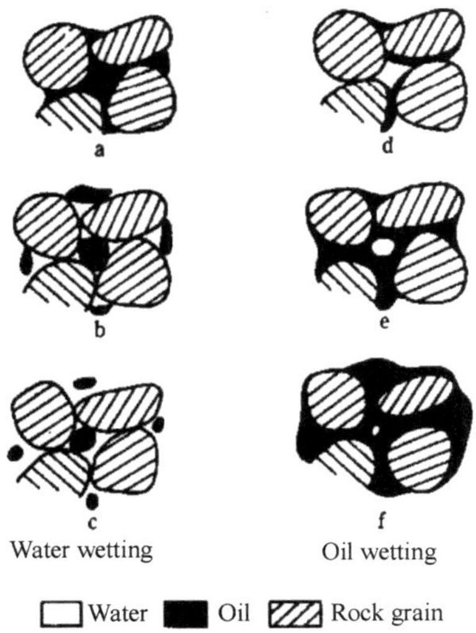

a :water-pendular, oil-funicular; b :water, oil-funicular;c :water-funicular, oil-solitary; d :water-funicular, oil-pendular;e :water, oil-funicular; f :water-solitary, oil-funicular

Fig. 4.35 The sketch of oil and water distributions in water-wet and oil-wet rocks pores

The water rings cannot contact with each other due to low water saturation, therefore, water cannot flow and exists in irreducible water state. Meanwhile, oil is in the middle position of pores in a continuous funicular distribution state, and will form channel flow due to the pressure difference. It is noted in Fig. 4.35b that as water saturation increases, water rings will increase until they connect together and become a kind of connate water. The pressure difference determines if the connate water can flow. If water saturation is greater than connate water saturation, then the water will form funicular flow. It is illustrated in Fig. 4.35c that oil will be not continuous any more with the increase of water saturation, and breaks into oil droplets, which is called solitary droplet distribution. Although oil droplets can be transported by water flow, but it can be easily blocked at the pore restriction, which will cause additional flow resistance. The water/oil displacement experiment of strongly water-wet cores carried by Moore shows that a considerable part of non-wetting phase is confined to branch structure, especially if oil saturation is close to S_{or}. Then the oil of branch system is trapped and isolated due to capillary force when the injected water bypasses the branch. In addition, the pore heterogeneity makes S_{or} increase as well, for the heterogeneity causes more captured oil.

When rock surface is oil-wet, the distribution of water–oil is contrary to the above, as shown in Fig. 4.35d–f.

The observation of water–oil flow in artificial porous medium indicates that oil and water flow through their own channel network separately, but the two channel networks are connected with each other at some points (Fig. 4.36).

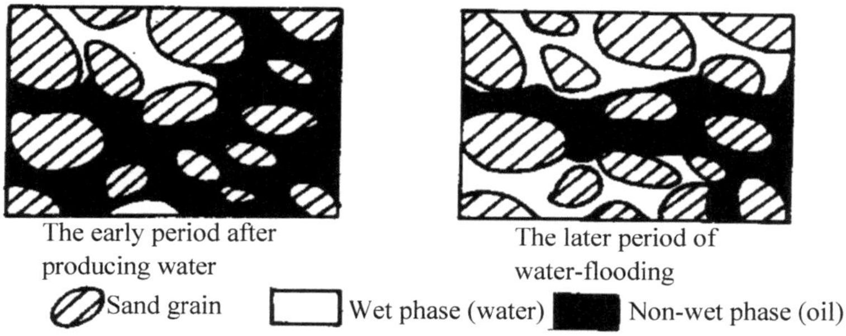

Fig. 4.36 Micrographs of flow regime (non-wetting phase displaces wetting phase)

The diameters of these channels may vary from less than a sand grain diameter to many times of a sand grain diameter. These channels are always surrounded by liquid–liquid interfaces or liquid–solid interfaces to form complex networks. The geometry of the flow channel network will also change when fluid saturation changes. In general, with water saturation increasing, the number of water flow channel increases; meanwhile, the number of oil flow channel reduces. Because of pore structure's connectivity, each flowing channel could find its own connective channel. In addition, the liquid flow in these channels observed is almost laminar flow.

What is the oil and water distribution during water injection? As for water-wet rock, with the increasing of water saturation, water is not only adheres to grain surface and corner, it also displaces oil through channels, yet residual oil occupies dead-end pores and fine connected throats, and some oil drops are surrounded by water. If reservoir rock is oil-wet, the non-wetting phase, water, will preferentially enter into large pores. With continuous injection, water will gradually flow into the small channels, and make narrow channels connected, then form new water flow channel. Some residual oil stays in small oil flow channels, and the other forms oil film on large pore wall. These oil films have high flow resistance, and it is difficult to be displaced by water. If rock is moderate hydrophilic, residual oil will occupy dead-end pores, and many oil droplets will adhere to the rock surface.

The above analysis shows that the distribution of oil and water in pores is related to not only oil and water saturation but also the direction of saturation change (wetting phase displaces non-wetting phase or non-wetting phase displaces wetting phase), which is shown in Fig. 4.37.

The process of non-wetting phase displacing wetting phase is usually called displacement process. The wetting phase saturation reduces and non-wetting saturation increases during this displacement process. The process of wetting phase displacing nonwetting phase is usually called imbibition process. The wetting phase saturation increases during the imbibition process. The distributions of oil and water in displacement process and imbibition process are, respectively, illustrated in Fig. 4.37a, b. Because of different saturation history, the oil and water distributions in pores are different even with same water saturation. It is illustrated in Fig. 4.37a

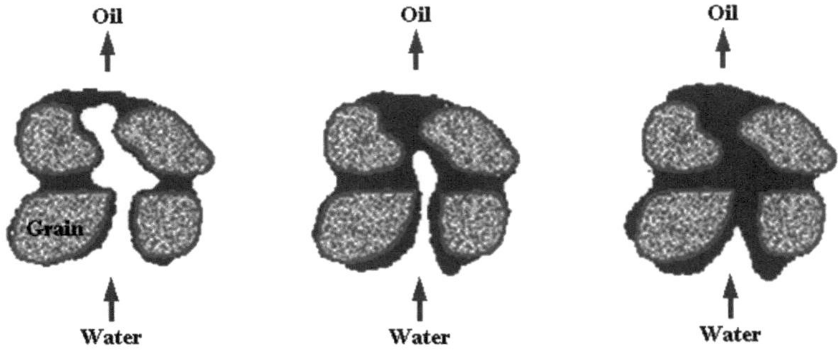

(a) The displacement process of water flooding in lipophilic rock

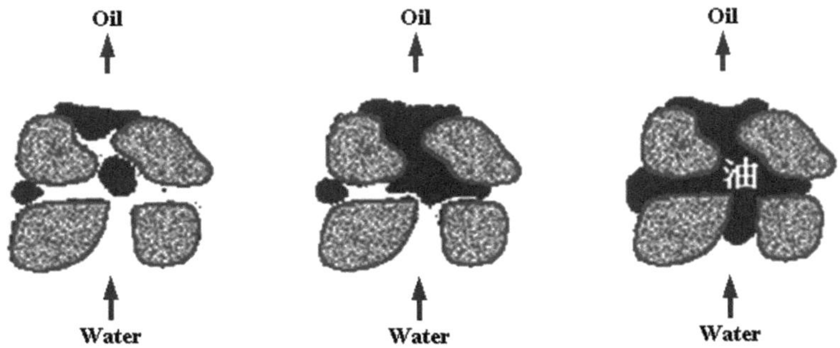

(b) The imbibition process of water flooding in hydrophilic rock

Fig. 4.37 The effect of wetting sequence on water injection during water flooding

that rock is saturated with oil first, then water is injected, so the saturation history is oil–water. It is noted in Fig. 4.37b that rock is saturated with water first, then oil displaces water, and a layer of water film is left on grain surface, then the oil is displaced by water, so the saturation history is water–oil–water. Saturation history represents the order of saturation process from the original to current state, wherein including static wetting hysteresis, for static wetting hysteresis is the hysteresis phenomenon caused by different wetting sequences.

4.2.4.1 Effect of Wettability on Capillary Pressure

There are water-wet pores and oil-wet pores in reservoir due to the difference of wettability. Contact angle is different with different wettability, and the shape and curvature of the interface is also different, which causes capillary pressure with different direction, as shown in Fig. 4.38. It is obvious that in water-wet capillary, the direction of capillary pressure p_c is consistent with the direction of displacement

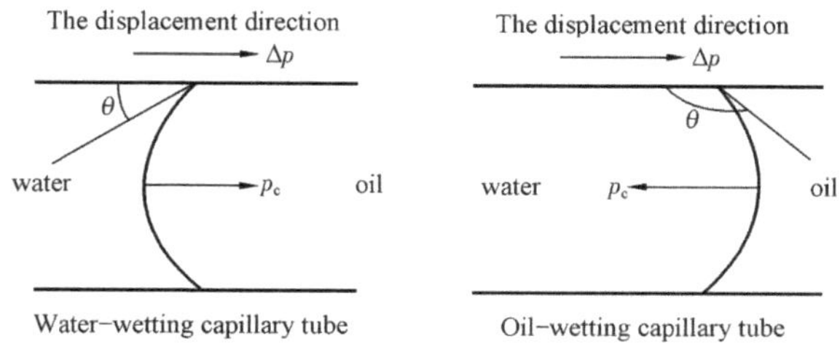

Fig. 4.38 Direction of capillary force in different wetting pores

pressure difference Δp, and p_c is driving force. Contrarily, in oil-wet capillary, the direction of capillary pressure P_c is opposite to the direction of displacement pressure difference Δp, and p_c is resisting force. The flow resistance directly affects the flow ability of oil and water. Capillary force plays an important role in displacing oil when production or injection pressure drop is very small.

4.2.4.2 Wettability Affects the Fines Migration in the Formation

During early stage of oilfield development, only oil flows in formation and only oil is produced while irreducible water does not flow. Water-wet particles protected by irreducible water film is not involved in oil flow, as illustrated in Fig. 4.39, and there is no fines migration in formation.

Once water flooding is started, oil and water can flow simultaneously when water displaces oil, and then particles protected by irreducible water also begin to flow. As long as water and oil flow simultaneously, water-wet particles will flow away with water without bridging at the pores restriction (as shown in Fig. 4.40).

For the mixed wettability rock, because of part of the grain surface is oil-wet while other part is water-wet (i.e., porphyritic wettability), the fine is on the

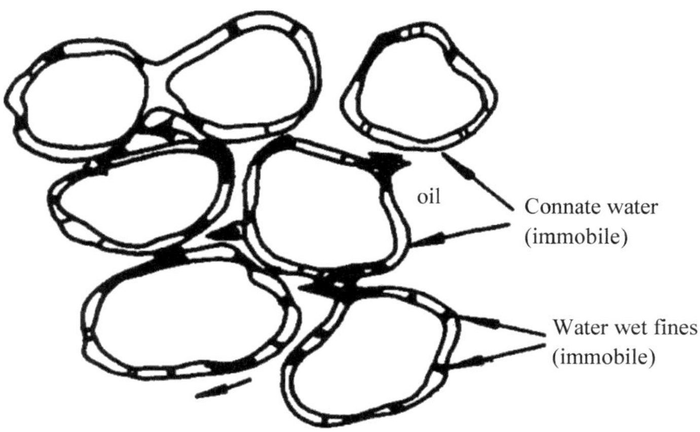

Fig. 4.39 Water-wet fines is immobile when water phase is immobile (Muecke [29])

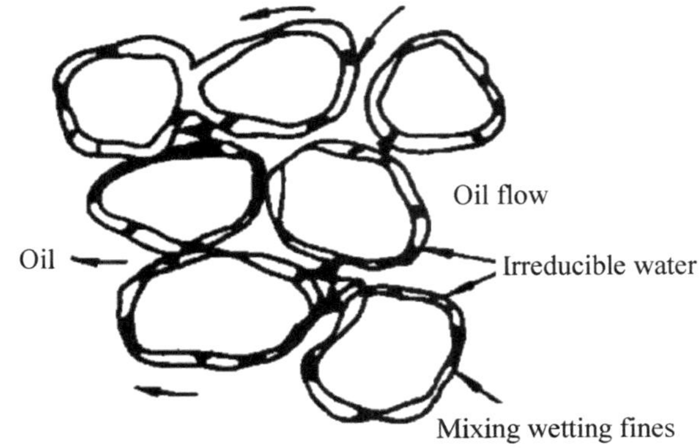

Fig. 4.40 Water-wet fines will not bridge at the narrow pores during two phase flow (Muecke [29])

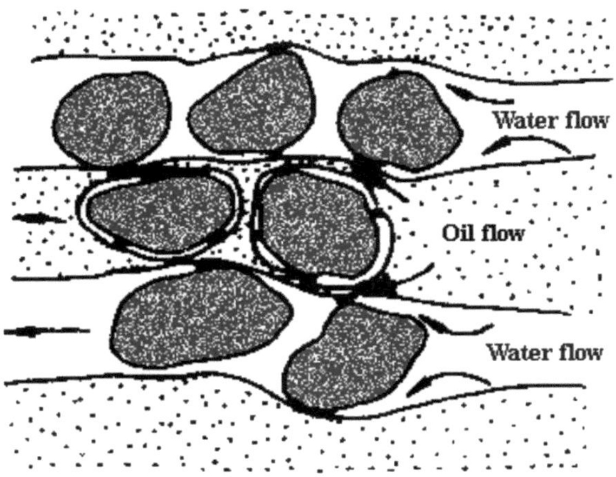

Fig. 4.41 Fines migration in the mixing wet system

interface of irreducible water film and oil phase, thus its movement is directly affected by both oil flow and irreducible water film. These fines cannot be carried away by oil flow, but just move in different directions on the interface. It is shown in Fig. 4.41.

Figure 4.42 shows the effect of oil–water mutual solvent on fines migration. With injection of mutual solvent or surfactant solution, Interface tension is reduced and even water–oil interface is disappeared so that the formation fines controlled by interface tension and wettability are released, causing them to migrate at high concentrations as shown in Fig. 4.42. The results are favorable for removing bridge plug, but a large number of released fines will also cause blockage and even damage reservoir. Therefore, it must be careful to use oil–water mutual solvent, and the current fines distribution in reservoir must be taken into consideration.

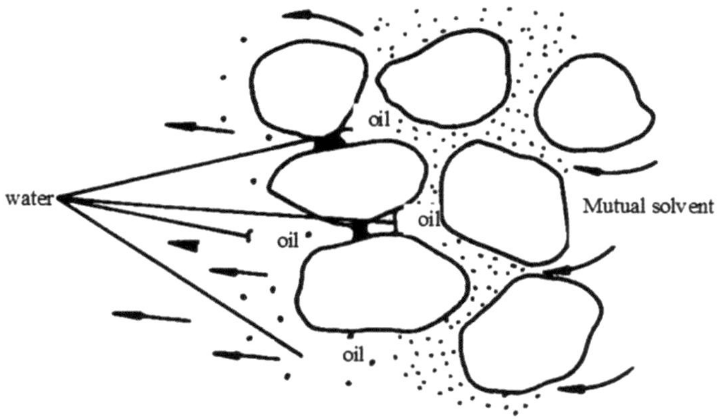

Fig. 4.42 Released fines migration controlled by interface tension and wettability during injection of chemicals for improved oil recovery

4.2.4.3 Effect of Wettability on Recovery

The oil recovery and displacement efficiency are determined by the wettability of the reservoir rock to a large extent during water flooding. Different wettability will lead to different oil recovery even when same pore volume (PV) water is injected into reservoir (Fig. 4.43).

When the PV is 0.7, the recovery of strongly oil-wet ($\theta = 180°$) rock is 37 %, but that of strongly water-wet rock is 57 %. It is generally acknowledged that water-wet reservoir is more effective than oil-wet reservoir in homogeneous wetting system, which indicates that the sweep region of water is larger in water-wet reservoir than that in oil-wet reservoir, and the production of oil on grain surface will be more if wetting of water works fully. Therefore, it is important to inject strongly water-wetting water. Besides, it should be taken into account that the effect of dynamic wetting hysteresis, oil, and the complexity of rock surface properties on wetting capacity of water.

Long-term injection will cause rock wettability change. The study on sealed coring from water flooded region in Daqing oil field shows that the wettability of most rock surface becomes slightly water-wet from slightly oil-wet as water saturation in reservoir is more than 40 %; and the wettability of all the rock is altered into water-wet as water saturation in reservoir is more than 60 %.

That is mainly because that after long-term flushing by injected water, the oil films on the rock surfaces are washed off and feldspar and quartz surface show their original water wet property. This wetting alternation is useful to improve water flooding efficiency, and has great practical significance to improve water injection efficiency and tertiary recovery after water flooding.

Warren and Calhoun studied the ultimate recovery of different wettability of artificial core after injecting a certain volume (20 PV), and drew a different conclusion that maximum recovery will be reached under the condition of

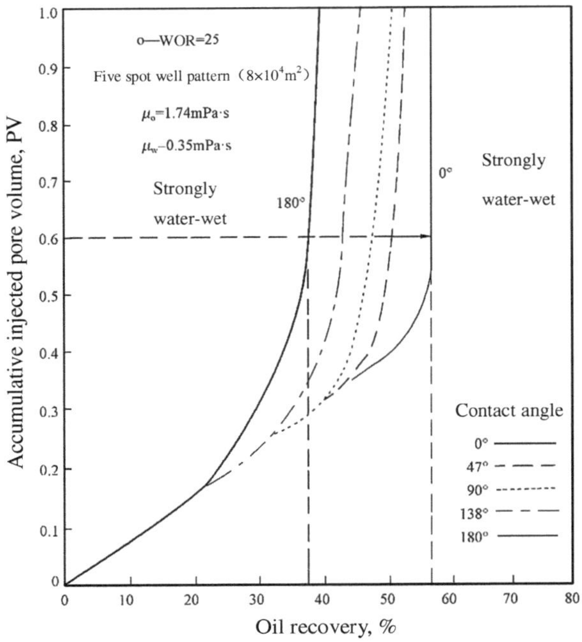

Fig. 4.43 The effect of wettability on oil recovery under water injection (Owens and Archer [30])

approximately intermediate wetting. It is because that the interfacial tension, which causes noncontinuous oil phase and trapped oil, is the minimum under that condition. In the strongly water-wet system, water prefers to flow in small pores, and bypasses some oil in large pores. Besides, large interface tension easily blocks the oil flow. In the strongly oil-wet system, water has a trend of fingering in large pores, and bypasses some oil. However, little water will bypass and trap oil when the rock is intermediate wettability.

Wettability has significant impact on oil recovery and many researchers still commit to study the relationship between oil recovery and wettability.

4.2.5 Measurement of Wettability

Evaluation of relative water/oil wetting of porous rocks is a very important aspect of petroleum reservoir characterization.

Changes in wettability have been shown to affect capillary pressure, relative permeability, waterflood behavior, dispersion of tracers, simulated tertiary recovery, irreducible water saturation (IWS), residual oil saturation (ROS), and electrical properties 23. Wettability has a decisive influence on oil production rates, the water/oil production ratio after water breakthrough, the oil production rates of enhanced oil production technologies, and the residual oil saturation of a reservoir at abandonment.

A large amount of research has therefore been conducted on wettability. Many different methods [31] have been proposed for measuring the wettability of a system, based on the observable characteristic interactions of water, oil, and rock. They include quantitative methods—contact angles, imbibition and forced displacement (Amott), and USBM wettability method—and qualitative methods—imbibition rates, microscope examination, flotation, glass slide method, relative permeability curves, permeability/saturation relationships, capillary pressure curves, capillarimetric method, displacement capillary pressure, reservoir logs, nuclear magnetic resonance, and dye adsorption. They are based on different principles, and every method has its advantages and disadvantages. These methods can be divided into two classes: direct measurement; indirect measurement. Two methods will be mainly introduced as follows.

4.2.5.1 Direct Measurement Contact Angle Method

The contact angle is the best wettability measurement method when pure fluids and artificial cores are used because there is no possibility of surfactants or other compound altering the wettability. The method is also used to determine whether a crude oil can alter wettability and to examine the effects of temperature, pressure, and brine chemistry on wettability. They include the tilting plate method, sessile drops or bubbles, vertical rod method, tensiometric method, cylinder method, and capillary rise method.

The method of measuring contact angle is mostly used to determine the wettability of a system, among which sessile drop method is simplest and most practical. A smooth, homogeneous surface is necessary for this test; a polished mineral surface is generally used to make contact angle measurements of water–oil system. Figure 4.44 illustrates the procedure where the polished mineral plate is suspended in the oil (or water) and a drop of water (or oil) whose diameter is about 1–2 mm is then placed on the top of the plate. Then use a certain optical instrument or microscope to amplify the drop and take a picture of the drop. The contact angle can be measured directly in the picture. Advanced angle and receding angle can be also measured if we incline the mineral plate or change the volume of the liquid drop.

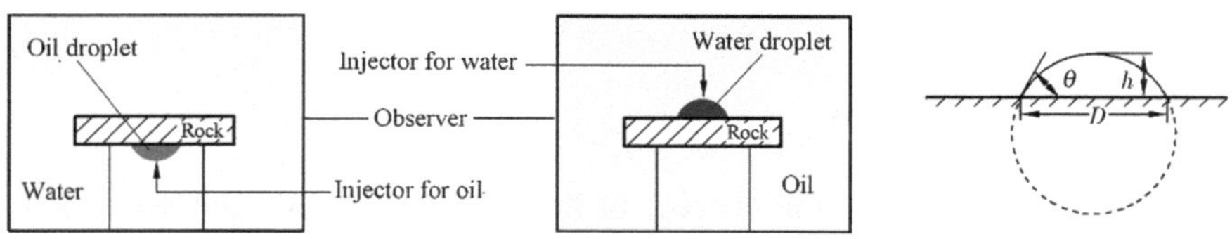

Fig. 4.44 Illustration of contact angle method [32]

To obtain a representative and real result, the fluid sample used in this method should be directly taken from reservoir. If it is not possible, simulated crude oil and formation water prepared based on formation water data can be used as well. However, mineral in rock can only be represented by polished mineral crystal. At present, the surface of sandstone and carbonate are usually replaced by the polished surface of ice land spar and calcite, respectively.

Mineral and fluid can be placed in cell of high pressure, then measurement can be carried under different formation temperature and pressure conditions in order to make the measurement condition to be closer to reservoir temperature and pressure. Generally, the effect of temperature on wettability is relatively large, and the effect of pressure is relatively small.

This method is simple and straightforward, but there are several problems: The measuring conditions are too strict or the contact angle will be inaccurate. Such as the mineral surface is required to be quite smooth and clean; temperature is required to be accurate, or slight error will affect measurement; the time for drop being stable is too long, may be several days or even several months. It will cause great error if stable time is insufficient. This method cannot directly measure the contact angle for reservoir rock. The mineral plate used in experiment contains main components of reservoir rock, but still not real reservoir rock. It is inevitable to lead to derivation by using single mineral to simulate the complex reservoir rock. So this method can only be used to evaluate qualitatively reservoir wettability.

However, many studies have shown that rock wettability is spot wetting, that is, rock wettability is heterogeneous both in macro- and microscopes. Surface of a same core may show different wettability. The method of measurement of contact angle cannot reflect spot wetting and mixing wetting, therefore, various indirect methods are developed on the basis of direct method.

4.2.5.2 Wilhelmy Plate Method

In recent years, Wilhelmy introduced a new method to solve the above problems. This measuring device is basically the same as Wilhelmy plate method of measuring interface tension (Fig. 4.9). The main difference is that the data recorder of tensiometer is connected to a computer so that the instantaneous value of the contact angle can be calculated. Oil–water interface tension, σ_{ow}, is measured first, in this method. As the plate is moved into and out of a liquid, the change in force, ΔF, due to the adhesion tension, can be measured. Then the values of cos θ and θ can be calculated by Wilhelmy equation.

4.2.5.3 Indirect Method

This method is based on the concept of "relative wettability" and spontaneous imbibition. If there are two kinds of fluids, one must have a better affinity for rock than the other, which means that the fluid with good affinity will penetrate into rock and displace the fluid with poor affinity. Therefore, the concept of relative wettability is the relative wetting ability of fluids.

Spontaneous imbibition method and imbibition centrifugation methods are used to measure the automatic substituting ability of two fluids. These two methods have been widely applied in domestic oil field.

The cores used in the experiment must be fresh that can represent the reservoir original condition or current condition. Oil and water should also represent the actual condition. For example, we can dilute the crude oil with neutral kerosene to prepare the oil sample for the experiment whose viscosity is similar to in situ crude oil.

Spontaneous Imbibition Method

The procedure for spontaneous imbibition method is as follows:

Immerse the core saturated with oil into the water, as shown in Fig. 4.45a. If the core is water-wet, the oil will be displaced by spontaneous imbibition of water due to the existence of capillary pressure. The displaced oil will occupy the top space of the instrument, and the volume of displaced oil can be directly read from the measuring scale. Spontaneous imbibition of water represents that the core is water-wet. Conversely, immerse the core saturated with water into the oil; if water is displaced by spontaneous imbibition of oil, then the core is oil-wet. The displaced water is collected at the bottom of the instrument, and the volume of displaced water can be directly read from the measuring scale.

Fig. 4.45 The device for measuring wettability of imbibometry

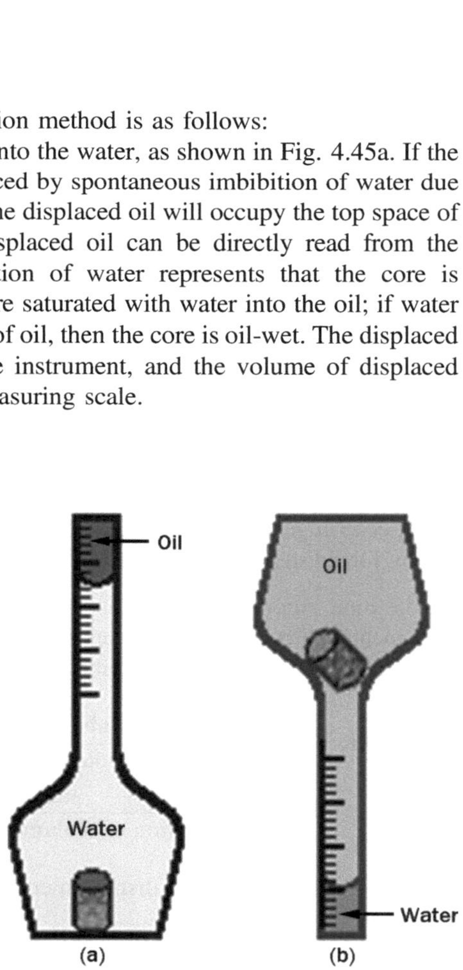

(a) Core saturated with oil

(b) Core saturated with water

Because of heterogeneous wettability of rock, a core is usually used to do both the water imbibition and oil imbibition experiments. Generally, rock is thought to be water-wet if water imbibition volume is more than oil imbibition volume, conversely rock is oil-wet. Water imbibition volume equals to oil imbibition volume means neutral wettability.

Spontaneous imbibition method is simple and good method for measuring wettability, its result is relatively close to real condition of reservoir. But this method has a disadvantage that it can only determine relative wettability. In addition, it is should be noticed in experiment that the pollution degree affects wettability greatly, so it is the key to ensure that the core is not contaminated during the processes of sampling and making sample, and improve measurement accuracy under the formation temperature and pressure condition.

Imbibometry centrifugation method is developed for measuring relative wettability quantitatively or semi-quantitatively, also known as Amott method.

Imbibition Centrifuge Method

After the imbibition experiment, a centrifuger is used to drive the remaining movable fluid in the core, and then the total amount of movable fluid in the core can be obtained. Based on the ratios of different volumes, the wettability can be judged.

The procedure for this method is as follows:

(a) The test begins at the residual oil saturation; therefore, the fluids are reduced to S_{or} by forced displacement of the oil.
(b) The core is immersed in oil for 20 h, and the amount of water displaced by spontaneous imbibition of oil, if any, is recorded. This volume also represents the amount of oil imbibition.
(c) A centrifuger is used to displace the water in the core, and the amount of water by forced displacement is also recorded.

The total amount of water displaced (by imbibition of oil and by forced displacement) represents the total capillary volume in oil flooding. Wettability can be determined by the ratio of oil imbibition to total capillary volume. That is as follows:

$$\begin{cases} \text{WPR} = \dfrac{\text{imbibition water volume}}{\text{imbibition water volume} + \text{forbicry imbibed water volume}} \\ \text{OPR} = \dfrac{\text{imbibition oil volume}}{\text{imbibition oil volume} + \text{forbicry imbibed oil volume}} \end{cases} \tag{4.24}$$

where WPR means Water displacement ratio and OPR means oil displacement ratio.

A value of WPR approaching 1 indicates a strongly water-wet sample, which means water can automatically imbibe into the sample even without centrifugal force. Therefore, the WPR is close to 1 and OPR is close to 0, in terms of the sample that preferentially water wet; it is opposite to sample that preferentially oil-wets, whose OPR is close to 1 and WPR is close to 0; and wettability is neutral if WPR and OPR are close.

It seemed that imbibition centrifuge method is more effective to evaluate relative wettability of core quantitatively or semi-quantitatively than spontaneous imbibition method.

However, centrifuger is not used widely in China, and imbibitions displacement method is recommended to measure rock wettability.

Imbibition Displacement Method

The principle of this method is similar with imbibition centrifuge method, while the difference is that core is placed in core holder and displaced by pressure difference, instead of being displaced by centrifugal force.

The principle of imbibition displacement is shown in Fig. 4.46. The experimental procedure is similar to that of imbibition centrifuge. Namely:

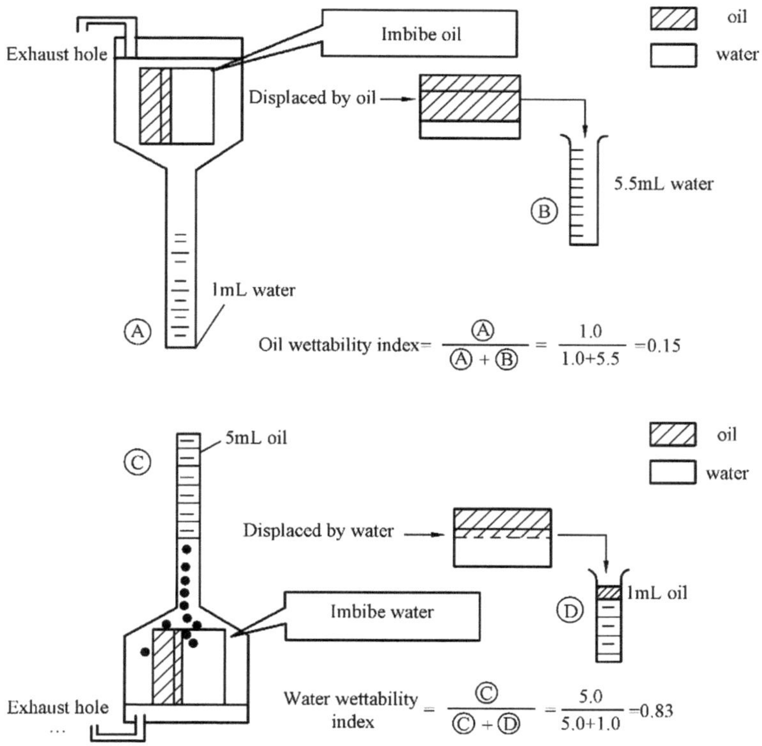

Fig. 4.46 Schematic diagram for water absorption

(a) immerse oil-saturated sample in water, and the water will imbibe into the core automatically, then oil volume displaced by imbibition of water, Ⓒ, can be measured;
(b) place sample in core holder for water flooding, then measure the volume of oil displaced by water flooding, Ⓓ;
(c) place water-saturated sample in oil, and then place the sample in core holder for oil flooding. The amount of water displaced by imbibition of oil, Ⓐ, and the amount of water displaced by oil flooding, Ⓑ, are measured respectively. Finally, Water wettability index and oil wettability index can be calculated through the following Eq. (4.25).

$$\begin{cases} \text{WWI} = \dfrac{\text{oil displaced by imbibition of water}}{\text{oil displaced by imbibition of water} + \text{oil displaced by water flooding}} \\ \text{OWI} = \dfrac{\text{water displaced by imbibition of oil}}{\text{water displaced by imbibition of oil} + \text{water displaced by oil flooding}} \end{cases} \tag{4.25}$$

Wettability can be evaluated by the two indexes, WWI and OWI, as shown in Table 4.4. Where WWI represents water wettability index and OWI represents oil wettability index.

In addition, there are other methods of determining the wettability of a rock, such as the method using capillary pressure curve, nuclear magnetic relaxation method and dye adsorption method, and so on. Nuclear magnetic relaxation method and dye adsorption method are not conventional. The ideas of these methods are as follows: from micro point of view, a porous medium may be wetted by any of the two fluids or be neutral. The determination of the wettability of the medium falls on the problem that how to determine the area of the medium wetted by the fluid. Greater value of wetting area indicates stronger wettability of this fluid. The details about these methods are omitted, but the new thoughts should be learned. In a word, how to determine wettability is still one of the subjects of petrophysics that needed be solved.

Whichever method is used to evaluate the wettability of the rock, there is a problem about how to store and deal with rock samples. The accumulation of asphalt and heavy components on rock surface will make rock more oil-wet, if preservation method is improper or samples are exposed to air for too long.

Table 4.4 The evaluation of wettability by wettability index

Wettability index	Wettability				
	Oil-wet	Slightly oil-wet	Neutral wettability	Slightly water-wet	Water wet
OWI	1–0.8	0.7–0.6	Two indices are close	0.3–0.4	0–0.2
WWI	0–0.2	0.3–0.4		0.7–0.6	1–0.8

Furthermore, natural core contains clay, so if the core is treated by solvents, the wettability may change which will cause clay swelling and dispersion, ion exchange, effective porosity change and surface conductivity change. Therefore, the maintenance of original wettability is the prerequisite of evaluating wettability accurately, so that natural cores must be protected carefully.

Wettability is evaluated with different methods and under different conditions, which may cause problem when compare different results. It should be careful when analyzing the experiment results. This is why more normalized test methods are needed.

The methods of determining wettability with capillary pressure curves, such as wettability index, apparent contact angle method, area per unit volume method or using the shape of relative permeability curve will be discussed in the next sections of this chapter.

4.3 Capillary Pressure Curve

The flow spaces in reservoirs are many meandrous, complex pores and channels, which have different sizes and connect with each other. These pores and channels can be seen as capillary tubes with variable cross sections and rough surfaces, thus reservoir rock can be treated as a multidimensional capillary network. Capillary tube is the basic flow path in reservoirs, so it is important to study the flowing behavior of oil, gas and water in capillary tubes.

4.3.1 Concept of Capillary Pressure

4.3.1.1 Determination of Various Additional Pressures on Curved Surfaces

The static liquid surface is flat in a large container. But under some special conditions, such as in the capillary tube, a curved interface will from at equilibrium. Curvature of the interface is the consequence of preferential wetting of the capillary walls by one of the phases. The interfacial tension on the curved interface is tangent to the interface. As for convex surface, there is a resultant force pointing in liquid direction, and the force in liquid inside is greater than the external force, so that an additional force acts on the convex surface. It is opposite for concave surface that concave surface is tended to be pull out of the interface, and there is an additional pressure (actually "pressure intensity", conventionally call it "pressure", or simply call it "force", such as capillary pressure calls capillary force for short) on concave surface.

The direction of this force is consistent with the concave direction (as indicated by the arrow in Fig. 4.47), and its value can be calculated by the Laplace Equation.

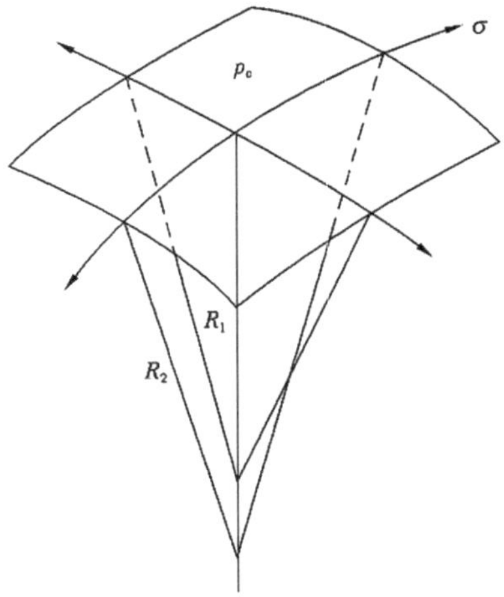

Fig. 4.47 Additional pressure on any curved interface

$$p_c = \sigma\left(\frac{1}{R_1} + \frac{1}{R_2}\right) \tag{4.26}$$

where p_c is additional pressure on meniscus (pressure intensity), Pa; σ is interfacial tension between the two fluids, mN/m; R_1, R_2 are the principle radius of curvature of the interface (i.e., curvature radius in the tangent plane of two perpendicular sections), mm.

Equation (4.26) is a basic formula to study capillary phenomenon, the key is to determine R_1 and R_2 for different meniscus.

Additional force on the interface can be ignored in a large container, while it will be not ignored in small capillary tube. This additional force is usually called capillary pressure.

Figure 4.48 indicates that curved interface in single capillary has two forms in reservoir. One kind is sphere and the other is cylindrical interface that is formed when the capillary tube is full of oil with water films on the wall. The following section will discuss the calculation of capillary pressure on different conditions (sphere, cylinder, curved surface in granular media).

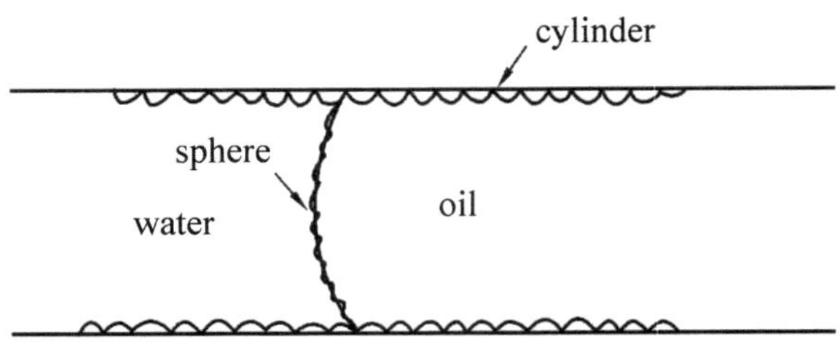

Fig. 4.48 Oil–water interface in water-wet capillary—sphere and cylinder

Spherical Interface

Cut the spherical surface with two mutually perpendicular planes, and we can get circles with curvature radius $R_1 = R_2 = R$. Substitute R_1, R_2 into Eqs. (4.26) and (4.27) can be get:

$$p_c = \sigma\left(\frac{1}{R_1} + \frac{1}{R_2}\right) = \frac{2\sigma}{R} \tag{4.27}$$

where p_c is additional pressure on meniscus (pressure intensity), Pa; σ is interfacial tension between the two fluids, mN/m; θ is wetting contact angle, (°); r is radius of capillary tube, mm; R is the principal radius of sphere surface, mm.

The following relationship can be obtained from Fig. 4.49:

$$\cos\theta = \frac{r}{R}$$

Substituting $\frac{1}{R} = \frac{\cos\theta}{r}$ in Eq. (4.27):

$$p_c = \frac{2\sigma}{R} = \frac{2\sigma\cos\theta}{r} \tag{4.28}$$

p_c points toward the non-wetting phase.

Equation (4.28) is the most important and commonly used formula for calculating capillary pressure. It indicates that there is an inverse relationship between p_c and the capillary radius; the smaller the capillary radius, the greater the capillary pressure is. The smaller the contact angle is, the greater the capillary pressure is.

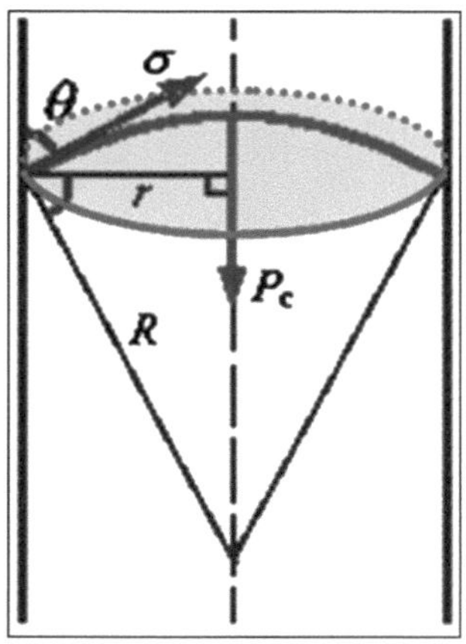

Fig. 4.49 The relationship between capillary radius and curvature radius

Cylindrical Interface

If capillary tube is water-wet, then the irreducible water will adhere to the tube wall and oil (or gas) will occupy the center of the tube, and the oil–water interface is cylindrical surface (as shown in Fig. 4.50).

Taking the plane passing through a generatrix of cylindrical surface as cross section (the first section), and cross section, and intersection of the first section and cylindrical oil–water interface gives a straight line, thus $R_1 = \infty$. Take the plane perpendicular to the first section as the second section, and intersection of the second section and cylindrical oil–water interface comes out to be a circle with curvature radius $R_2 = r$. Substituting $R_1 = \infty$ and $R_2 = r$ in Eq. (4.26) gives

$$p_{\mathrm{c}} = \sigma\left(\frac{1}{\infty} + \frac{1}{r}\right) = \frac{\sigma}{r} \tag{4.29}$$

where p_{c} is additional pressure on meniscus (pressure intensity), Pa; σ is interfacial tension between two phases, mN/m; r is radius of capillary tube, mm.

The direction of capillary force p_{c} points toward the center of capillary tube, and it can make water film thicken.

Then we discuss the calculations of capillary forces under the following two common circumstances.

Gradational Capillary Cross Section

Equation (4.28) only applies to the capillary tube with constant radius. But the radius of capillary tube always changes in reservoir rock. The effect of variable radius can be taken into account by considering the contact angle change when the wall of the capillary tube is inclined, such as the example of conical capillary tube shown in Fig. 4.51.

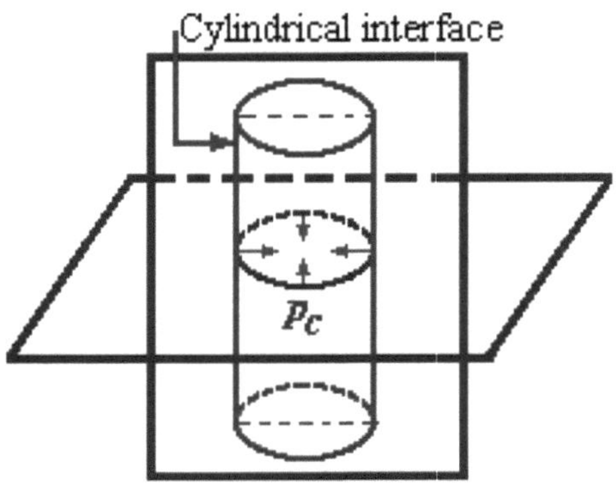

Fig. 4.50 The direction of additional force for cylinder

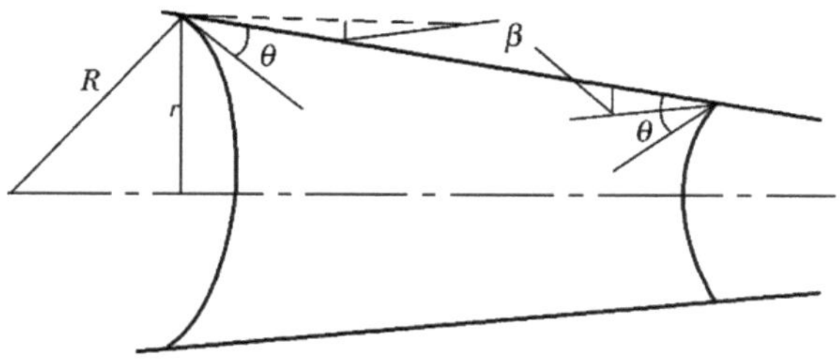

Fig. 4.51 The capillary pressure in conical capillary tube

The curvature radius of wider end is $R = r/\cos(\theta + \beta)$, and the curvature radius of thin end is $R = R/\cos(\theta - \beta)$, so the capillary force of conical tube is as follows:

$$p_c = \frac{2\sigma\cos(\theta \pm \beta)}{r} \tag{4.30}$$

where p_c is additional pressure on meniscus (pressure intensity), Pa; σ is interfacial tension between two phases, mN/m; β is the included angle between the wall of capillary tube and the centerline, (°), i.e., a half of cone angel; θ is the contact angle at equilibrium, (°); r is the radius of the thick end of capillary, mm.

Equation (4.30) indicates that capillary pressure changes with the change of radius. The capillary pressure is largest at the thin end of capillary tube, because the value of $\cos(\theta - \beta)$ reaches maximum at that point. Therefore, in flooding experiment, as long as driving pressure is greater than maximum capillary pressure at the thinnest end of capillary tube, the rest of wetting phase can be displaced out.

Interface in Fractures

Studies recently found that the pore throat of dolomite and limestone are mainly flaky, and oil and water flow between fractures. Then the capillary pressure formula of parallel plate can be used, i.e., assuming two-phase fluids flow between two parallel plates (as shown in Fig. 4.52), and the width of space between two parallel plates is W.

Fig. 4.52 The capillary pressure between two parallel plates

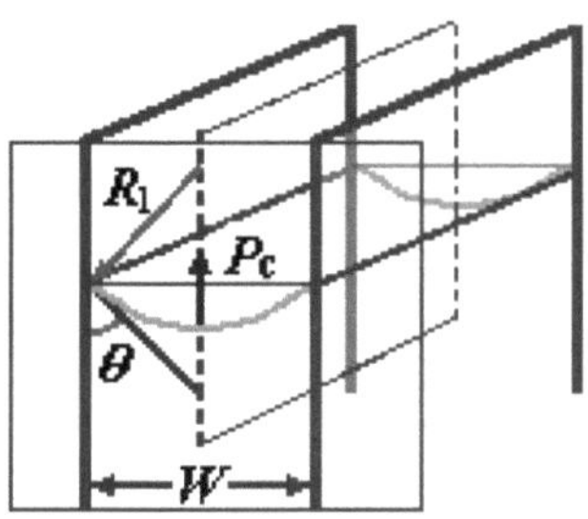

In this case, the meniscus between the two-phase fluids is semicircular cylinder. Similarly, the curvature radius of plane cut by normal section is R_1, and the two-phase interface cut by horizontal plane is a line, whose curvature radius $R_2 = \infty$. Substituting R_1 and R_2 in Eq. (4.26) and get p_c:

$$p_c = \frac{\sigma}{R_1} \tag{4.31}$$

It is noted that $\cos\theta = \frac{W/2}{R_1}$, i.e., $\frac{W}{2} = R_1 \cos\theta$.

So,

$$p_c = \frac{2\sigma\cos\theta}{W} \tag{4.32}$$

where p_c is additional pressure on meniscus (pressure intensity), Pa; σ is interfacial tension between two phases, mN/m; θ is the contact angle at equilibrium, (°); W is the width between two parallel plates, mm.

The above equation is used to calculate capillary pressure between two parallel plates. Its difference from capillary pressure formula of cylindrical capillary tube is that the capillary radius is replaced by the width between two parallel plates. Obviously, Eq. (4.32) can be used to study the capillary pressure of fractured reservoir rock. It is noted that the smaller the width, the greater capillary pressure is.

The following discussion is closer to the actual situation of rock.

Capillary Pressure in Ideal Rock

Ideal rock means that oil and water flow in the porous media made by equal size sands (shown in Fig. 4.53). Wetting phase occupies the contact points between

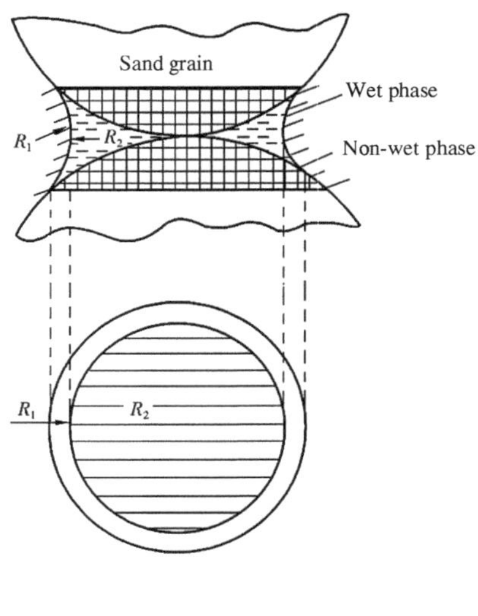

Fig. 4.53 The distribution of wetting phase and non-wetting phase in equivalent diameter circular sand grains

sands, and displays a circular distribution. Non-wetting phase is in the center of pores. There is a curved interface between two phases.

Cutting the oil–water interface by a perpendicular plane results a curvature radius R_1. Cut oil–water interface by a horizontal plane, and a section with curvature radius R_2 can be obtained (the centers of two curvature locate on the two sides of interface respectively), then the Laplace Equation can be get as the Eq. (4.26).

It is practically impossible to determine the values of R_1 and R_2, so they are generally replaced by the average radius of curvature R_m, and one can obtain

$$R_{\mathrm{m}} = \frac{1}{R_1} + \frac{1}{R_2} \tag{4.33}$$

Therefore,

$$p_{\mathrm{c}} = \frac{\sigma}{R_{\mathrm{m}}} \tag{4.34}$$

where R_m is the average curvature radius, which can be determined by other parameters (such as the equation $p_c = \Delta \rho g h = \sigma / R_m$), mm; p_c is the capillary pressure in ideal rock, Pa; σ is interfacial tension between two phases, mN/m.

It is indicated in Fig. 4.53 that with the amount of wetting-phase changes, the wetting-phase saturation S_w will change, as well as curvature radius R_1 and R_2, and corresponding average curvature radius R_m will change. For example, when wetting-phase saturation decreases with the decrease of wetting-phase amount, the curvature radius R_1, R_2, and R_m will decrease correspondingly, and capillary pressure p_c will increase. The following conclusions can be drawn from the discussion above: ① there is a functional relationship between average curvature radius Rm and wetting-phase saturation S_w: $S_w = S(R_m)$; ② there is a functional relationship between average curvature radius Rm and capillary pressure p_c: $p_c = p(R_m)$, therefore there is a functional relationship between wetting-phase saturation and capillary pressure as well

$$p_{\mathrm{c}} = f(S_{\mathrm{m}}) \tag{4.35}$$

where p_c is the capillary pressure in ideal rock, Pa; S_w is the saturation of wetting phase, decimal.

This is the theoretical base for capillary pressure curve.

Capillary Rise (or Descend)

When a capillary tube is inserted below the interface of a two-phase system, the meniscus of the immiscible fluids in the capillary will be either:

Concave with respect to the denser phase, which will rise above the interface between the two liquids outside the capillary;

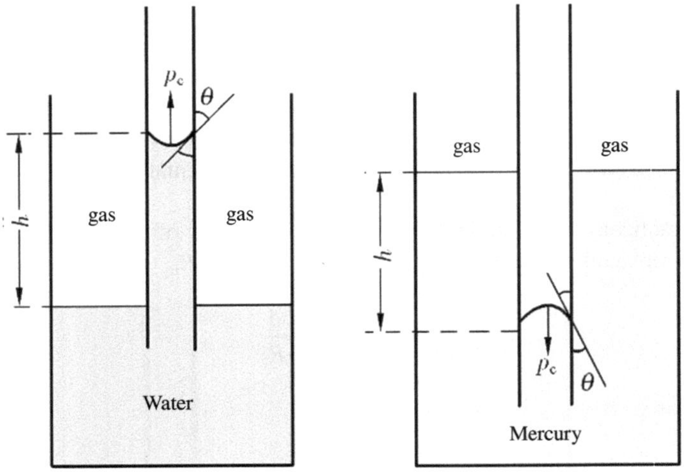

(a) Tube is placed in wetting phase

(b) Tube is placed in non-wettingphase

Fig. 4.54 The rising of wetting phase and the descending of non-wetting phase in capillary tube

Straight across the capillary and level with the bulk fluids interface; or

Convex with respect to the denser phase and below the bulk fluids interface as illustrated in Fig. 4.54.

This additional force that causes liquid rise or descend in a capillary tube is called capillary pressure.

If a capillary tube is inserted into a container filled with water and oil, the wetting phase, water, will rise to a height h in the capillary tube (see Fig. 4.55a). Set oil–water interface as σ, contact angle as θ, the densities of oil and water as ρ_o, ρ_w respectively. In the capillary tube, the oil phase pressure (point OB) and the water phase pressure (point WB) on the interface are assumed to be p_{ob} and p_{wb},

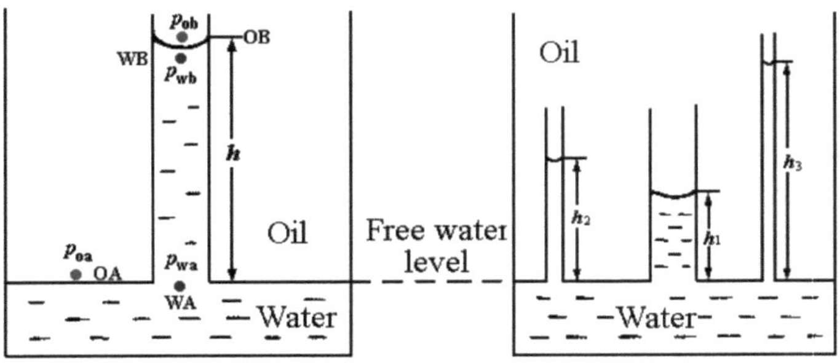

(a) single capillary tube

(b) capillary tube system

Fig. 4.55 The relationship between rising height and capillary force

separately. While in the container, the oil phase pressure (point OA) and the water-phase pressure (point WA) on the interface are assumed to be p_{oa} and p_{wa}, separately.

So one can get in oil phase:

$$p_{ob} = p_{oa} - \rho_o gh \quad (4.36)$$

In water phase:

$$p_{wb} = p_{wa} - \rho_w gh \quad (4.37)$$

Also,

$$p_{oa} = p_{wa} \quad (4.38)$$

where p_{ob} and p_{oa} are oil-phase pressure, Pa; p_{wb} and p_{wa} are water-phase pressure, Pa; ρ_o is the density of oil, kg/m^3; ρ_w is the density of water, kg/m^3; g is the acceleration of gravity, m/s^2; h is the rising height of wetting phase, mm.

Since the pressures at the same height in the connected pipes are equal, and the container is large enough, so the interface in the container (OA) is a horizontal plane and the capillary pressure there is zero.

Capillary pressure is defined as the difference in pressure between two immiscible fluids across a curved interface at equilibrium. Thus non-wetting phase pressure minus wetting-phase pressure is the value of capillary pressure.

Capillary pressure only exists on the interface, and may cause abrupt change of pressure. According to the above definition, we can get

$$p_c = p_{ob} - p_{wb} = (\rho_w - \rho_o)gh = \Delta\rho gh \quad (4.39)$$

where p_c is the capillary pressure on the interface, Pa; $\Delta\rho$ is density difference of two phases, kg/m^3; h is height of the column of water in the tube above that in the large vessel, m; g is acceleration of gravity, m/s^2.

Equation (4.39) is the basic formula for capillary equilibrium theory in reservoir, which indicates that the height of the liquid column depends on the value of capillary pressure. The greater the capillary pressure is, the higher the liquid column is.

By substituting capillary pressure $p_c = 2\sigma \cos \theta / r$ in Eq. (4.39), we have:

$$\Delta\rho gh = \frac{2\sigma \cos \theta}{r}$$

Therefore

$$h = \frac{2\sigma \cos \theta}{r \Delta \rho g} \tag{4.40}$$

where σ is oil-water interface tension, N/m; θ is contact angle, (°); r is radius of tube, m; $\Delta\rho$ is density difference of two phases, kg/m^3; h is height of the column of water in the tube above that in the large vessel, m; r is radius of capillary, m; g is acceleration of gravity, m/s^2.

Equation (4.40) can be used to calculate the rising height of liquid (such as water) in reservoir.

It is noted in Eq. (4.40) that as long as the other parameters are fixed, the rising height will not change as the capillary inclined in real reservoir. Figure 4.55b illustrates the effect of varying the radius of the capillary tube, and the smaller the radius of the tube, the higher the wetting phase column will rise. So conclusions can be drawn: the oil–water contact in reservoir is not a plane that separates the oil and water completely but a transition zone (or oil–gas transition zone) with a certain thickness. Generally, the thickness of oil–water transition zone is greater than that of oil–gas transition zone.

For the actual reservoir, if the rock is water wet ($\theta < 90°$, $\cos\,\theta > 0$), p_c is calculated by the equation $p_c = 2\sigma \cos \theta / r$. That means p_c is positive, and so does h, and the interface will rise. When a water-wet core is immersed into the oil, water can automatically imbibe into the core and displace the non-wetting phase (oil) due to the capillary pressure, and this process is called s imbibition process. The smaller the contact angle θ is, the stronger the hydrophilicity and the greater the ability of imbibition is. On the contrary, when $\theta > 90°$, rock cannot imbibe water automatically, and the capillary pressure p_c becomes resistance. It is necessary to exert an external force to overcome capillary pressure p_c in order to make water displace oil, which is called displacement process.

4.3.1.2 Capillary Hysteresis

Capillary hysteresis is a particular phenomenon occurs in capillary tubes. This phenomenon is similar to wetting hysteresis. For example, a single capillary tube is inserted into a container filled with wetting phase fluid (such as water) to do the imbibition experiment (Fig. 4.56), where the wetting phase (water) displaces the non-wetting phase (gas). The pressure in the air is P_a, and the liquid will rise to a certain height owing to the capillary pressure (as shown in Fig. 4.56a). Meanwhile, another tube of same size full of water is inserted into a container filled with water for the air flooding water experiment, in which the air pressure is still p_a. Then the water will fall to a certain height along capillary tube due to the gravity (in Fig. 4.56b). It is illustrated that even though the pressures in the air are the same, the heights of liquid column caused by displacement and imbibition is different because of the difference of saturated sequence and the liquid column of

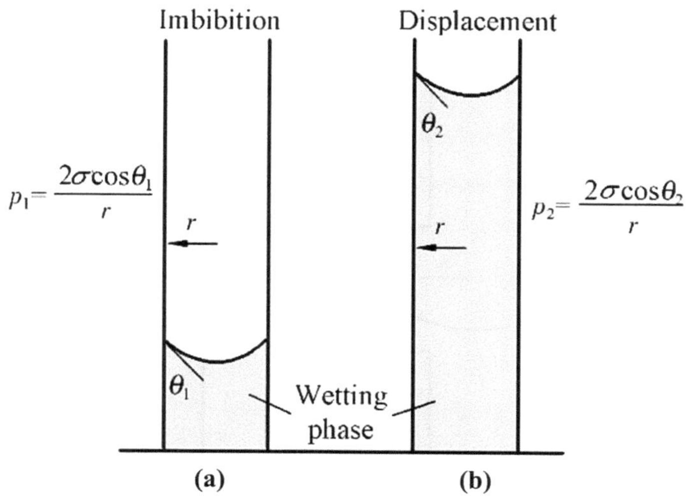

Fig. 4.56 Contact angle hysteresis

displacement is greater than that of imbibition, the phenomenon that is called as capillary hysteresis.

The causes of capillary hysteresis include wetting hysteresis and geometry of rock pores and throats. The previous researchers (such as Morrow, etc.) have done some studies on capillary hysteresis, and the causes of hysteresis are drawn:

Hysteresis Caused by Wetting Hysteresis (Also Called Contact Angle Hysteresis)

It is illustrated in Fig. 4.56 that with fixed radius of capillary tube, different wetting sequence leads to different contact angles in imbibition and displacement processes. The imbibition process will cause advancing angle θ_1, and the displacement process leads to receding angle θ_2. Because $\theta_1 > \theta_2$, so $\cos\theta_1 < \cos\theta_2$ which causes the capillary pressure P_{im} in imbibitions process less than the capillary pressure p_{dr} in displacement process. Therefore, under the same non-wetting phase pressure, the height of liquid column in imbibition process is less than that in displacement process. In other words, under same displacement pressure, the wetting phase saturation of displacement process is greater than that of imbibition process, which provides theoretical basis for investigating capillary pressure and pore structure with displacement or imbibition method. That is to say, if the rock is water-wet, displacement method (displace water by oil) can be used to obtain irreducible water saturation, while imbibition method (displace oil by water) can be used to obtain residual oil saturation, which has significance on oil recovery.

The capillary hysteresis is only related to contact angle hysteresis if the radius of the capillary tube is constant.

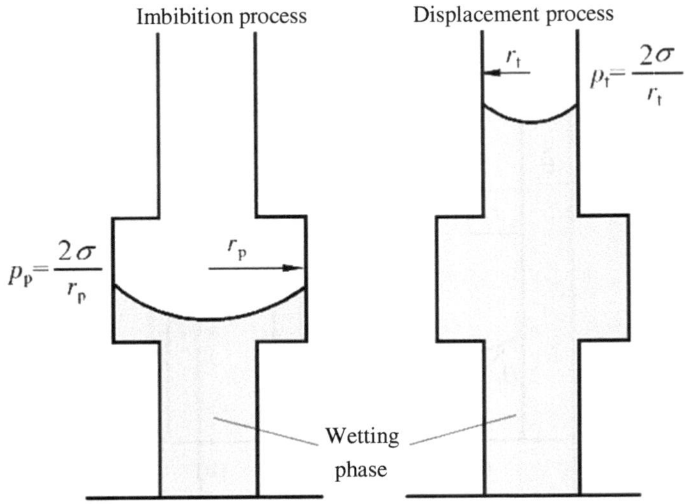

Fig. 4.57 The mutation of capillary radius

Hysteresis Caused by Capillary Radius Mutation (i.e., so Called "Inkpot" Effect)

As shown in Fig. 4.57, the radius of a capillary tube thickens abruptly in the middle part. The radius of upper small section (throat) is r_t, and the radius of middle large section (pore) is r_p, and $r_t < r_p$, so $p_t > p_p$. It is assumed that the non-wetting phase pressure equals p both in displacement and imbibitions processes. To simplify the problem, the contact angle θ is assumed to be 0°, thus the following equations can be obtained: $p_t = 2\sigma/r_t$ and $p_p = 2\sigma/r_p$. Then in the displacement process, the interface will decline and stay in the upper small part of the capillary tube. While in the imbibition process, the interface will rise and stay in the middle large part of the capillary tube. This difference leads to the result that wetting-phase saturation in imbibition process smaller than that in displacement process. Considering contact angle $\theta = 0°$, the capillary hysteresis is only related to the change of capillary radius.

Hysteresis Caused by Gradational Capillary Radius

Take conic capillary tube shown in Fig. 4.58 for example. In this case, capillary hysteresis is affected by both contact angle change and capillary radius change. Contact angle hysteresis is neglected, and the effect of capillary radius change is similar to the second case discussed above, i.e., $p_t = 2\sigma/r_t$ and $p_p = 2\sigma/r_p$. The change of contact angle is caused by the tilt of capillary wall. According to the Eq. (4.31), $p_c = 2\sigma \cos(\theta \pm \beta)/r$, contact angle should minus a half cone angle β

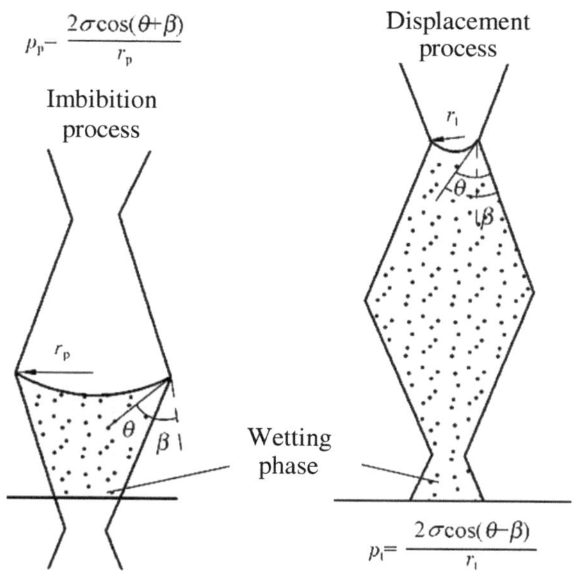

Fig. 4.58 Gradational capillary radius

when calculating displacement capillary pressure p_t, however, contact angle should plus a half cone angle β, when calculating imbibition capillary pressure p_p.

Imbibition process

$$p_p = \frac{2\sigma\cos(\theta + \beta)}{r_p} \tag{4.41}$$

Displacement process

$$p_t = \frac{2\sigma\cos(\theta - \beta)}{r_t} \tag{4.42}$$

where σ is interface tension, N/m; r_t is the radius of upper small section (throat), mm; r_p is the radius of middle large section (pore), mm; p_p is displacement pressure, Pa; p_t is imbibitions pressure, Pa; θ is contact angle, (°); β is semi-cone angle, (°).

It is illustrated in Fig. 4.58 that the effect of capillary radius and contact angle is consistent, $p_p < p_t$, and the wetting phase saturation in imbibitions process is less than that in displacement process.

In a real rock, the roughness of pore surfaces and the variation in the cross sections of pores (Fig. 4.59) also have evidently effects on the capillary hysteresis. That is to say, the effects of contact angle hysteresis (advancing angel θ_1 and receding angle θ_2), variable cross section ($\theta_1 + \beta$ and $\theta_2 - \beta$), pore radius variation (porous radius r_t, and throat radius r_p) and the roughness of pore surface should all be taken into consideration simultaneously. However, it is impossible to directly

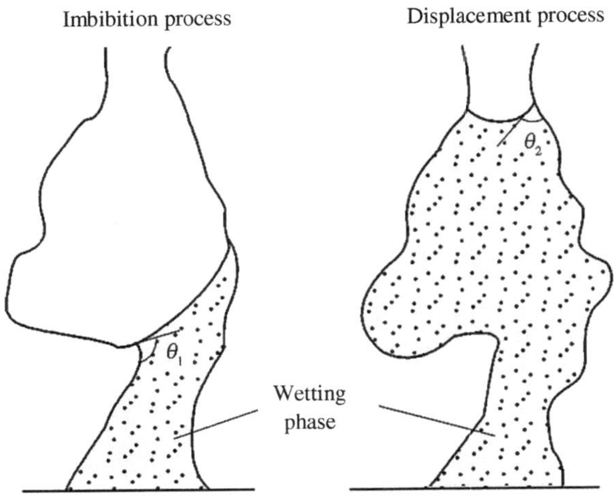

Fig. 4.59 Hysteresis caused by contact angle, gradational radius and rough surface in actual capillary

evaluate the effect of roughness on advancing angle and receding angle, which can only be determined by capillary pressure data collected from imbibition and displacement experiments. The apparent advancing angle and apparent receding angle are defined as $\theta_A = \theta_1 + \beta$ and $\theta_R = \theta_2 - \beta$, separately. Then the capillary pressure is:

Displacement process

$$p_{dr} = \frac{2\sigma\cos\theta_R}{r_t} = \frac{2\sigma\cos(\theta_2 - \beta)}{r_t} \tag{4.43}$$

Imbibition process

$$p_{im} = \frac{2\sigma\cos\theta_A}{r_p} = \frac{2\sigma\cos(\theta_1 + \beta)}{r_p} \tag{4.44}$$

where p_{dr}, p_{im} are displacement pressure and imbibition pressure respectively, MPa; σ is interface tension, N/m; θ_R, θ_A are apparent receding angle and apparent advancing angle respectively, (°); β is semi-cone angle, (°); r_t is the radius of upper small section (throat), mm; r_p is the radius of middle large section (pore), mm.

It is noted in Eqs. (4.43) and (4.44) that capillary pressure is calculated based on throat radius in displacement process, while it is calculated based on pore radius in imbibition process. In experiment, the quantitative value of pore and throat and their grain size distribution can be obtained by the capillary pressure-saturation curve in displacement and imbibition process.

Based on the above analysis, the capillary hysteresis in rock results in that wetting phase saturation in displacement process is greater than that in imbibition process under the same displacement pressure.

4.3.2 Measurement of Capillary Pressure Curve

Capillary pressure curve is the relation curve of capillary pressure and wetting phase (non-wetting phase) saturation in rock, as shown in Fig. 4.60.

Capillary pressure curve can be used to determine the pore size distribution, and to infer characteristics of pore geometry. The methods of determining capillary pressure curve should be discussed at first in order to recognize quantitative and qualitative characteristics and their various applications.

There are many methods to determine capillary pressure curve, and the following three methods are commonly used: Semi-permeable membrane method, mercury injection method and centrifuge method, whose basic principles are similar, and the differences include the fluid working media, way of pressurizing and required length of time, and so on. These three methods have their own advantages and are used widely.

4.3.2.1 Semi-permeable Membrane Method

The equipment for semi-permeable membrane method is shown in Fig. 4.61, and its principle also represents the measurement principle of mercury injection method

Fig. 4.60 The typical capillary curve

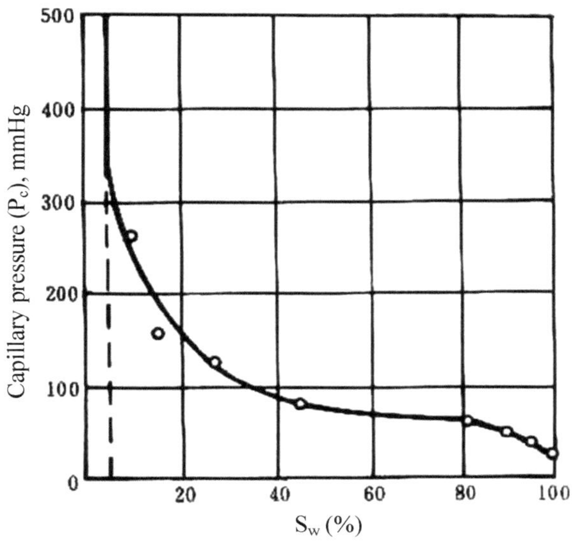

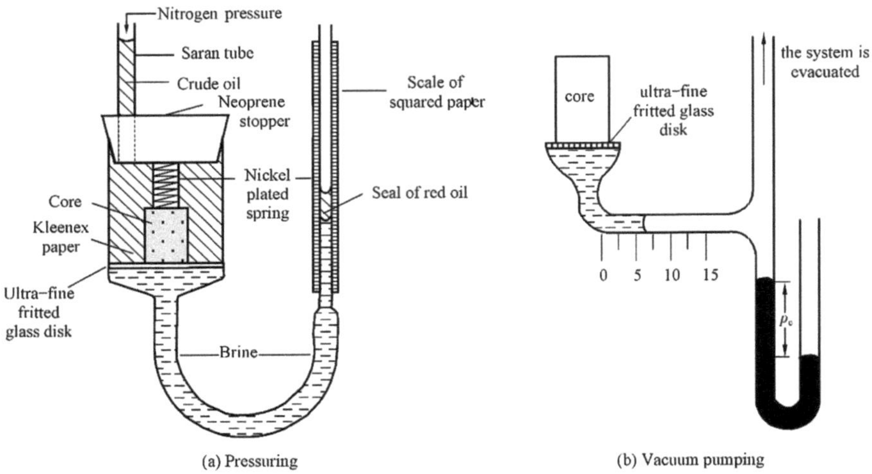

Fig. 4.61 Semi-permeable diaphragm capillary pressure device

and centrifuge method. In a displacement process, only when external pressure equals or is greater than the capillary pressure in a throat, can non-wetting phase pass through the throat and drive wetting-phase in pore out. At this time, the external pressure is equivalent to the capillary pressure.

The important part of pressuring method (Fig. 4.61a) for measuring rock capillary pressure curve is a metal (or glass) hopper with semi-permeable membrane, i.e., core chamber. Various materials including porcelain or glass, sintered powder metal, carburizing multi-rubber, and others have been used successfully as diaphragms with different wettability. It should be noted that the pores of the diaphragm should be smaller than that of the core. The membrane is saturated with the wetting phase fluid. Only wetting phase is allowed to penetrate through the diaphragm until the external pressure equals or exceeds the capillary pressure caused by the throat of diaphragm, and that is why it is called semi-permeable membrane. Beneath the diaphragm is filled with wetting phase (such as water) and air is evacuated, and the diaphragm is saturated with water beforehand. The test sample saturated with wetting-phase fluid is placed on the semi-permeable membrane, and a spring is used to make the sample is clinged to the membrane tightly. In addition, a layer of filter paper is usually put between them. The metal pipe above the core chamber is connected to the pressure source (nitrogen tank), which could fill the core chamber with non-wetting fluid (oil or gas). If exerting expulsion pressure on non-wetting phase, it will overcome the capillary pressure and then enter the core, and drive the wetting-phase (water) out. The displaced water will penetrate through the diaphragm and flow into the U shape tube, and the amount of the displaced water can be directly read from the graded tube. The non-wetting phase oil cannot pass through the diaphragm until the external pressure reaches the minimum capillary pressure. Meanwhile, in order to prevent displaced water from evaporating, a drop of oil in the graded tube may be helpful.

The pressure of the displacing fluid is increased in small increments from the minimum value during the experiment. With the increase of displacement pressure, non-wetting phase oil can pass through the increasingly fine throats and drive more and more water out. That is, the non-wetting phase saturation increases and the wetting phase saturation decreases, as the displacement pressure increases.

At each predetermined pressure, record the value of pressure and the amount of cumulative displaced water when the system reaches equilibrium (both the value of pressure and the interface in the tube keep stable). Then increase the pressure to the next point and repeat the record work until the pressure reaches the predetermined maximum pressure.

Finally, a series of pressure and corresponding cumulative volume of displaced water can be obtained (Table 4.5). According to the volume of displaced water and the initial volume of water saturated in the sample (1.365 cm^3 in Table), which equals the total pore volume V_p, the water saturation (or wetting phase saturation) at each pressure point can be calculated by the following equation:

$$S_w = \frac{V_p - \sum V_w}{V_p} \tag{4.45}$$

where S_w is wetting phase saturation (or water saturation), %; V_p is pore volume, i.e., initial water volume saturated in the sample, mL; $\sum V_w$ is cumulative displaced water volume, mL.

Based on the measured data, the relation curve of pressure and saturation can be drawn, and this is capillary pressure curve obtained from displacement process, as shown in Fig. 4.62 (curve ①).

The maximum capillary pressure that could be measured by semi-permeable membrane method depends on the pore size of diaphragm. The pressure at which non-wetting phase starts penetrating the diaphragm is the allowed maximum pressure. The smaller the pore size of diaphragm, the greater the valve pressure and

Table 4.5 Data for capillary pressure curve

No.	Capillary pressure (p_c) (mmHg)	Values read from graduated tube (cm^3)	Water volume in core (cm^3)	Water saturation (%)
1	20	0	1.365	100
2	40	0.075	1.29	94.6
3	50	0.15	1.215	89
4	60	0.25	1.115	81.8
5	80	0.75	0.615	45
6	120	1	0.365	27
7	160	1.125	0.24	17.9
8	260	1.225	0.14	10.7
9	390	1.285	0.08	6.3
10	>390	1.285	0.08	6.3

Fig. 4.62 Capillary pressure curve

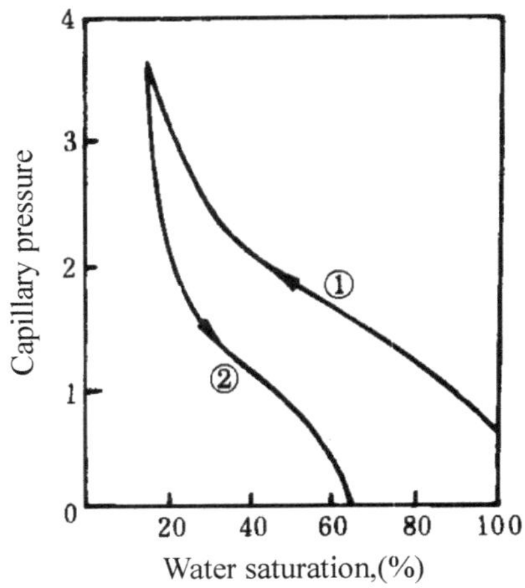

① displacement curve (mercury-injection curve); ② imbibition curve (mercury-withdrawal curve)

the test range. The valve pressure of domestic diaphragm can be as high as 0.7 MPa.

When maximum pressure is reached in displacement process, then decrease the pressure stage by stage. Water (wetting phase) will imbibe into the core automatically because of capillary pressure, and non-wetting phase will be displaced. This process is called imbibition process. Record the soakage of wetting phase at each pressure, and wetting phase saturation can be obtained. Then the capillary pressure curve obtained from displacement process can be plotted, as shown in Fig. 4.62 (curve ②).

A modification for this method is to vacuumize the tube under the membrane to cause pressure difference, as shown in Fig. 4.61b. Water in the sample can penetrate through the semi-permeable membrane and enter into the graded tube below. Different vacuum degree will cause different pressure drop, but the maximum pressure drop cannot exceed 0.1 MPa. The principles of the above two methods are almost the same, and the only difference is that the way of generating pressure drop.

The most disadvantage of semi-permeable membrane method is that the test time is too long, and one test takes a few hours or even a few days to reach equilibrium. If gas is chosen as the non-wetting phase, the time to equilibrium will be shorter. At present, the core chamber is designed to hold tens of samples at the same time, which greatly improves the testing efficiency.

Although semi-permeable membrane has the above disadvantage, but the testing condition is very close to the real reservoir condition with high accuracy whether it is gas/water displacement, gas/oil displacement, oil/water displacement or water/oil displacement. It is a standard and classic method and can be used as a criterion to compare with other methods.

4.3.2.2 Mercury Injection Method

Mercury injection method is initially applied to petrochemical industry, then it is introduced in petroleum industry and has been used and developed widely. The basic principles of mercury injection equipment are the same, and the differences are the way of pressurizing and measurement. The ways of pressurizing include metering pump, oil pump, and high pressure vessel. Metering pump directly adds pressure on the mercury, while oil pump and high pressure vessel transmit pressure to mercury through oil or nitrogen. The methods of determining the volume of mercury injection mainly include metering pump and dilatometer. Figure 4.63 shows the mercury injection equipment that pressurizes mercury by high pressure nitrogen vessel and metering pump. The main parts of the equipment include metering pump and core chamber, and the metering pump with mercury in is connected to the core chamber. The laboratory procedure is as follows:

(a) To conduct a test, a core is cleaned, dried, and placed in the sample chamber of the mercury injection equipment. Then tighten the gland.
(b) Evacuate the sample chamber and keep the mercury level lower than the lower scale mark on the lucite window at the bottom.
(c) Inject mercury until the mercury level reaches the lower scale mark on the lucite window at the bottom. The graduated scale reads zero.
(d) Inject mercury until the mercury level reaches the upper scale mark on the lucite window on the top. Record the reading on the graduated scale, V. The volume of the core chamber is known as V_E, so the external volume of sample is $V_f = V_E - V$.
(e) Turn off the vacuum system and prepare the high pressure gas source. Set a pressure which is balanced with a certain capillary pressure, p_c. Under this

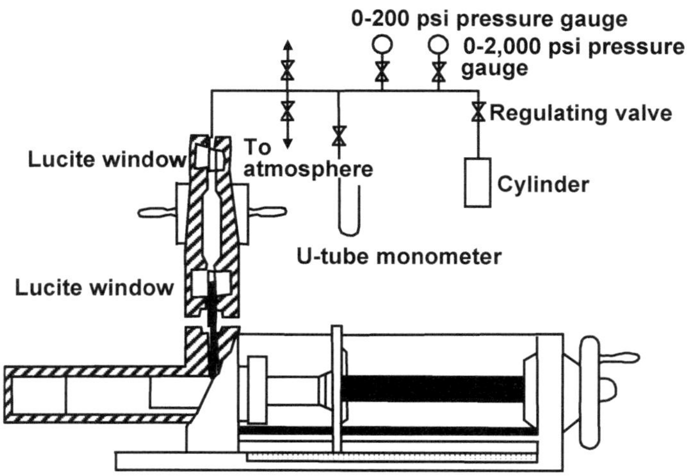

Fig. 4.63 Equipment for mercury Injection Capillary pressure measurement

pressure, mercury is force into the core, and the mercury level will fall below the upper scale mark. Then inject mercury until the mercury level reaches the upper scale mark again, and the volume of mercury forced into the core, V_{Hg}, can be read from the graduated scale.

(f) The saturation of mercury in the core can be calculated by the porosity φ and bulk volume V_f of core, that is $S_{Hg} = V_{Hg}/V_f$.

(g) According to the p_c and S_{Hg} data obtained by continuous mercury injection, the injection capillary pressure curve can be drawn (Fig. 4.62, curve ①). When the injection pressure reaches a limit, a mercury withdrawal capillary pressure curve can be obtained by decreasing the pressure in increments and recording the volume of mercury withdrawn (Fig. 4.62, curve ②).

Dilatometer is widely used to determine the volume of mercury injected abroad. Dilatometer is a stainless steel pipe, in which there is a platinum wire (Fig. 4.64). The change of mercury volume in pipe will lead to the change of resistance of platinum gold wire, so that the dilatometer calibrated beforehand can determine the volume of mercury injected at each pressure according to the change of resistance. The measurement accuracy of dilatometer is higher than metering pump.

The mercury injection method is simple to conduct and rapid and has no restrictions on the shape and size of the sample. It takes only 1–2 h to test one sample, and it can determine capillary pressure of irregular debris.

The mercury injection method also has two disadvantages:

Mercury is forced in the core under the vacuum circumstance, which is quite different from the real reservoir condition.

Mercury vapor is toxic, so strict safety precautions must be followed when using mercury. In addition, after mercury is injected into a core, it cannot be used for any other tests because mercury cannot be safely removed.

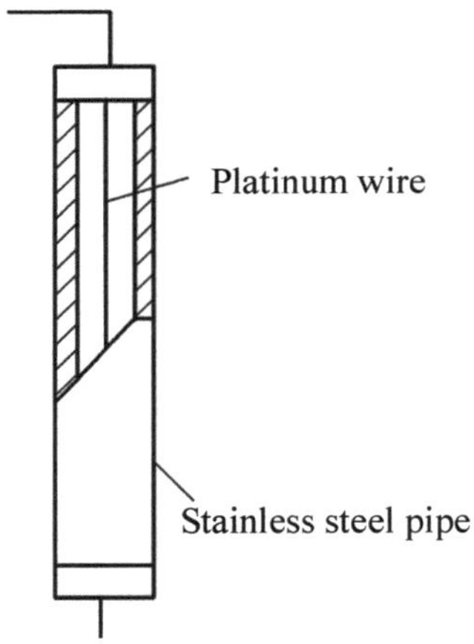

Fig. 4.64 Dilatometer

4.3.2.3 Centrifuge Method

This method uses the centrifugal force generated by a high speed centrifuger to replace the external displacement pressure to displace wetting phase from the core by non-wetting phase. A step-by-step procedure is as follows: A core is saturated with wetting phase fluid (water) and then placed in the core holder, which is filled with non-wetting phase fluid (oil) to cover the core. The core holder is placed in the centrifuge, and the centrifuge rotates at a certain angular speed (Fig. 4.65). Oil and water will suffer from different centrifugal forces, even under the same rotation radius and angular rate, due to the density difference. The centrifugal force difference is balanced by capillary pressure. Water is displaced out of the core and oil will occupy the pore space. Express centrifugal force with centrifugal pressure, the centrifugal pressure difference between the two phases (oil and water is) the driving pressure that non-wetting phase displaces wetting phase.

The volume of displaced water at equilibrium can be read through the observation window, and then the saturation under this centrifugal pressure can be calculated. As the rotational speed increases, the corresponding capillary pressure needed to balance the centrifugal force difference will also increase. Record the volume of displaced fluid, and the relation curve of capillary pressure and wetting phase (water) saturation can be obtained.

Capillary pressure and the saturation of wetting-phase fluid can be calculated by following methods.

Average Method

The standard formula of the oil industry standards is usually used, that is:

$$p_c = 0.50656 \Delta\rho n^2 H \left(r_e - \frac{H}{4} \right) \times 10^{-8} \tag{4.46}$$

Fig. 4.65 Centrifuge method

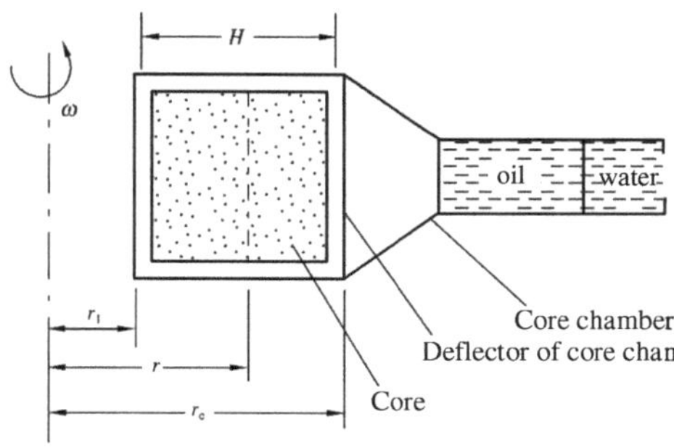

r_1 is the distance between core internal end and shaft,
r is the distance between core center and shaft
r_e is the distance between core external end and shaft

$$\bar{S}_{\mathrm{w}} = \frac{(V_{\mathrm{wi}} - V_{\mathrm{wc}})}{V_{\mathrm{p}}} \tag{4.47}$$

where p_c is capillary pressure, MPa; $\Delta\rho$ is density difference, g/cm^3; n is rotational speed of centrifuge; H is sample length, cm; r_e is external rotation radius of sample, cm; $\bar{S}_w$ is average water saturation of sample, %; V_{wi} is initial water saturation of sample, cm^3; V_{wc} is cumulative volume of displaced water, cm^3; V_p is effective pore volume of sample, cm^3.

First Approximation Method

$$p_{\mathrm{c}} = \frac{1}{2}\Delta\rho\omega^2\left(r_{\mathrm{e}}^2 - r_1^2\right) \tag{4.48}$$

where p_c is capillary pressure, Pa; ω = angular rate, r/min; $\Delta\rho$ is density difference, g/cm^3; r_1 is the internal rotation radius of sample, m.

$$S(p_{\mathrm{ci}}) = \frac{(d\bar{S}p_{\mathrm{ci}})}{dp_{\mathrm{ci}}} \tag{4.49}$$

The wetting phase and non-wetting phase fluids used in the experiment can be changed according to different requirements, such as gas/water displacement or gas/oil displacement, water/oil displacement or oil/water displacement, etc. When water displaces oil (Fig. 4.66a), displaced oil will float on the proximal end of centrifuger shaft because the density of oil is less than that of water. When oil displaces water, the displaced water will sink in the distal end of centrifuger shaft, which is opposite to water/oil displacement. It is shown in Fig. 4.66b. One centrifuger can do both the water/oil displacement experiment and oil/water displacement experiment with different rotary spinners.

The advantages of this method are that the capillary pressure curves of oil-water, oil-gas and gas-water, etc., can be gained directly, and it is convenient and fast to determine capillary pressure curve by displacement and imbibition. Its disadvantages are the troubles in calculation and complex equipment. The rotational speed of current centrifuger can be as high as 21,000 r/min, but its maximum pressure is only 1.4 MPa, which is too low for tight rocks (such as limestone). The above disadvantages also limited the application of this method.

Conversion of Capillary Pressure Curve

Capillary pressure curve is measured under laboratory condition, which is more or less different from the actual reservoir condition. For example, different methods (mercury injection method and Semi-permeable membrane method) use different

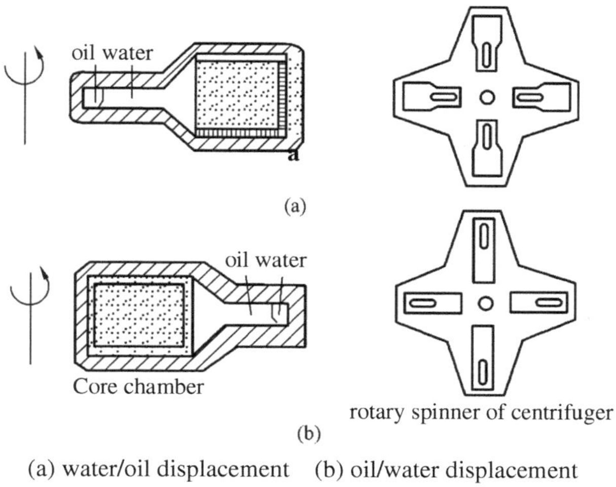

Fig. 4.66 The position of centrifuge tube in water/oil displacement and oil/water displacement

fluid system in laboratory. Thus the interfacial tension and contact angle are also different, which leads to different capillary pressure. So it is necessary to convert the capillary pressure curve to actual reservoir condition when applying the laboratory results to subsurface systems.

With one same sample, then the capillary pressure and capillary radius can be expressed as follows under Laboratory conditions:

$$p_{\mathrm{cL}} = \frac{2\sigma_{\mathrm{L}} \cos \theta_{L}}{r} \rightarrow r = \frac{2\sigma_{\mathrm{L}} \cos \theta_{\mathrm{L}}}{p_{\mathrm{cL}}} \tag{4.50}$$

And under reservoir conditions:

$$p_{\mathrm{cR}} = \frac{2\sigma_{\mathrm{R}} \cos \theta_{\mathrm{R}}}{r} \rightarrow r = \frac{2\sigma_{\mathrm{R}} \cos \theta_{\mathrm{R}}}{p_{\mathrm{cR}}} \tag{4.51}$$

where p_{cL} is additional pressure on meniscus (pressure intensity) under laboratory conditions, mN/cm^2; σ_{L} is interface tension under laboratory conditions, mN/cm; θ_{L} is contact angle under laboratory conditions, (°); p_{cR} is additional pressure on meniscus (pressure intensity) under reservoir conditions, mN/cm^2; σ_{R} is interface tension under reservoir conditions, mN/cm; θ_{R} is contact angle under reservoir conditions, (°), r is the radius of capillary tube, cm.

The r in the above two equations are equal because of the same sample, thus the following conversion formula can be obtained.

$$p_{cR} = \frac{\sigma_R \cos \theta_R}{\sigma_L \cos \theta_L} p_{cL} \tag{4.52}$$

Capillary pressure data determined by different methods can be converted with Eq. (4.52). Three cases of the conversion of capillary pressure data between different methods are illustrated as follows.

Example 4.1 Convert capillary pressure p_{Hg} determined by mercury injection method into the oil-water capillary pressure p_{ow} under reservoir conditions. Mercury surface tension σ_{Hg} = 480 mN/m, and θ_{wg} = 140°, and oil-water interface tension σ_{ow} = 25 mN/m, and θ_{ow} = 0° are known.

Solution

According to Eq. (4.52), we have

$$p_{ow} = \frac{\sigma_{ow} \cos \theta_{ow}}{\sigma_{Hg} \cos \theta_{Hg}} p_{Hg} = \frac{25 \times \cos 0^\circ}{480 \times \cos 140^\circ} p_{Hg} \approx \frac{1}{15} p_{Hg} \tag{4.53}$$

where the in situ capillary pressure p_{ow} is 1/15 of that determined by mercury injection method, p_{Hg} is capillary pressure determined by mercury injection method, mN/cm^2; p_{ow} the oil-water capillary pressure under reservoir conditions, mN/cm^2.

Example 4.2 Convert capillary pressure p_{wg} determined by semi-permeable membrane method (water–air system) into the oil–water capillary pressure p_{ow} under reservoir conditions. The water surface tension σ_{wg} = 72 mN/m, and contact angle θ_{wg} = 0° are known.

Solution

According to Eq. (4.52), we have

$$p_{ow} = \frac{\sigma_{ow} \cos \theta_{ow}}{\sigma_{wg} \cos \theta_{wg}} p_{wg} = \frac{25 \times \cos 0^\circ}{72 \times \cos 0^\circ} p_{wg} \approx \frac{1}{3} p_{wg} \tag{4.54}$$

where p_{wg} is capillary pressure determined by semi-permeable membrane method (water-air system), mN/cm^2; p_{ow} is the oil–water capillary pressure under reservoir conditions, mN/cm^2; the in situ capillary pressure p_{cow} is 1/3 of that determined by semi-permeable membrane method (Fig. 4.67).

Example 4.3 Convert capillary pressure p_{wg} determined by mercury injection method into the gas–water capillary pressure p_{wg} determined by semi-permeable membrane method.

Solution

Based on Eq. (4.52), following expression can be obtained:

$$p_{wg} = \frac{\sigma_{wg} \cos \theta_{wg}}{\sigma_{Hg} \cos \theta_{Hg}} p_{Hg} = \frac{72 \times \cos 0^\circ}{480 \times \cos 140^\circ} p_{Hg} \approx \frac{1}{5} p_{Hg} \tag{4.55}$$

where p_{wg} is capillary pressure determined by semi-permeable membrane method (water-air system), mN/cm^2; p_{Hg} is capillary pressure determined by mercury injection method, mN/cm^2.

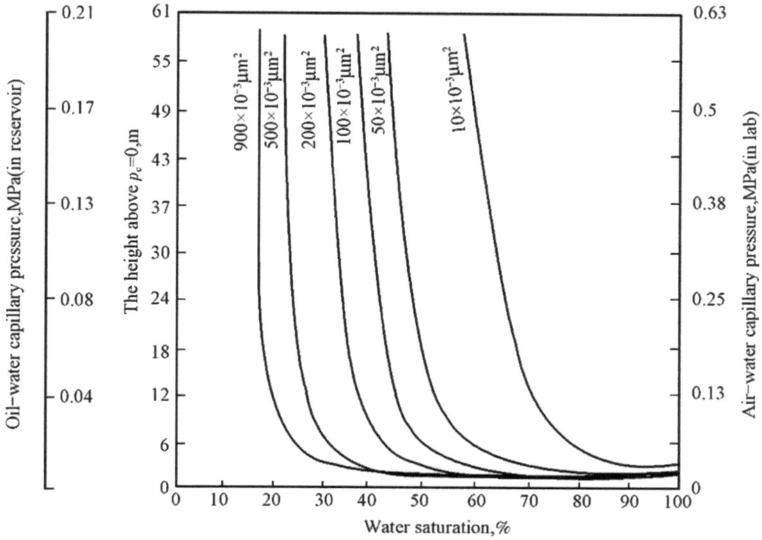

Fig. 4.67 Capillary pressure curve conversion (Frick, SPE-AIME)

Result shows that the capillary pressure curve determined by mercury injection can be compared with that determined by semi-permeable membrane after it is scaled down 5 times. It is generally believed that the semi-permeable membrane method is closer to the actual reservoir condition, and it can be used as the standard curve compared against all other methods due to its high accuracy. Practice has proven that results obtained by different methods are in good agreement with that obtained by semi-permeable membrane method.

An approximate pore size distribution of rocks can be obtained from capillary pressure curves. The pore throat size, r_c, and wetting phase height, h, under different pressures can be calculated from the capillary pressure curve.

Sometimes the throat radius and wetting phase height is directly marked on the vertical axis of the capillary pressure curve. The relationship between p_c, r_c and h, is as follows:

According to $p_c = 2\sigma \cos \theta / r$, and set $\sigma = 480$ mN/m, $\theta = 140°$ in mercury injection method, so the relationship between mercury injection capillary pressure p_{Hg} and throat radius r is:

$$p_{Hg} = \frac{2\sigma_{Hg} \cos \theta_{Hg}}{r} = \frac{0.75}{r} \quad \text{or } r = \frac{0.75}{p_{Hg}} \tag{4.56}$$

where P_{Hg} is capillary pressure obtained from mercury injection method, MPa; r is the throat radius corresponding to capillary pressure p_{Hg}, μm.

If capillary pressure equals 1 MPa, then the corresponding pore radius is 0.75 μm. Convert the units of Eq. (4.56) to SI system units, and convert laboratory capillary pressure data to reservoir capillary pressure p_{cR} base on Eq. (4.39), then

$$h = \frac{100 p_{cR}}{\sigma_w - \sigma_o} \tag{4.57}$$

where h is the height of wetting phase column above oil–water contact, m; p_{cR} is reservoir capillary pressure, MPa; ρ_w, ρ_o are the reservoir water and oil density, respectively, g/cm^3.

4.3.3 Basic Characteristics of Capillary Pressure Curve

As mentioned above, the capillary pressure of ideal porous media is a function of wetting phase (or non-wetting phase) saturation, but for actual reservoir rock, the influence factor of capillary pressure is more than that. Capillary pressure is not only a function of wetting phase (or non-wetting phase) saturation, but it is also affected directly by the pore-size of reservoir rock, pore sorting, fluid properties, mineral composition of rock, and capillary hysteresis, etc., so the measured capillary pressure curves can be quite different. Do different capillary pressure curves have something in common? What characteristic parameters should be studied to recognize capillary pressure curves?

4.3.3.1 Features of Capillary Pressure Curve

One typical capillary pressure curve is shown in Fig. 4.68. According to the shape, the capillary pressure curve is usually divided into three sections: the initial section, intermediate flat section, and the final upward section (Fig. 4.68).

For the initial steep section, the non-wetting-phase saturation increases slowly with capillary pressure increases. And the increase of non-wetting-phase saturation is mostly due to the uneven rock surface and large pores, which does not mean that the non-wetting phase has really entered the core. Sometimes, only a part of non-wetting phase is into the core, the rest is consumed to fill the concave and large pores.

The intermediate flat section is the main liquid inlet section, during which most of the non-wetting phase enters the core, so the non-wetting-phase saturation increases greatly but the capillary pressure does not change significantly. The shape and position of the intermediate curve is very important to analyze the pore distribution of the core. Longer intermediate flat section indicates well sorted grains, while lower position of the intermediate flat curve indicates larger throat size.

The final upwarp section shows that the displacement velocity gradually decreases to zero with rapid pressure increase. If the end of the curve is parallel to the vertical axis, non-wetting phase saturation will not change with increasing pressure.

Fig. 4.68 Typical capillary pressure curve

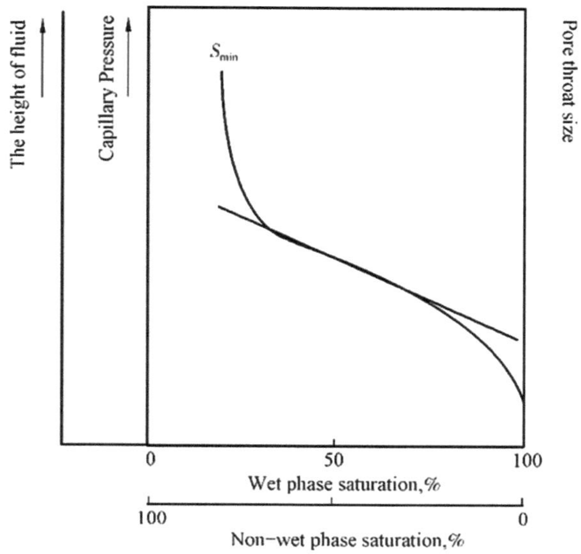

4.3.3.2 Characteristic Parameters of Capillary Pressure Curve

The following parameters, usually, are taken to describe the quantitative characteristics of capillary pressure curve (Fig. 4.69).

Expulsion Pressure (Valve Pressure)

Expulsion pressure means the pressure at which the non-wetting phase starts to enter the maximum throat in the core, which is also called inlet pressure, threshold pressure or valve pressure. Threshold pressure equals the capillary pressure in the maximum throat in the core. There are many methods to determine expulsion pressure. One commonly used method is as follows: extend the intermediate flat section in the curve to zero non-wetting phase saturation, then intersects with the vertical axis is expulsion pressure.

Expulsion pressure is one of the main parameters for evaluating the reservoir storage capacity and permeability. Good permeability leads to small expulsion pressure, and vice versa. In addition, with expulsion pressure, p_T, both the maximum throat radius and wettability of the core can be obtained (details will be discussed in the next section "Applications of capillary pressure curve").

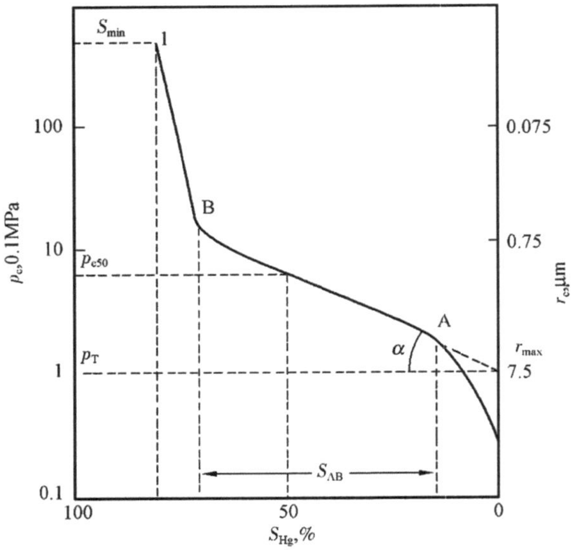

Fig. 4.69 Quantitative characteristics of capillary pressure curve

Median Capillary Pressure p_{c50}

Median capillary pressure p_{c50} refers to the capillary pressure necessary to fill 50 % of the pores with the non-wetting phase. The corresponding throat radius r_{50} is median saturation throat radius.

Obviously, the smaller the value of p_{c50}, the greater the r_{50} is, which indicates good storage capacity and permeability, and vice versa. Sometimes because of poor rock properties, the value of p_{c50} may exceed the maximum limit of the equipment and cannot be obtained from the capillary curve. The r_{50} can be roughly regarded as the average throat radius of the core because the majority of the pore size distribution of rock is close to normal distribution.

Minimum Saturation *of* Wetting Phase S_{min}

Minimum wetting phase saturation S_{min} refers to the irreducible saturation of wetting fluid. If the core is water-wet, then the S_{min} means irreducible water saturation, while if the core is oil-wet, S_{min} means irreducible oil saturation. Minimum wetting phase saturation represents the percentage of inaccessible pores at the maximum pressure of the equipment. The higher S_{min}, the more small throats are.

S_{min}, in fact, is an indicator of pore structure and permeability of rock. According to different lithology, porosity and permeability, S_{min} may vary between 0 and 90 %. The better the reservoir physical property is, the smaller S_{min} is.

The value of S_{min} also depends on the maximum pressure of equipment. It should be noted that when the end of the capillary pressure curve is not parallel to the pressure axis, S_{min} no longer represents irreducible water saturation. Regarding it as irreducible water saturation will lead to error, especially for low porosity, low permeability cores.

4.3.4 Applications of Capillary Pressure Curve

The initial purpose of determining the capillary pressure of a reservoir rock is to determine the irreducible water saturation of certain formation. With the development of capillary curve investigation, it can be used for the determination of many reservoir parameters, such as irreducible water saturation, residual oil saturation, porosity, absolute permeability, relative permeability, wettability, specific surface and pore throat size distribution, and other parameters being proposed for reservoir evaluation. Therefore, the capillary pressure data have been widely used in oil and gas production.

4.3.4.1 Rock Pore Structure

Capillary pressure curve means the curve of capillary pressure and saturation. A certain capillary pressure is corresponding to a certain throat radius ($r = 2\sigma \cos \theta / p_c$), so the capillary pressure curve actually represents pore size distribution of the core. In Fig. 4.69, the pore radius is marked on the right vertical axis of capillary pressure curve.

In order to make better use of capillary pressure curve to study pore size distribution, various pore size distribution diagrams are usually plotted, such as pore size frequency distribution diagram (Fig. 4.70) and cumulative pore size frequency

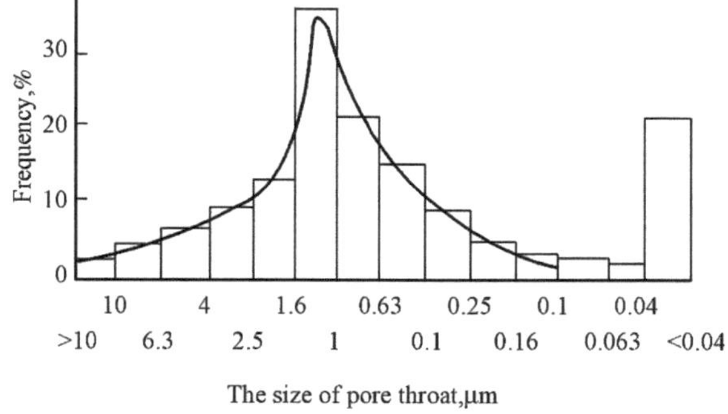

Fig. 4.70 Frequency distribution histogram and frequency distribution curve for pore-throat

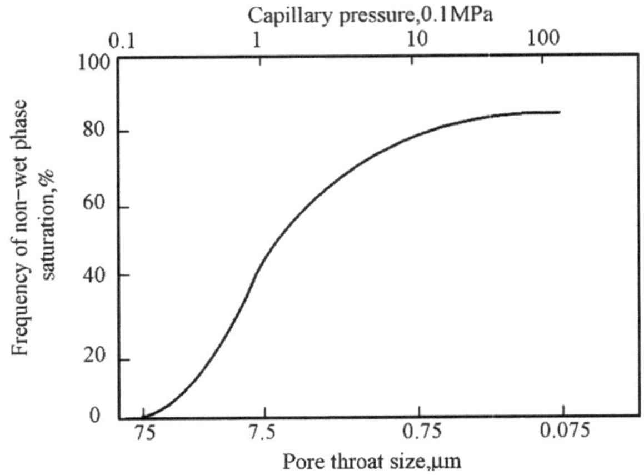

Fig. 4.71 Cumulative frequency distribution curve for pore throat

distribution diagram (Fig. 4.71). These curves can determine the size of main throat in rock. Pore size distribution is basic data for determining the grain size of mud temporary blocking agent, screening the high molecular compound of polymer flooding, and searching the optimal distribution matches.

4.3.4.2 Evaluate Reservoir Storage Capacity with Capillary Pressure Curve

The shape of capillary pressure curve mainly depends on the pore throat sorting and throat size. Sorting means the dispersion degree (or concentration degree) of the throat size. The more concentrative the throat size is, the better the sorting is, the longer the intermediate flat section of capillary pressure curve and the more parallel to the lateral axis is. The pore throat size and concentration degree mainly affects the skewness of curve. The greater the throats are, the closer the curve to the down left direction of coordinate, which is called large skewness. In contrary, the curve is closer to the upper right direction of coordinate, which is called small skewness. Figure 4.72 gives six ideal typical capillary pressure curves. It is obvious that *a* represents a reservoir with good rock properties, while *e* represents a reservoir with good rock properties.

The evaluation results of core samples based on the capillary curve can be used to identify reservoir flow units and nonflow units. The results are used to calibrate well logs, after which well logs can be used to map flow units throughout a field [33]. Those are shown in Fig. 4.73.

As for carbonate rocks, because of its own fracture structure, it is not sufficient to describe the storage and permeable capacities by using porosity and permeability merely. Capillary pressure curve is used to analyze the carbonate reservoir properties. In addition, the dual porosity of the carbonate reservoirs is clearly reflected on capillary pressure curve. Figure 4.74 presents the mercury injection capillary pressure curve of typical dual porosity media.

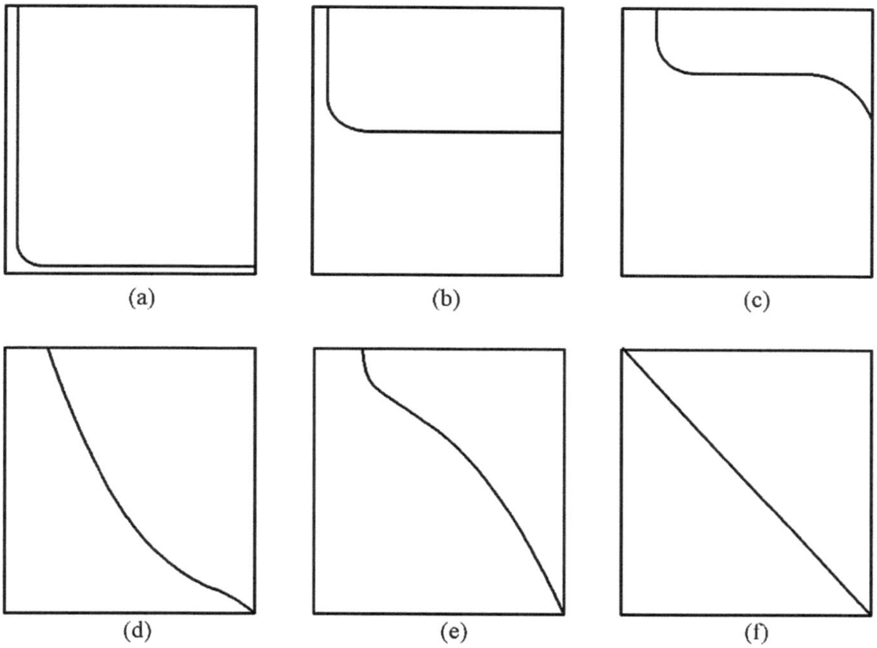

Fig. 4.72 Simplified capillary pressure curve

4.3.4.3 Determine Average Capillary Pressure of Reservoir—$J(S_w)$-Function

Capillary pressure curves obtained by experiments are based on small core samples, such as sidewall core, drilling cutting or a few representative cores, which represent an extremely small part of the reservoir. When apply them to the real reservoirs, all the capillary pressure curves from different formations should be reduced to an average capillary pressure curve. Leverett proposed the *J*-function of a specific reservoir which describes the heterogeneous rock characteristics, more adequately by combining porosity and permeability in a parameter for correlation [34].

Within a more limited field, Leverett thought that different samples from the same sediment have linked properties so that there exists an invariant specific to the sediment, He has introduced the function $J(S_w)$ known as the capillary pressure

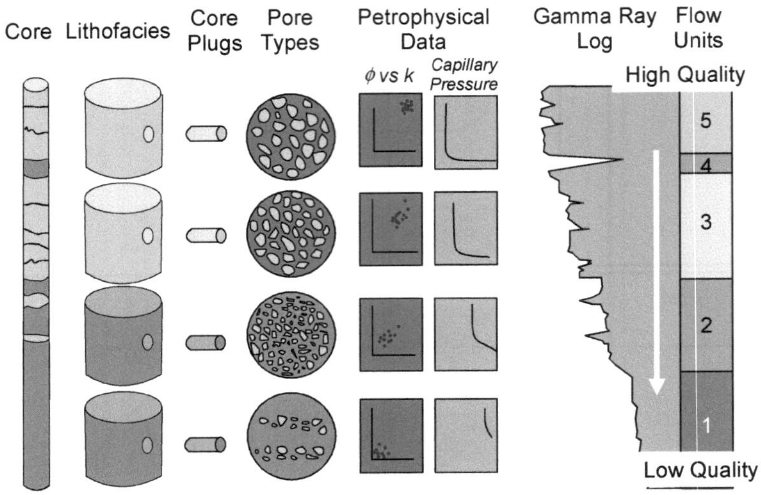

Fig. 4.73 Petrophysical analyses of core samples using to identify reservoir flow units [33]

Fig. 4.74 Dual media capillary pressure curve

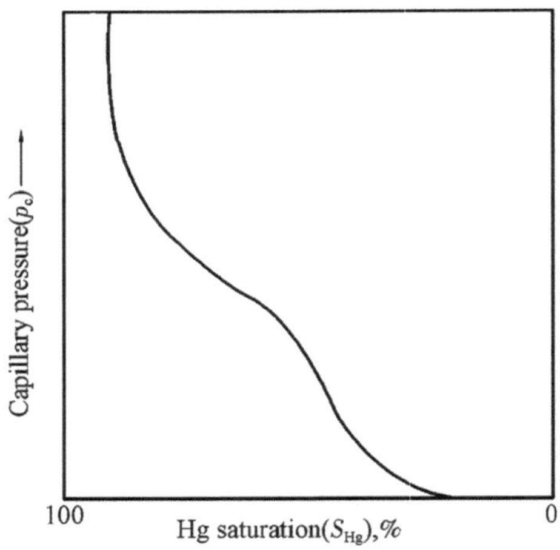

function which is a dimensionless grouping of the physical properties of the rock and the fluids saturating it. The *J*-function accounts for changes of permeability, porosity, and wettability of the reservoir as long as the general pore geometry remains constant. Therefore, different types of rocks exhibit different *J*-function correlations. The capillary pressure data for each core from a specific formation usually can be reduced to a single *J*-function versus the saturation curve, which is helpful for reservoir evaluation and comparison. The *J*-function which is a function of saturation can be expressed as follows:

$$J(S_{\rm w}) = \frac{p_{\rm c}}{\sigma\cos\theta}\left(\frac{K}{\phi}\right)^{\frac{1}{2}} \tag{4.58}$$

where $J(S_w)$ is *J*-function, dimensionless; p_c is capillary pressure, MPa; σ is interfacial tension, mN/m; ϕ is fractional porosity, decimal; θ is wetting contact angle, (°); K is air permeability, D.

The term $\cos\theta$ can be regarded as constant, which will not affect the shape of *J*-function curve. Thus the above equation can be simplified as:

$$J(S_{\rm w}) = \frac{p_{\rm c}}{\sigma}\left(\frac{K}{\phi}\right)^{\frac{1}{2}} \tag{4.59}$$

The *J*-function is a semi-empirical dimensionless function which is derived based on dimensional analysis. It has been proven valuable for correlating capillary pressure data within a lithology. The results (Fig. 4.75) obtained by Brown et al. also indicate that the *J*-function is useful for averaging capillary pressure data from same rock type from a given reservoir.

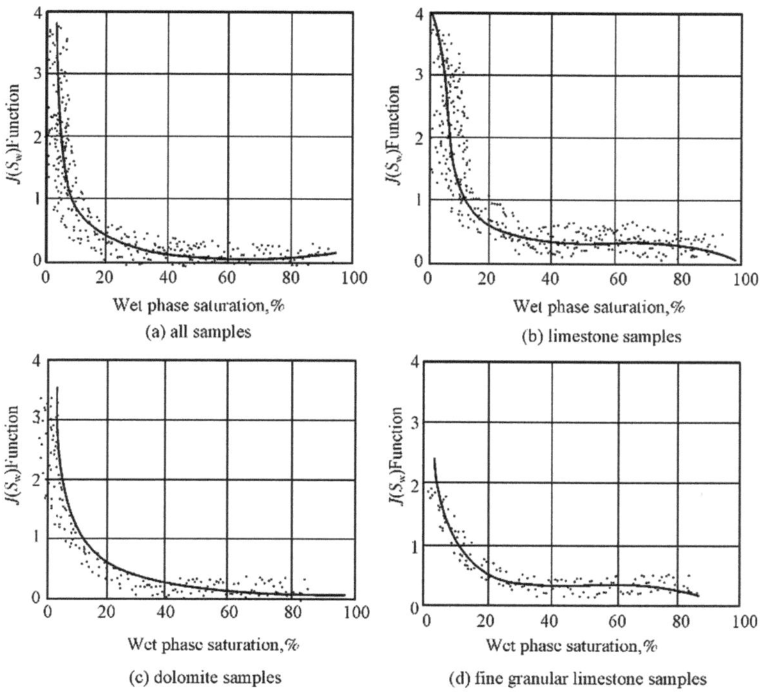

Fig. 4.75 *J*-function correlation of capillary pressure curve

Figure 4.75a shows the *J*-function curve obtained from all the capillary pressure curves from a certain reservoir. Although there is a general trend can be seen in the figure, there are still some scattered data points. Figure 4.75b, c show *J*-function curves for limestone and dolomite samples respectively. The *J*-function curve for limestone samples shows great agreement with the actual data. But the data points of limestone samples display a scattered distribution when water saturation is relatively low. The *J*-function curve for fine granular limestone samples is shown in Fig. 4.75d, which also indicates scattered date points.

Two results are obtained by using J-function: (a) the average capillary pressure data for one same type of rock (limestone or fine granular limestone) is obtained; (b) the rock physical characteristics of different types of rock are identified.

Usually, *J*-function curve varies from reservoir to reservoir, even in the same reservoir, a general *J*-function curve cannot be obtained if the permeability has strong heterogeneity. Therefore, the *J*-function method, usually, is used in relatively homogeneous formation. It will cause big error if it is used to strongly heterogeneous reservoirs. Whereas some people (such as Dullien) believe that it is improper to introduce K and Φ into *J*-function, whose value cannot reflect the differences of pore structure between different samples, which may be the main cause for the conditional use of *J*-function.

4.3.4.4 Calculate the Height of Transition Zone

As was discussed before, there is no clear interface between oil and water, but a thick oil–water transition zone. In the transition zone, water saturation gradually decreases from 100 % to the irreducible water saturation from bottom to top. The relationship between elevation and water saturation in transition zone is established in this part.

Figure 4.76 shows the oil (water) saturation change in the transition zone. According to capillary pressure data measured, the relationship between elevation and water saturation can be obtained as follows:

$$h(S_{\mathrm{w}}) = \frac{100 p_{\mathrm{c}}(S_{\mathrm{w}})}{\sigma_{\mathrm{w}} - \sigma_{\mathrm{o}}} \tag{4.60}$$

where $h(S_{\mathrm{w}})$ is the height of transition zone, m; p_{c} is capillary pressure, Pa; ρ_{w} is density of water, g/cm^3; ρ_{o} is density of oil, g/cm^3.

The above equation can be used to calculate the water and oil saturations at any height above the free liquid surface if the capillary pressure versus saturation data is available. It is usually used combined with relative permeability curves (see Fig. 4.100 for details).

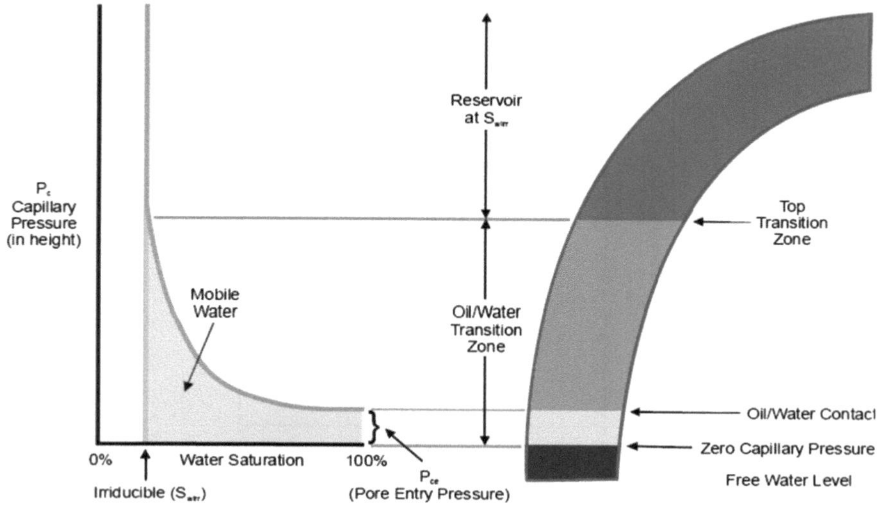

Fig. 4.76 The oil and water saturation change schematic diagram in transition zone [35]

4.3.4.5 Calculate Fluid Recovery by Capillary Pressure Loop Method

In the experiment on measuring capillary pressure curve, the capillary pressure curve obtained from a displacement process, in which non-wetting phase displaces wetting phase in core with the help of external pressure, is called a displacement capillary pressure curve, displacement curve for short. The capillary pressure curve obtained during the process of pressure release, in which non-wetting phase is displaced by wetting phase, is called imbibition capillary pressure curve, imbibition curve for short. In the mercury injection method, they are called injection curve and withdrawal curve separately.

In the past decade, many researchers have studied the withdrawal curves and found samples with similar injection curves can have different withdrawal curves, which indicates the differences and complexity of rock pore structure. And this is one of the reasons that the injecting withdrawal curves are used widely to investigate the pore structure of rocks, especially for the carbonate.

A simple method to get injection withdrawal capillary pressure curves is mercury injection method. During the test, mercury is first forced into the core and the injection curve can be obtained. Then the withdrawal curve can be obtained by decreasing pressure. A third capillary curve also can be obtained if mercury is re-injected by increasing the pressure. The above process can be repeated many times. As shown in Fig. 4.77, the measurement of capillary pressure closed loop includes at least three parts: injection, withdrawal and re-injection. The closed loop of the capillary pressure curves is used to study the pore structure and its influence on fluid distribution, flow mechanism and recovery. There are primary injection curve (curve ①), withdrawal curve (curve ②) and re-injection curve (curve ③) in Fig. 4.77.

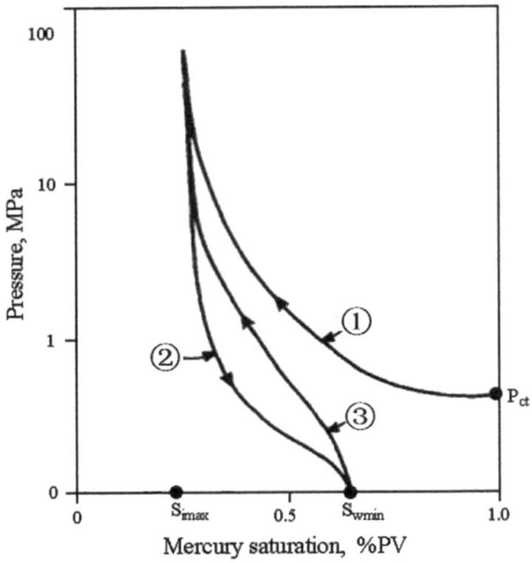

①primary injection curve; ②withdrawal curve; ③Secondary injection curve. Mercury-gas capillary pressure curves showing the initial injection curve with its threshold pressure and the hysteresis loop. Note that very high pressures are required for mercury injection.

Fig. 4.77 Injection and withdrawal capillary pressure curve

During the primary injection process, the pressure increases from zero to maximum pressure, and wetting-phase saturation decreases from 100 % to $S_{\min}$ (minimum wetting-phase saturation), while non-wetting-phase saturation increases from zero to $S_{\max}$. In the withdrawal process, the pressure decreases from maximum value to zero, but the non-wetting phase (mercury) is not totally displaced, and a small part of mercury remains in the core due to capillary force (residual mercury saturation S_R). The difference of mercury withdrawal curve and mercury injection curve is caused by capillary pressure hysteresis, which includes trap hysteresis and drag hysteresis.

Trap hysteresis is caused by internal pore structure of rock, thus trap hysteresis phenomena in different rocks with different pore structures are also different, which is shown in the difference of mercury injection curve and mercury reinjection curve. Drag hysteresis is caused by the change of wetting angle in mercury withdrawal process and the decrease of interfacial tension due to the pollution of mercury in core, which is shown in the difference of mercury withdrawal curve and mercury reinjection cure.

The minimum wetting phase saturations obtained by injection and reinjection curves are the same, $S_{\min}$. The closed loop of the withdrawal and reinjection curves is the characteristic capillary pressure hysteresis loop (the ②–③ loop in Fig. 4.77). Mercury is a non-wetting fluid therefore, the hysteresis loop exhibits a positive pressure for all saturations-that is, the hysteresis loop is above the zero pressure line [36].

Wardlaw and Taylor (1976) [36] defined the ratio of the volume of withdrawn mercury after pressure release and the volume of total injected mercury before the decrease of the pressure as withdrawal efficiency E_w, which is determined by the following equation:

$$E_w = \frac{S_{max} - S_R}{S_{max}} \times 100\,\% \tag{4.61}$$

Withdrawal efficiency, actually, is the recovery of non-wetting phase. For water-wet formation, the withdrawal efficiency is the recovery of non-wetting phase, i.e., oil recovery. This provides a new method to study oil recovery, and to discuss the relationship among recovery and pore structure and fluid properties. The difference between displacement curve and imbibition curve should be taken into account in laboratory simulation tests of secondary and tertiary oil recovery.

4.3.4.6 Determine Wettability of Reservoir Rock

Two main methods will be discussed below.

USBM Method

Donaldson et al. developed a method for determining wettability of a rock from the hystersis loop of capillary pressure curves. The test is known as the *United States Bureau of Mines* (USBM) method [37].

Figures 4.78 and 4.79 show the capillary pressure curves are obtained by alternately displacing water and oil from small cores using a centrifuge. The dashed line I in the figure is the capillary pressure curve obtained by displacement of water by oil, which is used to simulate the irreducible water saturation. Curve II is the capillary pressure curve (when the rock is water wet, curve II is called imbibition curve) obtained by the displacement of oil by water after the oil/water displacement. Curve III is the capillary pressure curve obtained by the re-displacement of water by oil. The areas under the capillary pressure curves represent the thermodynamic work required for the respective fluid displacements. A_1 is the area under the curve III, and A_2 is the area under the curve II.

If $A_1 > A_2$, the energy required to displace water by oil is larger than that required to displace oil by water, which means the reservoir rock is water wet. On the contrary, if $A_1 < A_2$, then the reservoir rock is oil-wet. It is illustrated in Fig. 4.78 that $A_1 = A_2$ indicates neutral wettability.

Usually, the logarithm of the area ratio of oil displacing water A_1 to water displacing oil A_2 is used as a convenient scale to determine the wettability of reservoir rock:

$\log\frac{A_1}{A_2} > 0$, increasing positive values to $+\infty$ indicate increasing preferential water wetting to infinite water wettability;

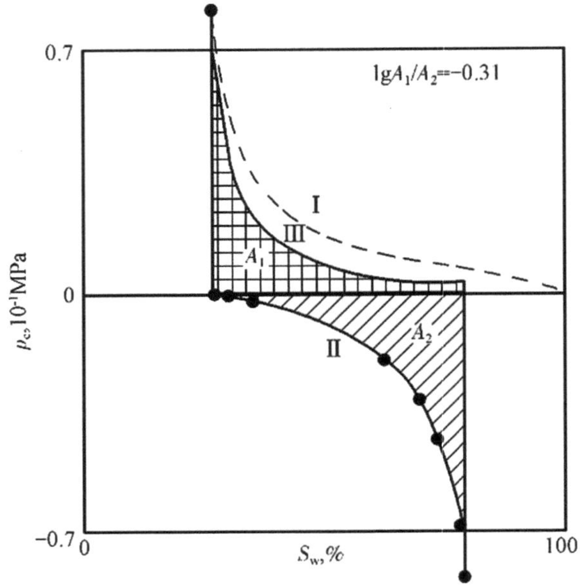

Fig. 4.78 The area ratio of capillary pressure curve for oil-wet rock

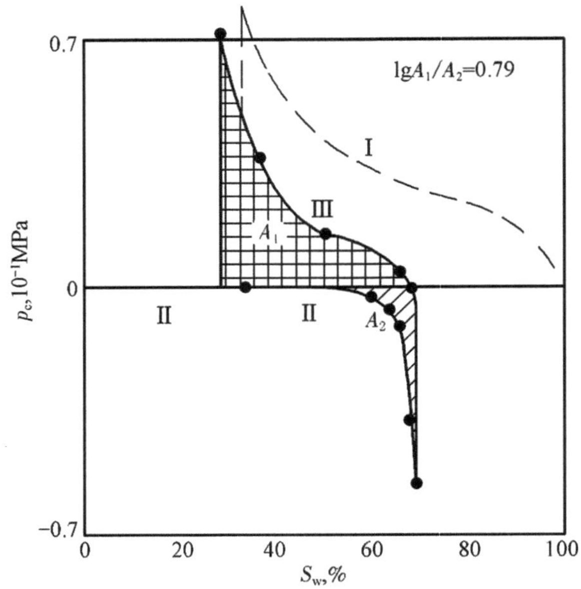

Fig. 4.79 The area ratio of capillary pressure curve for water-wet rock

$\log\frac{A_1}{A_2} < 0$, increasing negative values to $-\infty$ indicate increasing preferential oil wetting to infinite oil wettability;
$\log\frac{A_1}{A_2} = 0$, neutral wettability.

The USBM method does not depend on spontaneous imbibition, and, therefore, is sensitive to wettability throughout the range from complete water-wetting ($+\infty$) to complete oil-wetting ($-\infty$). For example, if a water–oil–rock system being tested repeatedly becomes progressively more water-wet, A_1 will become larger while A_2 will decrease. Eventually, A_2 will vanish as the hysteresis loop rises above the line representing $p_c = 0$. In this case, A_2 is zero and the wettability index $\log A_1/A_2$ is infinite, meaning 100 % wetting of the surface by water. Infinite oil wettability also is possible, in which case $A_1 = 0$ and the hysteresis loop is below the line where $P_c = 0$.

It should be pointed out that inverting the capillary pressure curve II is more convenient to compare and calculate.

Wettability Index and Apparent Contact Angle

The theoretical background for the method is as follows: Set the wetting degree of rock by oil in the oil–gas system as a criterion, and then compare it with the wetting degree of rock by water in the water–oil system, then the wetting degree of rock by water in the water-oil system can be determined.

The procedure of this method is as follows: A core is cut into two equal parts, and one is saturated with oil and then the oil in the core is displaced by gas, while the other one is saturated with water and then the water in the core is displaced by oil. Two capillary pressure curves can be obtained, and two threshold pressures can be determined by the curves, i.e., P_{Tog} and P_{Two}.

The wetting index W is defined as:

$$W = \frac{\cos\theta_{\mathrm{Two}}}{\cos\theta_{\mathrm{Tog}}} = \frac{p_{\mathrm{Two}}\sigma_{\mathrm{og}}}{p_{\mathrm{Tog}}\sigma_{\mathrm{wo}}} \tag{4.62}$$

where θ_{Two} is contact angle between water and oil in core, (°); θ_{og} is contact angle between air and oil in core, (°); P_{Two} is threshold pressure of the core saturated with water and displaced by oil, MPa; P_{Tog} is threshold pressure of the core saturated with oil and displaced by gas, MPa; σ_{og} is interfacial tensions between air and oil, mN/m; σ_{wo} is interfacial tensions between oil and water, mN/m.

Liquid has a better wettability for solid than air, therefore, oil has a better wettability on rock compared to gas, so $\theta_{\mathrm{Tog}} \approx 0$, and $\cos\theta_{\mathrm{Tog}} = 1$. So the threshold pressure needed to displace oil from the core by gas, P_{Tog}, is larger. Assume water has good wettability on rock, and then $\cos\theta_{\mathrm{Two}}$ will be greater, so that greater threshold pressure P_{Two} is needed to displace water from the core by oil.

Therefore, when W calculated by Eq. (4.62) equals 1, which indicates that the wetting ability of water on rock in the water–oil system is equivalent to the wetting ability of oil on rock in the oil-gas system, i.e., perfect wettability of water. If *W* equals zero, which indicates that water cannot wet rock completely, while oil can enter rock easily, i.e., perfect wettability of water. If the value of *W* is between 0 and 1, the closer to 0 the *W*, the more lipophilic the rock is, and the closer to 1 the *W*, the more hydrophilic the rock is.

Comparing oil and air, oil has a better affinity for rock, so set $\theta_{\mathrm{Tog}} = 0$, cos $\theta_{\mathrm{Tog}} = 1$, according to Eq. (4.62) apparent contact angle can be defined as:

$$\cos\theta_{\mathrm{Two}} = \frac{p_{\mathrm{Two}}\sigma_{\mathrm{og}}}{p_{\mathrm{Tog}}\sigma_{\mathrm{wo}}} \text{ or } \theta_{\mathrm{Two}} = \arccos\left(\frac{p_{\mathrm{Two}}\sigma_{\mathrm{og}}}{p_{\mathrm{Tog}}\sigma_{\mathrm{wo}}}\right) \tag{4.63}$$

The parameters in the right hand side of above equation can be measured directly, so apparent contact angle θ_{wo} of oil–water system can be calculated by Eq. (4.63), which can be regarded as an index for wetting degree of water and oil to rock. The closer to 0° the θ_{wo} is, the more hydrophilic the rock is, and the closer to 90° the θ_{wo} is, the more lipophilic the rock is.

Therefore, according to the threshold pressure and oil–water and water–oil interfacial tensions obtained by capillary pressure curve, the wettability index *W* or apparent contact angle θ_{wo} can be used to determine the wettability of rock.

4.3.4.7 Calculate Absolute and Relative Permeability of Reservoir

This part will be discussed in the section of relative permeability of this chapter.

4.3.4.8 Evaluate the Formation Damage Caused by Drilling Fluid and Effectiveness of Stimulation Treatment

If formation is damaged, the capillary pressure curve will show great threshold pressure and high irreducible water saturation, i.e., the curve moves toward upper right in the coordinate system. Whether there is formation damage caused by drilling fluid and effectiveness of stimulation treatments can be evaluated by comparing capillary pressure curves obtained before and after the operation. This method is provided by J.O. Amaefule et al. in 1984 [38].

4.4 Microcosmic Resistances During Oil-Water Displacing

The water flooding mechanism refers to the internal change of reservoir during exploitation. For example, the flooding medium water (or gas) should continue overcoming resistance in order to drive crude oil out in the process of water flooding oil, meanwhile, the amount and distribution of oil, gas and water will change continuously. The original continuous distribution oil will grade gradually into isolated oil droplets or oil columns. The redistribution of fluids contained in the pay formation is the result of mutual restriction of various resistances produced in the oil displacement process; meanwhile the circumstances after changing determines the energy and resistance statuses in continuous flooding process. Therefore, studying the resistance change in flooding process can better explain some phenomena that occur in production, so that measures can be taken to produce crude oil more effectively according to the reservoir performance.

4.4.1 Non-piston Characteristic of Water Displacing Oil

During early stage of water injection development, A view once imagined the process of water displacing oil is like the process of piston moving in cylinder, so all oil will be driven out (Fig. 4.80). However, the view was denied by various phenomena in actual production. For example, oil-bearing formation will produce oil and water simultaneously for a longtime after water breakthrough, which indicates that oil and water will flow simultaneously for a longtime; if a row of wells produce oil simultaneously, the time of water breakthrough will be far different from each other, etc. The non-piston characteristic of water displacing oil was found through experiment and further analysis, so the process of water displacing oil is not as shown in Fig. 4.80, but formation of three different flow regions: water flow region, oil–water mixed flow region and oil flow region (Fig. 4.81).

It has been known that the pore structure of reservoir rock is very complicated, and the wettability of pore surface is changeable according to the research in Chap. 2 and this chapter. As a result, when oil and water flow in the formation with serious heterogeneity, the driving energy and resistances are different in pores and throats. Usually, the energy of every pore should be equal as the external driving pressure differential maintains consistent. However, the capillary force is driving energy for water wet pores in fact, so they will gain additional energy; it is reverse to oil-wet pores that the capillary force is additional resistance. No matter capillary force is driving force or resistance, the driving force or resistance generated during the flow of water and oil in different pores will be different consequentially due to different pore sizes, so is the flow speed. Moreover, the oil–water viscosity difference will enlarge the flow rate difference between oil and water, as a result of

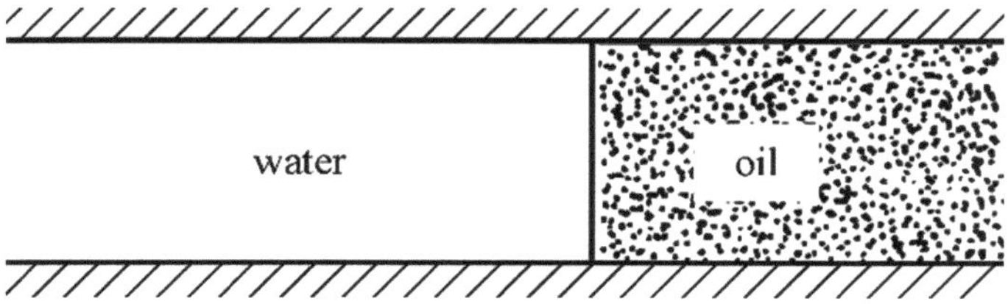

Fig. 4.80 Piston displacement of water displacing oil

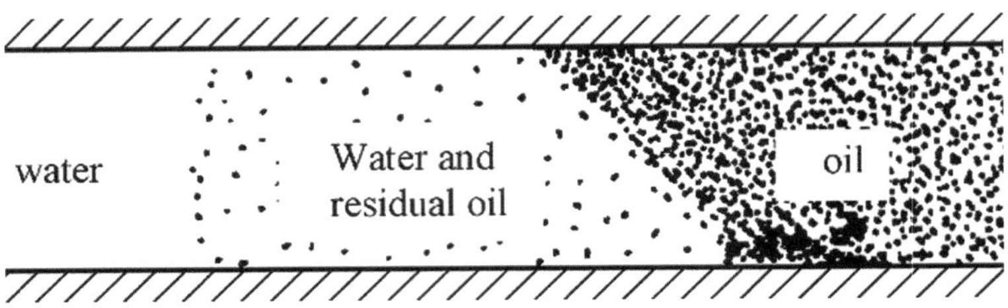

Fig. 4.81 Non-piston displacement of water displacing oil

different oil–water interfaces in various pores, so there is a mixed region that oil and water flow simultaneously.

It is true that different flow rate in various pore is one of the causes that leading to oil–water mixed region, but there are more complicated causes that bring about different flow rate, come into being mixed region and part of crude oil cannot be produced by further studying various resistances.

4.4.2 *Various Resistances in Capillary Tube*

Various resistances caused by capillary force will be discussed according to several common Scenarios in the reservoir.

4.4.2.1 Oil Column (or Bubble) in Static State

It is illustrated by the solid line in Fig. 4.82 that the radius of oil droplet is greater than that of capillary tube, and then the oil droplet becomes oil column, which will produce an extrusion force on wall.

The force on cylindrical interface points to the center of capillary tube, and it can be calculated based on Eq. (4.29):

$$p' = \frac{\sigma_{\mathrm{OW}}}{r} \tag{4.64}$$

Fig. 4.82 The additional force that oil droplet flows in capillary tube

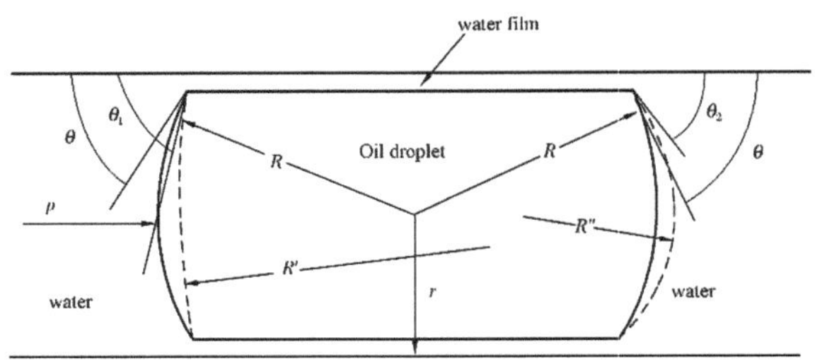

where p' is capillary pressure on cylindrical surface, Pa; σ_{ow} is interfacial tension between oil and water, mN/m; r is the radius of cylindrical tube, mm.

The force on spherical interface (which is meniscus painted by the solid line in the figure) points to the center of capillary tube, and it can be calculated by Eq. (4.65)

$$p'' = \frac{2\sigma}{R} = \frac{2\sigma_{\text{ow}} \cos \theta}{r} \tag{4.65}$$

where P'' is capillary pressure on spherical surface, Pa; σ_{ow} is interfacial tension between oil and water, mN/m; r is the radius of cylindrical tube, mm; θ is contact angle, (°).

P'' can be neglected due to they are equal and opposite in direction on the horizontal direction, but P'' can exert to the capillary tube wall, which is interpreted as the principle of transmission of liquid pressure, i.e., the Pascal's law (Pascal's law: pressure exerted anywhere in a confined incompressible fluid is transmitted equally in all directions throughout the fluid such that the pressure ratio (initial difference) remains the same), as shown in Fig. 4.83.

Therefore, the spherical capillary pressure p'' exerted on wall is opposite to the cylindrical capillary pressure p' exerted on tube center, when liquid column is in static state. The capillary pressure p'' makes liquid film thin, and the capillary pressure p' makes liquid film thick, so the liquid film maintains a certain thickness under the two forces. Ultimately, when oil droplet is static, the final resistance p_{I} (pointing to the wall of capillary tube) is

$$p_{\text{I}} = \frac{2\sigma \cos \theta}{r} - \frac{\sigma}{r} = \frac{2\sigma}{r}(\cos \theta - 0.5) \tag{4.66}$$

It is noted in Eq. (4.66) that the greater the capillary effect exerted on liquid film is, and the faster the thickness of liquid film will decrease before balance with the

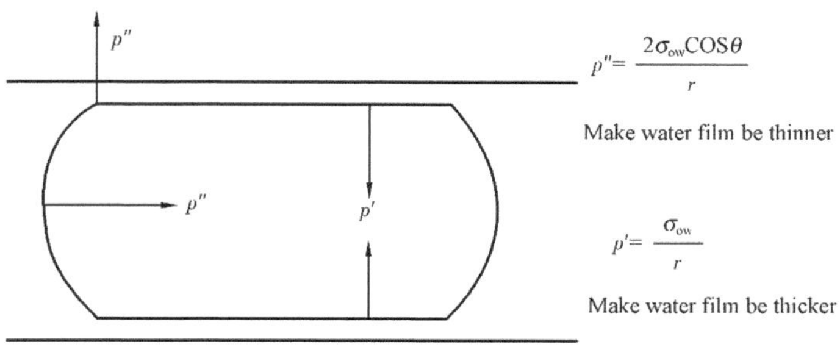

Fig. 4.83 The exerted point and direction of P', P''

greater the oil-water (or oil-gas) interfacial tension is, the smaller the capillary radius is.

It is worthwhile to note that the film on spherical surface has abnormal features: high viscous force and high intension. Therefore, there must be great enough differential pressure to overcome p_{I} and the friction pressure produced by film, so that the oil column can move.

4.4.2.2 Oil Column Is About to Move Under Differential Pressure

First, the meniscus of oil column deforms (see the dashed line shown in Fig. 4.82) due to wetting hysteresis, and the curvature radius of two ends of meniscus are unequal, and the capillary pressures of two ends are

$$p' = \frac{2\sigma}{R'} \tag{4.67}$$

$$p'' = \frac{2\sigma}{R''} \tag{4.68}$$

The second additional resistance p_{II}:

$$p_{\mathrm{II}} = p'' - p' = 2\sigma\left(\frac{1}{R''} - \frac{1}{R'}\right) \tag{4.69}$$

$$p_{\mathrm{II}} = \frac{2\sigma}{r}(\cos\theta_2 - \cos\theta_1) \tag{4.70}$$

The direction of p_{II} is opposite to the direction of liquid drop moving, and p_{II} is a resistance, so the pressure differential "$\Delta p > p_{\mathrm{I}} + p_{\mathrm{II}}$ + film resistance" must be exerted to the capillary tube in order to make liquid drop move. It is noted that the relation of p_{I} and p_{II} is not direct adding, because p_{I} exerts on the wall of capillary tube, and the friction factors of oil column and water film should be taken into consideration for the pressure in the flow direction.

4.4.2.3 The Resistance Encountered When the Bubble Is Passing Through a Narrow Throat

It is shown in Fig. 4.84 that liquid droplet or gas bubble deforms while passing through narrow pores, the curvature radii of the front and end of meniscus becomes unequal, and resistance increases, so the third additional resistance p_{III}

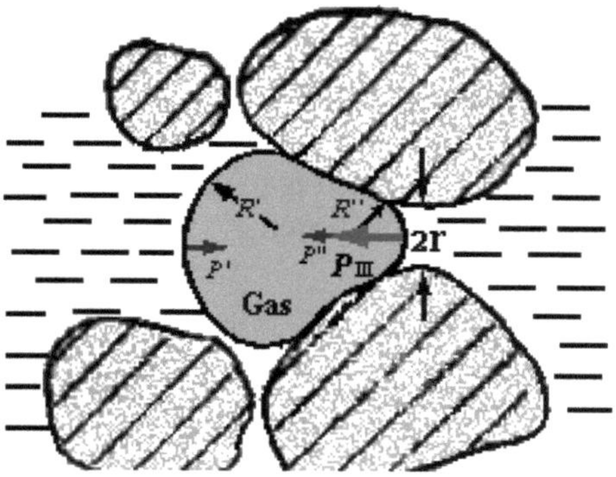

Fig. 4.84 Droplet bubble deforms when it passed through narrow pore

$$p_{\mathrm{III}} = 2\sigma\left(\frac{1}{R''} - \frac{1}{R'}\right) \tag{4.71}$$

How much the pressure difference should be exerted to make the bubble pass through the narrow channel? So when p_{III} will reach the max should be considered. And the fact is: p_{III} reaches the max when the front of oil drop deforms to the size of the narrowest channel. At this time, $R'' = r$, if the curvature radius of the end of oil drop is taken into account, $R' = \infty$, so the max of p_{III} is:

$$p_{\mathrm{III}} = 2\sigma\left(\frac{1}{r} - \frac{1}{\infty}\right) \tag{4.72}$$

Generally, liquid resistance effect refers to various effects of additional resistances which are resulted from the deformation of liquid droplet while it passes through narrow pores, and gas blocking effect refers to various effects of additional resistances which are resulted from the deformation of bubble while it passes through narrow pores, which is called Jamin effect, too. Solid resistance effect refers to various effects of additional resistances when solid fine passes through narrow pores, in the same way. Daqing Oilfield popularly call gas, liquid and solid blocking effect as gas, liquid and solid blockage (what the solid blocking different from gas and liquid blocking effect is that solid fine may not deform).

It is shown that various resistances are evident when two phases flow simultaneously, especially the one is continuous, but the other is not and it disperses in the continuous phase, furthermore, the pores change in diameter. So as less drilling mud as possible should be avoided entering oil-bearing layer, and the residual acid should be flow backed as much as possible, and the formation pressure should not be lower than saturation pressure that will lead oil to degas. Low fluid loss muds should be used when drilling or coring with water based mud to minimize or prevent water blocks in oil wells. Oil-based or inverted oil emulsion mud should be used during completion operations. In gas wells, air is recommended for drilling and coring. Using oil-based mud in gas wells may cause severe permeability damage [39].

The various additional resistances often happen in the oil–water mixed region. Besides, the other resistances such as viscous force should be studied. Several fluids flowing in simplified pore models will be discussed for study and understand some physical concepts better.

4.4.3 Fluid Flowing in Simplified Pore Models

The simplest case of single capillary tube, single phase, and only viscous force is considered at first. Then the case that taking multi-capillary, two-phase, and viscous force and capillary force in account simultaneously is studied on.

4.4.3.1 Single Phase Flow in Single Capillary Tube

At the early field life, there is only single phase with no degasification and no injection, so only viscous force is considered. Consider a single capillary tube, according to Poiseuille's equation, the flow rate through the capillary tube is calculated According to Poiseuille's equation:

$$q = \frac{1}{100} \times \frac{\pi r^4 \Delta p}{8\mu L} \quad or \ v = \frac{r^2 \Delta p}{8\mu L} \tag{4.73}$$

where L is the length of straight capillary tube, cm; r is the radius of straight capillary tube, mm; q is the flow rate through capillary tube, cm^3/s; Δp is pressure difference over length L, dyn/cm^2; μ is fluid viscosity, mPa s; v *is* flow velocity, cm/s.

It can be known that if pressure difference Δp, viscosity μ, and the length of two capillary tubes is equal, the flow velocity u is proportional to the square of the capillary radius.

If capillary radius is r_1 and r_2, respectively, and $r_1 > r_2$, then the velocity ratio as follows:

$$\frac{v_1}{v_2} = \left(\frac{r_1}{r_2}\right)^2 \tag{4.74}$$

Therefore

$$v_1 = \left(\frac{r_1}{r_2}\right)^2 v_2 \tag{4.75}$$

Equation (4.75) indicates that if there is a 10-time difference between two capillary radii, the flow velocity will have a 100-time difference. Therefore, fluid flows through dominating large pores under exerted pressure difference. The measured permeability is just a mean statistical value that describes the capacity of fluid flowing through a rock by analogy. In fact, large pores do the greatest contribution to seepage and permeability. So reservoir protection means the large pores should be protected first to a large extent.

4.4.3.2 Two-Phase Flow in Single or Multiple Noncommunicating Capillary Tubes

Both viscous force and capillary pressure are considered simultaneously in this case. To make the assumption: capillary radius r, crude oil viscosity μ_2, water (or filtered fluid) viscosity μ_1, water is wetting phase and oil is non-wetting phase (Fig. 4.85). Then, how does the flow velocity of two-phase interface change along the way?

(a) Regard the flows on both sides of the meniscus as single phase flow, the flow velocity of single phase can be calculated according to Eq. (4.73) when time is t:
Water phase

$$v_1 = \frac{\mathrm{d}x}{\mathrm{d}t} = \frac{r^2(p_1 - p_1')}{8\mu_1 x} \tag{4.76}$$

Oil phase

$$v_2 = \frac{\mathrm{d}x}{\mathrm{d}t} = \frac{r^2(p_2' - p_2)}{8\mu_2(L - x)} \tag{4.77}$$

(b) If liquid flow is a continuous flow, $v_1 = v_2$

$$\frac{\mathrm{d}x}{\mathrm{d}t} = \frac{r^2(p_1 - p_1')}{8\mu_1 x} = \frac{r^2(p_2' - p_2)}{8\mu_2(L - x)} \tag{4.78}$$

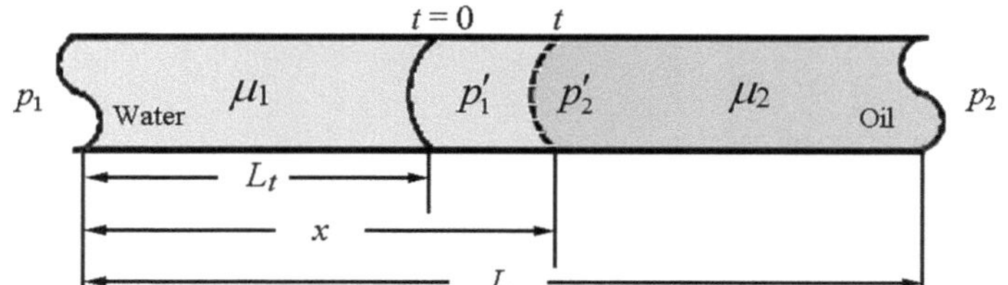

Fig. 4.85 Two-phase flow in capillary tube

$$\frac{\mathrm{d}x}{\mathrm{d}t} = \frac{r^2(p_1 - p_1' + p_2' - p_2)}{8[\mu_1 x + \mu_2(L - x)]} \tag{4.79}$$

And $p_2' - p_1' = p_c$, the differential equation is established

$$\frac{\mathrm{d}x}{\mathrm{d}t} = \frac{r^2(p_1 - p_2 + p_c)}{8[\mu_1 x + \mu_2(L - x)]} \tag{4.80}$$

(c) through separation of variables and integral

$$v = \frac{r^2(p_1 - p_2 + p_c)}{\sqrt{(\mu_2 L)^2 - (\mu_2 - \mu_1)\left[\frac{r^2 t}{4}(p_1 - p_2 + p_c) + 2\mu_2 L L_2 - L_t^2(\mu_2 - \mu_1)\right]}} \tag{4.81}$$

where p_1, p_2 are endpoint pressure of capillary tube, dyn/cm^2; p_c is capillary pressure, dyn/cm^2; L is the length of capillary tube, cm; μ_1 is the viscosity of water, mPa s; μ_2 is the viscosity of oil, mPa s; L_t is the position of oil–water interface at t time, cm, r is the radius of capillary tube, mm, t is displacement time, s; v is flow velocity, cm/s.

It is noted in Eq. (4.81) that velocity v is related to driving pressure and resistances (such as Δp, p_c, μ_1, μ_2, $(\mu_2 - \mu_1)$, r, L, L_t). Velocity is different in different pores, so does in the same size pore because it depends on viscosity difference $(\mu_2 - \mu_1)$. If $\mu_1 < \mu_2$, flow rate will be greater and greater like the water displacing oil, as a result of micro- and macro-fingering phenomenon in parallel pores with different diameters, namely in the formation with various permeable zones, which refers to viscous finger advance (Fig. 4.86).

The more serious the heterogeneity, the greater the pore size difference and the viscosity difference between injected fluid (such as water) and crude oil, i.e., μ_1 is much lower than μ_2, the more serious the viscous fingering phenomenon is, the

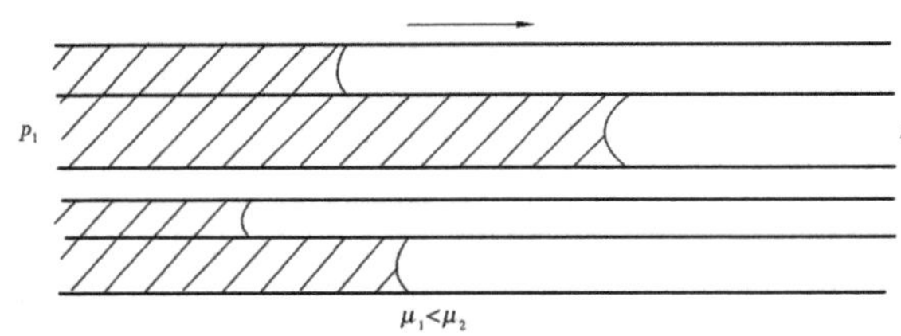

Fig. 4.86 Micro-finger advance

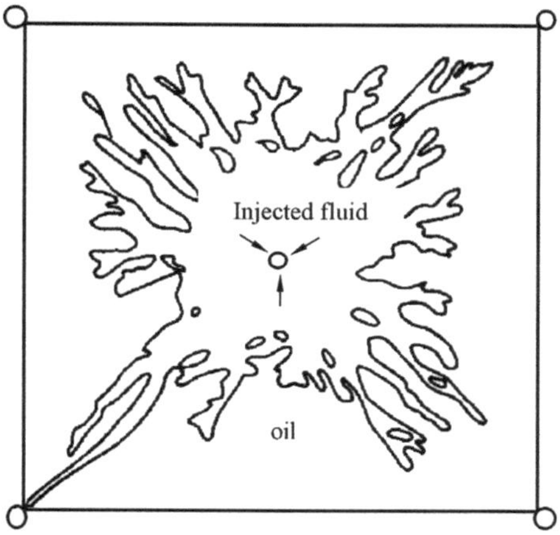

Fig. 4.87 Viscous fingering phenomenon on the plane

earlier the water breaks through, and the longer the time that oil and water are produced simultaneously is. Fingering phenomenon on the plane is shown in Fig. 4.87.

4.4.3.3 Two Phase Flow in Two Parallel Capillary Tubes with Different Diameters

It is illustrated in Fig. 4.88 that can be regarded as the common basic unit in the formation pore network and a most simplified model, which represents that pores with different sizes are interlaced or in parallel or in series. Under this condition, is the flow rate in large capillary tube greater than that in small capillary tube? It is

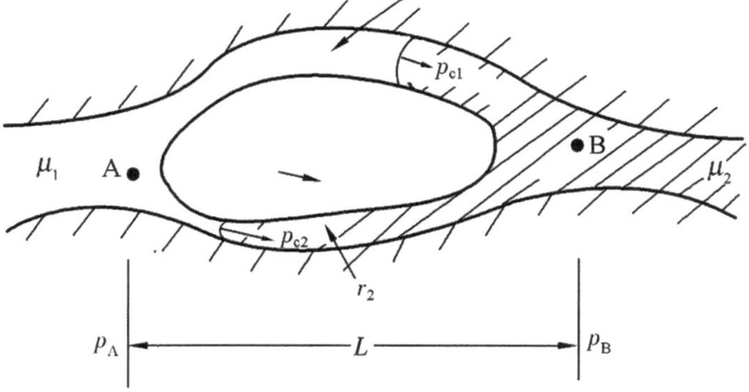

Fig. 4.88 Two-phase flow in two parallel capillary tubes with different diameters

known from the above analysis that the flow velocity of oil and water is affected by complex effects of driving energy and resistances.

To make the assumption: the radius of large pore r_1, flow rate q_1; the radius of small pore r_2, flow rate q_2; the length of both two capillary tubes L, and two pores are both water wetting. If the viscosities of oil and water are μ in two pores, and the total flow rate is q, the viscous force and capillary force can be obtained:

$$\Delta p_1 = \frac{8q_1\mu L}{\pi r_1^4}, \quad p_{c1} = \frac{2\sigma\cos\theta}{r_1} \tag{4.82}$$

$$\Delta p_2 = \frac{8q_2\mu L}{\pi r_2^4}, \quad p_{c2} = \frac{2\sigma\cos\theta}{r_2} \tag{4.83}$$

where, Δp_1, Δp_2 is pressure difference caused by viscous force in large and small pore, respectively, Pa; p_{c1}, p_{c2} is capillary pressure in large and small pore, respectively, Pa; σ is surface tension, mN/m; θ is wetting angle, (°); r_1 is the radius of small pore, mm; r_2 is the radius of large pore, mm; L is the length of straight capillary tube, mm.

The pressures of large and small pore are equal at point A and B because the capillaries are connected with each other. So the equilibrium correlation of pressure is

$$p_A + p_{c1} - \Delta p_1 = p_B; \quad p_A + p_{c2} - \Delta p_2 = p_B \tag{4.84}$$

or

$$p_{c1} - \Delta p_1 = p_{c2} - \Delta p_2 \tag{4.85}$$

Substituting Eq. (4.82) and (4.83) in Eq. (4.85), and $q = q_1 + q_2$
Then

$$q_1 = \left[\frac{4\mu Lq}{\pi r_2^4} - \left(\frac{1}{r_2} - \frac{1}{r_1}\right)\sigma\cos\theta\right] \Big/ \left[\frac{4\mu L}{\pi}\left(\frac{1}{r_1^4} + \frac{1}{r_2^4}\right)\right] \tag{4.86}$$

$$q_2 = \left[\frac{4\mu Lq}{\pi r_2^4} - \left(\frac{1}{r_2} - \frac{1}{r_1}\right)\sigma\cos\theta\right] \Big/ \left[\frac{4\mu L}{\pi}\left(\frac{1}{r_1^4} + \frac{1}{r_2^4}\right)\right] \tag{4.87}$$

If Substituting $q_1 = v_1\pi r_1^2$, $q_2 = v_2\pi r_2^2$ in Eqs. (4.86) and (4.87) respectively, the flow velocity ratio can be obtained

$$\frac{v_1}{v_2} = \frac{\frac{4\mu Lq}{\pi r_2^2} - r_2^2\left(\frac{1}{r_2} - \frac{1}{r_1}\right)\sigma\cos\theta}{\frac{4\mu Lq}{\pi r_1^2} - r_1^2\left(\frac{1}{r_2} - \frac{1}{r_1}\right)\sigma\cos\theta} \tag{4.88}$$

where q is total flow rate, cm^3/s; the other symbols are same as before.

Several cases will be discussed as follows:

(a) Capillary pressure does not work, i.e., when $\theta = 90°$, $\cos 90° = 0$, there is no capillary pressure on interface, so the velocity ratio is

$$v_1 = \left(\frac{r_1}{r_2}\right)^2 v_2 \tag{4.89}$$

If $v_1 > v_2$, oil column will remain in small pore when large pore is water out, and oil and water will flow simultaneously for a longtime.

The total flow rate q is low and capillary pressure $p_c \neq 0$ (i.e., $\cos\theta \neq 0$). At this time, velocity v_1 and v_2 depends on the complex effect of Δp and p_c. Total flow rate q is low, actually, external pressure difference Δp is low, and the value of Δp depends on σ, $\cos\theta$ and r.

(b) The two-phase interfaces in the two pores reach the exit simultaneously, i.e., $v_1 = v_2$, which is the most ideal state of water displacing oil. The total flow rate q_o is calculated by:

$$q_0 = 100 \times \frac{\pi\sigma\cos\theta\left(r_1^2 + r_2^2\right)r_1 r_2}{4\mu L(r_1 + r_2)} \tag{4.90}$$

where q_o is total flow rate, cm^3/s; θ is wetting angle, (°); r_1 is the radius of small pore, cm; r_2 is the radius of large pore, cm; L is the length of straight capillary tube, cm; μ is the viscosity of water and oil, mPa s; σ is surface tension, mN/m.

For example, given that $r_1 = 2 \times 10^{-4}$ cm, $r_2 = 1 \times 10^{-4}$ cm, $L = 5 \times 10^{-4}$ cm, $\mu = 1$ mPa s, $\sigma = 30$ mN/m. If $\theta = 0°$ and $v_1 = v_2$, the total flow rate is determined by Eq. (4.90) as follows:

$$q_0 = 100 \times \frac{\pi\sigma\cos\theta\left(r_1^2 + r_2^2\right)r_1 r_2}{4\mu L(r_1 + r_2)} = 1.6 \times 10^{-5}(\mathrm{cm}^3/\mathrm{s})$$

The results of further research indicate that:

(a) If $q < q_o$, the flow velocity of the oil-water interface in small pore is greater than in large pore. In this case, capillary pressure p_c dominates the flow, the oil–water interface in small pore thus arrives the exit first and the residual oil remains in the large pore.

(b) If $q > q_o$, the flow velocity of interface in large pore is greater than in small pore. Because pressure difference Δp between the two pores plays a leading role, the oil–water interface in large pore arrives the exit first, and the residual oil remains in the small pore. As a result, the additional resistance toward the flow will occur in narrow pore, and thus consumes more driving energy, which finally leads a lower oil recovery.

The above analysis shows that only a reasonable injection flow rate (or pressure difference, or flow velocity) can be used to displace cleaner when one phase fluid displace the other one. However, it is difficult to determine a reasonable flow rate in actual experiment. Therefore, a theory that low flow velocity first and then increasing flow velocity gradually is proposed to establish the irreducible water saturation in core in the test process of rock flow.

4.4.3.4 Two-Phase Flow in Single Capillary Tube with Variable Sections

The real porous medium system consists of pores with variable radii, and they are connected with each other in parallel or in series, and then forms capillary network. This type of network is more complicated than single pore model with consistent radius and dual pores model. When liquid–liquid interface (such as oil–water interface) moves in irregular pore, the shape of interface will change with the size of pore section, as shown in Fig. 4.89, which shows the variable section of pore. The interface is always in transient imbalance with the variable section, and the curvature of interface will change point by point, and then the capillary pressure on both sides of interface will change, and interface always deforms toward achieving the smallest energy. The shape of interface is always changing, and if exerted pressure difference is not great enough to overcome capillary pressure caused by the shape of interface, the interface will stop moving forward; if exerted pressure difference is great, the interface will jump forward. This phenomenon indicates that fluids does not flow evenly through porous medium, but in jumps, which is discovered by W.B. Haines, so it is called Haines jumps, which is an important

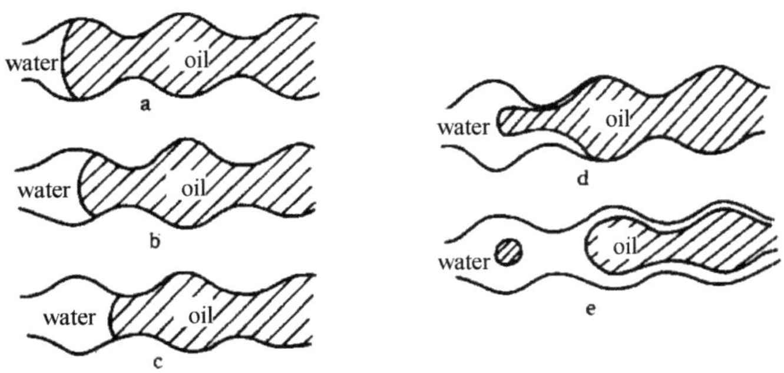

a, b and c: meniscus is in transient imbalance in variable pore;

d and e: snap-off of oil trapped by wetting phase (water) remains in pore

Fig. 4.89 Oil–water interface passes through variable sections of pore

characteristic of micro seepage mechanism in the capillary system of porous medium.

Figure 4.89c, d show that continuous oil thread is snapped off by wetting phase water when bulk oil is flowing through the narrow throat of porous medium, and then single oil droplet remains in pore. Snap-off of oil is a major mechanism causing residual oil in porous media, especially for strongly water-wet systems [40]. Chatzis et al. estimated that 80 % of the trapped oil in Berea sandstone cores occurred in snap-off geometries [41]. There are many narrow throats in actual formation and these oil drops will connect into beads. So, what features will these beads of single droplet bubble show when they flow in capillary tubes?

4.4.3.5 Mixing Fluid Flow in Capillary Pore

This case can be thought that there are water droplets and bubbles in pores, as shown in Fig. 4.90.

To make the assumption: pore length L, radius r_o, and there are bubbles distributing evenly in oil (or water). If pore is filled with bubbles (or oil drops), bubble radius is set as r. Bubbles will not deform and have no relative motion with dispersed medium (i.e., no slippage) when they flow in pore, which is equivalent to beadlike movement. If the viscosity of dispersed medium is μ (the pressure difference of two ends is $p_1 - p_2$), the closer to the wall of capillary tube dispersed medium is, the lower the flow velocity is in laminar flow, therefore, its velocity is related to the radius r of bubble (or oil drop) and the pore radius r_o.

How much is the flow velocity of liquid flow v_L when there are two flow patterns in tube: full pipe liquid flow and circulation flow (there are droplets and bubbles in pore).

Full pipe liquid flow:

$$v_0 = \frac{r_0^2 \Delta p}{8\mu L} \tag{4.91}$$

In terms of circulation flow, the liquid flow velocity v_L of fluid around droplets and bubbles is:

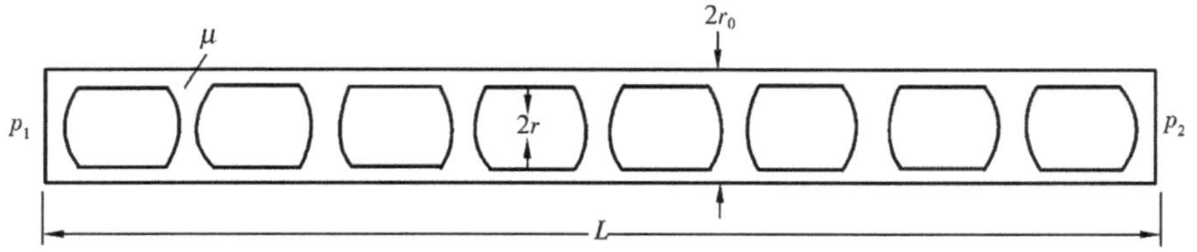

Fig. 4.90 Mixing fluid flow in capillary pore

$$v_L = \frac{(r_0^4 - r^4)\Delta p}{8\mu L r_0^2} \tag{4.92}$$

Dividing Eq. (4.92) by (4.91), and then the velocity ratio of circulation flow to single phase flow can be obtained

$$\frac{v_L}{v_0} = 1 - \left(\frac{r}{r_0}\right)^4 \tag{4.93}$$

$$v_L = \left[1 - \left(\frac{r}{r_0}\right)^4\right] v_0 \tag{4.94}$$

According to Eq. (4.93), it can be figured out that the velocity ratio (v_L/v_o) of the mixed multiphase flow to the single-phase flow in the same pore is 0.35, 0.04 and 0.004 respectively if the ratio of bubble radius to pore radius equals 0.9, 0.99 and 0.999.

Therefore, the small the difference between the radii of bubbles and the pore is, the faster the velocity of the mixed multiphase flow decreases. Besides, there is adsorption layer with abnormal viscosity on the pore surface, so that oil drops or bubbles are filled seriously with or block pores. Therefore, it will lead to increasing the oil flow resistance, difficult oil testing and low production when well control fluid invades oil-bearing layer and mud water loss.

The additional resistances and oil displacement case of capillary model previously discussed is the theory basis of seepage mechanism, studying and analyzing core flow experiments and various phenomena in production and experiments. It is difficult to calculate displacement rate and additional resistances by the above equations because of the complexities and varieties of pore structure and multiphase. Also, it is impossible to calculate the resistance caused by droplet bubble, abnormal viscosity of liquid film, and the resistances caused by capillary pressure. So the basic study methods of petrophysics will be used, i.e., the mean and macro flow resistance (or flow ability) shown in the process that fluid flows in pores can be determined by multiphase fluids flow experiment, which is the most effective method currently that will be discussed, and the concepts of effective permeability and relative permeability are introduced.

4.5 Effective Permeability and Relative Permeability Curves of Reservoir Rock

Routine permeability measurements are made with a single fluid filling the pore space. This is seldom the case in the reservoir situation except in water zones. Generally, two and sometimes three phases are present, i.e., oil, water, and occasionally gas as well. It has been known that the wettability of rock, various interfacial resistances and pore structures will affect the mobility of oil and water, that is, fluids will interact with each other in multiphase fluid flow. Here one would expect the permeability to either fluid to be lower than that for the single fluid since it occupies only part of the pore space and may also be affected by interaction with other phases. What parameters can be used to describe the mutual effect between different phases (such as rock–oil–water)? The concept used to address this situation is called relative permeability, which is a dynamic characteristic parameter of the interaction between rock and fluids, also one of the most important parameters during field development.

4.5.1 *Effective Permeability and Relative Permeability*

Darcy's law, an empirical equation describing the laminar flow of incompressible fluids, is largely used for calculation of fluid flow through porous media. It relates the flux of a fluid of known viscosity to the pressure gradient by a proportionality factor called absolute permeability, expressed in darcies. Generally, the fluid flow in hydrocarbon reservoirs involves more than one fluid, in which case the ability of each fluid to flow is reduced by the presence of other fluids. Effective permeability is a measurement of the conductance of a porous medium for one fluid phase when the medium is saturated with more than one fluid. The effective permeability of oil, gas and water can be expressed as K_O, K_q, K_w, respectively. Darcy's equation has been extended to such situations using the concept of effective permeability, which is the apparent permeability of a fluid at a given saturation.

4.5.1.1 Absolute Permeability

There is a rock sample whose length is 3 cm, and sectional area is 2 cm^2. The sample is 100 % saturated with the brine whose viscosity is 1 mPa s. The flow rate is 0.5 cm^3/s when pressure difference is 0.2 MPa, so the absolute permeability of the sample is:

$$K = \frac{Q\mu L}{A\Delta p} \times 10^{-1} = \frac{0.5 \times 1 \times 3}{2 \times 0.2} \times 10^{-1} = 0.375(\mathrm{D})$$

If the sample is 100 % saturated with the brine whose viscosity is 3 mPa s, and the flow rate is 0.167 cm^3/s under 0.2 Mpa pressure difference, and then the absolute permeability of the sample is:

$$K = \frac{Q\mu L}{A\Delta p} \times 10^{-1} = \frac{0.167 \times 3 \times 3}{2 \times 0.2} \times 10^{-1} = 0.375(\text{D})$$

It is indicated that absolute permeability is an intrinsic property of rock. Absolute permeability does not change with the properties of fluids through rock as long as there are no physical and chemical reactions between fluids and rock.

4.5.1.2 Effective Permeability

For the same sample, which is saturated with brine ($S_w = 70\ \%$) and oil ($S_o = 30\ \%$), and the saturations do not change during the flowing. If pressure difference is the same as the above, the flow rate of brine is 0.30 cm^3/s, and the flow rate of oil is 0.02 cm^3/s, then how to determine the effective permeability of oil and water?

Although Darcy's law is developed for single phase flow, it has been applied widely to multiphase flow. In multiphase flow, each phase can be regarded as single phase flow in the medium composed of solid and other phases, so that Darcy's law holds, but the permeability is replaced by effective permeability of the phase. The various resistances in multiphase flow can be reflected on effective permeability of the phase. Thus, the effective permeability of brine is

$$K_w = \frac{Q_w \mu_w L}{A\Delta p} \times 10^{-1} = \frac{0.3 \times 1 \times 3}{2 \times 0.2} \times 10^{-1} = 0.225(\text{D})$$

The effective permeability of oil is

$$K_o = \frac{Q_o \mu_o L}{A\Delta p} \times 10^{-1} = \frac{0.02 \times 3 \times 3}{2 \times 0.2} \times 10^{-1} = 0.045(\text{D})$$

The sum of the effective permeabilities of oil and brine is $K_w + K_o = 0.2709\text{D}$, which is less than absolute permeability of the rock $K = 0.375\text{D}$, that is to say, the sum of the effective permeability for all phases is always less than the absolute permeability of the rock because of the interference between fluids that share the same channels. That is because there are capillary pressure, adhesive force and other additional resistances during multiphase fluids flow. Therefore, effective permeability reflects not only the intrinsic property of reservoir rock but also the properties of fluids, the distribution of oil and water in reservoir rock and the interactions between them, that is why effective permeability is a dynamic property to describe rock–fluid interaction.

4.5.1.3 Relative Permeability

Another useful concept in describing the multiphase flow in porous media is relative permeability, which is defined as the ratio of the effective permeability of a fluid to the absolute permeability of the rock. Relative permeability has a first-order dependency on saturation level. It can measure the flow ability through the porous medium for one phase when the medium is saturated with more than one fluid.

The relative permeabilities of oil, gas, and water are separately defined as follows:

$$K_{ro} = \frac{K_o}{K} \tag{4.95}$$

$$K_{rg} = \frac{K_g}{K} \tag{4.96}$$

$$K_{rw} = \frac{K_w}{K} \tag{4.97}$$

where K_{ro}, K_{rg} and K_{rw} are the relative permeabilities of oil, gas and water in multiphase flow, respectively; K is the absolute permeability of the rock in which oil, gas, and water flow.

For example, the relative permeability of water in the above example is

$$K_{rw} = \frac{K_w}{K} = \frac{0.225}{0.375} = 0.60$$

The relative permeability of oil is

$$K_{ro} = \frac{K_o}{K} = \frac{0.015}{0.375} = 0.12$$

Although $S_w + S_o = 100\ \%$, but $K_{rw} + K_{ro} = 72\ \%$, which is less than 100 %. The conclusion on the relative permeability is universal. The sum of the relative permeability of each phase during multiphase flow in a rock is always less than 1 or 100 %.

The above example also shows that the water saturation is 70/30 = 2.33 times as big as oil saturation, but the relative permeability of water is 0.60/0.12 = 5 times as big as the relative permeability of oil. So what if the water saturation increases by 10 % ($S_w = 80\ \%$, $S_o = 20\ \%$), is the relative permeability of water is still 5 times as big as the relative permeability of oil or not? Many experiments have shown that the relative permeability is a function of water saturation, which can be measured by experiments, and the correlations curves of relative permeability and saturation can be obtained. Many interstitial fluid distributions are possible for each level of saturation, depending on the direction of saturation changes, Thus, values of relative permeability versus saturation obtained for drainage (reduction of wetting-phase saturation) maybe different from those for imbibition (increase in wetting-phase saturation). This phenomenon is called hysteresis.

4.5.2 Characteristics and Factors Affecting Relative Permeability Curves

4.5.2.1 Characteristics of Relative Permeability Curve

Typical oil–water (or oil or gas) relative permeability curves are shown in Fig. 4.91.

The vertical coordinate are the relative permeability for oil and water, and the horizontal coordinate is water saturation that increases from 0 to 1, wherein oil saturation decreases from 1 to 0.

In practical applications, the numerator of the relative permeability ratio is usually the effective permeability of a fluid, but the denominator has more than one choice:

Absolute permeability obtained by using gas as a test fluid (K or K_a)

The permeability of rock 100 % saturated with formation water (K_{sw})

EndPoint permeability: the oil phase permeability at irreducible water saturation K_o^0.

Two-Phase Relative Permeability Curves

Figure 4.91 is the oil–water relative permeability curve for water-wet rock. It shows the characteristics of two-phase relative permeability curve and can be used to study

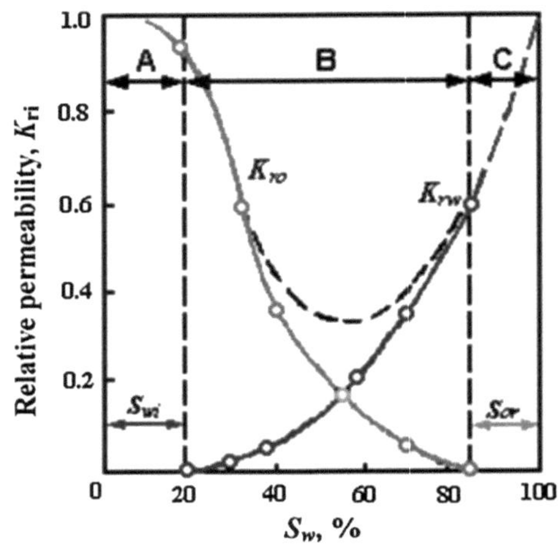

Fig. 4.91 Relative permeability curves of oil and water

oil (gas)–water distribution and fluid flow. According to the characteristics of the curve, it can be divided into three zones.

Oil Phase Flow Region (Region A)

S_w is small, and $K_{rw} = 0$; while S_{ro} is very large, K_{ro} decreases slightly. This characteristic depends on the oil–water distribution and flowing situation in the rock. When the water saturation is very small (e.g., $S_w < S_{wi} = 20\ \%$), water occupies the surface of the rock, the edges, corners, and narrow parts of the pore because the rock is water-wet, while oil is easy to flow in large pores. So, water flow has little effect on oil, and the relative permeability of oil decreases slightly. Water on the rock surface, the edges and corners cannot flow, so the relative permeability of water is zero. S_{wi} is the irreducible water saturation or known as the balanced saturation, the water saturation that cannot decrease any more (if water saturation is smaller than S_{wi}, water cannot flow), the coexisted water saturation and residual wetting phase saturation.

Although non-wetting phase saturation does not reach 100 % due to a little wetting phase exists in the rock (such as on the grain surface or in the dead-end pores), the effective permeability of the non-wetting phase nearly equals absolute permeability because the flow channel of non-wetting phase is almost the same as that when only non-wetting phase exists. With the further increase of the wetting phase saturation in the rock, the flow space of non-wetting phase is affected by the increasing wetting phase even though it does not flow. So, the relative permeability of non-wetting phase shows a slight decrease.

Two-Phase Flow Region (Region B)

In this zone, with increase of water saturation, the increase of K_{rw} and the decrease of K_{ro} are very obvious. Water begins to flow under pressure difference when water saturation reaches a certain value which is called the lowest wetting phase (water) saturation. If wetting phase (water) saturation is greater than the value, water will occupy more and more pores, and its flow network gradually expands, so that K_{rw} gradually increases. At the same time, non-wetting-phase (oil) saturation decreases, and the flow channels of oil is gradually occupied by water. When the non-wetting phase saturation reduces to a certain extent, not only the original oil flow channel occupied is by water, but also the liquid resistance occurs due to the complexity of pore structure, which will affect greatly oil (and water) flow. The simultaneous flow of oil and water will cause the oil–water interaction and interference. Therefore, oil–water flow area is the area with most obvious flow resistances. The sum of oil and water relative permeability, $K_{rw} + K_{ro}$, will have its minimum value in this zone (the dotted line in Fig. 4.91).

Single Phase Water Flow Region (Region C)

In this zone, non-wetting phase (oil) saturation is less than the residual oil saturation (the oil saturation corresponds to $K_{ro} = 0$). $K_{ro} = 0$, and K_{rw} changes significantly. Non-wetting phase (oil) has lost its continuity and dispersed into globules distributed in the wetting phase (water), and finally trapped in the pore. These oil droplets cause great resistance to the water flow due to the Jamin effect, which can be seen from Fig. 4.91 that when the oil saturation increases from 0 to 15 % (residual oil saturation), the relative permeability of water-phase decreases from 100 to 60 % (falls by 40 %), which indicates the resistance of the dispersed oil droplets on the flow.

A large number of experiments have shown that, whether oil-gas, gas-water or oil–microemulsion, water–microemulsion system, or unconsolidated sand or consolidated sandstone and so on, the characteristics of the curves of relative permeability and saturation are consistent with the general characteristics. Figure 4.91 actually represents the common characteristics of relative permeability curves of the wetting phase and non-wetting phase system. To sum up, they are:

(a) Whether the wetting phase or non-wetting phase, there is a minimum saturation when one just begins to flow (also known as the balanced saturation), when the fluid saturation is less than the minimum saturation, it cannot flow. The general rule is: The minimum saturation of wetting phase is greater than that of the non-wetting phase, such as the irreducible water saturation of hydrophilic rock is greater than the residual oil saturation.
(b) When the non-wetting phase saturation is less than 100 %, its relative permeability can be almost close to 100 % (or 1); however, only the wet phase saturation reaches 100 %, its relative permeability can be 100 %.
(c) When two phases flow simultaneously, the sum of two-phase relative permeability must be less than 1, and the minimum sum of relative permeability will occur.

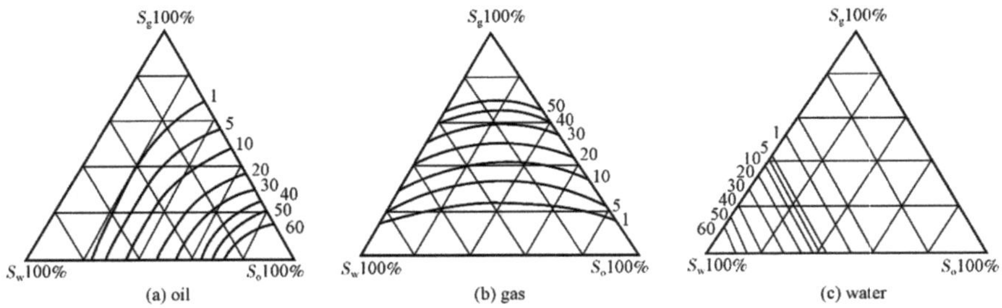

Fig. 4.92 The relative permeability curve of three phases (water wet media) (from Leverett and Lewis 1941 [42])

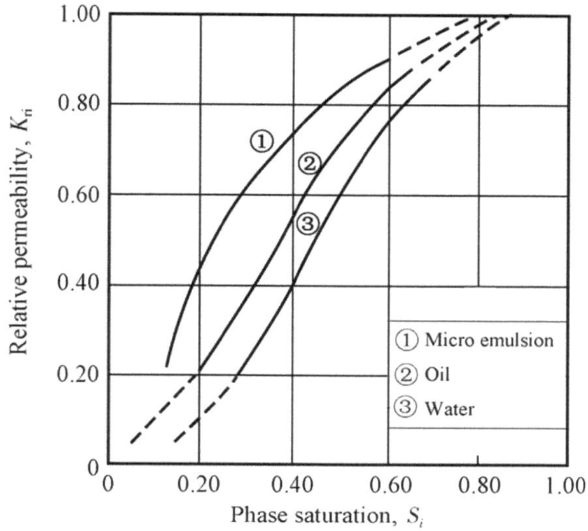

Fig. 4.93 Relative permeability curves of three phases (Delshad and Pope [43])

Three-Phase Relative Permeability

During field development, there may two phases or even three phases coexist with the development of water flooding and crude oil degassing, Three-phase flow occurs when the water saturation is higher than the irreducible level, and oil and gas are also present as mobile phases. How to determine the relative permeability when three phases are present?

When oil, gas, and water in the reservoir have certain saturation, whether all of the three phases flow, this must be determined by the correlations between the relative permeability of three phases and saturation. Three-phase relative permeability curves can be represented by using ternary diagram (Fig. 4.92) and Cartesian coordinate diagram (Fig. 4.93).

Leverett and Lewis obtained three separate triangular graphs showing lines of constant relative permeability ("isoperms") to the three phases; these were plotted against the saturations of the three fluids, as shown by Fig. 4.92.

Figure 4.92a–c, is the curve diagram of the relative permeability and saturation of oil, gas, water phase, respectively. And it shows that the relative permeability to wetting phase (water) is only related to the saturation of wetting phase in Fig. 4.92c. When the other two saturation values are equal, it has a same effect on permeability to water phase.

The points, at which the value of the relative permeability for each phase equals 1 %, is taken as the starting point that phase flow, and drawing the contour line of each phase relative permeability that equals 1 % in the same ternary diagram, as shown in

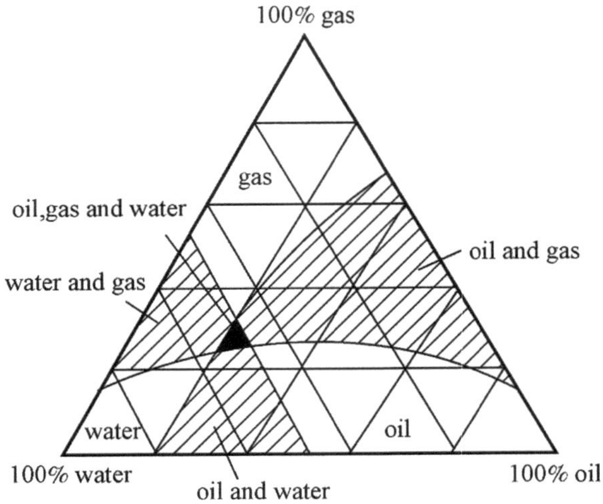

Fig. 4.94 The curves of flow behavior and saturation with three phases present (Leverett and Lewis [42])

Fig. 4.94. It may be used to determine that a single-phase, two-phase or three-phase flow will occur at some saturation. The single-phase and two-phase flow areas are larger, and three phases flow simultaneously is very small. Therefore, the two-phase relative permeability curves can meet reservoir engineering calculation. In the past, the use of three-phase relative permeability data for conventional reservoir engineering calculations has seldom been necessary. For example, Stone et al., in 1970, established a method to represent three-phase by using two-phase data.

When the gas content is low, the system can regarded as water-hydrocarbon two-phase (water and oil) system, and water is the wetting phase and hydrocarbon is a non-wetting phase.

When the water content is low, the system can regarded as gas–hydrocarbon two-phase (gas and oil) system, and liquid is wetting phase, gas is the non-wetting phase.

The method that two phases are attributed to the wetting phase and non-wetting phase greatly facilitates the practical applications of engineering.

It should be noted that the three-phase relative permeability curves are measured in the hydrophilic medium by Leverett, et al. in 1941 [42]. There will be different curves due to different rocks and fluids, and then what factors will affect the value of the relative permeability and the relative position of the curve at all.

4.5.2.2 Factors Affecting Relative Permeability Curve

Numerous studies show that: the relative permeability is not the unique function of the saturation for a given core, and it is strongly affected by the wettability, as well as the saturation history, pore structure, fluids used in the experiments, the experimental temperature and displacement pressure. The relative permeability curves measured in experiment are the final outcome affected by all the factors.

Effects of Rock Wettability on Relative Permeability

The wettability of a rock will strongly affect its water flooding behavior and the relative permeability of fluids flowing in the rock. Wettability affects relative permeability because it is a major factor in the control of the location, low, and distribution of fluids in a porous medium. The general trend of the relative permeability affected by rock wettability is that oil relative permeability tends to decrease as the rock wettability turns into complete lipophilicity from the complete hydrophilicity. This phenomenon can be explained according to the different wettability and oil–water distribution in the rock pores: for hydrophilic rock, water is often distributed in small pores, the stagnant pores or on the surface of grains; this distribution pattern has little effect on oil permeability. Water phase in the form of droplets or continuous phase in the pore channels impedes the flow of oil in the lipophilic rock at the same saturation conditions. The oil is attached to the grain surface or in the small pores with oil film, thus relative permeability of oil will decrease at the same oil saturation. This trend can be further observed from following two types of experimental results.

The series of relative permeability curves is measured by using natural cores and adding different concentrations of surfactants to the oil–water system to change the contact angle as well as change the wettability in the first class of experiments, as shown in Fig. 4.95. It can be seen from the figure that the oil relative permeability increases and that of water decreases when the wettability of rock changes from strong lipophilicity (θ = 180°, curve 5) to strong hydrophilicity (θ = 0°, curve 1),

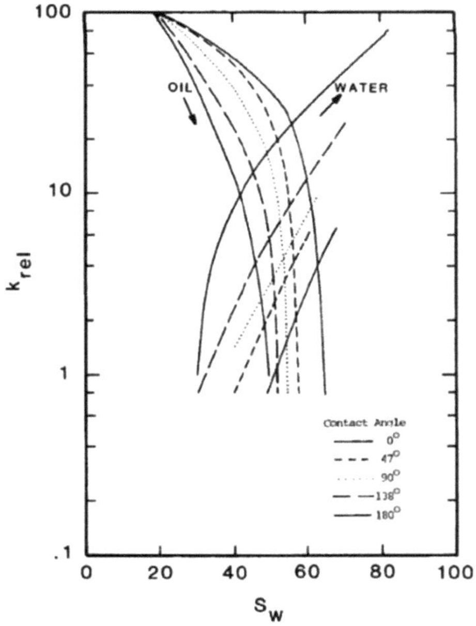

Fig. 4.95 The effects of wettability on relative permeability during water flooding (Owens and Archer [30])

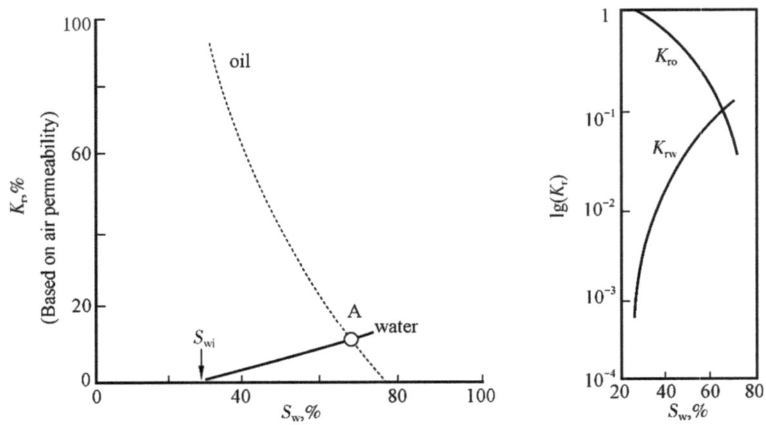

Fig. 4.96 Relative permeability to oil and water for strongly water-wetting rock

Table 4.6 Correlation of wetting contact angle and the oil relative permeability

Contact angle	0	47	90	130	180
Relative permeability to oil-phase at endpoint (K_{ro})	0.98	0.83	0.80	0.67	0.633

and the intersection point moves toward right gradually. This is consistent with the relative permeability curves (Fig. 4.96) of strongly water-wetting rock. The relative permeability curves are expressed in a semi-logarithmic coordinate system, which is conducive to observe and analyze the characteristics of relative permeability curves closed to the end saturation.

The second class of experiments carried by Owens et al. for measuring the relative permeability to oil phase at endpoint (i.e., the oil permeability when water saturation equals irreducible water saturation at the beginning of experiment), and the results are shown in Table 4.6. As can be seen from the table, the rock lipophilic becomes strong with the increase of the contact angle, and effective permeability of the oil gradually decreases, relative permeability at the endpoint of the oil phase decreases.

The wettability of reservoir rock is often determined based on the characteristics of relative permeability curve. Craig presented several rules of thumb, which are shown in Table 4.7, that indicate the differences in the relative permeability characteristics of strongly water-wet and strongly oil-wet cores.

The differences in relative permeability measured in strongly water-wet and strongly oil-wet systems are caused by the differences in the fluid distributions. Consider a strongly water-wet core. At the irreducible water saturation, the water is located in the small pores, where it has very little effect on the flow of oil. Because the water does not significantly block the oil flow, the oil effective permeability is relatively high, often approaching the absolute permeability [44]. In contrast, the

Table 4.7 Craig rules to determine the wettability of rock [23]

Feature point	Water wetting	Oil wetting
Irreducible water saturation (S_{wi})	>20–25 %	<15 %
Saturation (S_i) at which oil and water relative permeability are equal	>50 % S_w	<50 % water saturation
The water relative permeability at max water saturation (K_{rw})	<30 % (commonly)	0 %
The oil relative permeability at irreducible water saturation (K_{ro})	0 %	>50 % and approaching 100 %

effective water permeability at residual oil saturation is very low, because some of the residual oil is trapped as globules in the centers of the larger pores, where it is very effective in lowering the water permeability. Therefore, the water permeability at residual oil saturation is much less than the oil permeability at irreducible water saturation, with a ratio of less than 0.3 for a strongly water-wet core.

The above analysis tells us that the experimental results can be truly representative of the flow properties of the layer only when the wettability of rock is indeed the wettability of the rock from sampling layer in the determination of relative permeability curves of a certain reservoir rock. Therefore, the original wettability of rock must be preserved from the formation to laboratory.

Effects of Fluid Saturation History

The relative permeability measured in a drainage process and that in an imbibition process are difference because of effects of saturation history. Different process will affect the distribution of fluids in pores, and will exhibit hysteresis phenomenon in the wetting characteristics and capillary pressure characteristics. So it is essentially believed that the hysteresis will affect the relative permeability curves for drainage and imbibition process.

The relative permeability of a porous medium to a fluid at a given saturation depends on whether that saturation is obtained by approaching it from a higher value or a lower one. In a displacement process where the wetting phase saturation is approached from a lower value, the resulting relative permeability curve is referred to as an imbibition curve (an increase in the wetting phase). Examples of imbibition processes are the injection of water during water flooding and coring a water-wet rock with a water-base mud. On the other hand, in a displacement process where the wetting phase saturation is approached from a higher value, the resulting relative permeability curve is referred to as a drainage curve. Examples of drainage processes are the displacement of oil by expansion during primary depletion of a reservoir and the accumulation of hydrocarbons in oil and gas reservoirs; another example would be water flooding an oil-wet reservoir [45].

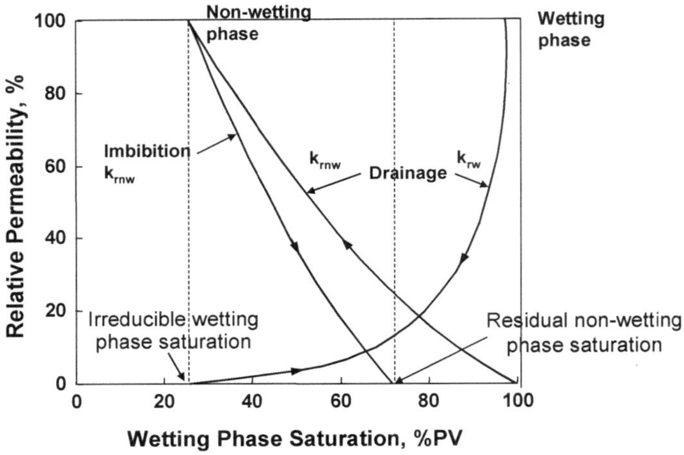

Fig. 4.97 Relative permeability in drainage and imbibition process

Figure 4.97 is the curve of relative permeability and wetting phase saturation in displacement and imbibition process. It is worth noting that the relative permeability to the wetting phase depends on the saturation of the wetting phase alone, which indicates that the imbibition and drainage wetting-phase relative permeability of a consolidated or unconsolidated rock are retraced under a succession of imbibition and drainage cycles. This case can be seen in a highly preferential wetting system (extreme water wetting or extreme oil wetting). However, a distinct path is traced by the non-wetting phase relative permeability curve to residual non-wetting phase saturation in a reversal of the saturation change from drainage to imbibition. The relative permeability to non-wetting phase at any saturation in imbibition process is always lower than the value in displacement process.

However, the conclusions obtained by other researchers (such as Osoba, et al. in 1951) is inconsistent with the above view, and their results (Fig. 4.98) show that whether it is wetting phase or non-wetting phase, its relative permeability will be affected by the saturation history. But there is a same view that the effect of saturation history on the relative permeability to non-wetting phase is much greater than that to wetting phase, and the relative permeability curves for wetting phase measured by displacement and imbibition have little difference.

The relative permeability curves measured by displacement and imbibition have difference, which is known as hysteresis, and the hysteresis is caused by capillary hysteresis. As was mentioned previously, there are many causes of capillary hysteresis, such as wetting hysteresis and the change of capillary radius and so on. These hysteresis phenomena will be expressed on the relative permeability curves measured by displacement and imbibition. The effect of saturation history on relative permeability (especially for non-wetting phase permeability) is great; therefore, either displacement process or imbibition process for determining relative permeability curve will be used for determination of relative permeability curve in

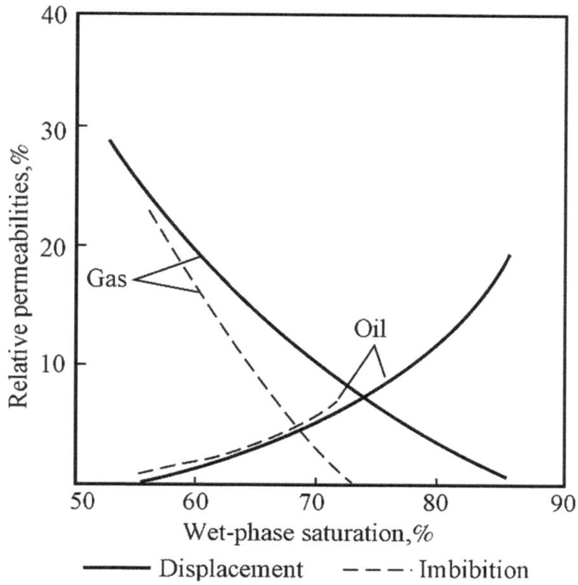

Fig. 4.98 Effects of saturation history on relative permeability

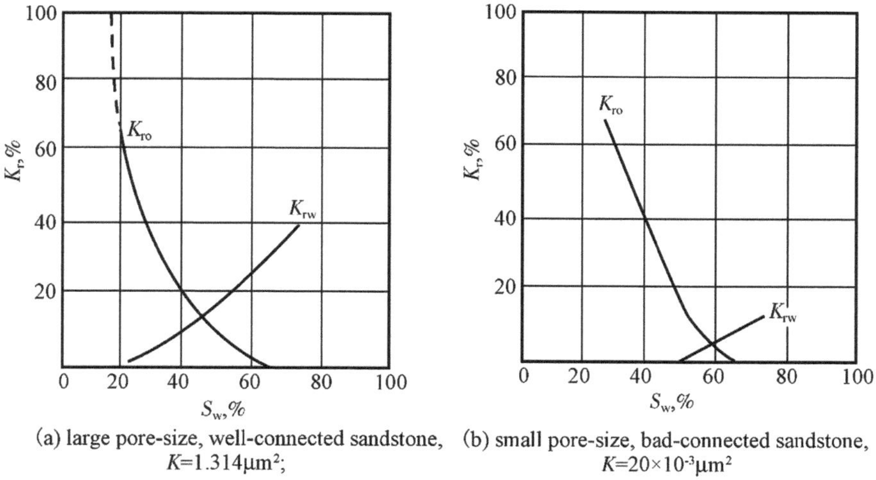

Fig. 4.99 Relative permeability to oil and water for different pore size sandstone

laboratory according to field development as much as possible. On the other hand, the actual physical processes of formation and exploitation of reservoir should be referred to choose the relative permeability curves measured by either displacement or imbibition in the calculation of the relative permeability curves.

Effects of Geometry of the Pore Spaces and Pore Size Distribution

The distribution of fluid saturation and the channels of fluid flow are directly related to the pore size and pore size distribution, therefore, the relative permeability curves that reflect the flow resistance of each phase must be affected by them. The relative permeability (Fig. 4.99) to oil and water for two sandstone cores with different pore structure and permeability measured by Morgan and Gordon (1970) [46] shows the effect of different pore size (permeability) on relative permeability. It can be seen that the extent of simultaneous flow of oil and water for sandstone with high permeability and high porosity is broad, and the irreducible water saturation is low; which is inverse to sandstone with low permeability and small pore size. This is because the flow channel in large pore is greater than that in small pores, and there is little small pores in which oil and water cannot flow.

Effects of Temperature

The effect of temperature on relative permeability is crucial to study on the seepage and displacement process of thermal recovery. In the past 40 years, the effects of temperature on relative permeability have been being studied in the world, and many experiments have been carried out, and the analytical expression of temperature and relative permeability is given by using mathematical methods.

The test results that temperature affects permeability and the corresponding curve are illustrated in Fig. 4.100. The impacts of temperature on relative permeability are globally controversial, but the characteristics of curves in Fig. 4.100 indicate the prevailing view on the correlation of temperature and relative permeability: irreducible water saturation increasing with temperature is an important feature of the correlation of temperature and relative permeability. It is widely known that: the polar substances adhered to oil-wet rock surface will release from rock surface at high temperature, and then plenty of water will be adhered to the rock surface; Rock becomes more hydrophilic, the contact angle decreases, and the oil-bearing pores that are isolated originally by water films change into water-bearing pores. In addition, the increase of temperature will lead to thermal expansion of rock, so that pore structure will change, so will permeability. However, some people believed that the impact of temperature on relative permeability results from viscosity ratio changing with temperature.

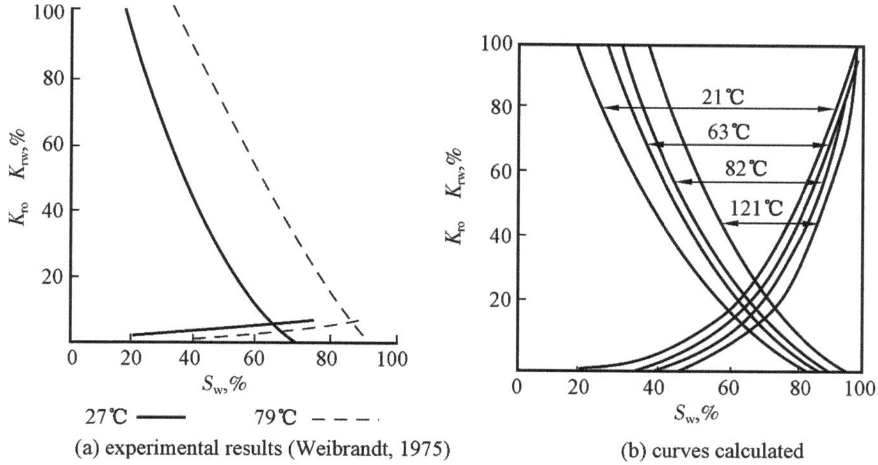

Fig. 4.100 Effect of temperature on relative permeability to oil and water (Nakornthap and Evans [47])

Effects of Other Factors

In addition to the above four main factors of affecting relative permeability, there are many other factors, fluid viscosity and π dimensionless number ($\pi = \sigma / K\Delta p$) will be mainly discussed.

Effects of Fluid Viscosity

By the 1950s, the viscosity of two-phase fluid is generally believed that has no effect on relative permeability. Later, however, it was discovered that when the viscosity of non-wetting phase is very great, and much greater than the viscosity of wetting phase, the relative permeability to non-wetting phase increases with viscosity ratio (non-wetting phase/wetting phase); and the relative permeability to wetting phase has nothing to do with viscosity ratio. These can be explained: the part of wetting phase that is adsorbed on solid surface can be regarded as a layer of wetting film, and the flow of non-wetting phase on it, to some extent, can be thought of slipping due to the lubrication of wetting film. Therefore, permeability to non-wetting phase increases with the decrease of the resistances of non-wetting phase. The effect of viscosity ratio on relative permeability decreases with the increase of radius, for example, when rock permeability is greater than 1 μm^3, the viscosity ratio is negligible.

Effects of π Dimensionless

There is a similar dimensionless number π in the similar simulation of laboratory experiment and actual reservoir, wherein $\pi = \sigma/K\Delta p$ or $\pi = \sigma/\mu v$, the physical concept of which expressed the effect of the ratio of microcapillary pressure gradient and driving pressure gradient. The numerical value of π includes influence of experimental pressure difference Δp, rock permeability K and the interfacial tension σ between fluids. In a experiment, the value of K and σ is fixed as long as the used core and fluids are certain, and the value of π is directly related to experimental pressure difference Δp. Experimental pressure difference Δp is generally believed that has no effect on relative permeability as long as it does not make the flow rate to cause inertial effect. On the other hand, when the magnitude of π changes, for example, when the interfacial tension decreases 2–8 order of magnitude, the fluid flow is ultralow interfacial tension flow, and the corresponding π has an effect on the relative permeability to two-phase fluid, such as the residual saturation of non-wetting phase decreases.

There are many factors affecting relative permeability, whether test conditions and reservoir conditions are consistent should be noted in the actual analysis and applications of curves. Therefore, determination of relative permeability should be measured under formation conditions such as rock wettability, fluids, temperature, pressure, and displacement process can truly show the rule of reservoirs flow.

4.5.3 Applications of Relative Permeability Curves

Relative permeability curves are the basis of studying multiphase flow. They are indispensable to some calculation about oil exploitation, performance analysis, determination of the distribution of oil, gas and water, and various calculations related to water flooding. There are three examples as follows.

4.5.3.1 Determination of Oil Production and Mobility Ratio

The corresponding K_{ro} and K_{rw} on relative permeability curves can be read if oil and water saturation is given during oil and water flow at the same time. The flow rate of oil and water Q_o, Q_w can be calculated by the following formula:

$$Q_o = \frac{K_{ro} K A \Delta p}{\mu_o L} = \frac{K_o A \Delta p}{\mu_o L} \tag{4.98}$$

$$Q_w = \frac{K_{rw} K A \Delta p}{\mu_w L} = \frac{K_w A \Delta p}{\mu_w L} \tag{4.99}$$

where K_o, K_w are effective permeability of oil and water, respectively, D; Q_o, Q_w are flow rate of oil and water, respectively, cm^3/s; μ_o, μ_w are viscosity of oil and water, respectively, mPa s; L is Length of rock sample, cm; Δp is pressure difference between the two end of the rock, 10^{-1} MPa; K_{ro}, K_{rw} are relative permeability of oil and water, respectively; K is absolute permeability, D.

It should be noted that if the relative permeability is the ratio of effective permeability to the permeability under formation water saturation, the K is the permeability under formation water saturation; if the relative permeability is the ration of effective permeability to the permeability under irreducible water saturation, the K should be the permeability under irreducible water saturation.

The water–oil ratio during oil–water simultaneous flow can be calculated based on the equation:

$$\frac{Q_w}{Q_w} = \frac{\frac{K_{rw} K A \Delta p}{\mu_w L}}{\frac{K_{ro} K A \Delta p}{\mu_o L}} = \frac{\frac{K_w}{\mu_w}}{\frac{K_o}{\mu_o}} = \frac{\lambda_w}{\lambda_o} = M \tag{4.100}$$

If the viscosity of oil and water is given, the ratio of oil and water production depends on the ratio of the effective permeability (or relative permeability) of oil and water, and mobility (λ) is introduced to describe the flow ability of fluid in multiphase flow.

Mobility is defined as the ratio of the effective permeability of one fluid to its viscosity during multiphase flow. The mobilities of oil and water are separately expressed as follows:

$$\lambda_o = \frac{K_o}{\mu_o} \tag{4.101}$$

$$\lambda_w = \frac{K_w}{\mu_w} \tag{4.102}$$

In general, the greater the mobility of a fluid is, the easier the fluid flows in a rock.

Mobility ratio is defined as the ratio of the mobility of the displacing fluid to the mobility of displaced fluid. During the process of oil production, the mobility rate of water displacing oil can be expressed as:

$$M = \frac{\lambda_w}{\lambda_o} \tag{4.103}$$

Based on the example of relative permeability above, we can see the great difference between mobilities of oil and water. From the basic information of the example, we know that there are three times difference between the viscosities of oil and water (μ_o = 3 mPa s, μ_w = 1 mPa s) and the water relative permeability differs 5 times (K_w = 0.225D, K_o = 0.045D) as the oil phase because of the 2.33 times difference of saturation (S_w, = 70 %, S_o = 30 %). The difference between flow rates of oil and water is then determined as follows:

$$\frac{Q_w}{Q_o} = \frac{K_w \cdot \mu_o}{K_o \cdot \mu_w} = \frac{0.225 \times 3}{0.045 \times 1} = 15$$

According to Eq. (4.100), we know that the mobility rate of water displacing oil is 15. It means the difference between the flowability of oil and water is 15 times. Therefore, mobility ratio is very important to predict the swept area and oil recovery in petroleum engineering investigation.

4.5.3.2 Determine Oil–Water Contact and Closure Height of Oil Reservoir

The irreducible water saturation, residual oil saturation, and relative permeability under different saturation can be calculated according to the relative permeability curves, and the heights over free water levels corresponding to different oil and water saturation can be gotten by capillary pressure curves. Therefore, the distribution of oil and water in the reservoir can be determined by combining relative permeability curve with capillary pressure curve in the homogeneous reservoir, and then the zones of 100 % oil production, simultaneous production of oil and water and 100 % water production can be recognized.

Figure 4.101 is the schematic diagram of determining oil–water contact and productivity using the relative permeability curves and capillary pressure curves. In the diagram, the area of a reservoir above point A only contains irreducible water and thus is the zone of single-phase oil production. In the area between point A and B, oil and water coexist in the reservoir. It is thus the zone of simultaneous production of oil and water. In the area between point B and C, water and residual oil coexist in the reservoir. The area is thus the zone of single-phase water production but containing residual oil in the rock. In the area below point C, water saturation is 100 %. It is known as the aquifer region.

If capillary pressure is expressed by the liquid column over oil–water contact, the height of the liquid column corresponding to the capillary pressure of point A represents the minimum closure height of the oil reservoir where oil produces only. If the actual closure height of a reservoir is greater than this value, the reservoir will only produce single-phase oil. The larger the closure height is, the greater the thickness of the formation producing single-phase oil is. If the actual reservoir closure height is smaller than this value, conversely, the reservoir will produce simultaneously oil and water, and it possibly has no industrial exploitation value.

The difference of Figs. 4.101 and 4.76 is: the position of point A and B in capillary pressure curves can be determined accurately because of combining relative permeability curves and capillary pressure curves, and then the height of oil–water contact and the thickness of the zone producing oil and water simultaneously.

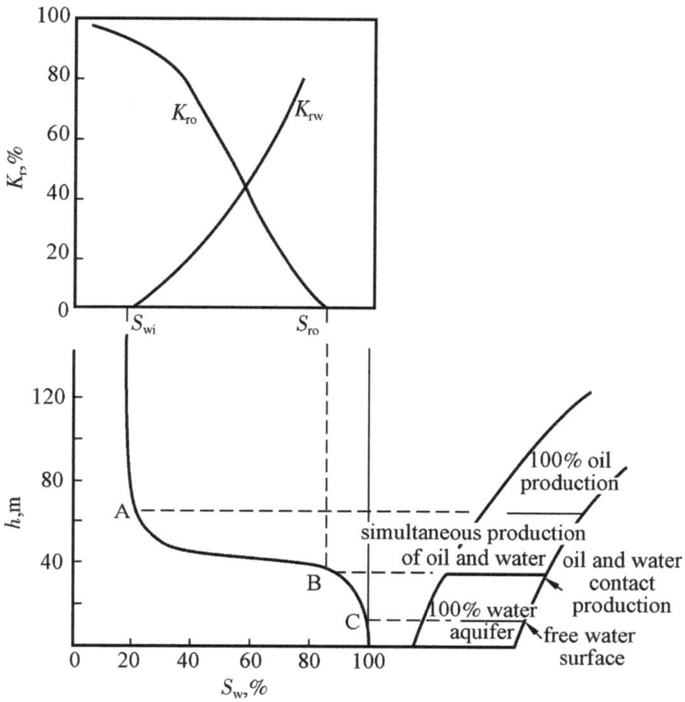

Fig. 4.101 determine water-oil contact and productivity

4.5.3.3 Analyze the Producing Water Rule

Fractional flow can be predicted from capillary pressure data and relative permeability curves. Capillary pressure data gives the saturations expected, and the relative permeability curves provide the values for Krw and Kro at that saturation. Water cut can then be calculated. In addition, the rules of fractional water cut varying with water saturation can be researched based on the fractional flow equation.

Water cut is the ratio of water production Q_w to total liquid production ($Q_w + Q_o$), that is:

$$f_w = \frac{Q_w}{Q_w + Q_o} \tag{4.104}$$

From Eqs. (4.98) and (4.99):

$$f_w = \frac{Q_w}{Q_w + Q_o} = \frac{Q_w}{Q_o} = \frac{\frac{K_{rw} K A \Delta p}{\mu_w L}}{\frac{K_{rw} K A \Delta p}{\mu_w L} + \frac{K_{ro} K A \Delta p}{\mu_o L}} \tag{4.105}$$

So,

$$f_w = \frac{K_w/\mu_w}{K_w/\mu_w + K_o/\mu_o} = \frac{1}{1 + \left(\frac{K_o}{K_w}\right)\left(\frac{\mu_w}{\mu_o}\right)} \tag{4.106}$$

Thus, fractional flow equation is obtained:

$$f_w = \frac{1}{1 + \left(\frac{K_o}{K_w}\right)\left(\frac{\mu_w}{\mu_o}\right)} \tag{4.107}$$

where f_w is water cut, %.

Equation (4.107) is known as fractional flow equation. The viscosity ratio of water to oil (μ_w/μ_o) is certain for a given reservoir, the water cut thus depends on the ratio of the relative permeabilities of oil and water. Because the relative permeability ratio of oil to water is a function of the water saturation (S_w) in the reservoir, the water cut (f_w) is thus a function of water saturation as shown in Fig. 4.102. In Fig. 4.102, we can see that even if water cut is 100 %, the water saturation in reservoir is less than 100 %. It indicates that there is still residual oil in the reservoir even oil wells produce 100 % water.

In Eq. (4.107), the K_o/K_w can be expressed as a function of water saturation S_w so as to apply conveniently. Convert the K_{ro} and K_{rw} in Fig. 4.91 produce the curve as shown in Fig. 4.103. The correlation of K_{ro}/K_{rw} and water saturation (S_w) is the curve which is curved at both ends and is a straight line in the middle segment. The straight line means exactly the ratio range of relative permeability corresponding to the saturation of two-phase flow. Most of the relative-permeability ratio curves have the similar characteristics. The straight line segment on the curve can be expressed as:

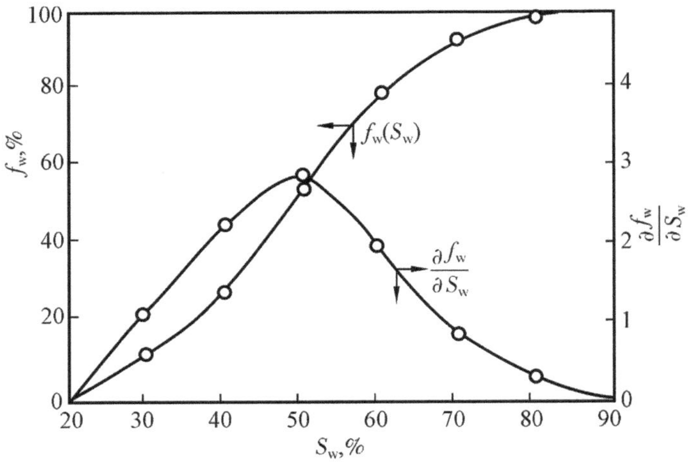

Fig. 4.102 Water cut and correlation of water cut varying with water saturation

$$\frac{K_{\mathrm{ro}}}{K_{\mathrm{rw}}} = ae^{-bS_{\mathrm{w}}} \tag{4.108}$$

where e is the base of natural logarithm, its value is about 2.71828; a is intercept of straight line; b is slope of straight line.

The coefficients "a" and "b" depend on properties of rock and fluid. Rock permeability, pore-size distribution, fluid viscosity, interfacial tension and wettability are different, so is "a" and "b".

The value of "a" and "b" can be obtained by graphic method, as shown in Fig. 4.103.

$$\text{Intercept: } a = 1222$$

$$\text{Slope: } b = \frac{2 \times 2.303}{0.335} = 13$$

where 2.303 is the conversion factor of common logarithm and natural logarithm, 2 represents 2 cycles that $K_{\mathrm{o}}/K_{\mathrm{w}}$ changes from 10-1 to 101, 0.355 is the change of water saturation ΔS_{w} during the two cycles.

Fig. 4.103 The correlation of the ratio of oil and water relative permeability and water saturation

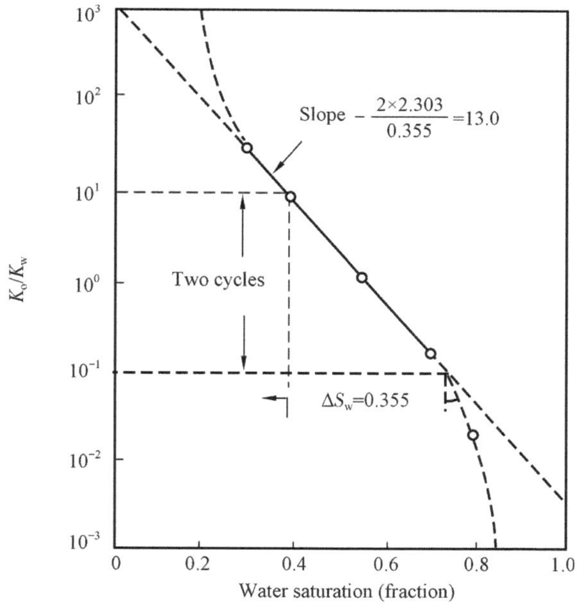

The value of a, b can also be calculated by solving simultaneous equations, for example: $S_w = 0.3$, $K_o/K_w = 25$; or $S_w = 0.7$, $K_o/K_w = 0.14$, according to Eq. (4.108):

$$25 = ae^{-0.3S_w}$$

$$0.14 = ae^{0.7b}$$

Solving the above simultaneous equations, $a = 1222$ and $b = 13$ are obtained that are same as the graphical method.

If the equation $K_o/K_w = ae^{-bS_w}$ is obtained, and substituting it in Eq. (4.107), and then the relation of f_w and S_w can be obtained as shown in Eq. (4.109). Thus there are several conclusions from Eqs. (4.107) or (4.109):

$$f_w = \frac{1}{1 + \left(\frac{\mu_w}{\mu_o}\right) ae^{-bS_w}} \quad (4.109)$$

(a) f_w increasing with two-phase mobility ratio, the denser the oil ($\mu_o \gg \mu_w$), the greater the water cut f_w. Therefore, water cut will increase quickly once water breaks through in heavy oil reservoir;
(b) f_w increasing with water saturation S_w, water cut will increase. S_o the water cut is different in different locations of oil–water transitional zone. The correlation of increasing velocity of water cut ($\partial f_w / \partial S_w$) with water saturation can be obtained by the partial derivative of S_w in Eq. (4.109)

$$\frac{\partial f_w}{\partial S_w} = \frac{(\mu_w/\mu_o) bae^{-bS_w}}{[1 + (\mu_w/\mu_o) ae^{-bS_w}]^2} \quad (4.110)$$

$\partial f_w / \partial S_w$ with S_w by graphical method, and the physical meaning of S_w is the increasing percent of water cut f_w with unite increase (i.e., 1 %) of water saturation S_w. The curve shows that the gradual flooding of reservoir, the water cut rises slowly at the early stage, and then it rises rapidly as water saturation increases; however, the water cut rising rate will low down again when water saturation is high at the later stage. Understanding the rule of water cut rising is helpful to take measures to prevent early water out, which can explain why the middle section of water flooding characteristic curve ($\Sigma Q_w - \Sigma Q_o$ curve) is a straight line, and both sides of curve deviate from the straight line.

4.5.3.4 Relative Permeability Data Treatment

As above mentioned, relative permeability and relative permeability curves are widely used in petroleum engineering. There are many factors affecting relative permeability, such as rock wettability, fluids, temperature, pressure and displacement process, etc., so interpretation and use of relative permeability data to predict individual well or reservoir performance become complicated. Thus, although the laboratory tests can adequately describe the behavior of a particular test plug, modeling of a well or reservoir performance may require modified relative permeability data. Correlations of overall curve shape, crossover points, recovery at a given produced volume, brine permeability at residual oil saturation, etc., must all be made with reference to lithology, permeability, and initial fluid saturations [48].

Choice of test method will be governed by application of data, i.e., high flow rate for near the well bore, and low flow rates away from the well bore. As stressed previously, no one test method can fully describe a system and choice of data will be influenced by laboratory scale limitations; in particular, end effects in oil-wet cores and problems sometimes caused by wettability alterations and mobile fines.

Good petroleum engineering reports should highlight any experimental difficulties encountered and indicate the most reliable data. However, it is often impossible to assess service company data since flow rates are seldom constant and D_p/D_{pi} versus Q_i curves is not reported. If you are ever in a position to commission this type of work, ensure that provision of this information is part of the contract.

Exercise

Question

1. Give a definition of adhesion tension in an oil–water system
2. What are the factors affecting adhesion tension?
3. What are the “imbibition process” and “drainage process”? Illustrate the two processes by example.
4. What are the “hysteresis” occurring in capillary pressure curve and relative permeability curves? Why does the phenomenon exhibit?
5. Explain the porous diaphragm method for determining capillary pressure in the laboratory.
6. What methods are there for the determination of capillary pressure in the laboratory?
7. What are the factors affecting the capillary pressure curve?
8. Give a qualitative definition of effective permeability to the wetting phase. And list four factors affecting effective permeability–saturation relationships.
9. Give a definition of relative permeability to a non-wetting phase.
10. Illustrate, using diagram, the effect of hysteresis on the relative permeability curve of a non-wetting phase, and explain the reason in terms of the relationship between capillary and viscous forces.
11. The equation shown below is used to determine the average capillary pressure of a rock.

$$J(S_w) = \frac{p_c}{\sigma\cos\theta}\left(\frac{K}{\phi}\right)^{0.5}$$

Determine the constant, and its units to calculate $J\ (S_w)$, when capillary pressure, P_c, is in Mpa; interfacial tension, σ, is in dyn/cm; the permeability, K, is in md; and the porosity, ϕ, is expressed as a fraction.

12. Consider the balance of forces at the interface between oil and water in a small capillary tube of radius, r, inserted vertically across the oil–water contact in container. Derive the equation for the height of water rise, h, in the tube. Include a sketch for this case where water, the denser phase, is the wetting phase.
13. Match the names in Table 4.8 Column 1 with the definitions in column 3. Place column 3 in Column 2.
14. What are the effects of fluid saturation history on relative permeability?
15. Define the term "interfacial tension" and explain the changes of interfacial tension with gas solubility in reservoir fluid, pressure and temperature.
16. What is adsorption? Explain the causes of wetting and give the factors affecting rock wetting.
17. How to describe rock wetting? How to determine the wettability of an oil-wet-rock system?
18. What is Jiamin effect? Why does it happen?

Table 4.8 Concepts and their definitions

Column 1	Column 2	Column 3
Wettability		(a) The pressure required to force non-wetting fluid into larger pores
Drainage		(b) The difference between two solid-fluid interfacial tensions
Capillary pressure		(c) Fluid flow process in which the saturation of the wetting phase increases and the non-wetting phase saturation decreases
Irreducible wetting phase saturation		(d) The lagging of an effect behind its cause
Adhesion tension		(e) The pressure difference existing across the interface separating two immiscible fluids in capillaries
Displacement pressure		(f) Tendency of one fluid to spread on or adhere to a solid surface in the presence of other immiscible fluids
Hysteresis		(g) The limiting value in reduction of the wetting phase saturation
Interfacial tension		(h) Fluid flow process in which the saturation of the wetting phase decreases and the non-wetting phase saturation increases
Imbibition		(i) The energy per unit area (force per unit distance) at the surface between phases

19. How to determine saturation distributions of oil and water, oil–water contact and closure height of oil reservoir during only producing oil base on relative permeability? Give an example.
20. Descript the characteristics of relative permeability and factors affecting relative permeability curves
21. Descript the characteristics of capillary pressure curve and illustrate the applications of capillary pressure curve in petroleum engineering.
22. Studies have shown that rock wettability has an impact on water flooding process; and rock wettability can be changed in turn after longtime water flooding.

    (a) Give three or more aspects about the effect of wettability on water flooding.
    (b) Interpret the reason that rock wettability can be changed during water flooding.

23. Describe briefly the variation of fluid saturations in the oil–water transition zone. Why the oil–water transition zone is thicker than the oil–gas transition zone? Why the oil–water transition zone becomes thicker in a heavy oil reservoir?
24. How do the oil and water distribute in the pore space of a reservoir rock? What are the factors affect the distribution of fluids in the reservoir pores?
25. What factors may result in the wetting hysteresis?

Calculation

1. The height of a fluid rise in a capillary can be given by: $h = \frac{\sigma \cos \theta}{\rho g r}$
   Calculate the height, in cm, for an air–mercury system where the interfacial tension is 480 dynes/cm, the contact angle is 140°, the density of mercury is 13.6 g/cm^3, and the radius of the capillary is 0.01 cm. Interpret the result of your calculations.
2. See the following sketch. Calculate $p_{c,ow}$ and $p_{c,go}$ in dynes/cm^2 at points A and B if $h_1 = 800$ cm, $h_2 = 1000$ cm, $\rho_w = 1.05$ g/cm^3, $\rho_o = 0.8$ g/cm^3, $\rho_g =$ 0.15/cm^3 (Fig. 4.104).
3. The laboratory capillary pressure curves, shown below, were measured using air/water displacement tests (interfacial tension of 72 dynes/cm and contact angle of zero degrees) for a core plug having porosity of 0.22 and permeability of 135 md. Estimate the initial water saturation at a point 10 ft above the free water level in the reservoir for the reservoir is water wet. Other parameters are given in Table 4.9 (Fig. 4.105).
4. Sketch the curves of relative permeability of the wetting and non-wetting phases versus wetting phase saturation. Mark the hysteresis effect. Clearly label the axes of your graph(s) and endpoints. Suppose the following data (Table 4.10) on permeability (mD) are available from a drainage laboratory flow test.
   Calculate the relative permeabilities in following table (Table 4.11) to oil and water if the "base permeability" is the absolute permeability.

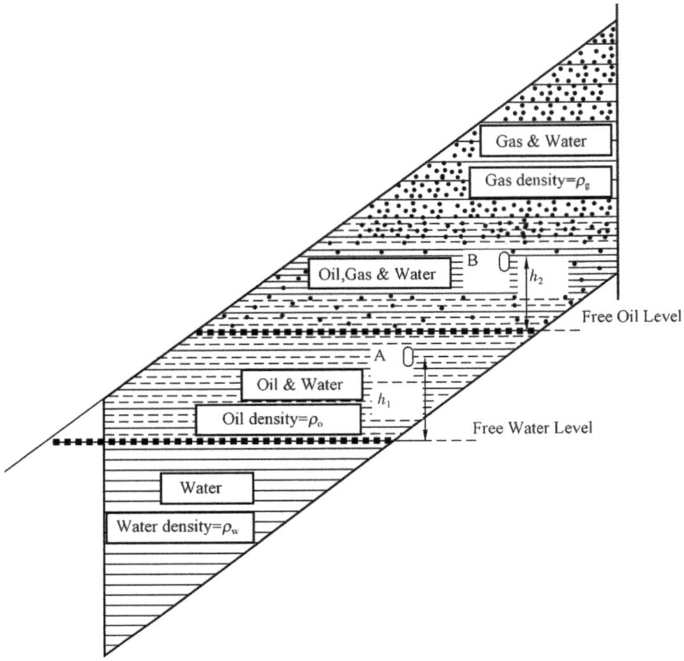

Fig. 4.104 The sketch map of oil–water transition zone

Table 4.9 Basic parameters for water saturation calculation

Porosity	0.17
permeability	110 md
Interfacial tension	30 dynes/cm
Contact angle	8°
Oil density	53.0 lb_m/ft^3
Water density	63 lb_m/ft^3

5. Table 4.12 shows the oil-water relative permeability data obtained on a sand rock at reservoir conditions.

   (a) Plot the oil–water relative permeability curves on the Cartesian coordinate paper.
   (b) Plot the relationship curves of the relative permeability ratio K_{ro}/K_{rw} versus water saturation on a semi-logarithmic coordinate system.
   (c) Determine the constant a and b in the equation $\frac{K_{ro}}{K_{rw}} = ae^{-bS_w}$
   (d) The oil viscosity is assumed to be 3.4 mPa s, the water viscosity is 0.68 mPa s, formation volume factor of o/l is 1.5 and formation volume factor of water is 1.05. A well is drilled in the oil-water transition zone. If the

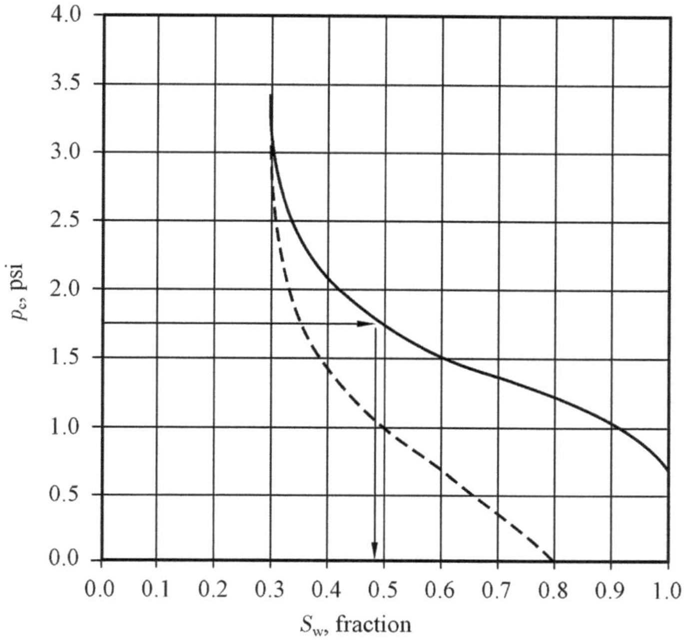

Fig. 4.105 The laboratory capillary pressure curves

Table 4.10 Relative permeability data

S_w	K_{rw}	K_{ro}
0.2	0	80
1.0	100	0

Table 4.11 Water saturation data

S_w	K_{rw}	K_{ro}
0.2		
1.0		

water saturation at the perforation point is 50 %, calculate the water cut of this well at surface and reservoir conditions

6. There is a rock sample whose length is 10 cm, and sectional area is 6.25 cm^2. The sample is 100 % saturated with the brine whose viscosity is 0.8 mPa s. The flow rate is 0.04 cm^3/s when pressure difference is 5 atm.

 (a) Calculate the absolute permeability of the sample.
 (b) Calculate the effective permeability of the brine.
 (c) What can you conclude from above results?

Table 4.12 Oil-water relative permeability data

S_w (%)	0	10	20	30	40	50	60	70	75	80	90	100
K_{rw}	0	0	0	0	0.04	0.11	0.2	0.3	0.36	0.44	0.68	1
E_{ro}	1	1	1	0.94	0.8	0.44	0.16	0.045	0	0	0	0

7. For the same sample in exercise last question, which is saturated with brine (S_w = 40 %) and oil (S_o = 60 %). If pressure difference is the same as the above, the oil flow rate is 0.04 cm^3/s and oil viscosity is 2.5 mPa s. the air permeability is 0.1D.
   (a) Calculate the effective permeability of the brine and oil.
   (b) Calculate the relative permeability of the brine and oil.
   (c) Calculate the mobility of the brine and oil.
8. The oil–water relative permeability data are obtained using steady-state method. The length of the rock sample is 15 cm and the cross-sectional area is 4.9 cm^2. The pressure difference is 2×10^5 Pa. The oil viscosity is 3 mPa s and the water viscosity is l mP.s. The flow rates of water and oil under various water saturations are listed in Table 4.13.
   (a) Plot the relative permeability curves, given that the absolute permeability of the rock sample is l.2D.
   (b) Determine the irreducible water saturation and residual oil saturation.
   (c) Determine the wettability of the rock sample.
9. The air–water capillary pressure curve is obtained by semi-permeable diaphragm method in the laboratory, When the water saturation is 50 %, the capillary pressure is measured to be $p_{cL} = 0.6 \times 10^5$ Pa The surface tension of water is 72 mN/m in surface conditions. While in the reservoir conditions, the interfacial tension between water and oil is 24 mN/m. The water density is ρ_w = 1.088 g/cm^3 and the oil density is ρ_o = 1.088 g/cm^3. The altitude of free water level is −1000 m. The reservoir rock is water-wet, and the contact angle between water and reservoir rock is assumed to be the same as that at surface condition.
   Calculate:
   (a) The distance of the water level where water saturation is 50/70 to the free water level.
   (b) The altitude of the water level where water saturation is 50 %.

Table 4.13 Flow rates of water and oil

S_w (%)	80	70	60	50	40	30	20
Q_w (cm^3/s)	0.549	0.336	0.202	0.101	0.045	0.011	0.000
Q_o (cm^3/s)	0.000	0.026	0.067	0.109	0.149	0.213	0.261

References

1. Professor H, Yidirim E (2006) Surface chemistry of solid and liquid interfaces. Blackwell Pub
2. Aveyard R, Haydon DA (1973) An introduction to the principles of surface chemistry. University Press
3. He GS, Tang H (2011) Reservoir physics. Petroleum Industry Press, Beijing
4. Danesh A, University HW, Edinburgh AS (1998) PVT and phase behaviour of petroleum reservoir fluids. In: Developments in Petroleum Science
5. Katz DL, Monroe RR, Trainer RP (1943) Surface tension of crude oils containing dissolved gases. Pet Tech 1–10
6. Hocott CR (1938) Interfacial tension between water and oil under reservoir conditions. Trans AIME J 132(1):184–190
7. Livingston HK (1938) Surface and interfacial tensions of oil-water systems in texas oil sands. Petrol Technol J 1(4):1–13
8. William C. Lyons (2009) Working guide to reservoir engineering, Gulf Professional Publishing
9. Hjelmeland OS, Larrondo LE (1986) Experimental investigation of the effects of temperature, pressure, and crude oil composition on interfacial properties [interfacial between crude oil and brine]. SPE Reservoir Eng J 1(4):321–328
10. Competition Science Vision (2004) 7(78):136
11. Glossary (2009) The brownfields and land revitalization technology support center. Retrieved 21 Dec 2009
12. Gurdeep R (2002) Surface chemistry. Krishna Prakashan Media
13. Banker GS, Siepmann J, Rhodes C (2002) Modern pharmaceutics. Dekker
14. Oura K, Lifshits VG, Saranin AA, Zotov AV, Katayama M (2003) Surface science, an introduction. Springer, Berlin
15. Rouquerol J, Rouquerol F, Sing KSW (1999) Adsorption by powders and porous solids: principles, methodology and applications. Academic Press
16. Sivasankar (2008) Engineering chemistry. Tata McGraw-Hill Education
17. Rouquerol F, Rouquerol J, Sing K (1999) Adsorption by powders and porous solids
18. Tadros TF (2005) Applied surfactants: principles and applications
19. Criddle DW (1960) Rheology, theory and applications. In: Eirich FR (ed) Vol 3, Chapter 11. Academic Press, New York
20. Fertl WH, Chilingarian GV (1989) evaluation and monitoring of enhanced oil recovery projects based on geophysical well logging techniques. Dev Pet Sci 17:451–494
21. Laurier LS (2000) Surfactant: fundamentals and applications in the petroleum industry
22. Donaldson EC, Thomas RD (1971) Microscopic observations of oil displacement in water-wet and oil-wet systems. Fall Meeting of the Society of Petroleum Engineers of AIME
23. Craig FF (1971) The reservoir engineering aspects of water flooding, Monograph series, SPE, Richardson, TX, p 3
24. Raza SH, Treiber LE, Archer DL (1968) Wettability of reservoir rocks and its evaluation. Prod Mon (United States) 32:4
25. Anderson GW (1986) Wettability literature survey-part 1: rock/oil/brine interactions and the effects of core handling on wettability. J Petrol Technol
26. Denekas MO et al (1959) Effect of crude oil components on rock wettability. Trans AIME J 216:330–333
27. Morrow NR, McCaffery FG (1976) Displacement studies in uniformly wetted porous media. Proc, Soc Chem Ind Inti Symposium on Wetting, Loughborough, U.K. (Sept. 27–29), pp 289–319
28. Johansen RT, Dunning HN (1958) Relative wetting tendencies of crude oils by the capillarimetric method
29. Muecke TW (1979) Formation fines and factors controlling their movement in porous media. J Pet Technol 31(2):144–150

30. Owens WW, Archer DL (1971) The effect of rock wettability on oil-water relative permeability relationships. SOC Petrol Eng JPT 71:873–878
31. Anderson WG (1986) Wettability literature survey-part 3: the effects of wettability on the electrical properties of porous media. J Petrol Technol J 38(12):1371
32. Yang SL et al (2004) Reservoir physics. Petroleum Industry Press, Beijing
33. Piper L (2003) Reservoir petrophysics. Spring
34. Leverett MC (1941) Capillary behavior in porous solids. Trans AIME J 142:152–169
35. Holmes M (2002) Capillary pressure & relative permeability: petrophysical reservoir models. Prepared by Michael Holmes. Digital Formation, Inc. Denver, Colorado, USA
36. Wardlaw NC, Taylor RP (1976) Mercury capillary pressure curves and the interpretation of pore structure and capillary behavior in reservoir rocks. Bull Can Petrol Geol J 24(2):225–262
37. Donaldson EC, Lorenz PB, Thomas RD (1969) Wettability determination and its effect on recovery efficiency. SOC Petrol Eng J 9(1):13–20
38. Amaefule JO, Masuo ST (1986) Use of capillary pressure data for rapid evaluation of formation damage or Stimulation. SPE Production Engineering, SPE 12475
39. Keelan DK, Koepf EH (1977) The role of cores and core analysis in evaluation of formation damage. Soc Petrol Eng JPT 482–490
40. Wardlaw NC (1980) The effects of pore structure on displacement efficiency in reservoir rocks and in glass micro-models. In: Paper SPE 8843 presented at the first joint SPE/DOE Sym on EOR held in Tulsa, OK, USA, April 20–23
41. Chatzis I, Morrow NR, Lim HT (1983) Magnitude and detailed structure of residual oil saturation. SPE 10681
42. Leverett MC, Lewis WB (1941) Steady flow of gas-oil-water mixtures through unconsolidated sands. Trans AIME 142:107
43. Delshad M, Pope GA (1989) Comparison of the three-phase oil relative permeability models. Transp Porous Media 4(1):59–83
44. Geffen TM et al (1951) Experimental investigation of factors affecting laboratory permeability measurements. Trans AIME J 192:99–110
45. Treiber LE, Archer DL, Owens WW (1972) A laboratory evaluation of the wettability of fifty oil producing reservoirs. Soc Pet Eng J 12(6):531
46. Morgan JT, Gordon DT (1970) Influence of pore geometry on water-oil relative permeability. J Pet Technol 22(10):1199–1208
47. Nakornthap K, Evans RD (1986) Temperature-dependent relative permeability and its effect on oil displacement by thermal methods (Spe11217)
48. https://zh.scribd.com/doc/187702058/Formation-Evaluation-MSc-Course-Notes-Paul-Glover

Chapter 5
Principles of Enhanced Oil Recovery

Fayang Jin

Enhanced oil recovery (EOR) processes are very important for us to maintain the production of oilfields. This chapter first introduces the definition and characteristics of EOR. Classification and description of EOR processes are presented. Then, thermal recovery processes are introduced, such as cyclic steam stimulation, steamdrive, and in situ combustion. Then the chapter discusses several kinds of miscible gas floodings, design procedures, and criteria of a miscible displacement process. Then it focuses on the most widely used EOR process in China, named chemical flooding, especially polymer flooding. Finally, the chapter introduces MEOR (Microbial Enhanced Oil Recovery) process, which is believed to be profound in the future.

5.1 Fundamental Concepts and Principles of EOR

5.1.1 Concept of EOR

Approximately 60–70 % of the oil in place cannot be produced by conventional methods. Enhanced oil recovery methods gain importance in particular with respect to the limited worldwide resources of crude oil. Enhanced oil recovery is oil recovery by the injection of materials not normally present in reservoir. All of currently available EOR is based on one or more of two principles: increasing the capillary number and/or lowering the mobility ratio, compared to their waterflood values. Increasing the capillary number means, practically speaking, reducing oil–water interfacial tension. The injectant mobility can be reduced through increasing water viscosity, reducing oil viscosity, reducing water permeability, or all of the above.

F. Jin (✉)
Southwest Petroleum University, Chengdu, Sichuan, China
e-mail: jfy975@swpu.edu.cn

X. Hu et al. (eds.), *Physics of Petroleum Reservoirs*, Springer Geophysics,
DOI 10.1007/978-3-662-55026-7_5

Traditionally, oil recovery operations have been subdivided into three stages: primary, secondary, and tertiary. Historically, these stages describe the production from a reservoir in a chronological sense. Primary production, the initial production stage, results from the displacement energy naturally existing in reservoirs. Secondary recovery, the second stage of operations, usually is implemented after primary production exhausted. Traditional secondary recovery processes include waterflooding, pressure maintenance, and gas injection. Tertiary recovery, the third stage of production, is that obtained after waterflooding. Tertiary processes used miscible gases, chemicals, and/or thermal energy to displace additional oil after the secondary recovery process became uneconomical.

In other situations, the so-called tertiary process might be applied as a secondary operation in lieu of waterflooding. This action might be dictated by such factors as the nature of the tertiary process, availability of injectants, and economics. For example, if a waterflood before application of the tertiary process diminishes the overall effectiveness, then the waterflooding stage might reasonably be bypassed.

Because of such situations, the term "tertiary recovery" fell into disfavor in petroleum engineering literature and the designation of "enhanced oil recovery" (EOR) became more accepted. Another descriptive designation commonly used is "improved oil recovery" (IOR), which includes EOR but also encompasses a broader range of activities, e.g., reservoir characterization, improved reservoir management, and infill drilling, etc. [2]. Oil recovery processes now are classified as primary, secondary, and EOR processes. A classification scheme is clearly useful, in which a basis for communication among technical persons is established.

Primary recovery refers to the use of natural energy presenting in reservoirs as the main source of energy for the displacement of oil to producing wells [2, 4]. These natural energy sources are solution–gas drive, gas cap drive, natural water-drive, fluid and rock expansion, and gravity drainage. The particular mechanism of lifting oil to the surface, once it is in the wellbore, is not a factor in the classification scheme.

Secondary recovery results from the augmentation of natural energy through injection of water or gas to displace oil toward producing wells [2, 4]. Gas injection, in this case, is either into a gas cap for pressure maintenance and gas cap expansion or into oil-column wells to displace oil immiscibly according to relative permeability and volumetric sweepout considerations. Gas processes based on other mechanisms, such as oil swelling, oil viscosity reduction, or favorable phase behavior, are considered EOR processes. An immiscible gas displacement is not as efficient as a waterflood and is used infrequently as a secondary recovery process today. Today, waterflooding is almost synonymous with the secondary recovery classification.

EOR results principally from the injection of gases or liquid chemicals and/or the use of thermal energy. Hydrocarbon gases, CO_2, nitrogen, and flue gases are the gas sources used in EOR processes. A number of liquid chemicals are commonly used, including polymers, surfactants, and hydrocarbon solvents. Thermal processes typically consist of the use of steam or hot water, or rely on in situ generation of thermal energy through oil combustion in reservoir rock [2, 3].

EOR processes involve the injection of a fluid or fluids of some type into reservoirs. The injected fluids and injection processes supplement the natural energy present in reservoirs to displace oil to producing wells. In addition, the injected fluids interact with the reservoir rock/oil system to create conditions favorable for oil recovery. These interactions might, for example, result in low IFT, oil swelling, oil viscosity reduction, wettability modification, or favorable phase behavior. The interactions are attributable to physical and chemical mechanisms and to the injection or production of thermal energy. Simple waterflooding or the injection of dry gas for pressure maintenance or oil displacement is excluded from the definition.

EOR processes often involve the injection of more than one fluid. In a typical case, a relatively small volume of an expensive chemical (primary slug) is injected to mobilize the oil. This primary slug is displaced with a larger volume of a relatively inexpensive chemical (secondary slug). The purpose of the secondary slug is to displace the primary slug efficiently with as little deterioration as possible of the primary slug. In some cases, additional fluids of even lower unit cost are injected after a secondary slug to reduce expenses. In such a case of multiple fluid injection, all injected fluids are considered to be part of the EOR process, even though the final chemical slug might be water or dry gas that is injected solely to displace the fluids injected earlier in the process.

5.1.2 Characteristics of EOR Process

5.1.2.1 Microscopic and Macroscopic Displacement

The overall displacement efficiency of any oil recovery displacement process can be considered conveniently as the product of microscopic and macroscopic displacement efficiencies [2]. In equation form:

$$E = E_D E_V, \tag{5.1}$$

where E is the total displacement efficiency expressed as a decimal; E_D is the microscopic displacement efficiency expressed as a decimal, and E_V is the macroscopic (volumetric) displacement efficiency expressed as a decimal. Microscopic displacement relates to the displacement or mobilization of oil at the pore scale. That is, E_D is a measure of the effectiveness of the displacing fluid in moving (mobilizing) the oil at those places in the rock where the displacing fluid contacts the oil. E_D is reflected in the magnitude of the residual oil saturation, S_{or}, in the regions contacted by the displacing fluid.

Four factors generally control how much of a reservoir will be contacted by a displacement process: the properties of the injected fluids, the properties of the displaced fluids, the properties and geological characteristics of the reservoir rock, and the geometry of the injection and production well pattern.

Macroscopic displacement efficiency relates to the effectiveness of the displacing fluid in contacting the reservoir in a volumetric sense. Alternative terms conveying the same general concept are sweep efficiency and conformance factor. E_V is a measure of how effectively the displacing fluid sweeps out the volume of a reservoir, both areal and vertically, as well as how effectively the displacing fluid moves the displaced oil toward production wells. Both areal and vertical sweeps must be considered, and it is often useful to further subdivide E_V into the product of areal and vertical displacement efficiencies. E_V is reflected in the magnitude of average or overall residual oil saturation, $\overline{S}_{or}$, because the average is based on residual oil in both swept and unswept parts of the reservoir.

Consider the magnitude of these efficiencies in a typical waterflood. For an example in which initial oil saturation, S_{oi} is 0.80 and S_{or} in the swept region is 0.40

$$E_D = \frac{S_{oi} - S_{or}}{S_{oi}} = \frac{0.80 - 0.40}{0.80} = 0.50 \tag{5.2}$$

A typical waterflood sweep efficiency, E_V, at the economic limit 0.7. Therefore,

$$E = E_D E_V = 0.50 \times 0.70 = 0.35 \tag{5.3}$$

Thus, for a typical waterflood, the overall displacement efficiency is on the order of one-third, which also represents the oil recovery efficiency. This one-third figure is by no means a universal result applicable to all reservoirs. Individual reservoirs yield higher or lower recovery efficiencies, depending on the oil and reservoir characteristics.

It is desirable in an EOR process that the values of E_D and E_V, and consequently E, approach 1.0. An idealized EOR process would be one in which the primary displacing fluid (primary slug) removes all oil from pores contacted by the fluid ($S_{or} \rightarrow 0$), and in which the displacing fluid contacts the total reservoir volume and displaces oil to production wells. A secondary fluid slug used to displace the primary slug would behave in a similar manner and would displace the primary slug efficiently both microscopically and macroscopically.

Several physical/chemical interactions occur between the displacing fluid and oil that can lead to efficient microscopic displacement. These include miscibility between the fluids, decreasing the IFT between the fluids, oil volume expansion, and reducing oil viscosity. The maintenance of a favorable mobility ratio between displaced and displacing fluids also contributes to better microscopic displacement efficiency. EOR processes are thus developed with consideration of these factors: Fluids used as primary displacing slugs have one or more of the favorable physical/chemical interactions with the oil. Fluids used to displace the primary slug ideally should also have similar favorable interactions with the primary slug. The goal with an acceptable EOR fluid is to maintain the favorable interaction as long as possible during the flooding Process.

Macroscopic displacement efficiency is improved by maintenance of favorable mobility ratios between all displacing and displaced fluids throughout a process. Favorable ratios contribute to improvement of both areal and vertical sweep efficiencies. An ideal EOR fluid is the one that maintains a favorable mobility ratio as the fluid being displaced. Another factor that is important to good macroscopic efficiency is the density difference between displacing and displaced fluids. Large density differences can cause in gravity segregation, i.e., the underriding or overriding of the fluid being displaced. The effect is to bypass fluids at the top or bottom of a reservoir, reducing E_V. Reservoir geology, and in particular geologic heterogeneity, is an important factor in the consideration of macroscopic displacement efficiency. The effects of mobility and density differences can be amplified or diminished by the nature of the geology.

5.1.2.2 Volumetric Displacement Efficiency

Volumetric sweep efficiency can be considered conceptually as the product of the areal and vertical sweep efficiencies. Consider a reservoir that has uniform porosity, thickness, and hydrocarbon saturation, but that consists of several layers. For a displacement process conducted in the reservoir, E_v can be expressed as follows [2]:

$$E_V = E_A E_I, \tag{5.4}$$

where E_A is the areal sweep efficiency in an idealized or model reservoir, area swept divided by total reservoir area, expressed as a decimal; and E_I is the vertical sweep efficiency, pore space invaded by the injected fluid divided by the pore space enclosed in all layers behind the location of the leading edge (leading areal location) of the front, expressed as a decimal.

All efficiencies are expressed as fractions. E_I is the volumetric sweep efficiency of the region confined by the largest areal sweep efficiency in the system.

For a real reservoir, in which porosity, thickness, and hydrocarbon saturation vary areally, E_A is replaced by pattern sweep efficiency, E_P [2]:

$$E_V = E_P E_I, \tag{5.5}$$

where E_p is the pattern sweep efficiency, hydrocarbon pore space is enclosed behind the injected-fluid front divided by total hydrocarbon pore space in the pattern or reservoir.

In essence, E_p is an areal sweep efficiency that has been corrected for variations in thickness, porosity, and saturation. In either case, overall hydrocarbon recovery efficiency in a displacement process may be expressed as follows [2]:

$$E_V = E_P E_I E_D \tag{5.6}$$

While Eqs. 5.4 and 5.5 are conceptually correct, application to practical problems generally is difficult. To use Eqs. 5.4 and 5.5 to determine E_V, independent estimates of E_p (or E_A) and E_I are required. Such estimates are difficult to obtain because, for displacements in 3D systems, E_A and E_I typically are not independent. In the absence of vertical effects, areal sweep can be approximated from correlations developed from scaled physical models or mathematical models. Additionally, there are methods to estimate vertical sweep efficiency, but the availability of model studies is very limited. In practice, E_V is usually determined by the application of appropriate correlations or mathematical models based on 3D systems and not by independent calculation of E_A and E_I. Nonetheless, it is useful to consider E_V as the product of E_A and E_I to understand the parameters that affect volumetric sweep.

5.2 Classification of EOR Processes

Using the nomenclature adopted by the Oil and Gas Journal, EOR processes are divided into four categories: thermal, gas, chemical, and other. Table 5.1 summarizes the main processes within each category. The processes are typically defined by the nature of their injected fluid. For instance, gas EOR includes hydrocarbon miscible/immiscible and CO_2 miscible and immiscible processes.

Table 5.1 EOR categories and processes

Category	EOR method
Thermal EOR processes	Steamflooding
	Cyclic steam stimulation
	In situ combustion
	Hot waterflooding
	Steam-assisted gravity drainage
Gas EOR processes	Hydrocarbon miscible/immiscible
	CO_2 miscible
	CO_2 immiscible
	Nitrogen
	Flue gas (miscible and immiscible)
Chemical EOR processes	Micellar-polymer
	Polymer
	Alkaline
	Surfactant/ASP
Other EOR processes	Microbial
	Electromagnetic heating

5.3 Thermal Recovery Processes

5.3.1 Introduction

Thermal recovery processes rely on the use of thermal energy in some form both to increase the reservoir temperature, thereby reducing oil viscosity, and to displace oil to a producing well. Three processes have evolved over the past 30 years to the point of commercial application. These are cyclic steam stimulation, steamdrive, and forward in situ combustion.

The motivation for developing thermal recovery processes was the existence of major reservoirs all over the world that were known to contain billions of barrels of heavy oil and tar sands that could not be produced with conventional techniques. In many reservoirs, the oil viscosity was so high that primary recovery on the order of a few percent of original oil in place (OOIP) was common. In some reservoirs, primary recovery was negligible.

Thermal recovery processes are the most advanced EOR processes and contribute significant amounts of oil to daily production. Most thermal oil production is the result of cyclic steam injection and steamdrive. In 1993, worldwide production from cyclic steam and steamdrive was more than 700,000 bbl/d. Consequently, more emphasis is placed on these processes. Reservoirs under treatment by these processes are shallow (generally less than 3500 ft in depth).

In situ combustion has been tested extensively in a variety of reservoirs with mixed results. Daily production from in situ combustion in 1993 was approximately 22,000 bbl/d [2].

Although steam injection and in situ combustion have been used successfully in the same reservoir, steam injection has been the process of choice for reasons other than process efficiency. In situ combustion is the only thermal recovery process that can be used in deep, high-pressure reservoirs.

Thermal recovery processes are applicable to a wide range of reservoirs. Table 5.2 summarizes the criteria for thermal recovery processes. These criteria are to be used as a guide in selecting candidates for thermal recovery process. Exceptions to the criteria may be found in specific reservoirs.

5.3.2 Cyclic Steam Stimulation

Historically, primary production from heavy-oil reservoirs before the development of thermal recovery techniques was 5 % OOIP or less. Production rates were low, declining with time as the reservoir energy was depleted. Using downhole heaters to increase oil production have been attempted and largely discarded [2].

Cyclic steam stimulation was discovered by accident in the Mene Grande field in Venezuela in 1959 when steam broke out behind casing in a steam-injection well.

Table 5.2 Screening parameters for thermal recovery processes [2]

Screening parameters	Thermal recovery steam	In situ combustion
Oil gravity (°API)	10–34	10–35
In situ oil viscosity (μ, cp)	≤ 15,000	≤ 5000
Depth (D, ft)	≤ 3000	≤ 11,500
Pay-zone thickness (h, ft)	≥ 20	≥ 20
Reservoir temperature (T_r, °F)	–	–
Porosity (φ, fraction)	≥ 0.20	≥ 0.20
Average permeability (k, mD)	250	35
Transmissibility (kh/μ, mD-ft/cp)	≥ 5	≥ 5
Reservoir pressure (P_r, psi)	≤ 1500	≤ 2000
Minimum oil content at start of process ($S_o \times \varphi$, fraction)	≥ 0.10	≥ 0.08
Salinity of formation brine[total dissolved solids (TDS)] (ppm)	–	–
Rock type	Sandstone or carbonate	Sandstone or carbonate

This well, which did not produce any oil previously, flowed at rates of 100–200 bbl/d when the well was blown down.

Cyclic steam stimulation process is shown in Fig. 5.1. In this method, steam is injected into a production well for a period of 2–4 weeks. The well is shut in and allowed to "soak" before returning to production. The initial oil rate is high because of the reduced oil viscosities at the increased reservoir temperatures. There is also some acceleration from increased reservoir pressure near the wellbore. With time, the heated zone temperature declines as a result of heat removed with the produced

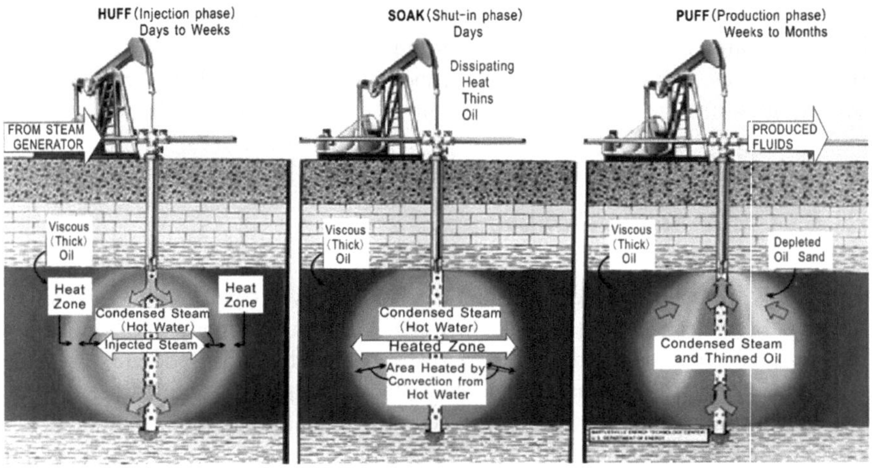

Fig. 5.1 Cyclic steam stimulation process

fluids and conduction losses to over and underlying formations. Oil rates decline as the heated zone temperature and oil viscosity decrease. When the production declines to a predetermined level, another cycle of steam injection is initiated. In some reservoirs, up to 20 cyclic have been carried out.

Cyclic steam stimulation is a precursor to steamdrive in most reservoirs. Cyclic steam stimulation is preferred when the natural reservoir energy has not been depleted. Steamdrive is used when the reservoir energy is depleted. Cyclic steam injection heats the reservoir rock around the wellbore and permits this region to remain at an elevated temperature for long periods of time.

The essential requirement for successful cyclic steam stimulation is a source of natural reservoir energy. Reservoir energy may be available in the form of fluid expansion by solution–gas drive or reduction in reservoir pressure, natural water-drive, gravity drainage, or compaction.

5.3.3 Steamdrive

The second general method of steam application is the steamdrive or steamflood, process shown in Fig. 5.2. In this method, steam is injected through the injection wells and the fluids are displaced toward production wells that are drilled in specified patterns.

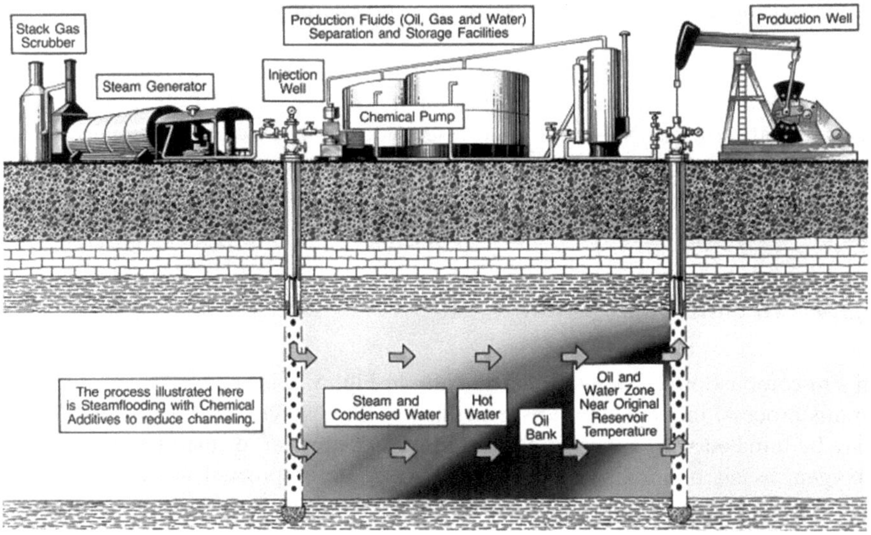

Fig. 5.2 Steamflooding process

Recovery mechanisms in this method are also based on viscosity reduction, oil swelling, steam stripping, and steam–vapor drive. As the steam loses energy in its movement through reservoir, condensation to liquid water occurs. Therefore, the process consists of a hot-water flood in the region of condensation followed by steam displacement. The process has been applied primarily to low API gravity, high-viscosity oils but it is also applicable to lighter crudes.

A major problem with steam processes is that the steam density is much lower than that of oil and water, and therefore the steam tends to move to the top of a reservoir, overriding a large part of the oil body. This is compensated partially by heat conduction away from the zone of actual contact by the steam, however, and the heated portion of a reservoir can be a high percentage of total reservoir volume. The heated volume depends significantly on the reservoir structure. Mobility control is also a problem with the steamdrive process because steam viscosity is small compared with the viscosities of liquid water and oil. Other points of concern include heat loss, equipment problems from operating at high temperatures, and pollutant emissions resulting from surface steam generation.

A steamdrive is a complex process. In the simplest representation presented, the steam zone was assumed to occupy the entire heated region until the critical time was reached. Then, a hot-water zone formed that preceded the steam zone. Oil is assumed to be displaced to residual values in each zone.

Willman et al. identified the principal displacement mechanisms for a steamdrive from the results of a set of laboratory experiments on short and long cores. Crude oil viscosity at ambient temperature varied from 5.4 to 6500 cp. Waterfloods were conducted at ambient temperature (80 °F) and 330 °F. Steamfloods were conducted at 330 and 520 °F. Typical results are shown, recovery from steam injection was significantly higher than that for either hot or cold waterflood. Furthermore, less PV of injected fluid was required for steam injection compared with hot or cold water injection.

Now, steamflood is a mature EOR method and is practiced in many reservoirs. As the technology matured, strategies for the management of steamfloods have been hypothesized and tested in field tests and expansion of commercial projects. Some of the strategies worked and others did not. Reservoir management strategies continue to develop as the steamflood is expanded all over the word.

5.3.4 In Situ Combustion

In situ combustion, as shown schematically in Fig. 5.3, is another thermal process. In this process, thermal energy is generated in the reservoir by combustion, which may be initiated with either an electric heater, gas burner, or may be spontaneous. Oxygen, as air or in a partially purified state, is compressed at the surface and continuously injected (dry process), often together with water (wet process). In the heating and combustion that occur, the lighter components of the oil are vaporized and move ahead. Depending on the peak temperature attained, thermal cracking

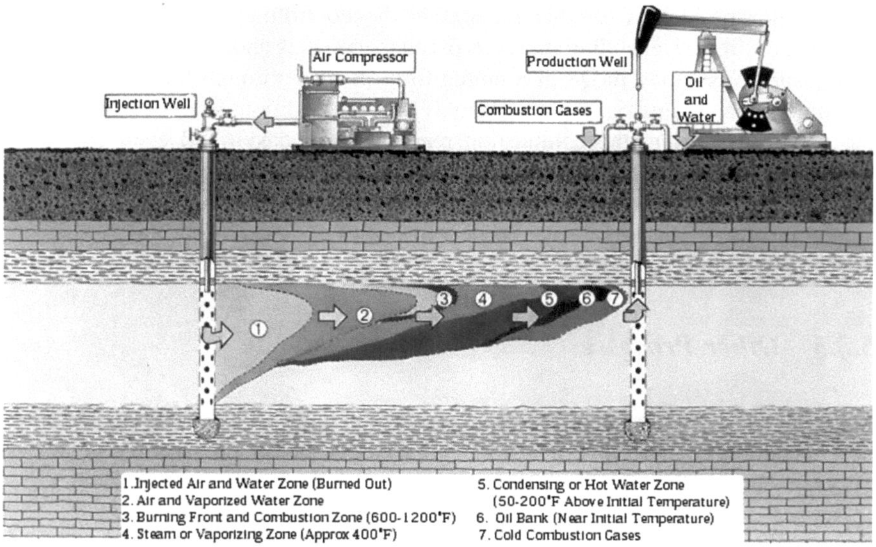

Fig. 5.3 In situ combustion process

may occur, and vapor products from this reaction also move downstream. Part of the oil deposits as a coke-like material on the reservoir rock, and this solid material serves as the fuel in the process. Thus, as oxygen injection continuing, a combustion front slowly propagates through the reservoir, with the reaction components displacing vapor and liquids ahead toward production wells.

Recovery mechanisms include viscosity reduction from heating, vaporization of fluids, and thermal cracking. Injected gases and water pick up energy as they pass through the burned zone and move toward the combustion front. Ahead of the combustion front, a steam plateau exists, i.e., a region of condensing steam in which the temperature is almost constant at the steam saturation temperature corresponding to the reservoir pressure A hot-water flood essentially exists in this region, much in the same manner as in a steamdrive process. Ahead of the steam plateau, the temperature decreases to the original reservoir temperature.

There are two types of forward combustion: dry combustion and wet combustion. In wet combustion, water is injected along with air. The water effectively picks up energy in the burned zone behind the front. It also has beneficial effects on the combustion process and reduces the combustion zone temperature. In another variation, not often applied, the combustion is carried out in a reverse manner. Combustion is started at the production wells. Oxygen is still injected at injection wells and so the combustion zone moves in the direction opposite to the fluid flow.

The in situ combustion process effectively displaces oil in the contacted. Approximately 30 % of the oil in place is required as fuel in the burning. This percentage varies, of course, depending on the oil composition and saturation, combustion conditions, and rock properties.

A major problem with this method is the controlling of movement of the combustion front. Depending on reservoir characteristics and fluid distributions, the combustion front may move in a nonuniform manner through the reservoir, with resulting poor volumetric contact. Also, if proper conditions are not maintained at the combustion front, the combustion reaction can weaken and even cease. The process effectiveness is lost if this occurs. Finally, because of the high temperatures generated, significant equipment problems can occur at the wells. Pollutant emission control also can be of concern in some cases.

5.3.5 Other Processes

Many innovative concepts in heavy-oil production have been developed. The major new technologies that have positively affected the heavy-oil industry in the last 10 years are as follows:

(a) Horizontal well technology for shallow applications (<1000 m), often combined with gravity drainage approaches.
(b) Cold production using long horizontal wells that are often combined with multilaterals with aggregate lengths exceeding several kilometers.
(c) Gravity-driven processes, particularly steam-assisted gravity drainage (SAGD), vapor-assisted petroleum extraction, and inert gas injection, all using horizontal wells to establish stable, gravity-assisted oil recovery.
(d) Cold heavy-oil production with sand, or CHOPS technology, where sand production is encouraged and managed as a means of enhancing well productivity.
(e) Pressure pulse flow enhancement technology, both as a reservoir wide method and as a workover method.

5.4 Miscible Gas Flooding

5.4.1 Introduction

The primary objective in a miscible process is to displace oil with a fluid that is miscible with the oil at the conditions existing at the interface between the injected fluid and the oil bank being displaced. There are two major variations in this process. In one, called the first-contact miscible (FCM) process, the injected fluid is directly miscible with the reservoir oil at the conditions of pressure and temperature existing in the reservoir. Figure 5.4 illustrates the FCM process. A relatively small slug of a hydrocarbon fluid, such as liquified petroleum gas (LPG), is injected to displace the oil. The primary slug size would be about 10–15 % PV. The LPG slug

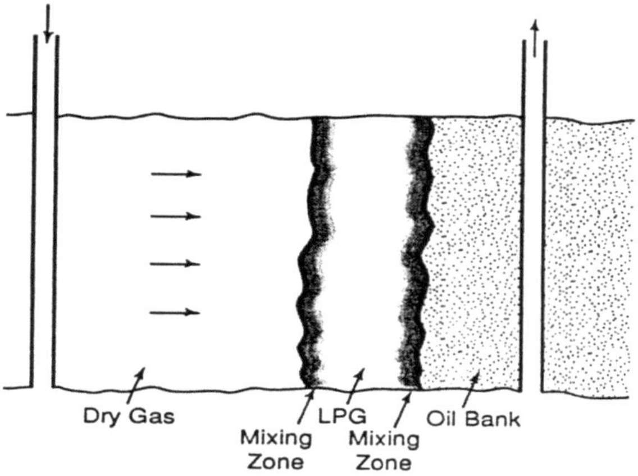

Fig. 5.4 FCM process with LPG and dry gas

is displaced by a larger volume of a less expensive gas that is high in methane concentration (dry gas). In some cases, water may be used as the secondary displacing fluid [2].

The process is effective primarily because of miscibility between the primary slug and the oil phase. Primary slug/oil interfaces are eliminated, and oil drops are mobilized and moved ahead of the primary slug. Miscibility between the primary slug and the secondary displacing fluid (dry gas in Fig. 5.4) is also desirable. Otherwise, the primary slug would be trapped as a residual phase as the process progresses.

The other variation of the miscible processes is the multiple-contact miscible (MCM) process. In this, the injected fluid is not miscible with the reservoir oil on the first contact. Rather, the process depends on the modification of composition of the injected phase, or oil phase, through multiple contacts between the phases in the reservoir and mass transfer of components between them. Under proper conditions of pressure, temperature, and composition this composition modification will generate in situ miscibility between the displacing and displaced phases.

The CO_2 miscible process illustrated in Fig. 5.5 is one such process [2].

A volume of relatively pure CO_2 is injected to mobilize and displace residual oil. Through multiple contacts between the CO_2 and oil phase, intermediate molecular weight hydrocarbons are extracted into the CO_2-rich phase. Under proper conditions, this CO_2-rich phase will reach a composition that is miscible with the original reservoir oil. From that point, miscible or near-miscible conditions exist at the displacing front interface. Under ideal conditions, this miscibility condition will be reached very quickly in the reservoir and the distance required to establish multiple-contact miscibility initially is negligible compared with the distance between wells. CO_2 volumes injected during a process are typically about 25 % PV.

Other gases are suitable for application as MCM displacement fluids in a manner similar to that described for CO_2. These include relatively dry hydrocarbon gases

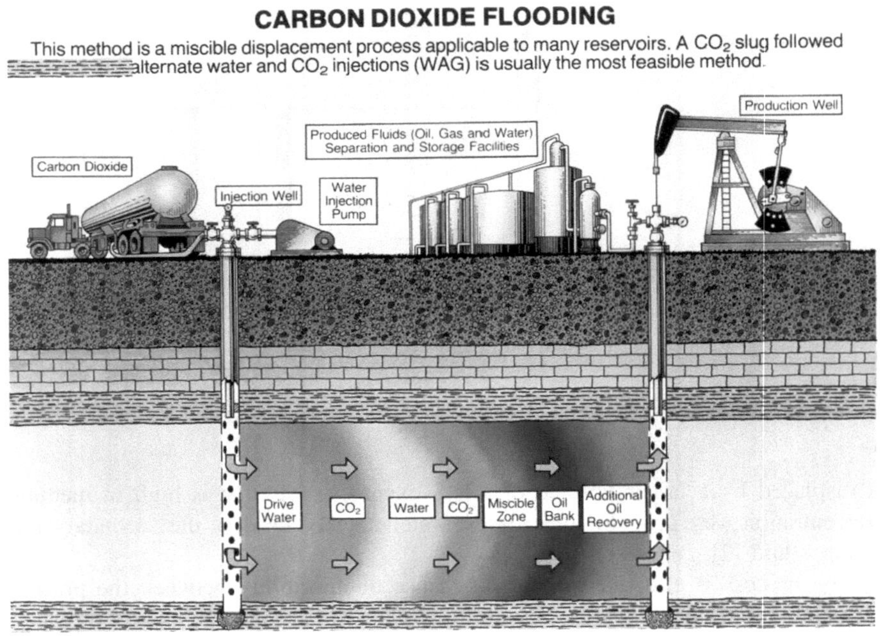

Fig. 5.5 CO_2 miscible process

(high CH_4 content), nitrogen, or flue gases. The difference is that these gases usually require much higher pressures to achieve miscibility than CO_2.

Another modification of the MCM process is the uses of a hydrocarbon fluid that is rich in components such as ethane and propane. In this process, these injected components condense into the oil phase, enriching the oil with the lighter components. Again, under proper conditions, the oil-phase composition can be modified so that it becomes miscible with the injected fluid and in situ generation of miscibility occurs.

5.4.2 *Description of Miscible Displacement*

In an immiscible displacement process the microscopic displacement efficiency, E_D, is generally much less than unity. Part of the crude oil in the places contacted by the displacing fluid is trapped as isolated drops, stringers, or pendular rings, depending on the wettability [2]. When this condition is reached, relative permeability to oil is reduced essentially to zero and continued injection of the displacing fluid is ineffective because the fluid simply flows around the trapped oil. The oil does not move in the flowing stream because of capillary forces, which prevent oil deformation and passage through constrictions in the pore passages.

This limitation to oil recovery may be overcome by the application of miscible displacement processes in which the displacing fluid is miscible with the displaced fluid at the conditions existing at the displacing-fluid/displaced-fluid interface. Interfacial tension (IFT) is eliminated. If two fluids do not mix in all proportions to form a single phase, the process is called immiscible.

Figure 5.6 shows schematically an idealized FCM displacement process that involves the injection of a specified volume of a solvent that is miscible with the crude oil. The solvent in Fig. 5.6 is a mixture of low molecular weight hydrocarbons (LPG). If the process is operated as a secondary recovery process, the oil is displaced efficiently ahead of the LPG slug, leaving little or no residual oil.

If the process is used as a tertiary process with the oil at waterflood residual saturation, the injected solvent must displace enough of the water phase to contact the residual oil and then displace the oil as a single-phase mixture with the solvent. Mixing of solvent and oil results in a mixture with a higher viscosity than the pure solvent, which makes water displacement more efficient. The mixing leads to the development of an oil bank, followed by a solvent/oil mixture bank that is rich in oil at the front and rich in solvent at the back. As the displacement proceeds, the oil bank continues to grow, and oil is displaced through the reservoir as long as the integrity of the injected solvent slug is maintained; that is, as long as the fluid bank displacing the oil is miscible with the oil. The resulting microscopic displacement efficiency is much greater than for immiscible processes and can approach 100 %.

In practice, solvents that are miscible with crude oil are more expensive than water or dry gas, and thus an injected solvent slug must be relatively small for economic reasons. For this reason, the primary solvent slug may be followed by a larger volume of a less expensive fluid, such as water or a lean gas, as shown in Fig. 5.6. Ideally, this secondary slug should be miscible with the primary slug, thus yielding an efficient displacement of the primary slug. Under proper conditions of pressure, temperature, and composition, a lean gas that is high in CH_4 concentration will be miscible with LPG, as Fig. 5.6 illustrates. Again, a mixing zone will develop at the interface between the primary slug (LPG) and the secondary slug (lean gas).

When water, which is not miscible with the solvent, is used as the chase fluid a residual solvent saturation will be retained in the rock and the primary slug will deteriorate as it is displaced toward a producing well by the water. The displacing water typically has a relatively low mobility, which improves sweep efficiency over

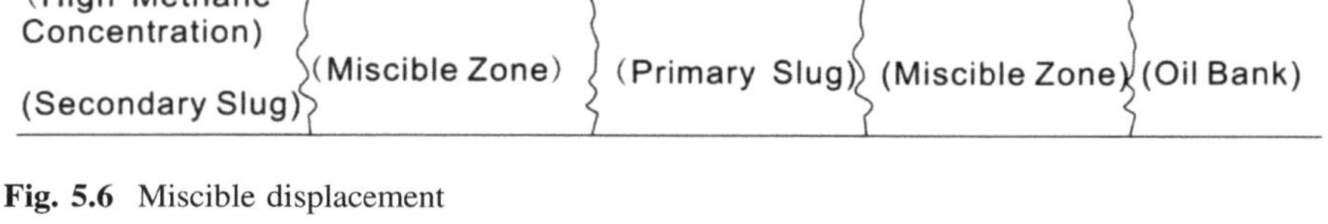

Fig. 5.6 Miscible displacement

that obtained with the solvent. For this situation, alternate injection of water and solvent, called a water-alternating-gas (WAG) process, is often used.

Figure 5.6 indicates that only hydrocarbon is flowing. If the process is applied as a tertiary process after waterflooding, however, water saturation will be above the waterflood residual saturation in at least part of the reservoir and water also be will flowing in that part of the reservoir.

In the MCM process, the oil and injected solvent are not miscible upon first contact at reservoir conditions. Rather, the process depends on modification of the oil or injected solvent compositions to such a degree that the fluids become miscible as the solvent moves through the reservoir. For example, under the proper conditions, injected methane, which is not miscible in all proportions with oil, might extract certain hydrocarbon components from oil as flow occurs through the reservoir. The enriched methane slug could thus become miscible with oil. In this process, miscibility does not exist initially but is dynamically developed as the process continues. Such processes are sometimes called dynamic miscible processes because of the manner in which miscibility develops.

Various gases and liquids are suitable for use as miscible displacement agents in either FCM or MCM processes. These include low-molecular-weight hydrocarbons, mixtures of hydrocarbons (LPG), CO_2, nitrogen, or mixtures of these. The application particular depends on the reservoir pressure, temperature, and compositions of the crude oil and the injected fluid.

5.4.3 Vaporizing Gas (Lean Gas) Displacement Process

The vaporizing process may be described on a pseudoternary diagram, as shown in Fig. 5.7[2]. The apexes of the diagram represent 100 % concentration of methane (C_1), intermediate hydrocarbons (C_2 through C_6), and heavy hydrocarbons (C_{7+}), respectively. Temperature and pressure are constant and equal to the displacement conditions in the reservoir. Phase Boundary VO is the saturated vapor curve and Phase Boundary LO is the saturated liquid curve. Point O is the critical or plait point.

Figure 5.7 is a pseudoternary diagram, because certain components in the system are combined to form pseudocomponents. These are C_2 through C_6 and C_{7+}. For this approximation to be valid, the individual components in each pseudocomponent should part into the different phases in the same proportions as they exist in the pseudocomponent. This does not happen precisely, but the approach to this behavior is often sufficiently close that the use of the pseudoternary diagram is justified. In particular, a pseudoternary diagram is useful in describing the development of miscibility in a multiple contact process.

The process descriptions that follow are based on the assumption that local thermodynamic equilibrium exists between different phases. This assumption is generally thought to be valid at reservoir displacement conditions where advance rates are very low. As discussed later, at high water saturations, the water can block

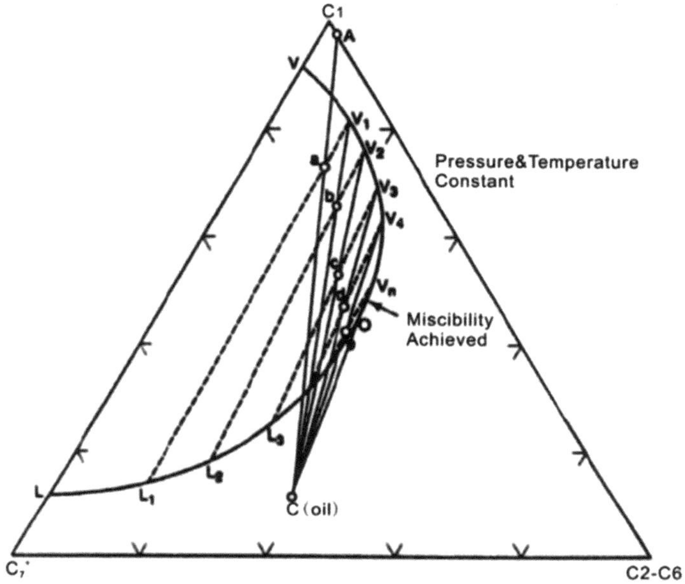

Fig. 5.7 The vaporizing gas displacement process on a pseudoternary diagram

contact between the solvent and crude oil in strongly water-wet rocks and mass transfer effects can be important.

Assuming that the reservoir crude oil is represented by Point C in Fig. 5.7. The injected fluid is shown as Point A and is a mixture consisting mostly of methane and much smaller amounts of the intermediate components. The injected fluid would be classified as a dry gas. Miscibility may be created in a dynamic fashion in the reservoir. The process operates conceptually as follows (see Fig. 5.7) [2].

(a) At the leading edge of the displacement front, gas A mixes with oil C. The resulting composition of the mixture is along line AC, say at point a.
(b) Mixture a is in the two-phase region and separates into a vapor, V_1,and a liquid, L_1.
(c) Vapor V_1 moves ahead of liquid L_1 and contacts fresh oil of composition C. The resulting mixture is along line V_1C, say at Point b.
(d) Mixture b separates into Vapor V_2 and Liquid L_2.
(e) The process continues with the vapor-phase composition changing along the saturated vapor curve, V_3, V_4, …, etc.
(f) Finally, at Point O, the vapor becomes miscible with Oil C because the mixing line lies entirely in the single-phase region.
(g) From this point, as the displacement continues the process will be miscible for an ideal process. In fact, as a result of mixing in the reservoir, miscibility may be successively lost and redeveloped.

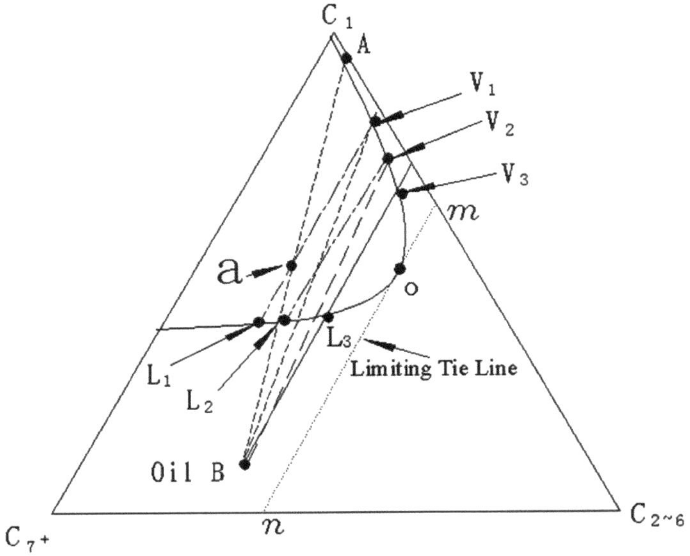

Fig. 5.8 Representation of the high-pressure process on a pseudoternary diagram (miscibility does not develop)

Miscibility will not be generated for all combinations of injected gases and crude oils. An example in which miscibility does not result is shown in Fig. 5.8.

For this example, the reservoir oil is assumed to have the composition at Point B. As before, assume Gas A mixes with Oil B along Line AB, yielding Mixture a, which gives Vapor V_1 and Liquid L_1. Vapor V_1 moves ahead and mixes with the new Oil B, which gives V_2 and L_2. In this case, however, as the process continues, the vapor will change by successive mixing only to Point V_3 and further enrichment will not occur. The composition change ceases because Line V_3L_3, when extended, passes through Point B, the oil composition. That is, mixtures of V_3 and B will be along the tie line and further enrichment of the vapor by the intermediate and heavy components will not occur. The process will continue as an immiscible process.

The distinction between processes that will and will not be miscible is determined by the so-called limiting tie line, illustrated by Line mn in Fig. 5.8. This line is tangent to the binodal curve at Critical Point O.

The process described is called a vaporizing-type process. The vaporizing term derives from the fact that the injected dry gas is enriched by vaporizing intermediate and heavy components from the oil. The general effect of pressure on the process is illustrated in Fig. 5.9.

As pressure is increased, the two-phase envelope becomes progressively smaller and the critical point shifts toward lower concentrations of intermediates. The limiting tie line thus shifts toward lower concentrations of intermediates in the oil, and more oils develop miscibility with a dry gas with specified composition.

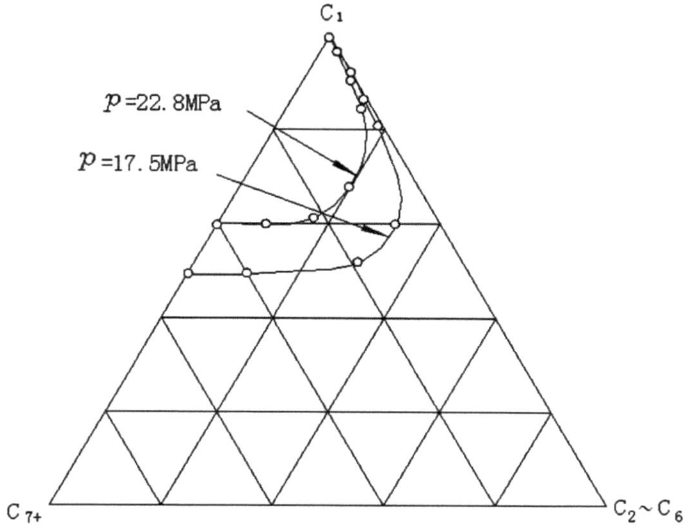

Fig. 5.9 Effect of pressure on the phase behavior

The minimum pressure at which the limiting tie line lust passes through the reservoir oil composition is called the MMP. That is, the MMP is the minimum pressure at which in situ miscibility can be achieved in the MCM process for a specified fluid system. In the vaporizing MCM process, the injected gas composition must lie to the left and the reservoir oil composition must lie on or to the right of the limiting tie line, as shown in Fig. 5.7. Relatively high pressures are required for most of oils and thus the processes is sometimes designated a high-pressure process. Typical applications of this process are at pressures well above 20 MPa.

This described development of miscibility is an idealized picture of the events at the leading edge of the displacement front. Behind the front, the situation is essentially the same. In the region of developing miscibility, the oil-phase composition at a fixed point moves along the lower part of the binodal curve toward higher concentrations of the C_{7+} components, as shown in Fig. 5.7. Thus, the oil at a fixed position is being depleted of its lighter components.

There is a transition zone in MCM displacement processes because a finite distance is required for the gas phase composition to be modified to create a miscible condition. Data from laboratory experiments suggest that this distance is short (no more than a few feet in laboratory cores or sandpacks) relative to reservoir dimensions.

5.4.4 Condensing (Enriched Gas) Displacement Process

In this process, the injected fluid contains significant amounts of intermediate components (C_2 through C_6). The process depends on the condensation of these components into the reservoir oil, thereby modifying the oil composition. The modified oil then becomes miscible with the injected fluid [2].

The process is illustrated in Fig. 5.10. The fluid to be injected is represented by Point A and the reservoir oil by Point C.

The idealized sequence of events for the condensing process is as follows:

(a) The injected Fluid A mixes with Oil C. The resulting composition is along Line AC, say at Point a.
(b) Mixture a is in the two-phase region and separates into Vapor V_1 and Liquid L_1.
(c) Vapor V_1 moves ahead of Liquid L_1. For ease of conceptualization, it might be viewed that L_1 is immobile at this point in the process. Liquid L_1 mixes with additional injected Gas A and the resulting fluid is along Mixing Line AL_1, say at Point b.
(d) Mixture b separates into Vapor V_2 and Liquid L_2.
(e) The process continues with the oil-phase composition changing from L_2, L_3, L_4, …, L_n.

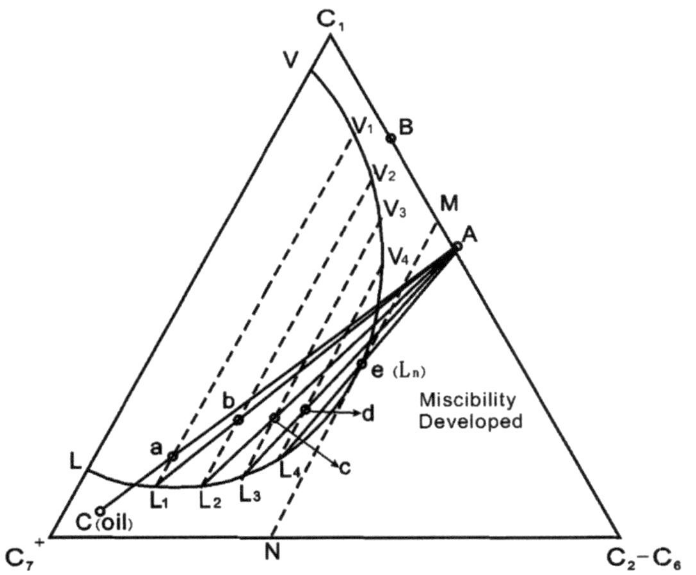

Fig. 5.10 Representation of the enriched gas (condensate) process

(f) For the tie line through Mixture e the immobile oil, L_n, becomes miscible with the injected fluid because Mixing Line AL_n, will lie entirely in the single-phase region.
(g) From this time, the process will become miscible and will continue as a miscible process because the mixture resulting from mixing along Line AL_n will be miscible with the fluids ahead of this point.

For this ideal conceptualization of the condensing type process, a composition transition zone exists ahead (downstream) of the point where miscibility achieves. This results because the injected gas continues to condense partially into the oil as the injected gas moves downstream. Once miscibility is generated, the transition zone stabilizes.

The limiting Tie Line MN is shown in Fig. 5.10. In the example, Fluids A and C are on opposite sides of Tie Line MN. If the injection fluid was at Point B, to the left of this line, miscibility would not be achieved. This is because, at some point in the process, extension of one of the L_nV_n lines would pass through Point B and further modification of the oil-phase composition would not occur. The process would then continue as an immiscible displacement.

If pressure were changed, the two-phase region would be modified and the limiting tie line would shift. For a given injection fluid/crude oil pair, the MMP in a condensing gas process is the minimum pressure at which the limiting tie line just passes through the injection fluid composition; i.e., the injection fluid composition is on the limiting tie line. Thus, in the condensing MCM process, the injected fluid composition must lie on or to the right and the oil composition must lie to the left of the limiting tie line.

In the condensing gas process, there is an alternative to increase the pressure in that the injection gas composition can be enriched to achieve miscibility. At a fixed pressure, the minimum enrichment at which the limiting tie line passes through the injection gas composition is called the minimum miscibility enrichment (MME).

More recent investigations of the enriched gas process have shown that an alternative mechanism, one that involves both condensation and vaporization, is often responsible for the efficient displacement in this process. Zick was the first to present a detailed description of the mechanism, although Stalkup indicated earlier that there were problems in trying to describe the enriched gas process on a pseudoternary diagram.

The problem of description arises from the assumption that a pseudoternary diagram is a valid representation of a multicomponent system having many more than three components. If specific components making up a pseudocomponent do not partition into the different phases in the same proportions as specified in the pseudocomponent then the pseudoternary diagram will not describe the system correctly. To the extent that the pseudocomponent concept fails, the description of the condensing process on a pseudoternary diagram for a multicomponent system may be incorrect.

5.4.5 Design Procedures and Criteria

A number of factors must be considered in the design of a miscible displacement process. This section discusses those that relate primarily to the laboratory and modeling studies that precede field testing. In addition to the design work discussed here, significant reservoir analysis is required. The reservoir analysis should focus on geology, fluid distributions, performance of primary production, analysis of any prior displacement process (such as waterflooding), injection rates, and pattern geometry effects. Gaining an understanding of reservoir heterogeneity is especially critical for the final design. An economic study, including a sensitivity analysis, also should be done.

In a miscible displacement process, a solvent slug that efficiently displaces the reservoir oil at the microscopic level must be used. Typically, ROS (residual oil saturation) to the solvent should be 10 % or less. Depending on reservoir fluids and conditions (P and T) and the solvents availability, either a first-contact or multiple-contact process can be applied. The solvent-slug size should be designed so that the miscible and immiscible portions of the process yield this acceptable recovery.

Meeting these criteria requires that certain laboratory and modeling studies be conducted in the design work. Phase-behavior studies relating to miscibility must be conducted. Methods of mobility control should be considered. This probably will require conducting laboratory displacements. Mathematical modeling should be used to examine such factors as optimum slug size and volumetric sweep efficiency.

5.4.5.1 Phase Behavior Selection of a Solvent

The selection of a solvent is governed by reservoir pressure and temperature, reservoir oil composition, and solvent availability and cost [2]. Reservoir temperature and especially pressure are related to reservoir depth. Pressures in excess of the fracture pressure cannot be used because fracturing would lead to poor sweep efficiency.

Miscibility of a solvent/oil system and efficiency of displacement in a phase-behavior sense are typically determined from slim-tube experiments. These experiments should be designed so that viscous fingering, gravity segregation, and heterogeneity effects are minimized. Under these conditions, the slim-tube apparatus is basically a multiple-contact device in an idealized porous medium. Recovery from an experiment represents the maximum displacement efficiency at the microscopic level that can be expected on the basis of the phase behavior of the solvent/oil system.

Slim-tube experiments or RBA (rising bubble apparatus) experiments can be used to measure the MMP for a specified solvent or the MME of a solvent at a specified pressure. The experiments are relatively simple and inexpensive to conduct, making the testing of several systems feasible.

In the device used for this test, multiple contacts between oil and solvent are made in a continuous process but a porous medium is not used. Phase behavior data that relate to the development of miscibility are obtained. Classic phase-behavior data taken in a PVT cell are also can useful. These experiments consist of allowing fixed amounts of solvent and oil to equilibrate in a cell at fixed pressure and temperature. Samples of each phase then can be used for measurements of compositions and other properties such as viscosity and density. PVT data are relatively time-consuming and expensive to obtain. The data also can be used to identify potential problems, such as asphaltene deposition.

5.4.5.2 Mobility-Control Considerations

Viscosities of solvents are relatively small compared with those of reservoir oils. For example, viscosities of hydrocarbon gases or CO_2 are on the order of 0.05–0.10 cp at usually encountered flooding conditions. Thus, the mobility ratio between the solvent slug and the displaced oil bank is typically very unfavorable.

Research is in progress on the use of additives to increase hydrocarbon or CO_2 solvent viscosity. The application of mobility-control agents, such as foams or gels, also has been examined. To date, however, these approaches are experimental and are not widely accepted technology in solvent flooding.

The most widely used method of mobility control is to alternate water and gas injection—the WAG injection process. In this process, slugs of solvent and water are alternately injected through the injection wells. The presence of a high water saturation reduces the effective permeability of the solvent and thus its mobility. An evaluation of the improved sweep resulting from improving the effective mobility ratio by use of the WAG process is probably best accomplished with a reservoir simulator. Relative permeability data for the reservoir rock are required for these calculations.

5.4.5.3 Corefloods

Reservoir corefloods are useful to estimate the displacement efficiency at the microscopic level. Effects on displacement of such phenomena as water blocking at high water saturations can be investigated. Data on relative permeabilities in cores can be used to estimate injectivities under field conditions if hysteresis in the relative permeabilities is carefully modeled.

Corefloods need to be designed carefully. Samples of reservoir rock suitable for making long cores for laboratory displacement investigations are difficult to obtain. Miscible displacements in short, small-diameter cores cannot be readily extrapolated to field dimensions. Quarried rocks, such as Berea sandstone, can be used, but the microscopic characteristics, especially wettability, would be quite different from actual reservoir rocks.

5.4.5.4 Mathematical Modeling

Mathematical modeling, or simulation, of a miscible displacement process can be done with a variety of models ranging from those that are relatively simple to complex, computer-based, numerical finite-difference simulators. The calculation is based on data from laboratory physical models, which is an example of a relatively simple approach. Results from such a model can be used to describe the general performance behavior and as a screening tool. The results, however, should be considered only approximate at best.

The most comprehensive models are numerical finite-difference simulators based on numerical solutions of partial differential equations that describe the fluid system. The literature contains many examples of the use of such simulators to describe miscible displacement processes.

Two types of models that have been widely used are modified black oil simulators and compositional simulators. These differ in the treatment of phase behavior and composition of each phase. Black oil simulators basically treat the different phases as individual components, i.e., oil, gas, or water. Compositional simulators account for the different components in each phase and use a more rigorous EOS (equation of state). Compositional simulators are more complex and thus generally more expensive and time-consuming to apply.

One problem commonly encountered in numerical finite-difference simulators used to describe miscible displacement processes is numerical dispersion. This is an artificial and often large mixing or dispersion that results from the solution of the equations in the model. However, dispersion caused by reservoir heterogeneity is significant, and simulators often incorporate numerical dispersion to approximate dispersion in a reservoir.

5.5 Chemical Flooding

5.5.1 Introduction

Chemical flood are identified by the chemical type that is injected. The most common processes are polymer (P), surfactant (S), and alkaline (A), and any combination of these processes. Emulsion and foam are more related to mobility control. These two processes are not discussed in detail because they are thermodynamically unstable processes and quite different from the stable processes we deal with here [2, 3].

Polymer flooding consists of adding polymer to the water of a waterflood to decrease its mobility. The water-soluble polymers used in EOR consist of high molecular weight chain-like molecules with molecular weights up to or exceeding 35 million. Polymers such as polyacrylamides and polysaccharides often are used as mobility-control buffers for permeability reduction and/or increasing viscosity. Polysaccharides sometimes are called biopolymers. Polymers increase the viscosity of the water and prevent it from running ahead of the oil. Increased resistance to flow, particularly in high-permeability zones, improves the volumetric reservoir sweep efficiency resulting in an increased oil recovery. Water-soluble synthetic polyacrylamides consist of high molecular weight chain-like molecules with $CONH_2$, COON, and COONa groups attaching to every other carbon atom on a carbon chain. Naturally occurring polysaccharides consist of cyclic carbohydrate monomers alternating in the polymer structure. These additives aid oil recovery by decreasing the flood water's mobility.

A fundamental chemical process is surfactant flooding in which the key mechanism is to reduce interfacial tension (IFT) between oil and displacing fluid. The mechanism, because of the reduced IFT, is associated with the increased capillary number, which is a dimensionless ratio of viscous-to-local capillary forces. Experimental data show that as the capillary number increases, the residual oil saturation decreases. Therefore, as IFT is reduced through the addition of surfactants, the ultimate oil recovery is increased. In alkaline flooding, the surfactant required to reduce IFT is generated in situ by the chemical reaction between injected alkali and naphthenic acids in the crude oil. However, more important mechanisms in alkaline injection are its synergy with surfactant and its function to reduce surfactant (even polymer) adsorption.

The synergy makes the alkaline-surfactant process more robust and results in a wider range of application conditions. In modern chemical EOR, the most important processes are to reduce the amount of injected chemicals and to fully explore the synergy of different processes. This effort has resulted in the alkaline-surfactant-polymer (ASP) process. Laboratory studies, pilot tests, and field applications have demonstrated the greatest potential for enhancing oil recovery. However, some problems, such as scaling and emulsion, have also emerged in practical applications. Although ASP has the greatest potential, the practical problems lead operators to consider chemical processes without alkaline injection. Other factors challenging chemical EOR processes include expensive water treatment such as filtering, softening, and post-filtering; disposal of produced chemical solution; initial capital investment for facilities and equipment; and so on [3].

5.5.2 Polymer Flooding

Polymer flooding is a process in which a small amount of polymer is added to thicken brine to reduce the water mobility, process shown in Fig. 5.11[2, 4]. Under current technology, most light-oil reservoirs are waterflooded unless a strong

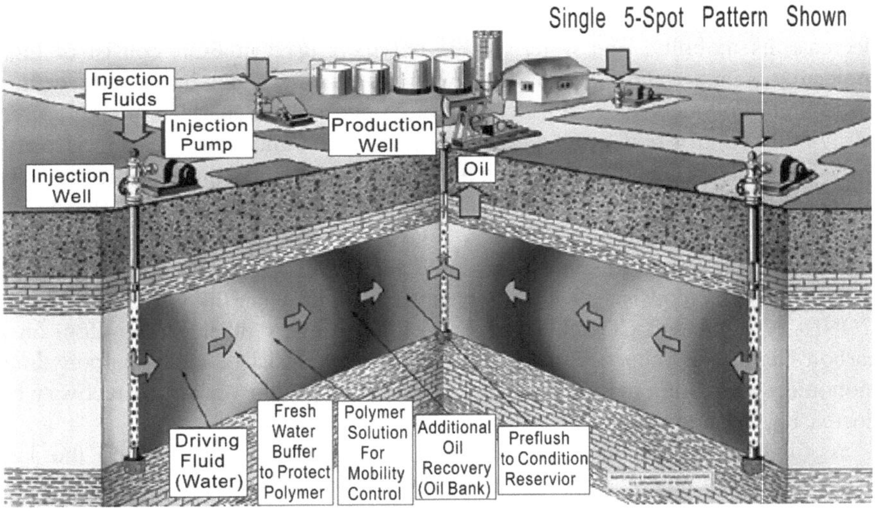

Fig. 5.11 Polymer flooding process

natural water drive or an effective gas cap drive is already taking place. The oil recovery that can be expected by waterflooding a given reservoir is controlled by two factors: volumetric coverage efficiency and the residual oil saturation. Generally, at least one-half of the original oil remains after waterflooding. Under some conditions, the volumetric coverage efficiency can be improved by "thickening" the injected water to improve the mobility ratio.

The mobility ratio of a waterflood can be reduced by adding a chemical such as polyacrylamide or XC biopolymer to the injected water. The effect of mobility ratio on vertical invasion efficiency and on pattern coverage efficiency was covered in the waterflooding section. Both of these efficiencies increase as the mobility ratio decreases. The main incentive for polymer is in flooding medium viscosity oil reservoirs. The mobility ratio of a waterflood can be effectively reduced by about a factor of 10 by adding polymer to the water. As a result, the most attractive oil reservoirs for polymer flooding will fall in the 20–200 cp viscosity range. As we will see later, water banks thickened with polymer are used to provide mobility control in pushing miscible or surfactant banks in tertiary recovery projects.

5.5.2.1 Types of Polymer

Several polymers have been considered for polymer flooding: xanthan gum (XC biopolymer), hydrolyzed polymer (HPAM), copolymers of acrylic acid and acrylamide, etc. But only two basic types of polymer: XC biopolymer and partially hydrolyzed polyacrylamides are widely used in field recovery projects. XC biopolymer is a natural polysaccharide produced by a microbial fermentation

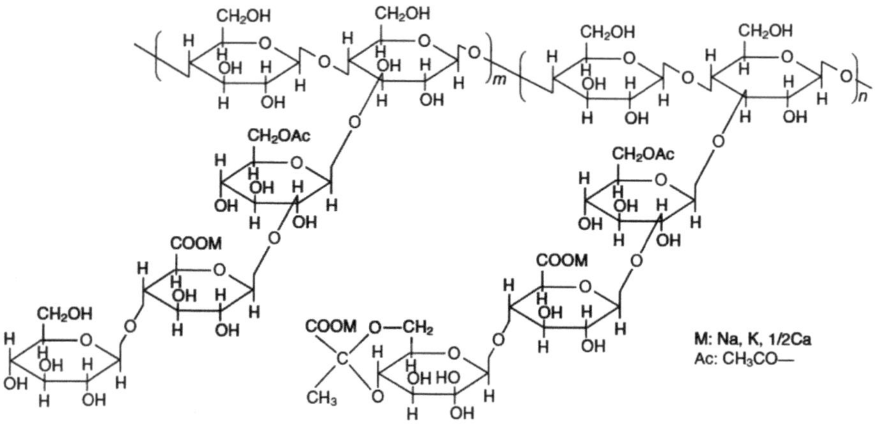

Fig. 5.12 The structure of xanthan biopolymer

process. XC biopolymer reduces the mobility ratio by simply increasing water viscosity. The structure of a xanthan biopolymer is shown in Fig. 5.12.

Polyacrylamides are synthetic chemical products. Polyacrylamides reduce water mobility mainly by reducing the formation permeability to water. The structure of hydrolyzed polyacrylamide is shown in Fig. 5.13.

Hydrolysis converts some of the amide groups ($CONH_2$) to carboxyl groups (COO^-), the degree of hydrolysis is the mole fraction of amide groups that are converted by hydrolysis. It ranges from 10 to 35 % in commercial products. Hydrolysis of polyacrylamide introduces negative charges on the backbones of polymer chains that have a large effect on the rheological properties of the polymer solution. At low salinities, the negative charges on the polymer backbones repel each other and cause the polymer chains to stretch. When an electrolyte, such as NaCl, is added to a polymer solution, the repulsive forces are shielded by a double layer of electrolytes; thus, the stretch is reduced.

5.5.2.2 Properties of Polymer Solutions

Polymer Viscosity

Viscosity is the most important parameter for polymer solution. In general, a polymer solution behaves like a pseudoplastic fluid. A fluid whose viscosity decreases with increasing shear rate ($\dot{\gamma}$) is shear thinning. The behavior of the polymer solution is caused by the uncoiling and unsnagging of polymer chains when they are placed in a shear flow. The polymer solution viscosity—shear rate relationship may be described by a power-law model.

Fig. 5.13 The structure of HPAM

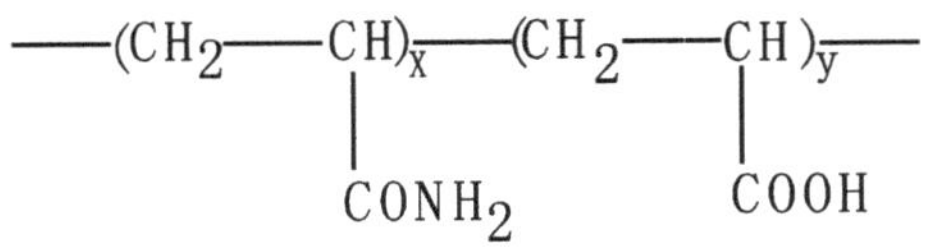

$$\mu_p = K\dot{\gamma}^{n-1}, \tag{5.7}$$

where μ_p is the viscosity of polymer solution, mPa.s, K is the flow consistency index, mPa.s^n, $\dot{\gamma}$ is shear rate, s^{-1}, and n is the flow behavior index, dimensionless. In the pseudoplastic regime $n \leq 1$, typically n ranges between 0.4 and 0.7. At different concentrations, n hardly changes, but K changes. For a Newtonian fluid, $n = 1$ and K is simply the constant viscosity, μ.

A more general model is the Carreau equation,

$$\mu_p = \mu_\infty + \frac{\mu_p^0 - \mu_\infty}{\left[1 + (\lambda\dot{\gamma})^\alpha\right]^{\frac{(n-1)}{\alpha}}}, \tag{5.8}$$

where μ_∞ is the limiting viscosity at the high (approaching infinite) shear limit and is generally taken as water viscosity μ_w, mPa.s; λ and n are polymer-specific empirical constants, dimensionless; and α is generally taken to be 2, dimensionless. μ_p^0 is the limiting viscosity at the low (approaching zero) shear limit, mPa.s. For intermediate shear rates, the Carreau equation represents a power-law relation. Practically, μ_p and μ_p^0 are much higher than μ_∞, and $(\lambda\dot{\gamma})^\alpha$ is much larger than 1. Then Eq. 5.8 becomes the power-law equation of the form $\mu_p = \mu_p^0(\lambda\dot{\gamma})^{n-1}$, which describes the viscosity at the intermediate and high shear rate regimes. At the low shear rate regime, $\mu_p = \mu_p^0$, as shown in Fig. 5.14. At the intersection of these two regimes, the viscosity and shear rate are the same. Then we must have $\dot{\gamma} = \lambda^{-1}$.

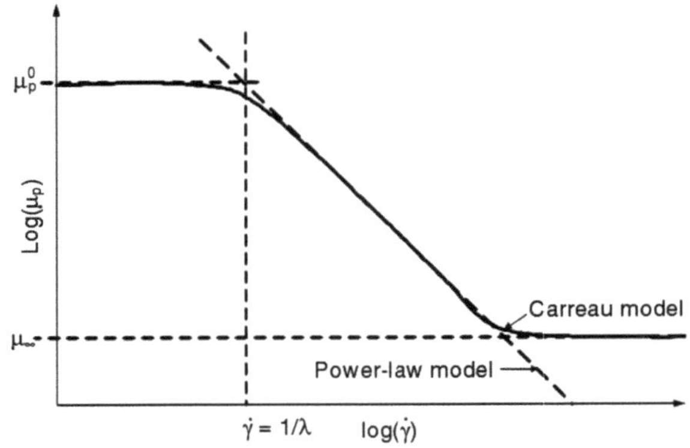

Fig. 5.14 Comparison of the Carreau and power models

This intersection is the first critical shear rate at which the fluid deviates from the Newtonian behavior.

Polymer Retention

Polymer retention is due to solid surface adsorption or mechanical trapping within porous media. Field measurement values of retention range from 20 to 400 lbm polymer/acre-ft bulk volume with desirable retention level being less than 50 lbm/acre-ft. Retention causes the loss of polymer from solution, which can cause mobility-control effect to be destroyed. Retention also causes a delay in the rate of polymer propagation.

Polymer retention includes adsorption, mechanical trapping, and hydrodynamic retention. Mechanical entrapment and hydrodynamic retention are related and occur only in flow-through porous media. Retention by mechanical entrapment is viewed as occurring when larger polymer molecules become lodged in narrow flow channels.

Permeability Reduction

Permeability reduction, or pore blocking, is caused by polymer adsorption. Polymer retention reduces the apparent permeability of the rock. Permeability reduction depends on the type of polymer, the amount of polymer retained, the pore size distribution, and the average size of the polymer relative to the pores in the rock. The reduction in permeability to water caused by the polymer is a big advantage since a relatively small polymer bank can be followed by brine without losing mobility control.

Mobility control is required for processes using miscible or surfactant banks. Because these banks remove the residual oil, the permeability to water behind the bank is greatly increased. The bank must be driven by thickened water to compensate for this relative permeability increase. This phenomenon is measured by the following factors: the resistance factor ($F_{\rm R}$, dimensionless), and residual resistance factor ($F_{\rm Rr}$, dimensionless).

$$F_{\rm R} = \frac{K_{\rm w}/\mu_{\rm w}}{K_{\rm p}/\mu_{\rm p}} \tag{5.9}$$

$$F_{\rm Rr} = \frac{K_{\rm wb}}{K_{\rm wa}}, \tag{5.10}$$

where $K_{\rm p}$ is the permeability of polymer solution, mD; $K_{\rm w}$ is the permeability of water phase, mD; $K_{\rm wb}$ is the permeability of water phase before polymer injection, mD; $K_{\rm wa}$ is the permeability of water phase after polymer injection, mD.

Screening Candidates for Polymer Flooding

Polymer flooding should not be considered for any reservoir unless the following guidelines are met:

(a) Oil viscosity should be less than 200 cp.
(b) Reservoir temperature should be less than 200 °F.
(c) Permeability should be greater than 20 mD.

5.5.3 *Micellar-Polymer Flooding*

In most situations, the micellar/polymer process is implemented as a tertiary displacement near the end of a watertlood. Figure 5.15 shows a tertiary process where the initial oil saturation is S_{orw}, ROS (residual oil saturation) after waterflooding.

A specified volume of micellar solution is injected. The volume of the slug is one of the process variables; however, typical volumes are about 3–30 % of the flood-pattern PV. The micellar solution has a very low IFT with the residual crude oil and mobilizes the trapped oil, forming an oil bank ahead of the slug. The micellar slug also has a relatively low IFT with the brine and thus displaces brine as well as oil. Both oil and water flow in the oil bank. Because oil is initially at residual saturation in a tertiary flood no oil production occurs until the oil bank breaks through at the end of the flow system.

The micellar solution must be designed so that a favorable mobility ratio exists between the micellar slug and the oil bank. The viscosity of the micellar solution is adjusted to accomplish this. Polymer is often added to the micellar solution to increase its apparent viscosity. Thus, the process has the potential to increase both volumetric sweep efficiency and microscopic displacement efficiency. This

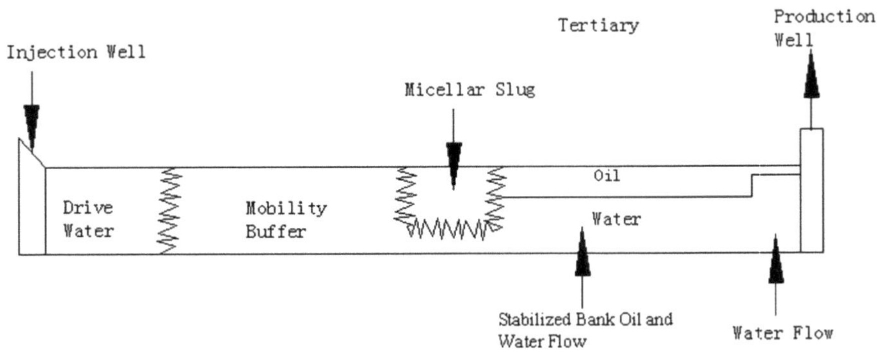

Fig. 5.15 Micellar/polymer displacement process

improvement is an advantage of the micellar/polymer process over mobility-control and miscible processes.

In some cases, a preflush is injected ahead of the micellar solution to adjust brine salinity or pH. The preflush solution may contain a sacrificial adsorbent,i.e., a relatively cheap chemical that will be adsorbed on the rock and occupy adsorption sites. The purpose is to reduce adsorption and loss of the surfactant contained in the micellar solution that will follow. In pilot tests, the volume of the preflush may be as large as 100 % of the flood-pattern PV. This amount is not practical in a full-scall field application where much smaller volumes must be used. Field experience has shown that preflushes are often not effective in controlling brine salinity and divalent ion content. Thus, a micellar solution generally must be designed to tolerate.

A micellar solution is relatively expensive, so a limited volume, or slug size, is used. The micellar slug must be displaced with a less expensive fluid. Water is not a suitable displacing fluid because an unfavorable mobility ratio usually exists between water (brine) and a micellar solution. Therefore, a mobility buffer slug that consists of a solution of polymer in water is used. The mobility buffer slug, which may range up to 100 % of the flood-pattern PV, is then followed by water. The polymer mobility buffer typically contains polymer at a concentration from 250 to 2500 ppm. Both polyacrylamide and biopolymer are used. The salinity of the polymer solution is a design variable. The solution often contains a biocide, especially when a biopolymer is used, because microorganisms can degrade the polymer and reduce the viscosity of the solution.

In laboratory corefloods, through this method the recovery efficiency can be quite high, approaching 90 % or more of the oil present at the start of displacement.

5.5.4 Alkaline Flooding

Alkaline flooding also called caustic flooding. In alkaline flooding, a high-pH chemical system is injected. If the reservoir crude oil has sufficient "saponifiable components," a reaction will occur in which surfactants are formed in situ. In most of the literature, these saponifiable components are described as petroleum acids, even though their structure is not known. Several mechanisms, including reduction of IFT, contribute to increased oil displacement efficiency as a result of the formation of the surfactant [1]. The process generally has been applied with crude oils of relatively low API gravity.

5.5.4.1 Chemicals Used and in Situ Formation of Surfactants

Several different alkaline agents have been used, including sodium hydroxide, sodium orthosilicate, sodium carbonate, ammonium hydroxide, and ammonium carbonate. The first three have been the most widely considered. The process is dependent on alkali reacting with petroleum acids in crude oil to form surfactants in situ. These petroleum acids are many in number and varied in composition. In the past, sodium hydroxide and sodium carbonate were used most often. However, owing to the emulsion and scaling problems observed in Chinese field applications, the tendency now is not to use sodium hydroxide.

Addition of the alkali chemicals results in a high pH because of the dissociation in the aqueous phase. NaOH, for example, dissociates to yield $[OH^-]$ as follows:

$$NaOH \rightarrow Na^+ + OH^- \tag{5.11}$$

Sodium carbonate dissociates as

$$Na_2CO_3 \rightarrow 2Na^+ + CO_3^{2-} \tag{5.12}$$

followed by the hydrolysis reaction:

$$CO_3^{2-} + H_2O \rightarrow HCO_3^- + OH^- \tag{5.13}$$

The dissociation of sodium silicate compounds is complex, involving formation of oligomeric species. Consequently, dissociation cannot be represented by a single chemical equation. The hydroxide ion must react with a petroleum acid from the crude oil to form a surfactant. A general mechanism is shown in Fig. 5.16 and described by the following.

Some of the petroleum acid in the crude oil partitions into the aqueous phase according to the solubility expression

$$K_D = \frac{[HA_o]}{[HA_w]}, \tag{5.14}$$

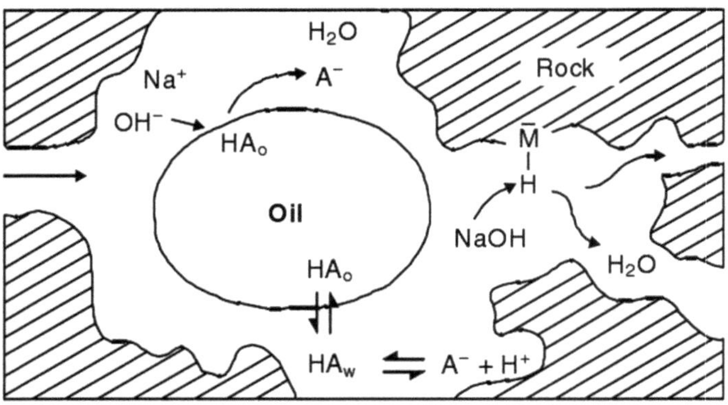

Fig. 5.16 Schematic of alkaline recovery process

where K_D is distribution or partition coefficient; HA_o and HA_w denote petroleum acid in the oil and water phases, respectively. This also can be expressed as

$$HA_o \Leftrightarrow HA_w \tag{5.15}$$

This process is as shown in Fig. 5.16 [3].

The petroleum acid dissociates in aqueous phase according to the expression

$$HA_w \Leftrightarrow H^+ + A^- \tag{5.16}$$

as governed by the equilibrium relationship

$$K_A = \frac{[H^+][A^-]}{[HA_w]} \tag{5.17}$$

The species A^- is an anionic surface-active agent. In effect, caustic uses up the hydrogen ion by the reaction

$$HA_w + OH^- \Leftrightarrow A^- + H_2O \tag{5.18}$$

According to Eq. 5.18, this results in an increase in $[A^-]$. The net effect of all the reactions is shown in the upper left corner of Fig. 5.16.

A measure of the potential of a crude oil to form surfactants is given by the acid number. This is the amount of KOH, usually given in milligrams, required to neutralize 1 g of the petroleum acid in the crude oil. Unfortunately, acid number does not always correlate with oil recovery.

5.5.4.2 Recovery Mechanisms

The effectiveness of alkaline flooding is attributed to a number of mechanisms. The four most commonly referenced are emulsification and entrainment, wettability reversal (oil-wet to water-wet), wettability reversal (water-wet to oil-wet), and emulsification and entrapment. Other mechanisms mentioned include emulsification with coalescence, wettability gradients, oil-phase swelling, disruption of rigid films, low IFT, and improved sweep resulting from precipitates altering flow. Johnson describes the first four mechanisms listed in detail with references to the original sources.

The first mechanism, emulsification and entrainment, results from reduction of IFT and the formation of an emulsion in which oil is entrained. If the emulsion is mobile, the oil saturation will decrease and oil will move through the reservoir.

Introduction of alkaline chemicals can cause a reversal of wettability from either oil-wet to water-wet or vice versa. The change in wettability and subsequent readjustment of fluids within the pores favorably affects the relative permeability to the oil phase. Discontinuous residual oil can be reconnected and made to flow. When this wettability reversal is coupled with IFT reduction, the ROS can significantly decrease.

Jennings et al. described recovery by emulsification and entrapment. In this mechanism, an emulsion formed by decreasing the IFT is subsequently trapped by the pore throats. This, in effect, causes a reduction of flow in high-permeability zones and results in an improvement of the effective mobility ratio between displacing and displaced fluids. Viscous fingering is diminished. A series of laboratory experiments in glass bead packs, where the displacement process could be photographed, was used to demonstrate this mechanism. In these experiments, no significant reduction in oil saturation below that resulting from a waterflood occurred. Improved oil recovery resulted from the improvement in mobility ratio.

In any specific alkaline chemical displacement, one or more of the different mechanisms may dominate the recovery efficiency. This depends on the chemical/rock system, but the process is not sufficiently understood to predict this behavior. Results from laboratory core experiments generally have shown that improved oil recovery is not directly correlated to the acid number of the crude oil. This suggests that other saponifiable constituents must exist that react with alkali to produce surface-active agents. The magnitude of IFT reduction is a better criterion. IFT values between the alkaline chemical solution and the oil of about 0.01 mN/m or less are required for significant recovery. Some disagreement exists in the literature about whether the pH of the alkaline system must be high (>11–12) or can be lower.

Addition of polymer to an alkaline chemical solution to increase the viscosity and to provide mobility control has been shown to be effective. Burk presented a series of corefloods using sodium carbonate, sodium hydroxide, and sodium orthosilicate, which were shown to be equally effective in decreasing IFT. In alkaline floods without added polymer, tertiary recoveries of oil were about 8–17 % of residual oil after waterflooding. With addition of polymer to the alkaline solutions, recoveries increased to about 88–97 % of residual oil. The increase is attributed to improved mobility ratios with polymer.

5.5.4.3 Application Conditions of Alkaline Flooding

Injection wells should be located within oil zones not in the peripheral aquifer to avoid alkali consumption caused by reaction with divalents. However, if the bottom water has a high divalent content, an alkaline solution may be injected to form precipitates so that bottom water coning may be mitigated by precipitates.

The reservoir does not have a gas cap. For reservoirs with oils having high acid numbers, alkaline flooding can be executed at any development stage. However, for reservoirs with oils having low acid numbers, alkaline flooding in an earlier stage performs better. In this case, remaining oil saturation should be higher than 0.4. There is no temperature limitation for alkaline flooding.

Alkaline consumption by chemical reaction and ion exchange is mainly due to the existence of clays. Thus, clay content should not exceed 15–25 %. Formation permeability should be greater than 100 mD.

Oil viscosity should be less than 50–100 cp. However, currently alkaline-surfactant injection into very high viscous oils has attracted more and more interest [4]. Formation water salinity should be less than 20 %, and the divalents in the injection water should be less than 0.4 %.

In the design of an alkaline flooding project, the following facts may be taken into account:

(a) Alkaline consumption in the field is higher than that in the laboratory because the contact time of alkalis with rocks in the former is much longer than that in the latter.
(b) Oil recovery factor in the field is generally lower than that in the laboratory.
(c) When alkaline flooding is combined with other methods, such as polymer flooding, surfactant flooding, hydrocarbon gas injection, or thermal recovery, much better effects will be obtained.
(d) Alkaline injection could cause scale problems in the reservoir, wellbore, and surface facility and equipment.
(e) When an alkaline solution makes contact with oil, stable emulsions may be formed. This will increase the cost to treat produced fluids on the surface.

5.5.5 *Surfactant Flooding*

Surface-active agents, or surfactants, are chemical substances that adsorb on or concentrate at a surface or fluid/fluid interface when present at low concentration in a system. They alter the surface tension, or IFT. In their most common form, surfactants consist of a hydrocarbon portion (nonpolar) and a polar, or ionic, portion. Figure 5.17 shows the structure of surfactant.

The hydrocarbon portion is often called the "tail" and the ionic portion the "head" of the molecule. The hydrocarbon portion can be either a straight chain or a branched chain. The nonpolar and polar portions are called lipophilic and

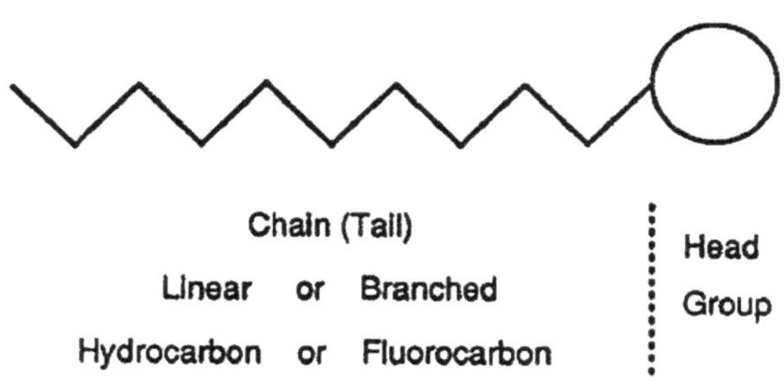

Fig. 5.17 Schematic of surface-active molecule

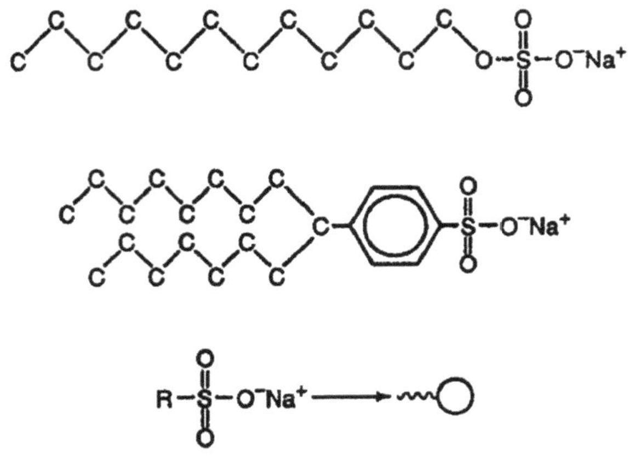

Fig. 5.18 Representative surfactant molecular structures

hydrophilic moieties, respectively. The entire molecule is sometimes called an amphiphile because it contains the nonpolar and polar moieties. Figure 5.18 shows examples of surfactants [2].

In surfactants of the type shown in Fig. 5.18, the hydrocarbon portion interacts very weakly with water molecules in aqueous solution. In fact, the water molecules act to try to "squeeze" the hydrocarbon out of the water. Thus, the tail is called hydrophobic. On the other hand, the head or polar portion interacts strongly with water molecules, undergoing solvation. This part of the surfactant is hydrophilic.

Basically, it is the balance between the hydrophilic and hydrophobic parts of a surfactant that gives it the characteristics associated with a surface-active agent. An empirical number that has sometimes been used to characterize surfactants is the hydrophilic/lipophilic balance (HLB). This number indicates relatively the tendency to solubilize in oil or water and thus the tendency to form water-in-oil or oil-in-water emulsions. Low HLB numbers are assigned to surfactants that tend to be more soluble in oil and to form water-in-oil emulsions.

5.5.5.1 Classification and Structure of Surfactants

Surfactants may be classified according to the ionic nature of the head group as anionic, cationic, nonionic, and zwitterionic.

Anionic — In aqueous solution, the molecule ionizes, and thus the surfactant has a negative charge. This surfactant is classified as anionic because of the negative charge on its head group.

Cationic — In aqueous solution, ionization occurs and the surfactant head group has a positive charge and is cationic.

Nonionic In this particular molecule, which does not ionize, the head group is larger than the tail group.

5.5.5.2 Displacement Mechanisms

Surfactant flooding can be grouped into dilute surfactant flooding and micellar flooding. Discussion of the displacement mechanisms may be made according to these two groups. The key mechanism for these surfactant floods is the low IFT effect. Surfactant flooding is principally an immiscible process in field applications, where slug size and surfactant concentration are limited by economics. Miscible displacement may occur in the early states of a flood, but the chemical slug quickly breaks down (multiple phases form) and the process becomes immiscible.

In dilute surfactant flooding a water-wet reservoir, when surfactant solution contacts residual oil droplets, the oil droplets are emulsified because of low IFT and entrained in surfactant solution. These entrained oil droplets are carried forward and are "pulled" to become long oil threads so that they can deform and pass through pore throats. When the salinity is low, oil-in-water (O/W) emulsions are formed. When the salinity is high, water-in-oil (W/O) emulsions are formed. These oil droplets are coalesced to form an oil bank ahead of the surfactant slug. As surfactant contacts rock surfaces, wettability may be changed.

In an oil-wet reservoir, oil sticks at pore throats or on pore walls. When surfactant solution flows through the pores, because of low IFT, oil droplets at pore throats are displaced. The oil droplets on pore walls are deformed and displaced along the walls by the dilute surfactant solution. These oil droplets are moved down to bridge with the oil droplets downstream to form continuous oil flow paths. They are pulled into long threads, and the oil threads flow downward. The oil threads could be broken during flow. Once broken, they become small droplets and are emulsified. These small droplets flow downward and lodge at the next throats to be coalesced with other oil droplets. Generally, these emulsions are W/O type. The emulsified oil and displaced oil coalesce to form an oil bank ahead. Wettability may be changed from oil-wet to water-wet owing to surfactant adsorption.

In dilute surfactant flooding, oil droplets must be able to deform to pass through pore throats. In other words, the deformation capability of oil droplets in dilute surfactant flooding is very important. This effect can be achieved by low IFT.

5.5.5.3 Screening Guideline

Following is a list of reservoir and fluid properties that are considered to be desirable for a surfactant flood. Guideline values are given for certain reservoir and fluid properties.

(a) Homogeneous sandstone reservoir.
(b) Sand should be at least 10 ft thick.
(c) Permeability should be at least 100 mD.
(d) Reservoir temperature should be less than 200 °F.
(e) Oil viscosity should be less than about 40 cp.
(f) Reservoir water salinity should be less than 100,000 ppm.
(g) Divalent ion concentration should be less than 500 ppm.
(h) Waterflood residual oil saturation should be greater than about 30 %.

5.6 Microbial Enhanced Oil Recovery (MEOR)

MEOR is a technique where microbes are used to generate products such as acids, gases, biomass, and biopolymers, as well as surfactants and solvents that aid oil production. The use of microorganisms and their metabolic products to stimulate oil production is now receiving renewed interest worldwide. This technique involves the injection of selected microorganisms into the reservoir and the subsequent stimulation and transportation of their in situ growth products in order that their presence will aid in further reduction of residual oil left in the reservoir after secondary recovery is exhausted.

Not all of these metabolites are needed at the same time, and deciding which products will be most useful for enhancing oil production will depend on the factors limiting petroleum production in a particular reservoir. Due to their size and ability to grow under the conditions present in petroleum reservoirs, only prokaryotes are considered promising candidates for MEOR, where as molds, yeasts, algae, and protozoa are not suitable. The MEOR concept was mooted as early as 1920, and pioneering field studies were carried out in the USA in the 1930s and 1940s by Claude ZoBell and his group. However, 80 years onward the oil industry commonly remains skeptical about the potential of MEOR to become a mainstream part of their EOR portfolio. This perception is nurtured by mixed successes and a general lack of a full scientific understanding of the processes. While it is clear that biocatalysis performed by microbes may promote beneficial chemical reactions such as the production of biosurfactants in a very specific and energy-efficient manner, a sound understanding of the underlying principles is important to predict site-specific effects of microbial activity on fluid flow in porous media and hence on the efficiency of oil production.

Microbial methods of flooding to enhance oil production include microbial flooding and cyclic microbial recovery.

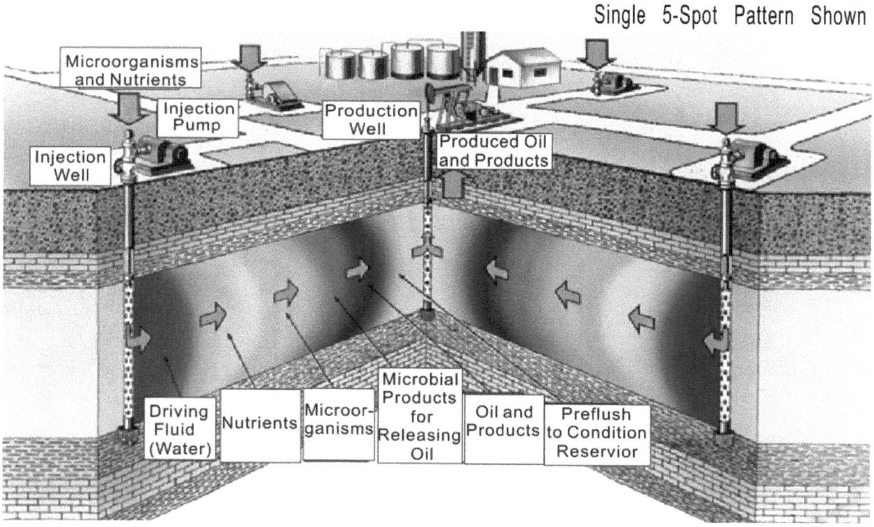

Fig. 5.19 Microbial flooding process

5.6.1 Microbial Flooding

Microbial flooding is performed by injecting a solution of microorganisms and a nutrient such as industrial molasses down injection wells drilled into an oil-bearing reservoir. As the microorganisms feed on the nutrient, they metabolically produce products ranging from acids and surfactants to certain gases such as hydrogen and carbon dioxide. These products act upon the oil in place in a variety of ways, making it easier to move the oil through the reservoir to production wells.

The microbial and nutrient solution and the resulting bank of oil and products are moved through the reservoir by means of drive water injected behind them, as shown in the Fig. 5.19.

5.6.2 Cyclic Microbial Recovery

Cyclic microbial recovery, one of the newest EOR methods, requires the injection of a solution of microorganisms and nutrients down a well into an oil reservoir. This injection can usually be performed in a matter of hours, depending on the depth and permeability of the oil-bearing formation. Once injection is accomplished, the injection well is shut in for days to weeks. During this time, known as an incubation or soak period, the microorganisms feed on the nutrients provided and multiply in number. These microorganisms produce substances metabolically that affect the oil in place in ways that facilitate its flow, making it easier to produce. Depending on

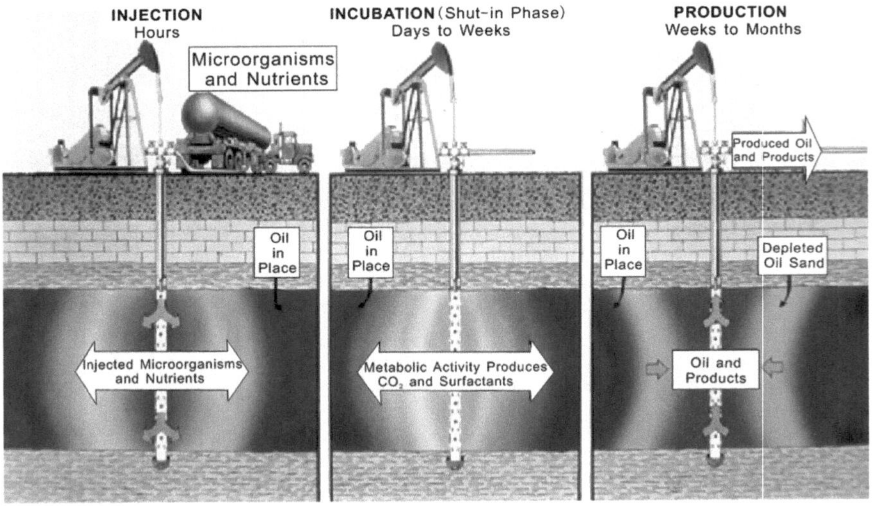

Fig. 5.20 Cyclic microbial recovery process

the microorganisms used, these products may be acids, surfactants, and certain gases, most notably hydrogen and carbon dioxide. At the end of this period, the well is opened, and the oil and products resulting from this process are produced.

This method eliminates the need for continual injection, but after the production phase is completed, a new supply of microorganisms and nutrients must be injected if the process is to be repeated, process shown in Fig. 5.20.

Although the basic concepts of MEOR are understood, the current state of knowledge lacks a detailed understanding of the scale and detail of the causes and effects of MEOR. Ambitious research programs are required to close these gaps. When this occurs, MEOR will have enormous potential to become a cost-effective mainstream tertiary recovery method. Such a research program should integrate findings in microbiology, geochemistry, and rock and fluid properties, including quantitative assessments of the interplay of these factors. A clear definition of reservoir production issues is critical for any MEOR study. New techniques and paradigms in molecular biology and microbiology offer an exciting opportunity to advance research in the MEOR domain.

Questions

1. How are EOR, IOR, E_D, and E_V defined?
2. A pseudoternary phase diagram for the fluid system is shown in Fig. 5.21. The oil composition is indicated on the diagram. A displacement test is conducted with a fluid that is 96 % CO_2 and 4 % C_2 through C_6. Will miscibility be achieved in this displacement? Explain and show the reason for your answer.
3. A laboratory study is needed to determine whether CO_2 flooding would be suitable for a reservoir. If you are to design the laboratory investigation, what

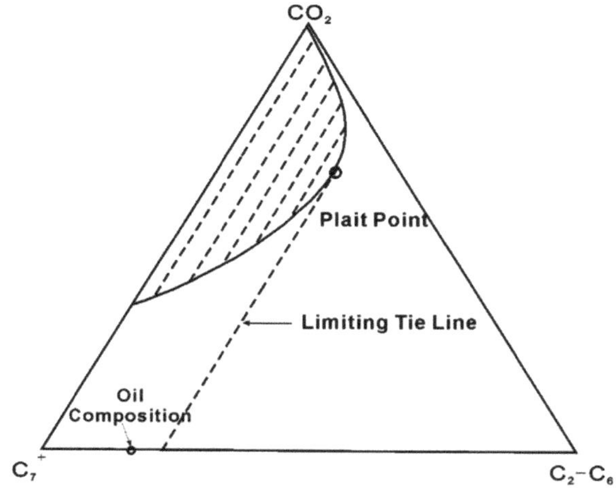

Fig. 5.21 Pseudoternary diagram

kinds of experiments would you conduct and what experimental data would you take and what would be your indications that CO_2 either is or is not suitable? Describe in a general way.

4. The viscosity of a 5000 ppm solution of partially hydrolyzed polyacrylamide was measured at several shear rates with a cone-and-plate rheometer. Table 5.3 gives experimental data as a function of shear rate. Plot the viscosity versus shear rate on log–log paper and determine the power-law exponent, n, and the power-law constant, K, from the experimental data using least squares analysis and correlate the viscosity as a function of shear rate. Compare values of the viscosity from the correlation with experimental data.

Table 5.3 Rheological date for 5000 ppm HPAM in 3 % NaCl

Shear rate (s^{-1})	Viscosity (cp)
1.75	6.921
2.083	7.364
2.733	7.111
3.546	6.938
4.661	6.676
6.353	6.205
8.624	5.785
11.95	5.298
16.56	4.838
23.31	4.351
33.04	3.883
47.22	3.441
68.04	3.026
97.68	2.669
139.5	2.363
199.5	2.094

5. An oil reservoir at waterflood ROS is to be considered as a candidate for a low-tension flood. The reservoir is a narrow, linear sand body that can be flooded with a linear pattern. The viscosity of the oil is 15 cp, and the viscosity of the reservoir brine is 0.9 cp. It is desired to conduct the flood at a mobility ratio of unity by adding polymer to the injected surfactant solution. The polymer increases the viscosity of solution but does not alter oil-recovery capability. Determine the viscosity of the chemical solution required to obtain mobility control between the oil bank and the chemical solution.
6. What is the difference of Cyclic Stem Stimulation, Steamdrive, and In Situ Combustion? Explain the concept and mechanism of these three kinds of thermal recovery processes.
7. Make a schematic representation of the steam drive process and explain the various zones formed in the reservoir.
8. Enumerate the mechanisms suggested for oil displacement by alkaline flooding and explain the special aspects to consider as screening criteria for alkaline flooding.
9. What are the products generated by bacterial activity and how can oil production be improved by them?

References

1. Campbell TC (1982) The role of alkaline chemicals in the recovery of low-gravity crude oils. J Petrol Technol 34(11):2510–2516
2. Green DW, Willhite GP (1998) Enhanced oil recovery. Henry L. Doherty memorial fund of AIME. Society of Petroleum Engineers, Richardson
3. Sheng J (2010) Modern chemical enhanced oil recovery: theory and practice. Gulf Professional Publishing
4. Terry RE (2001) Enhanced oil recovery. Encycl Phys Sci Technol 18:503–518

MIX
Papier aus verantwortungsvollen Quellen
Paper from responsible sources
FSC® C105338

If you have any concerns about our products,
you can contact us on
ProductSafety@springernature.com

In case Publisher is established outside the EU,
the EU authorized representative is:
Springer Nature Customer Service Center GmbH
Europaplatz 3, 69115 Heidelberg, Germany

Printed by Libri Plureos GmbH
in Hamburg, Germany